Table of Contents

Introduction i

Weatherability iii

Test Methods vii

Thermoplastics

ABS

ABS - Chapter 1
Textual Information 1
Tabular Information 3
Graphical Information 5

Acetal Resin

Acetal Resin - Chapter 2
Textual Information 21
Tabular Information 22

Acetal Copolymer - Chapter 3
Textual Information 23
Graphical Information 24

Acrylate Stryrene Acrylonitrile Polymer

Acrylate Styrene Acrylonitrile Polymer (ASA) - Chapter 4
Textual Information 31
Tabular Information 32
Graphical Information 33

Acrylic Resin

Acrylic Resin - Chapter 5
Textual Information 39
Tabular Information 41
Graphical Information 48

Acrylic Copolymer - Chapter 6
Graphical Information 53

Cellulosic Plastic

Cellulosic Acetate Butyrate (CAB) - Chapter 7
Textual Information 55
Tabular Information 57
Graphical Information 58

Fluoropolymer

Ethylene Chlorotrifluoroethylene Copolymer (ECTFE) - Chapter 8
Textual Information 61
Tabular Information 61

Ethylene Tetrafluoroethylene Copolymer (ETFE) - Chapter 9
Textual Information 63
Tabular Information 64

Polychlorotrifluoroethylene (CTFE) - Chapter 10
Textual Information 67
Graphical Information 67

Polyvinylidene Fluoride (PVDF) - Chapter 11
Textual Information 71
Tabular Information 72

Ionomer

Ionomer - Chapter 12
Textual Information 75
Tabular Information 76

Modified Polyphenylene Oxide

Modified Polyphenylene Oxide (PPO) - Chapter 13
Textual Information 81
Tabular Information 82
Graphical Information 83

Nylon

Nylon 12 - Chapter 14
Textual Information 87
Graphical Information 88

Nylon 6 - Chapter 15
Textual Information 91
Tabular Information 92
Graphical Information 93

Nylon 610 - Chapter 16
Textual Information 101
Tabular Information 102

Nylon 66 - Chapter 17
Textual Information 103
Tabular Information 105
Graphical Information 112

Nylon 6/6T - Chapter 18
Textual Information 119

Nylon MXD6 - Chapter 19
Graphical Information 121

Polyarylamide - Chapter 20
Textual Information 125
Graphical Information 126

Parylene

Parylene - Chapter 21
Textual Information 129

Polycarbonate

Polycarbonate (PC) - Chapter 22
Textual Information 131
Tabular Information 133
Graphical Information 139

Polyester

Polybutylene Terephthalate (PBT) - Chapter 23
Textual Information 147
Graphical Information 148

Polyethylene Terephthalate (PET) - Chapter 24
Textual Information 153
Tabular Information 154
Graphical Information 156

Glycol Modified Polycyclohexylenedimethylene Terephthalate (PCTG) - Chapter 25
Textual Information 159

Polycyclohexylenedimethylene Ethylene Terephthalate (PETG) - Chapter 26
Textual Information 161

Liquid Crystal Polyester (LCP) - Chapter 27
Textual Information 163
Tabular Information 163

Polyarylate - Chapter 28
Textual Information 165
Tabular Information 165
Graphical Information 166

Polyimide

Polyimide (PI) - Chapter 29
Textual Information 167
Graphical Information 168

Polyamideimide (PAI) - Chapter 30
Textual Information 171
Graphical Information 171

Polyetherimide (PEI) - Chapter 31
Textual Information 173
Graphical Information 173

Polyketone

Polyetheretherketone (PEEK) - Chapter 32
Textual Information 175
Tabular Information 175

Polyaryletherketone (PAEK) - Chapter 33
Textual Information 177

Polyolefin

Polyethylene (PE) - Chapter 34
Textual Information 179
Tabular Information 180

Low Density Polyethylene (LDPE) - Chapter 35
Graphical Information 183

High Density Polyethylene (HDPE) - Chapter 36
Textual Information 185
Tabular Information 189
Graphical Information 196

Ultrahigh Molecular Weight Polyethylene (UHMWPE) - Chapter 37
Textual Information 203

Polyethylene Copolymer - Chapter 38
Textual Information 205
Tabular Information 206

Polypropylene (PP) - Chapter 39
Textual Information 207
Tabular Information 207

Polypropylene Copolymer - Chapter 40
Graphical Information 211

Polymethylpentene - Chapter 41
Textual Information 215
Graphical Information 215

Polyphenylene Sulfide

Polyphenylene Sulfide (PPS) - Chapter 42
Textual Information 217
Tabular Information 218

Polystyrene

General Purpose Polystyrene (GPPS) - Chapter 43
Textual Information 219
Tabular Information 220
Graphical Information 220

Impact Polystyrene (IPS) - Chapter 44
Textual Information 223
Graphical Information 224

Polysulfone

Polysulfone (PSO) - Chapter 45
Textual Information 225
Tabular Information 225
Graphical Information 226

Polyethersulfone (PES) - Chapter 46
Textual Information 227
Graphical Information 227

Styrene Acrylonitrile Copolymer

Styrene Acrylonitrile Copolymer (SAN) - Chapter 47
Textual Information 229
Tabular Information 229
Graphical Information 230

Olefin Modified Styrene Acrylonitrile Copolymer (OSA) - Chapter 48
Textual Information 231
Tabular Information 231
Graphical Information 233

Styrene Butadiene Copolymer -

Styrene Butadiene Copolymer - Chapter 49
Textual Information 235

Styrene Butadiene Block Copolymer - Chapter 50
Textual Information 237

Vinyl Resin

Polyvinyl Chloride (PVC) - Chapter 51
Textual Information 239
Tabular Information 241
Graphical Information 250

Chloronated Polyvinyl Chloride (CPVC) - Chapter 52
Graphical Information 257

Thermoplastic Blends/Alloys

ABS Vinyl Resin Alloy

ABS Polyvinyl Chloride Alloy - Chapter 53
Textual Information 259
Tabular Information 259
Graphical Information 260

Acrylic Vinyl Resin Alloy

Acrylic Resin Polyvinyl Chloride Alloy - Chapter 54
Textual Information 261

Polycarbonate ABS Alloy

Polycarbonate ABS Alloy - Chapter 55
Graphical Information 263
Tabular Information 263

Biodegradable Thermoplastic Alloys

Starch Polyolefin Alloy

Starch Modified Low Density Polyethylene - Chapter 56
Graphical Information 265

Starch Modified Polyethylene Alloy - Chapter 57
Graphical Information 267

Starch Synthetic Resin Alloy

Starch Synthetic Resin Alloy - Chapter 58
Textual Information 269
Tabular Information 270

Thermosets

Diallyl Phthalate Resin

Diallyl Phthalate Resin (DAP) - Chapter 59
Tabular Information 271

Polyester

Thermoset Polyester - Chapter 60
Textual Information 273
Tabular Information 274
Graphical Information 275

Polyurethane

Polyurethane Reaction Injection Molding System (RIM PU) - Chapter 61
Textual Information 277
Tabular Information 278
Graphical Information 281

Thermoplastic Elastomers

Thermoplastic Elastomers

Chlorinated Polyethylene Elastomer (CPE) - Chapter 62
Textual Information 283

Olefinic Thermoplastic Elastomer (TPO) - Chapter 63
Textual Information 285
Tabular Information 290
Graphical Information 329

Polyester Thermoplastic Elastomer - Chapter 64
Textual Information 331
Tabular Information 333

Styrenic Thermoplastic Elastomer - Chapter 65
Textual Information 341
Tabular Information 343

Urethane Thermoplastic Elastomer (TPU) - Chapter 66
Textual Information ... 345
Tabular Information ... 347
Graphical Information ... 350

Vinyl Resin

Polyvinyl Chloride Polyol (pPVC) - Chapter 67
Tabular Information ... 353

Nitrile Thermoplastic Elastomer

Nitrile Thermoplastic Elastomer - Chapter 68
Textual Information ... 357
Tabular Information ... 358

Thermosetting Elastomers (Rubbers)

Butyl Rubber

Isobutylene Isoprene Rubber (IIR) - Chapter 69
Textual Information ... 359

Bromoisobutylene Isoprene Rubber (BIIR) - Chapter 69
Textual Information ... 359

Chlorosulfonated Polyethylene Rubber

Chlorosulfonated Polyethylene Rubber (CSM) - Chapter 70
Textual Information ... 361
Tabular Information ... 364
Graphical Information ... 373

Ethylene Propylene Copolymer

Ethylene Propylene Copolymer (EPM) - Chapter 71
Textual Information ... 375
Graphical Information ... 375

Ethylene Propylene Rubber

Ethylene Propylene Diene Methylene Terpolymer (EPDM) - Chapter 72
Textual Information ... 377
Tabular Information ... 382
Graphical Information ... 405

Fluoroelatomer

Fluoroelatomer - Chapter 73
Textual Information ... 407

Neoprene Rubber

Polychloroprene Rubber (CR) - Chapter 74
Textual Information 409
Tabular Information 411

Polybutadiene

Polybutadiene - Chapter 75
Tabular Information 419

Polyisoprene Rubber

Polyisoprene Rubber - Chapter 76
Textual Information 421
Tabular Information 422

Polyurethane

Polyurethane - Chapter 77
Textual Information 427
Tabular Information 427
Graphical Information 428

Silicone

Silicone - Chapter 78
Textual Information 429
Tabular Information 429

Indices and Appendices

Glossary of Terms 431
Graph Index 457
Table Index 465
Endnotes to Tables Index 473
Reference Index 475
Trade Name Index 479

Introduction

This reference publication is an extensive compilation of how the elements of weathering affect the properties and characteristics of plastics and elastomers. The basic physical characteristics of polymers are generally well defined by manufacturers. The effects of weathering methods, however, are not well compiled. This volume serves to turn the vast amount of disparate information from wide ranging sources (i.e. conference proceedings, test laboratories, materials suppliers, monographs, trade and technical journals) into useful engineering knowledge.

The information provided ranges from a general overview of the resistance of various plastics and elastomers to weathering (ultraviolet light, moisture, heat) to detailed discussions and test results. For users to whom the effects of both outdoor weathering and indoor exposure are relatively new, the detailed glossary of terms, including descriptions of test methods, will prove useful. For those who wish to delve beyond the data presented, source documentation is presented in detail.

In compiling data, the philosophy of Plastics Design Library is to provide as much information as is available. This means that complete information for each test is provided. At the same time, an effort is made to provide information for as many weathering tests and conditions (i.e. outdoor, outdoor accelerated, artificial accelerated, indoor, microbiologic attack, etc.) and material combinations as possible. Therefore, even if detailed test results are not available (i.e. the only information available is that the material is resistant or degrades), information is still provided. The belief is that some limited information serves as a reference point and is better than no information. Flexibility and ease of use were also carefully considered in designing the layout of this book.

How a material performs in its end use environment is a critical consideration and the information here gives useful guidelines. However, this or any other information resource should not serve as a substitute for actual testing in determining the applicability of a particular part or material in a given end use environment.

We trust you will greet this reference publication with the same enthusiasm as previous Plastics Design Library titles and that it will be a useful tool in your work. As always, your feedback on improving this volume or others in the series is appreciated and encouraged.

If you need more information:

Source documents for information presented in this volume are available. Software products and other publications dealing with additional subjects pertaining to plastics and elastomers are also available, some of which are outlined inside the back cover of this book. If we can be of further assistance please contact us at:

Plastics Design Library
P.O. Box 443
Morris, NY 13808-0443
Tel: 607-263-2318 Fax: 607-263-2446

How To Use This Book

This data bank publication presents the results of weathering exposure for more than 80 families of plastics and elastomers. Each chapter represents a single generic family. Data appears in textual, tabular, and graphical forms. Textual information is useful as it is often the only information available or the only way to provide an expansive discussion of test results. This is especially true in the case of weathering data where many results are qualitative.

Tables and graphs provide detailed test results in a clear, concise manner. Careful study of a table or graph will show how variations in material, exposure conditions and test conditions influence a material's physical characteristics. Endnotes associated with data in the tables are presented as an appendix and appropriately referenced in the tables.

Each table or graph is designed to stand alone, be easy to interpret and provide all relevant and available details of test conditions and results. The information's source is referenced to provide an opportunity for the user to find additional information. The source information might also help to indicate any bias which might be associated with the data.

Weatherability

Weather Defined

Webster defines weathering as "Noun. Action of the elements in altering the color, texture, composition, or form of exposed objects. Weather: to expose to air; to season, dry, pulverize, discolor, etc. by exposure to air." In essence, weathering is the natural tendency of materials to return - corrode, oxidize, chalk, permeate, delaminate, depolymerize, flex crack, etc. - to their elemental forms.

The sun's powerful ultraviolet rays combined with the effects of heat and moisture cause millions of dollars of damage every year to exposed materials. Ultraviolet light as sunlight, water as rain or dew, temperature and their combined effects are substantial stresses. Degradation also occurs from localized stresses such as microbiologic attack, air pollution and salt water.

Variations In Natural Weathering

Weather is variable - regional variations, local variations, seasonal variations, yearly variations, etc. Exposure to a tropical climate such as Florida can easily be twice as severe as exposure in northerly regions. This is due to the increase in ultraviolet radiation caused by the higher average sun angle and its exceptionally moist climate. On the other hand, Arizona may offer an increased ultraviolet radiation degradation, but it has a much lower rate of deterioration due to wetness.

Seasonal variations such as higher temperature and increased ultraviolet radiation (due to a higher sun angle in summer) can cause summer exposure to be 2 to 7 times as severe as winter exposure in the same place. Variations in weather can change from year to year making one year twice as severe as the last. Natural, accelerated, and artificial testing attempts to recreate weathering and its variability, usually under conditions more severe than normally encountered.

Testing For Weatheribility

Man attempts to prevent deterioration and discoloration due to weathering by examining the nature of the physics and chemistry at the interface of pigments, fillers and polymers and then creating the anti-corrosion, anti-oxidant, additive, or coating technology to extend the useful life of natural and synthetic products. However, the only way to reliably predict their durability is through comprehensive testing.

The primary purpose served by examining data on weathering of materials is to predict any potential changes of both physical properties and appearance of a part made from those materials. Data on the aging behavior of plastics are acquired through accelerated tests and/or actual weather exposure. These tests serve as a means for comparison of materials and can also be exploited to determine the ability of the material to serve its function when formed into a part and used in a particular environment. Comparisons between materials are made by measuring the retention of properties (e.g. impact strength, gloss, tensile strength, yellowness index, etc.) significant to the application as a function of exposure time. The demand for new products has shortened the time available for determining the durability of a particular material. Therefore, accelerated weathering is increasingly used in an attempt to predict the long term environmental effects in less time than the real time working life expectancy.

Natural sunlight isn't standard; there are variations in clouds, smog, angle of the sun, rain, industrial environments, etc. Likewise, opinions and approaches to the testing of plastic materials and their additives vary among individuals and change as theory becomes application. As a result, there are differing opinions as to the validity of testing, both with natural exposure and under accelerated or laboratory conditions. Nevertheless, testing and reliance on test results is necessary in the product development process to determine the durability of materials in particular applications. An accelerated weathering method correlates with real time exposure testing when specific defects can be generated in a material with an acceptable precision in a repeatable, lesser interval. Many manufacturers and suppliers worldwide find the correlations acceptable and have adopted test specifications for accelerated and artificial weathering.

The factors that influence the degree of weathering are:

- Solar radiation (usually ultraviolet)
- Moisture (dew, humidity, rain)
- Heat (surface temperature of the material)
- Pollutants (ozone, acid rain)
- Microbiologic attack
- Salt water

Because these factors vary so widely over the earth's surface, the weathering of materials is not an exact science. It is almost impossible to rank the degenerative power of temperature, moisture, and ultraviolet radiation. Most materials are weathered by a combination of these factors, but some are degenerated by moisture alone or ultraviolet radiation alone. Before one can determine the appropriate test procedure for a particular material, it is important to become familiar with the elements of natural weathering, how they work, and may work together to cause adverse effects on the durability of a given part.

Elements of Weather

Ultraviolet wavelengths from sunlight are an important component in outdoor degradation. The energy from sunlight is mainly visible light and infrared; ultraviolet makes up less than 5% of sunlight. However, the photodegradation of exterior materials is caused mainly by ultraviolet light. The solar ultraviolet radiation spectrum is divided into three ranges. UV-A is the energy in wavelengths between 400 nm and 315 nm. UV-B is the 315nm to 290 nm range. UV-C includes the solar radiation below 290 nm. The following table summarizes wavelength regions of ultraviolet radiation and their characteristics relative to degradation of materials.

Wavelength Regions of UV

UV-A 400 to 315 nm	Always present in sunlight; 400 nm upper limit for UV-A is the boundary between visible light and ultraviolet light; energy at the 315 nm boundary begins to cause adverse effects and pigmentation changes in human skin and some polymers.
UV-B 315 to 290 nm	Includes the shortest wavelengths found at the earth's surface; responsible for severe polymer damage; absorbed by window glass; ultraviolet light absorption by ozone varies with solar altitude (290 - 315 nm completely absorbed at altitudes below 14°, at 19° solar cutoff is 310 nm, at 40° solar cutoff is 303 nm, at solar altitudes between 60° and 90° maximum UV-B reaches the earth's surface with a solar cutoff at approximately 295 nm).
UV-C 290 to 100 nm	The UV-C 290 nm boundary is a sharp cut-off of solar radiation at the earth's surface due to complete absorption by ozone; found only in outer space.

UV-B wavelengths cause the most damage to polymeric materials.

Exposure of many plastics to ultraviolet radiation causes a loss in their mechanical properties and/or a change in their appearance. The mechanical property most severely affected is usually the ductility of the material. The embrittlement produced by ultraviolet radiation exposure can be evaluated by measuring the impact resistance (toughness) of the material.

Deterioration in appearance produced by ultraviolet radiation exposure can be evaluated by measuring color shift and loss of gloss. The observed color shift (ΔE) may be affected by the change in gloss. While the determination of gloss is straight forward, techniques for evaluating color shift may vary considerably.

Pigment systems influence to what extent materials are affected by ultraviolet radiation exposure. Some colors, such as black, may make the material less susceptible to ultraviolet radiation degradation than others.

While the effects of indoor and outdoor ultraviolet radiation exposure are similar, radiation frequencies do differ and the intensity of radiation is less indoors. Materials exposed indoors to fluorescent lamps are exposed to ultraviolet radiation and a color shift can occur over time. Indoor applications are also subject to ultraviolet radiation through window glass; however, much of the harmful radiation is filtered out by window glass.

Wetness

A high incidence of wetness has important implications on the durability of a part. It is often the case that water is not destructive in itself, but that water causes damage by bringing oxygen into intimate contact with the material and thereby promoting oxidation.

Generally, the potential for degeneration from dew exceeds that associated with rain. In Florida, materials are exposed to outdoor wetness an average of eight hours per day, or about 2,900 hours of wetness per year. Materials are wet from dew or condensation more frequently and for longer duration than from rain.

In order to condense dew, a material must be cooler than the dewpoint temperature of the air. This usually occurs in the night sky when solid objects lose their heat through radiation. The fact that materials are exposed primarily to dew and not rain affects the type of degradation that will occur. Dew is saturated with oxygen and lies on materials for hours at a time allowing the water to penetrate deep within the material to cause internal oxidation. This also allows time for soluble additives to be leached out from the material. In the case of paint films, it gives the water time to permeate through the coating and dissolve the solubles between the coating and the substrate. The severity of wetness attack increases dramatically with increasing temperature.

Surface Temperature

Surface temperature is the most variable factor in weather. An automobile driven at 55 mph on a highway will attain a surface temperature near ambient. The same car, locked and parked in direct sunlight, can reach a surface temperature 30°C above ambient. At night, with no wind and a clear sky, the surface temperature can drop 8°C below ambient.

Color is also a contributing factor in surface temperature. White paints typically attain a temperature 10°C to 15°C lower than black paints. It is difficult to match outdoor temperature differences between dark and light materials in the laboratory. For example, the introduction of air for heating and cooling will reduce temperature differences between colors. While temperature is an important factor in weathering it is also important to note that not all materials show increased degradation with increased temperature.

Test Methods

Outdoor Testing

Real-time weathering data from natural environment exposure programs remains the standard to which all other weathering data is compared. Two of the most commonly used harsh aging sites are Arizona and Florida. Arizona is important because of its high annual radiation and ambient temperature. Southern Florida is unique because of its high radiation combined with high rainfall and humidity. These two areas have become U.S. and international reference climates for gauging the durability of materials since they represent the worst case for applications in the Northern Hemisphere.

Accelerated Outdoor Tests

In outdoor tests the usual standard procedure calls for specimen exposure on racks facing due south at an angle of 45°. These are conditions that offer a maximum direct sunlight exposure and intensity. This tilt is also preferable to 0° as it allows for some drainage and wash off during rains. In a further attempt to accelerate outdoor effects, many studies are conducted in tropical and hot, dry climates, such as Florida and Arizona in the United States, Panama, Germany, and Japan to obtain the most wide ranging and severe environments possible. Outdoor accelerated weathering is a relatively recent technique. It relies heavily on technology to follow the track of the sun and to keep a constant temperature on the sample.

Conventional aging: This test method, which may occur in many different geographic locations (i.e., Florida, Arizona, Okinawa - Japan), is real time exposure at a 45° tilt from the horizontal. Direct exposures are intended for materials which will be used outdoors and subjected to all elements of weather. Exposure times are generally 6, 12, 24 and 48 months. Location is an important factor in the harshness of this test. The assumption being that test results from a hostile environment will prevail in more moderate conditions.

Conventional aging with spray: This test method, which may occur in many different geographic locations (i.e., Florida, Arizona, Okinawa - Japan), is real time exposure at a 45° tilt from the horizontal with a water spray used to induce moisture weathering conditions. The introduction of moisture plays an important role in improving both the relevancy and reproducibility of the weathering tests results. The purpose of wetting is two fold. First, the introduction of water in an otherwise arid climate induces and accelerates some degradation modes which do not occur as rapidly, if at all without moisture Second, a thermal shock causes a reduction in specimen surface temperatures, as much as 14°C (25°F). This results in physical stresses which accelerate the degradation process. Spray nozzles are mounted above the face of the rack at points distributed to insure uniform wetting of the entire exposed area. Distilled water is sprayed for 4 hours preceding sunrise to soak the samples, and then twenty times during the day in 15 second bursts. Direct exposures are intended for materials which will be used outdoors and subjected to all elements of weather. Exposure times are generally 6, 12, 24 and 48 months. Location is an important factor in the harshness of this test. The introduction of moisture is thought to play an important role in improving both the relevancy and reproducibility of the weathering test results.

With all outdoor tests it is important to be aware of bias introduced by the choice of location. Details about a few representative test sights are listed below.

Climatological Data						
Location	**Miami, Florida**		**New River, Arizona**		**Hiratsuka, Japan**	
Latitude	25°33'N		33°54'N		33°54'N	
Longitude	80°27'W		112°8'W		139°19'E	
Elevation	4.1 meters		610 meters		3.0 meters	
Temperature °C/°F	**Summer**	**Winter**	**Summer**	**Winter**	**Summer**	**Winter**
Average High	34°C/93°F	26°C /79°F	39°C /102°F	20°C /68°F	34°C /93°F	13°C /55°F
Average Low	23°C /73°F	13°C /55°F	24°C /75°F	8°C /46°F	19°C /66°F	0°C /32°F
Relative Humidity						
Annual Mean	74%		37%		73%	
Annual Precipitation						
Rain	188.1 cm/74 in.		25.5 cm/10 in.		156 cm/61 in.	
Solar Radiant Exposure						
Total	6588 Mj/m^2		8004 Mj/m^2		5000 Mj/m^2	
UV	280 Mj/m^2		333.5 Mj/m^2			

Equatorial Mount with Mirrors for Acceleration (EMMA): This is an accelerated outdoor weathering test method which uses natural sunlight and special reflecting mirrors to concentrate the sunlight to the intensity of about eight suns. The test apparatus is a follow the sun rack with mirrors positioned as tangents to an imaginary parabolic trough. The axis is oriented in a north-south direction, with the north elevation having the capability for periodic altitude adjustment. A blower which directs air over and under the samples is used to cool the specimens. This limits the increase in surface temperatures of most materials to 10°C (18°F) above the maximum service temperature that is reached by identically mounted samples exposed to direct sunlight at the same times and locations without concentration. Exposure periods of 6 and 12 months, have been correlated to about 2-1/2 and 5 years of actual aging in a Florida environment, respectively.

Equatorial Mount with Mirrors for Acceleration plus Water (EMMAQUA): This is an accelerated outdoor weathering test method which uses natural sunlight and special reflecting mirrors to concentrate the sunlight to the intensity of about eight suns. In addition to intensifying the power of the sun, water spray is used to induce moisture weathering conditions. The test apparatus is a follow the sun rack with mirrors positioned as tangents to an imaginary parabolic trough. The axis is oriented in a north-south direction, with the north elevation having the capability for periodic altitude adjustment. A blower which directs air over and under the samples is used to cool the specimens. This limits the increase in surface temperatures of most materials to 10°C (18°F) above the maximum service temperature that is reached by identically mounted samples exposed to direct sunlight at the same times and locations without concentration. Exposure to EMMAQUA is considered to be the harshest. Applications generally tested with EMMAQUA include adhesives, agricultural films, automotive exteriors, building materials, elastomers, glass, packaging, paints and coatings, plastics, roofing and sealants.

Artificial Accelerated Tests

Exposure to ultraviolet radiation can occur both outdoors and indoors. During outdoor exposure, both direct and reflected sunlight are sources of radiation. During indoor exposure, products are subjected to ultraviolet radiation from fluorescent lights as well as from glass-filtered ultraviolet rays transmitted through windows.

A number of variables must be considered when assessing the effect of indoor ultraviolet radiation exposure on a material. The type of light source, its energy flux and its distance from the specimens determine the intensity of the radiation impinging on the surface of the part.

For outdoor applications, geographical location is a major consideration. For outdoor testing it is critical to remember that interactions between ultraviolet radiation, water and temperature are the rule not the exception. Temperature can alter the rate of photochemical attack or oxidation reactions, and photochemical attack can alter the rate of oxidation during wetness. The total ultraviolet radiation exposure time will therefore have an effect.

The conditions necessary for reproducing natural weathering stresses in the laboratory are summarized in the following table:

Conditions for Reproducing Natural Weathering Stresses in the Laboratory

	UV Conditions	Water Conditions
Quality	UV-B emission with minimal emission below 290 nm	Condensed from vapor phase, pH approximately 4.0 to 6.0 saturated with 0_2.
Exposure Duration	No theoretical maximum or minimum. Practical minimum of 3 to 4 hr.	Time and temperature interact. 4 to 20 hr. are practical limits.
Temperature	55°C to 80°C as required to duplicate service temperature.	60°C sometimes causes abnormal effects. 50°C for 8 hr. can cause problems. 40°C is safe but slower

In automotive testing there are two methods that are predominately used: 1) xenon arc (Ci65 or Ct35) with a quartz-borosilicate filter combination for plastics and for interior textiles 2) exposure with Fluorescent UV/condensation devices (QUV). Xenon arc was used originally because the spectral distribution of a properly filtered xenon arc is a good match with sunlight. However, comparisons with sunlight and sunlight filtered through auto glass indicate that this filter combination may allow too much short wave emission for good correlation on some materials, particularly for automotive interiors. The use of QUV (FS-40, UV-B lamps) was once used for plastics, but over the last several years, testing indicates that their short wavelengths were inappropriate for plastics and textile coatings.

Atlas UV-Con or QUV: The QUV is an artificial accelerated weathering test to help predict the durability of materials subject to attack by sunlight or moisture inside and out. QUV testing simulates the effect of the sunlight with fluorescent ultraviolet (UV) lamps, while it simulates rain and dew with constant humidity. It is much stronger on the high energy, short wavelength region of the spectrum than normal sunlight. In a fluorescent UV apparatus, materials are alternatively exposed to ultraviolet light alone and condensation alone in a repetitive cycle.

The exposure conditions in fluorescent UV-condensation apparatus may be varied by selection of the fluorescent UV lamp, the timing of UV, condensation exposures, the temperature of the UV exposure and the temperature of the condensation exposure.

Carbon arc: Carbon arc devices generally use two lamps (twin arc) with the flame carbon arc open or encased in a borosilicate glass cover that acts as a filter for low wavelength radiation. The open flame carbon arc spectral distribution exhibits a significant amount of ultraviolet radiation below 300nm, the cut-off of radiation at the earth's surface. Generally the device is considered deficient compared to natural sunlight.

Xenon Arcs: Xenon arcs require a combination of filters to reduce unwanted radiation. SAE J1885, the current automotive interior test method, calls for xenon arc exposure with quartz inner and borosilicate outer filters. This filter combination transmits short wave ultraviolet radiation as low as 275 nm. While this may allow useful acceleration for some materials, the unnatural short wave ultraviolet radiation may cause unrealistic degradation in others.

FS-40 Lamp - F40-UVB: In the early 1970's the FS-40 became the first fluorescent lamp to achieve wide use. This lamp is currently specified for coatings (SAE J2020). It has demonstrated good correlation to outdoor exposures for gloss retention on automotive coatings and for material integrity of plastics. In these situations, short wavelength ultraviolet radiation allows great acceleration without affecting correlation. However, the output from fluorescent UV-B lamps that is below the solar cut-off often causes changes in color that are unrealistic when compared to those actually seen in service.

UVA-340 Lamp: This lamp simulates sunlight from about 365 nm down to a solar cut-off of 295 nm. The spectral distribution of the UVA-340 is similar to that of the xenon arc in the short wave region, except that its ultraviolet radiation cut-off matches that of sunlight and the xenon arc does not. It generally improves the correlation possible with the Fluorescent UV and condensation devices. It has demonstrated good correlation to Florida weathering on rigid vinyl, vinyl films, polyurethane films, polyethylene, polystyrene, urethane, and epoxy coatings and various pigments.

Glass Filtering: Glass of any type acts as a filter on the sunlight spectrum. The shorter, most damaging wavelengths are the most greatly effected. Ordinary window glass is essentially transparent to light above 370 nm. However, the filtering effect becomes more pronounced with decreasing wavelength. Windshield glass is thicker than window glass. It acts as a more efficient filter. Safety features associated with windshield glass, i.e., tinting and plastic adds to the filtering efficiency. The most damaging wavelengths below 310 nm are completely filtered out by windshield glass. Almost all ultraviolet is filtered out by windshield glass.

Variations In Testing and Results

Apparently similar test methods can yield test results that vary widely. When comparing results the user should be cognizant that factors such as test sights in similar locations, time of year, pollution counts, sample conditioning, to name a few can have a huge impact on test results. For instance, at Florida test sites, results can vary widely due to an increase in wetness caused by proximity to a pond or other source of moisture or dew, or by proximity to woods that shield the drying effects of wind. Another example is variation in sample mounting and effect on the period of degradation. Plywood backed samples, for instance, get much hotter in direct sunlight than unbacked samples and they are wet for up to twice as long.

The "synergistic" effect of ultraviolet radiation with moisture is an important area in assessing the validity of accelerated testing and its correlation to the natural weathering process. Much weathering literature focuses on the interaction between morning dew and ultraviolet radiation. While it is clear that this effect can be reproduced in a laboratory situation, it is not clear that it occurs under actual exposure conditions. The situation is reproduced with devices such as UV-CON and EMMAQUA. However, under normal outdoor exposure, sunlight likely will dry materials before the sun elevation reaches a point where UV-B is transmitted through the atmosphere. Laboratory conditions must be closely matched to actual weathering situations and it is recommended that you carefully consider your own situation in comparing it to the test conditions and data presented in this book.

Chapter 1

ABS

Outdoor Weather Resistance

BASF AG: Terluran 800 (modifier: butadiene acrylic rubber); Terluran 900 (modifier: butadiene rubber)

Solar radiation and atmospheric oxygen damage the butadiene elastomer in ABS during long periods of outdoor exposure. The consequences are yellowing and diminished impact resistance. Although the degradation can be inhibited by coloring the articles black, the outdoor performance will never be as good as that attainable with Acrylonitrile Styrene Acrylate (ASA).

Reference: *Terluran Product Line, Properties, Processing,* supplier design guide (B 567e/ (8109) 9.90) - BASF Aktiengesellschaft, 1990.

GE Plastics: Cycolac

Prolonged exposure to the weather, and especially direct sunlight, will cause significant changes in both the appearance and the mechanical properties of ABS plastics. The material will lose gloss, shift in color tone towards yellow and can surface craze in areas of high strain. Very severe weather exposure in some climates can also degrade the surface to a chalking condition.

The plastic will lose much of its impact resistance and ductility, particularly at low temperatures. Tensile and flexural strength values are maintained at normal temperatures, but these properties also drop appreciably at lower temperatures or higher strain rates. Modulus and hardness properties are not severely affected. These changes in properties are due to the formation of a very thin brittle layer on the exposed surface of the ABS. Any load sufficient to crack this veneer can, by "notch effect", cause the crack to propagate into the ductile core of the plastic.

Depending on the method of test or application of load, this fracture of the veneer can be very sudden in a localized area, as in the case of small projectile impact tests. In other cases, fracture of the veneer occurs over a period of time as progressive formation of multiple fractures (crazing) in the surface, as in the case of repeated flexing or vibration. In either situation, the fracture of the surface veneer is aggravated by conditions that include high strain rates, cyclic loading, or low temperature.

Reference: *Weatherability of Cycolac Brand ABS - Technical Publication P-405,* supplier technical report (8203-5M) - General Electric Company, 1982.

ABS

In contrast to ASA, ABS articles exposed outdoors undergo changes in mechanical properties and color after a comparatively short period of time. Solar radiation, particularly at the ultraviolet end of the spectrum, acts together with atmospheric oxygen to cause embrittlement and yellowing. These changes occur mainly in the butadiene elastomer.

Reference: *Luran S Acrylonitrile Styrene Acrylate Product Line, Properties, Processing,* supplier design guide (B 566 e / 11.90) - BASF Aktiengesellschaft, 1990.

Accelerated Artificial Weathering Resistance

ABS

The effect of exposure to ultraviolet radiation on the impact strength of ASA and ABS was plotted. Results were obtained by exposing one side of 50 x 6 x 4 mm standard bar specimens in the Xenotest 1200 and subsequently subjecting the specimens to the DIN 53453 impact test with the blow struck on the unexposed side. ABS exposed to such extreme ultraviolet radiation suffered a very rapid drop in impact strength. ASA retains its impact strength under these conditions over a very much longer period, i.e. about seven times as long.

Reference: *Luran S Acrylonitrile Styrene Acrylate Product Line, Properties, Processing,* supplier design guide (B 566 e / 11.90) - BASF Aktiengesellschaft, 1990.

Effect of Color Pigments on Weatherability

GE Plastics: Cycolac

Pigmented Cycolac ABS has somewhat better resistance to weathering than the natural unpigmented grades, both in appearance and in physical properties. When protective coatings cannot be used, a properly pigmented black grade should be chosen for best maintenance of physical properties.

Where colors must be used, the earthen tones are usually more satisfactory. Pastel blues or blue undertone colors which might shift towards green should be avoided. The best light-fast color pigments should be specified.

Reference: *Weatherability of Cycolac Brand ABS - Technical Publication P-405,* supplier technical report (8203-5M) - General Electric Company, 1982.

Effect of Surface Protectants on Weatherability

GE Plastics: Cycolac

End-use products requiring exterior long-life retention of color, gloss, and abuse resistance, should be designed with the surface protected with a pigmented protective film or coating.

Protective Film Laminates One of the most effective techniques for improving the weathering resistance of Cycolac ABS sheet products is the lamination of pigmented Korad acrylic film to the ABS. Pigmented Korad films (0.076 mm or more) can be laminated to Cycolac ABS sheet during the extrusion process; compatibility of the materials permits the use of regrind. Test exposures of up to 5 years in Arizona have shown this method to be extremely effective in protecting against both color shift and loss of gloss. In addition to providing excellent retention of aesthetic properties, Korad film has proven to be an effective means of reducing the rate of degradation of physical properties.

Protective Paints Specifically compounded paint systems based upon weather resistant resins, having sufficient flexibility to avoid brittle-veneer effects and can also be effective in minimizing weather degradation.

Reference: *Weatherability of Cycolac Brand ABS - Technical Publication P-405,* supplier technical report (8203-5M) - General Electric Company, 1982.

Indoor UV Light Resistance

BASF AG: Terluran 800 (modifier: butadiene acrylic rubber); Terluran 900 (modifier: butadiene rubber)

Terluran is stabilized against oxidative and high-temperature aging. As a result, articles produced from it can last for years indoors, even if they are exposed to considerable ultra-violet radiation (e.g. automobile interior trim). Terluran 800 resins display great resistance to yellowing when exposed to UV radiation.

Reference: *Terluran Product Line, Properties, Processing,* supplier design guide (B 567e/ (8109) 9.90) - BASF Aktiengesellschaft, 1990.

TABLE 01: Outdoor Weathering in Florida of White ABS.

Material Family	ABS
Features	white color
Reference Number	194

EXPOSURE CONDITIONS

Exposure Type	outdoor weathering
Exposure Location	Florida
Exposure Country	USA
Exposure Note	45° south
Exposure Time (days)	365

SURFACE AND APPEARANCE

ΔE Color	12.8
Yellowness Index	35.6 {q}
Δ Yellowness Index	21 {q}
Yellowness Index	30.7 {r}
Δ Yellowness Index	18.8 {r}
Whiteness Index	-16.7
Δ Whiteness Index	-61.1

TABLE 02: Outdoor Weathering in Ludwigshafen, Germany of ABS.

Material Family	ABS			
Reference Number	143	143	143	143

EXPOSURE CONDITIONS

Exposure Type	outdoor weathering			
Exposure Location	Ludwigshafen			
Exposure Country	Germany			
Exposure Note	45° angle south			
Exposure Time (days)	15	30	45	60

PROPERTY VALUES AFTER EXPOSURE

Impact Strength (kJ/m^2)	10.4 (struck on unexposed side, thus subjecting exposed side to dynamic loading)	11.3 (struck on unexposed side, thus subjecting exposed side to dynamic loading)	12.2 (struck on unexposed side, thus subjecting exposed side to dynamic loading)	10.2 (struck on unexposed side, thus subjecting exposed side to dynamic loading)

PROPERTIES RETAINED (%)

Notched Impact Strength	99.2 (notched side exposed to light)	98.4 (notched side exposed to light)	96.8 (notched side exposed to light)	95.3 (notched side exposed to light)

TABLE 03: Accelerated Indoor Exposure by HPUV of General Electric Cycolac ABS.

Material Family	ABS		
Material Supplier/ Grade	GE Plastics Cycolac V100	GE Plastics Cycolac V200	GE Plastics Cycolac VW300
Features	flame retardant; Hewlett Packard parchment white color	flame retardant; indoor UV resistance; Hewlett Packard parchment white color	flame retardant; indoor UV resistance; Hewlett Packard parchment white color
Reference Number	176	176	176
MATERIAL COMPOSITION			
Material Composition Note	non-PBBE additives	non-PBBE additives	non-PBBE additives
EXPOSURE CONDITIONS			
Exposure Type	Accelerated Indoor UV Light		
Exposure Apparatus	HPUV	HPUV	HPUV
Exposure Note	3 year office exposure simulation	3 year office exposure simulation	3 year office exposure simulation
SURFACE AND APPEARANCE			
ΔE Color	10.5	9	2

TABLE 04: Accelerated Indoor Expsoure to Fluorescent Light of General Electric Cycolac ABS.

Material Family	ABS									
Material Supplier/ Grade	GE Plastics Cycolac KJM					GE Plastics Cycolac KJW				
Features	tan color					tan color				
Reference Number	118					118				
EXPOSURE CONDITIONS										
Exposure Type	Indoor UV Exposure									
Exposure Apparatus	Fluorescent Light					Fluorescent Light				
Exposure Note	1.83 meters from 4, 40 watt fluorescent lights					1.83 meters from 4, 40 watt fluorescent lights				
Exposure Time (days)	30	91	182	274	365	30	91	182	274	365
SURFACE AND APPEARANCE										
ΔE Color Change	0.09 {a}	0.36 {a}	0.85 {a}	1.59 {a}	3.11 {a}	0.17 {a}	0.23 {a}	0.75 {a}	1.51 {a}	3.60 {a}

TABLE 05: Accelerated Indoor Exposure to Fluorescent Light of General Electric Cycolac ABS.

Material Family	ABS									
Material Supplier/ Grade	GE Plastics Cycolac KJB					GE Plastics Cycolac KJU				
Features	tan color					tan color				
Reference Number	118					118				

EXPOSURE CONDITIONS

Exposure Type	Indoor UV Exposure									
Exposure Apparatus	Fluorescent Light					Fluorescent Light				
Exposure Note	1.83 meters from 4, 40 watt fluorescent lights					1.83 meters from 4, 40 watt fluorescent lights				
Exposure Time (days)	30	91	182	274	365	30	91	182	274	365

SURFACE AND APPEARANCE

ΔE Color Change	0.07 {a}	0.26 {a}	0.44 {a}	0.77 {a}	1.18 {a}	0.08 {a}	0.13 {a}	0.26 {a}	0.41 {a}	0.53 {a}

GRAPH 01: Outdoor Weathering Exposure Time vs. Yellowness Index of ABS

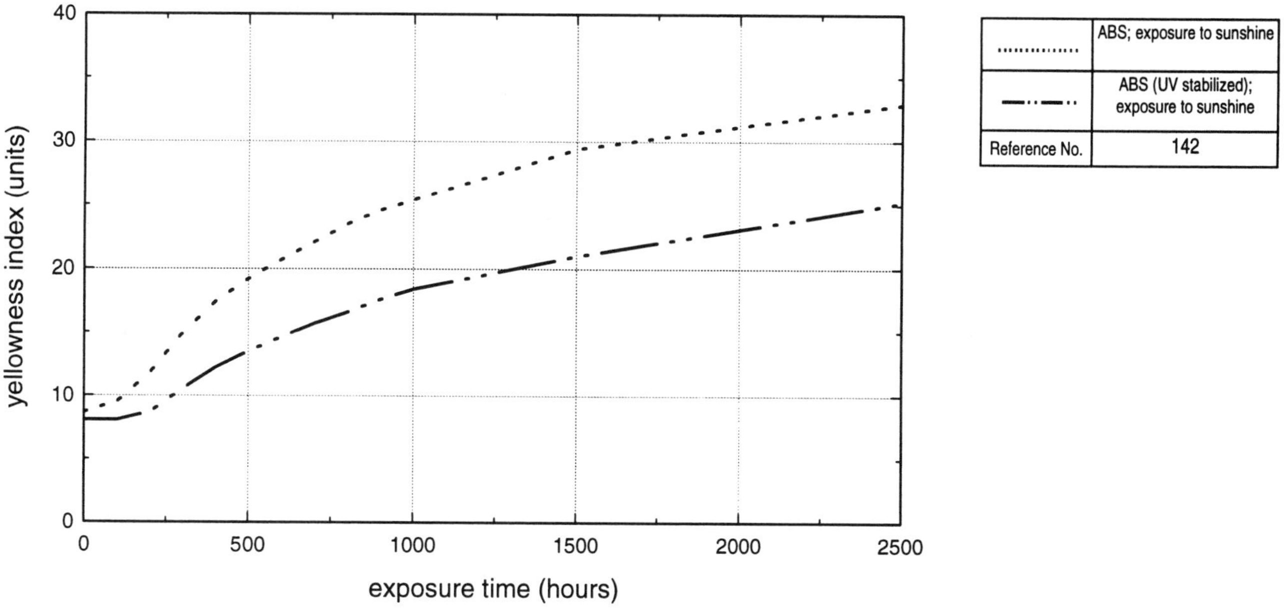

GRAPH 02: Arizona Outdoor Weathering Exposure Time vs. Drop Dart Impact Strength of ABS

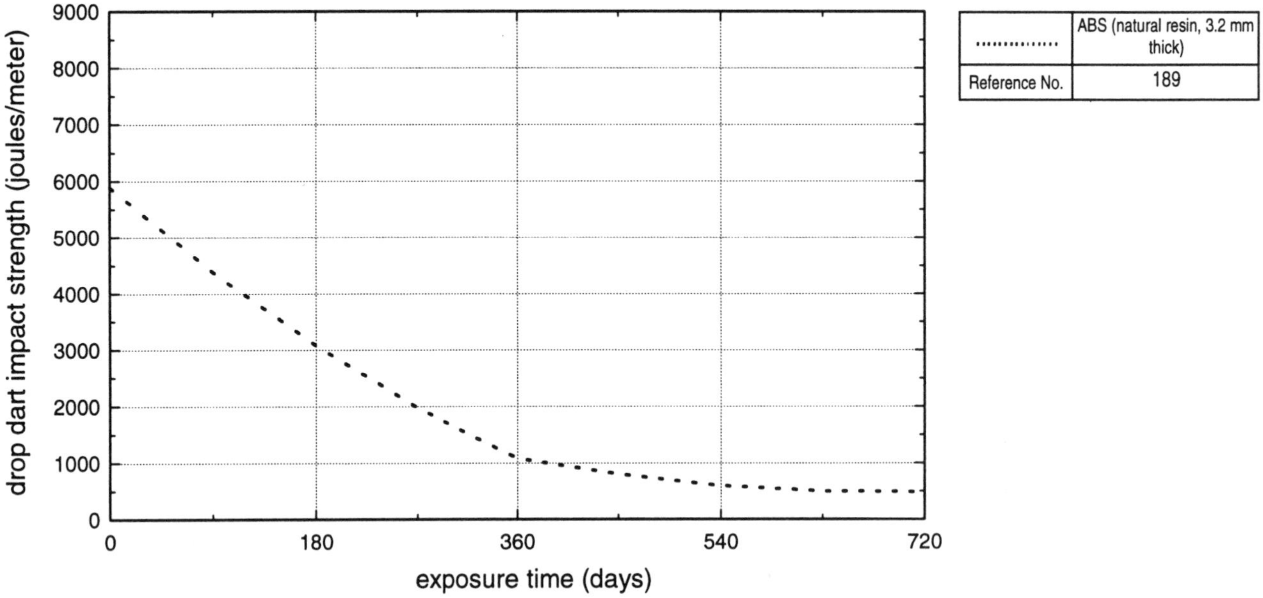

GRAPH 03: Arizona Outdoor Weathering Exposure Time vs. Elongation of ABS

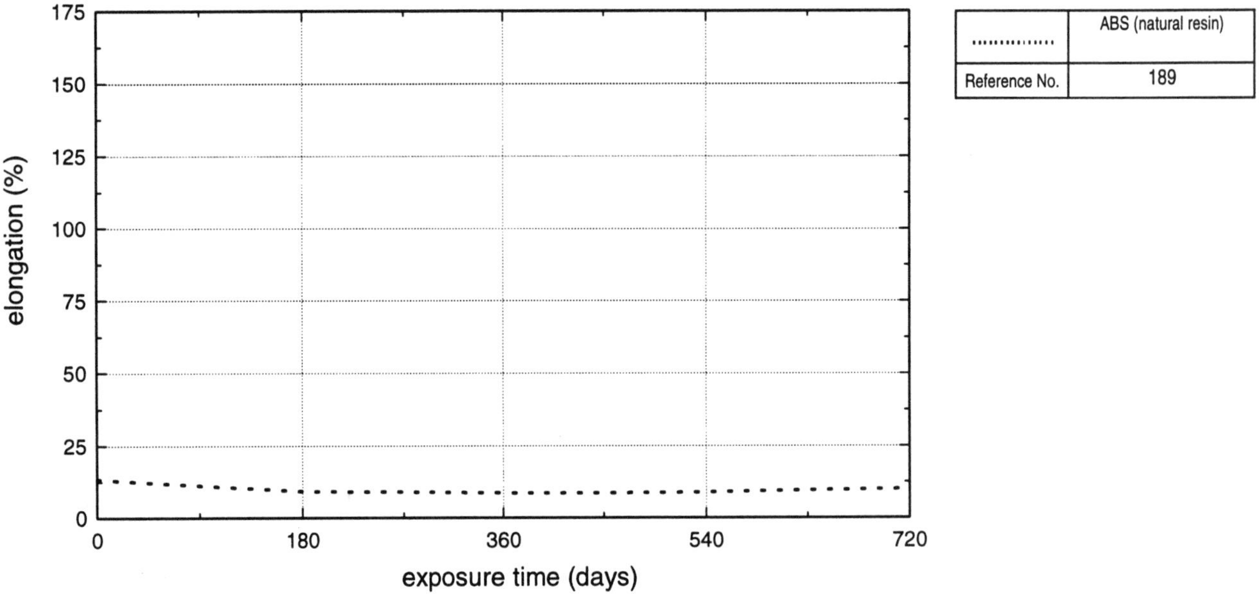

GRAPH 04: **Arizona Outdoor Weathering Exposure Time vs. Tensile Strength at Yield of ABS**

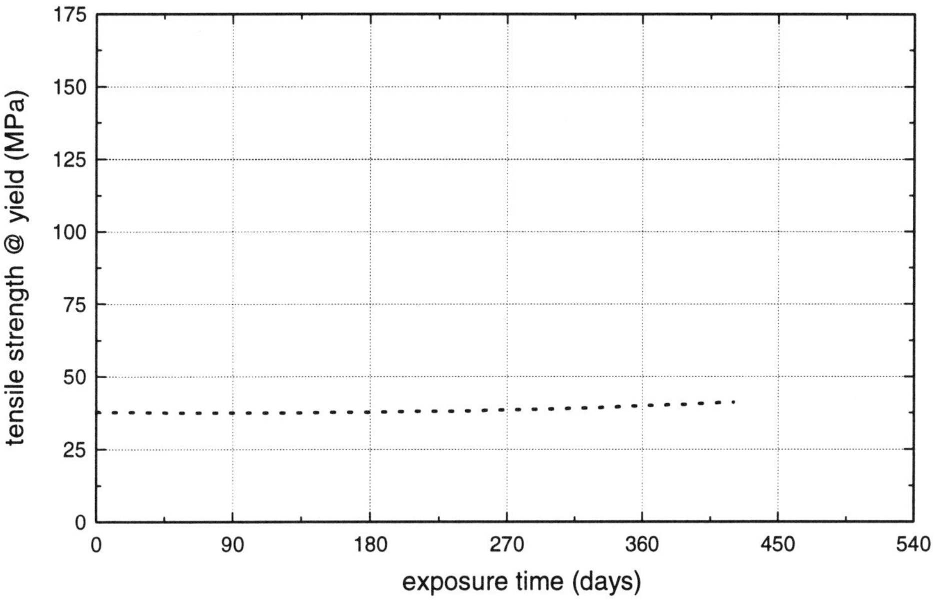

···············	ABS
Reference No.	189

GRAPH 05: **Arizona Outdoor Weathering Exposure vs. Delta E Color Change of ABS**

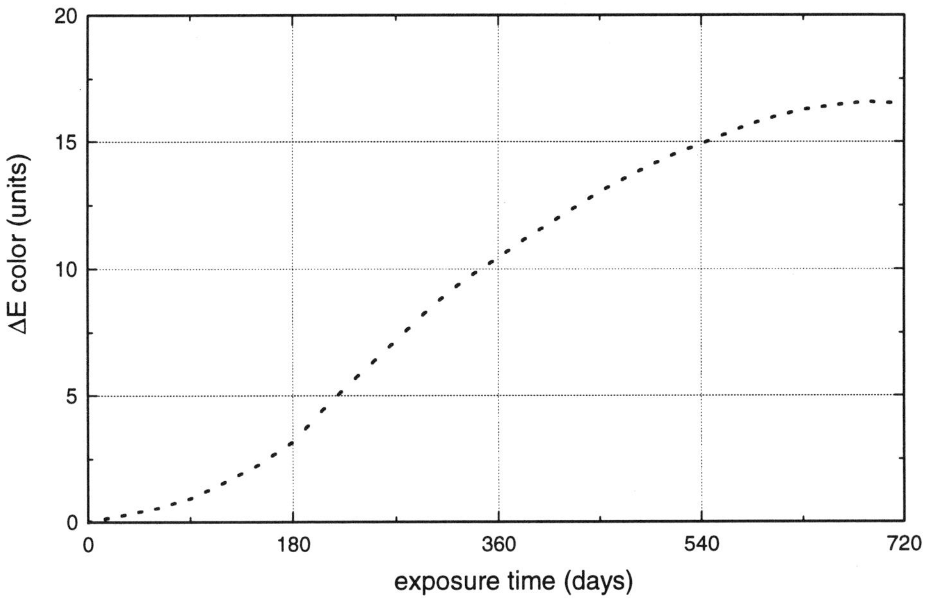

···············	ABS (natural resin); CIE lab color scale
Reference No.	189

GRAPH 06: **Arizona, Florida and Ohio Outdoor Weathering Exposure Time vs. Drop Dart Impact Strength of ABS**

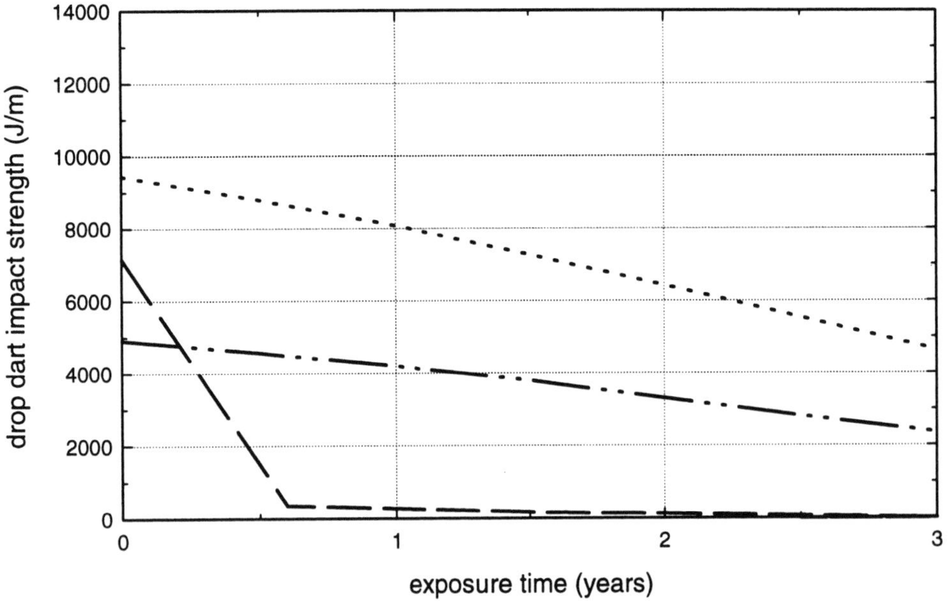

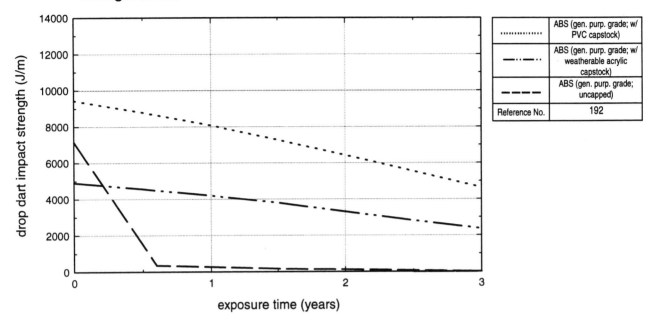

···············	ABS (gen. purp. grade; w/ PVC capstock)
—··—··	ABS (gen. purp. grade; w/ weatherable acrylic capstock)
— — — —	ABS (gen. purp. grade; uncapped)
Reference No.	192

GRAPH 07: **Florida Outdoor Weathering Exposure Time vs. Drop Dart Impact Strength of ABS**

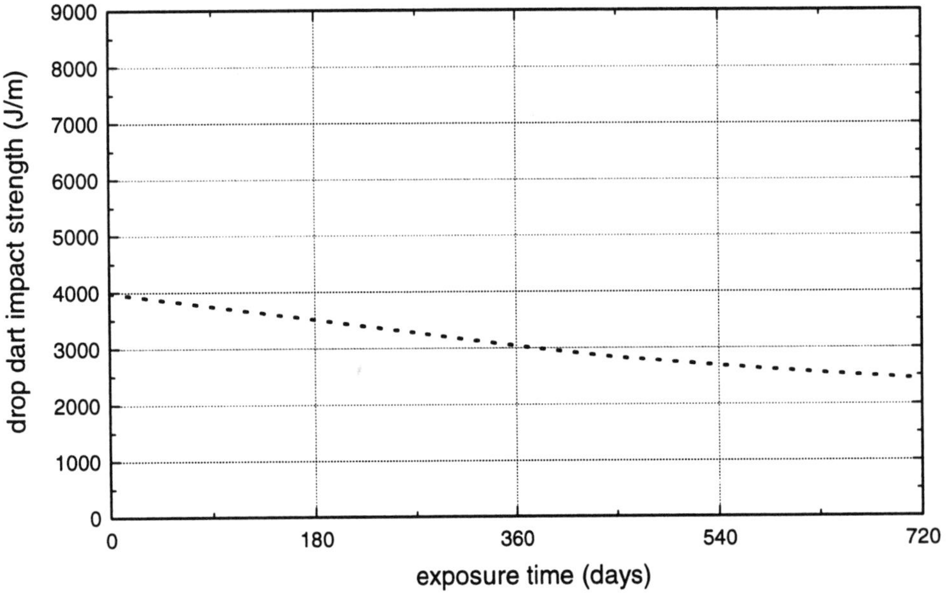

···············	ABS (natural resin, 3.2 mm thick)
Reference No.	189

GRAPH 08: Florida Outdoor Weathering Exposure Time vs. Drop Weight Impact of ABS

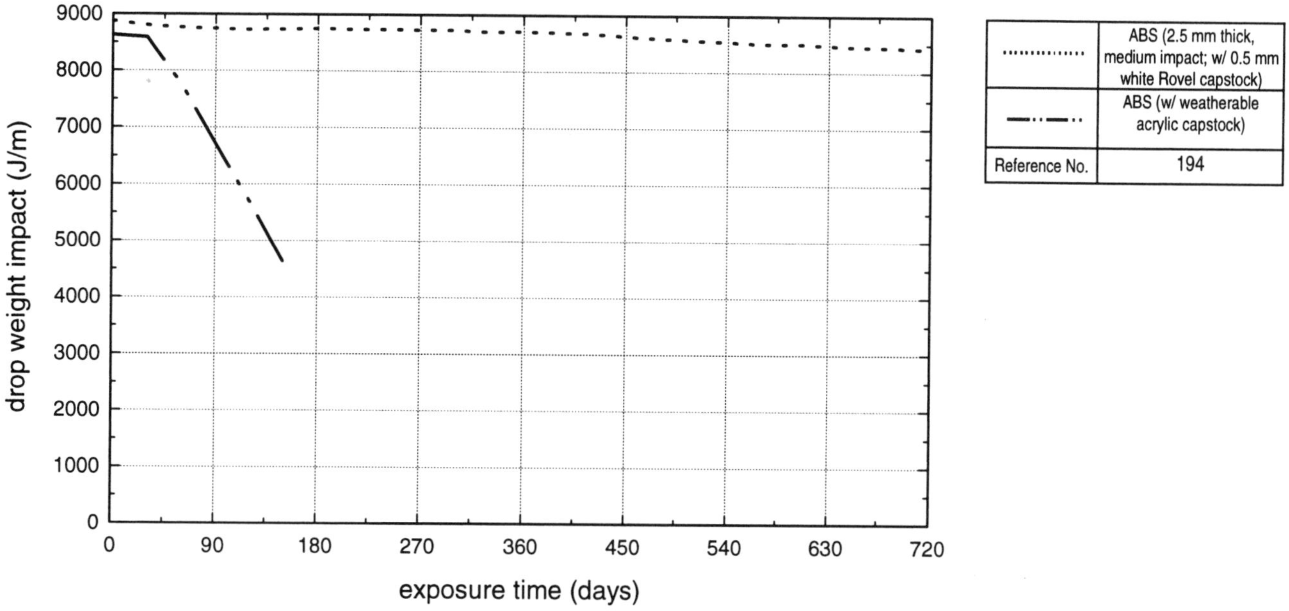

GRAPH 09: Florida Outdoor Weathering Exposure vs. Delta E Color Change of ABS

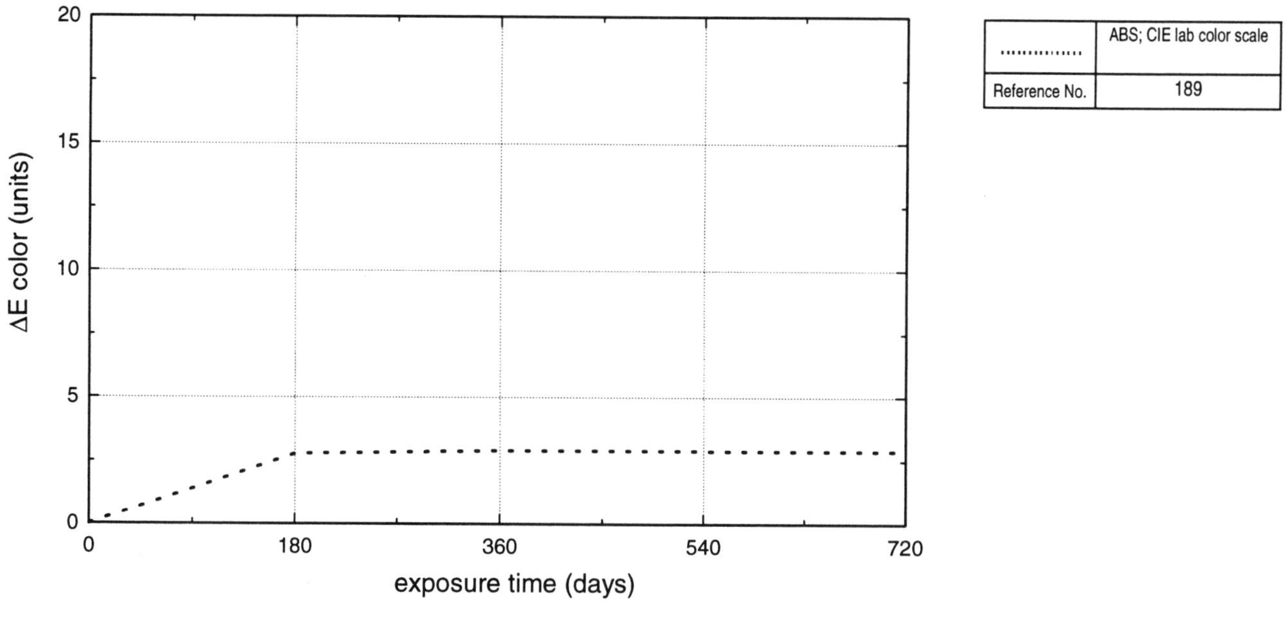

GRAPH 10: Florida Weathering Exposure Time vs. Chip Impact Strength of ABS

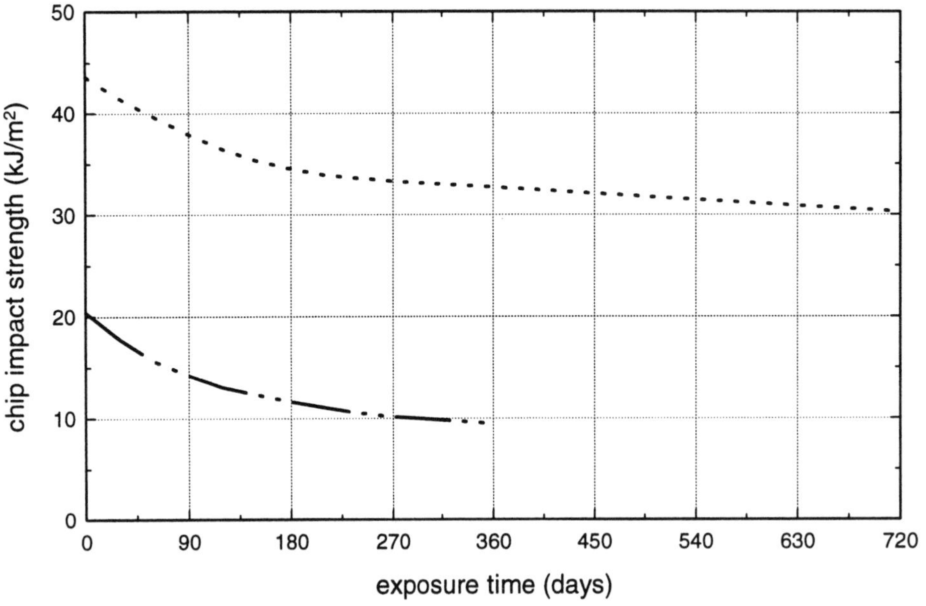

....	ABS (2.5 mm thick, medium impact; w/ 0.5 mm white Rovel capstock); 45° angle facing south
—·-—··	ABS (w/ weatherable acrylic capstock); 45° angle facing south
Reference No.	194

GRAPH 11: Florida Weathering Exposure Time vs. Chip Impact Strength of ABS

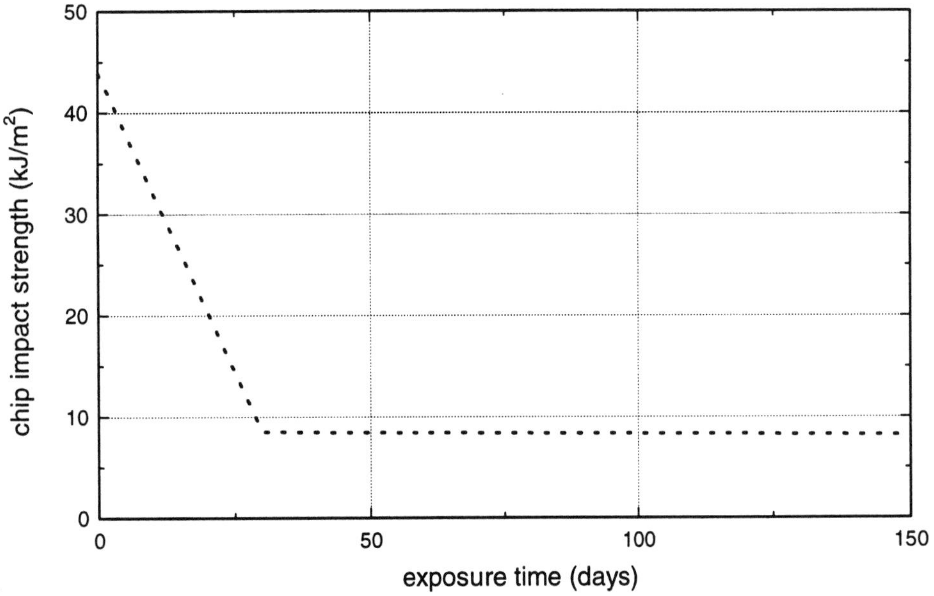

....	ABS (natural resin); 45° angle facing south
Reference No.	194

GRAPH 12: Ohio Outdoor Weathering Exposure Time vs. Delta E Color Change of ABS

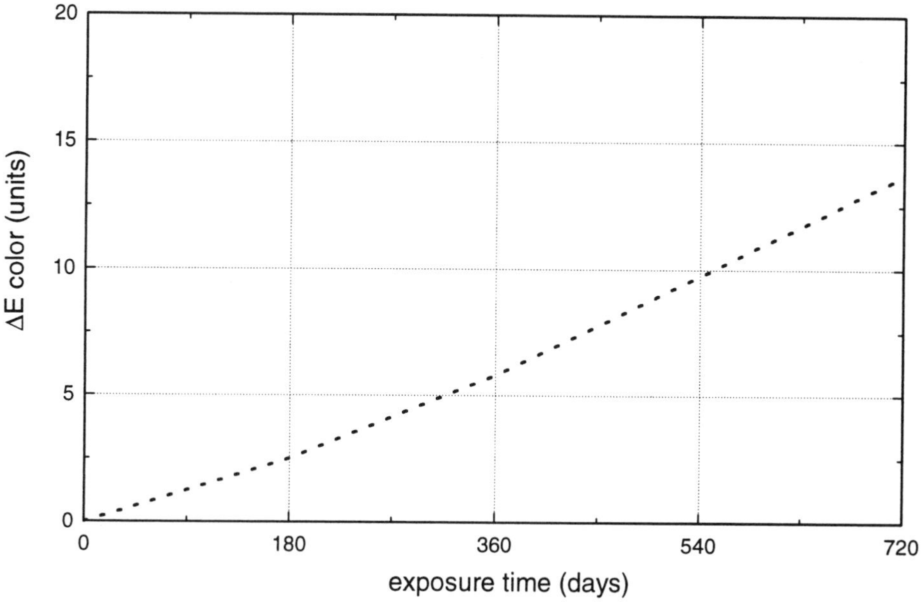

···············	ABS (natural resin); CIE lab color scale
Reference No.	189

GRAPH 13: Ohio Outdoor Weathering Exposure Time vs. Drop Dart Impact Strength of ABS

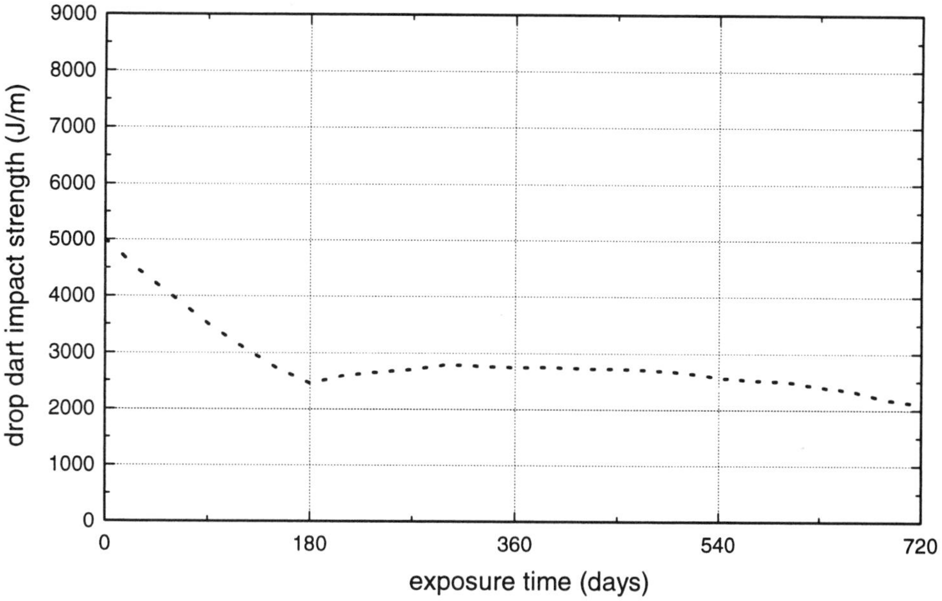

···············	ABS (natural resin, 3.2 mm thick)
Reference No.	189

GRAPH 14: **Okinawa, Japan Outdoor Weathering Exposure Time vs. Delta E Color Change of ABS**

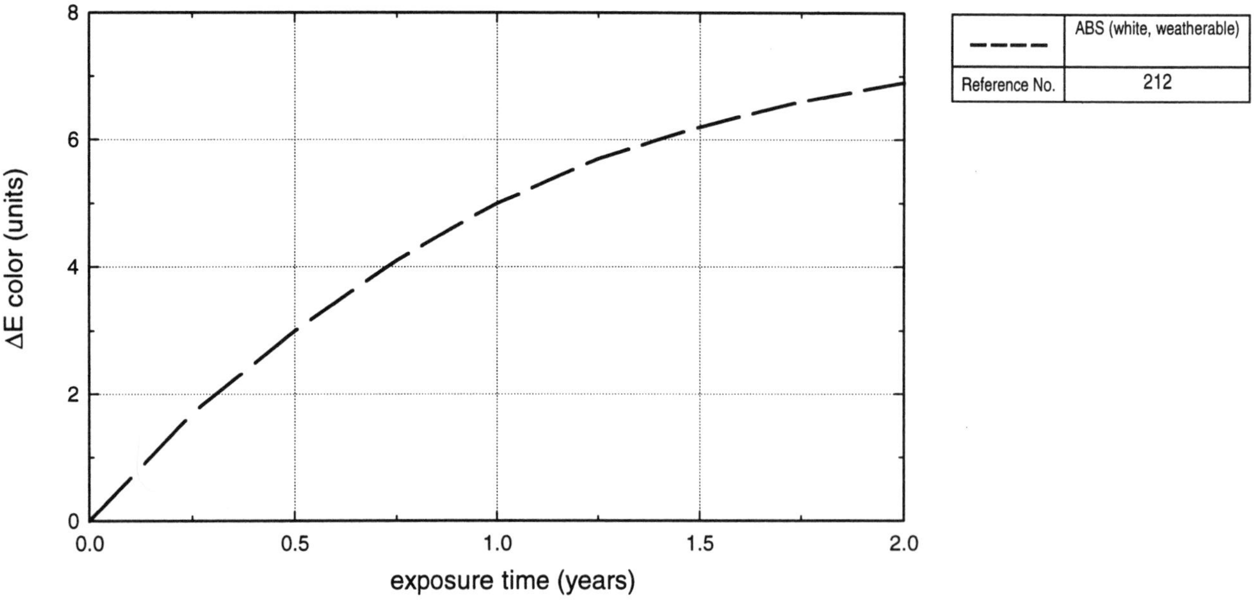

GRAPH 15: **Okinawa, Japan Outdoor Weathering Exposure Time vs. Dynstat Impact Strength Retained of ABS**

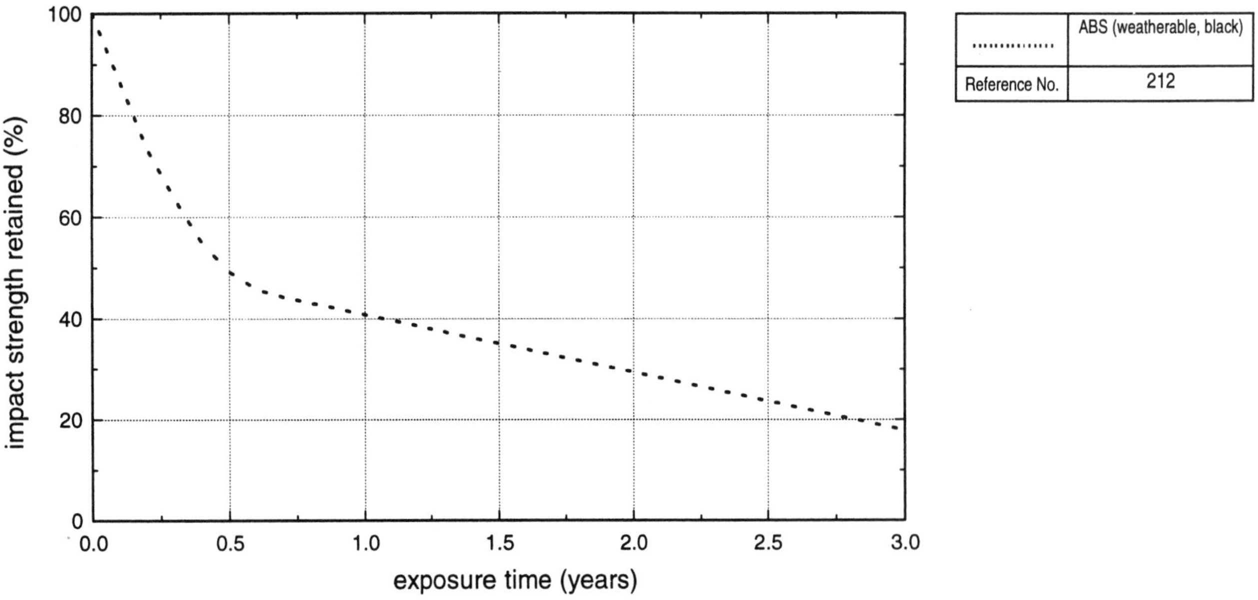

GRAPH 16: Okinawa, Japan Outdoor Weathering Exposure Time vs. Elongation at Break Retained of ABS

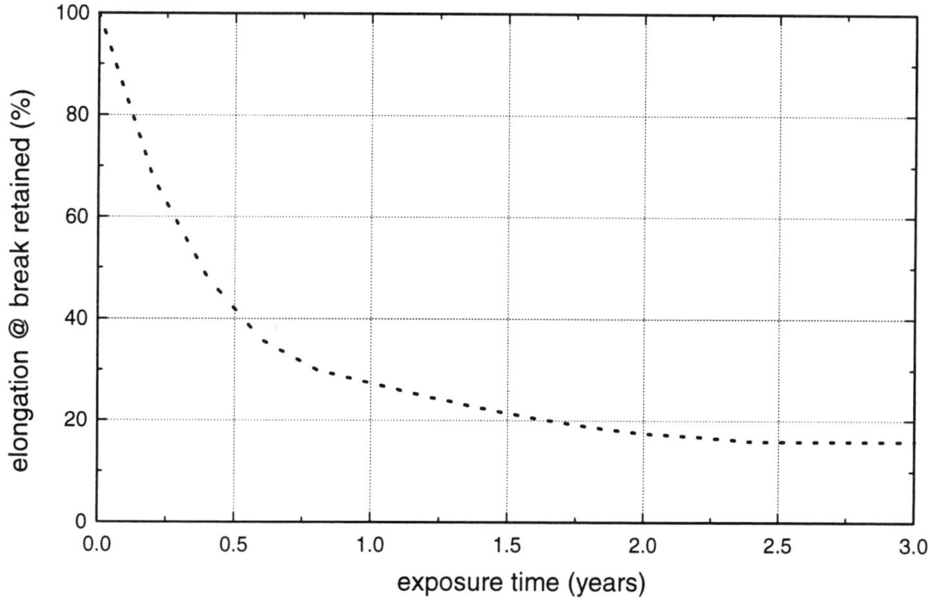

....	ABS (weatherable, black)
Reference No.	212

GRAPH 17: Okinawa, Japan Outdoor Weathering Exposure Time vs. Gloss Retained of ABS

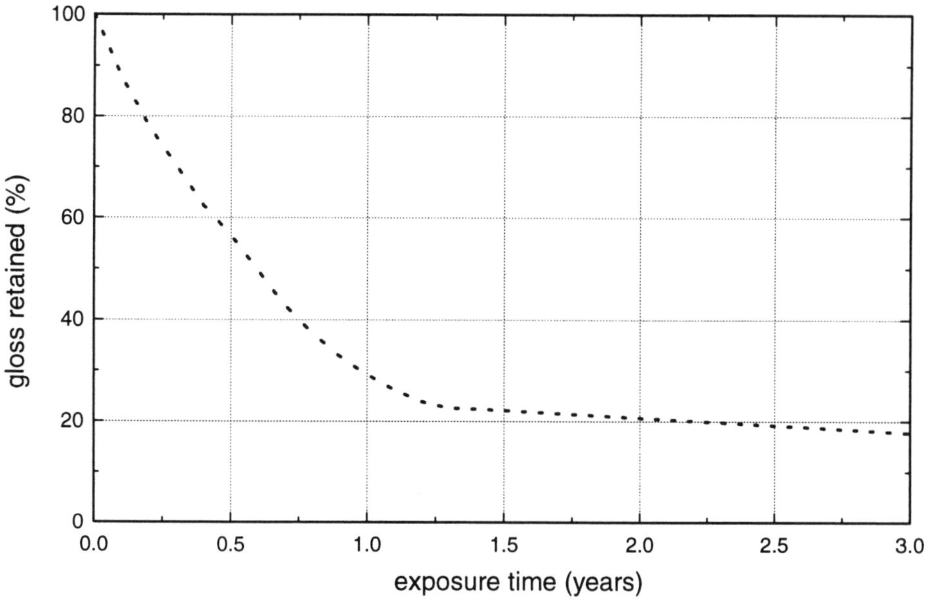

....	ABS (weatherable, black)
Reference No.	212

GRAPH 18: West Virginia Outdoor Weathering Exposure Time vs. Falling Dart Impact of ABS

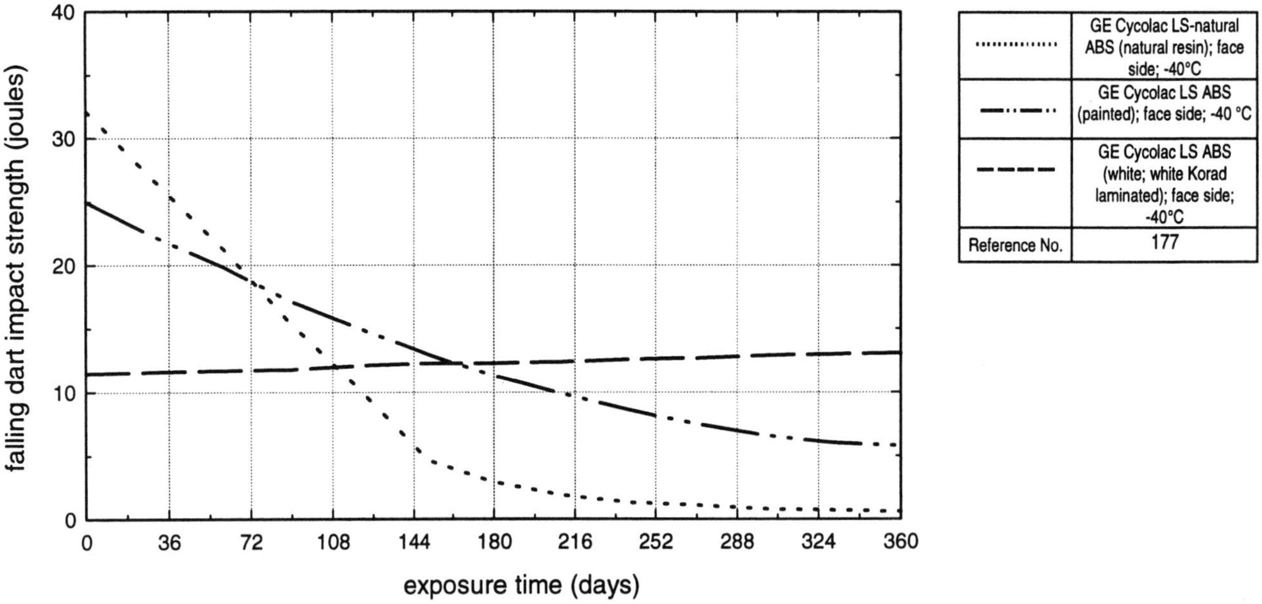

GRAPH 19: West Virginia Outdoor Weathering Exposure Time vs. Falling Dart Impact of ABS

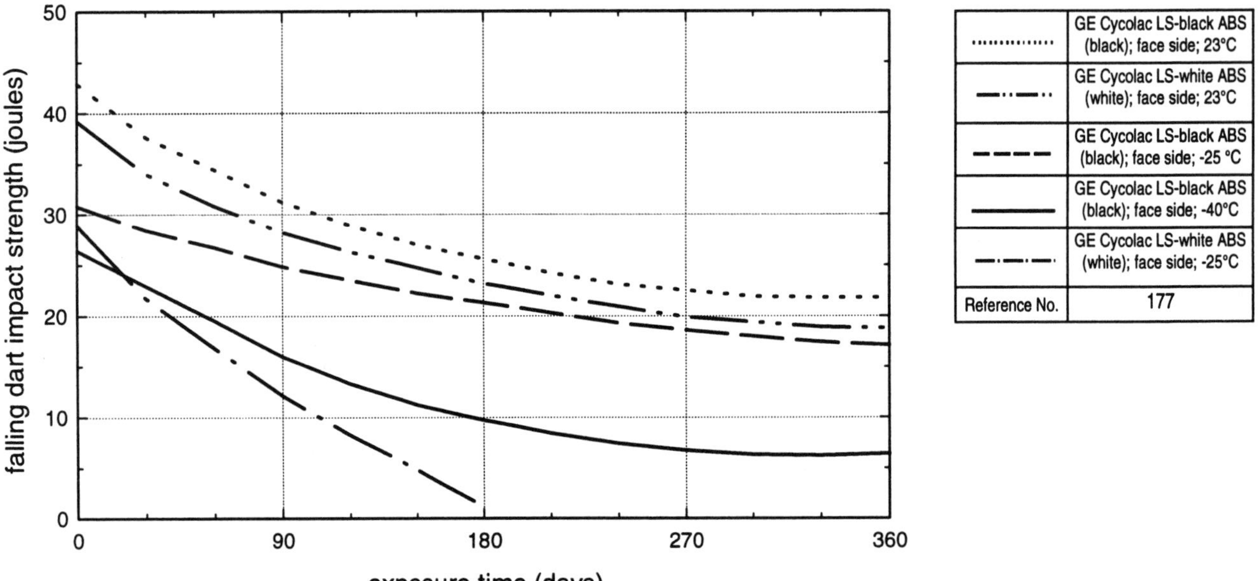

GRAPH 20: West Virginia Outdoor Weathering Exposure Time vs. Falling Dart Impact of ABS

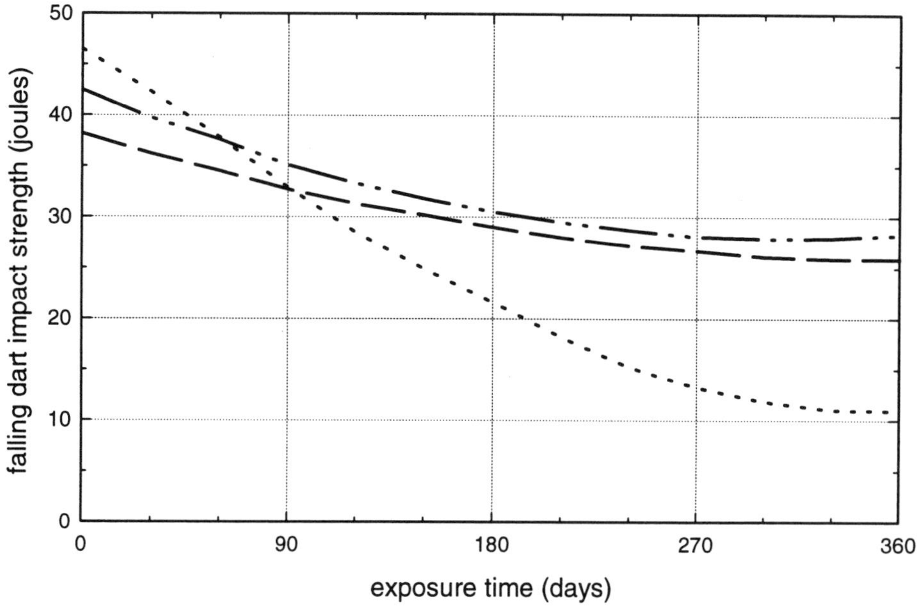

···············	GE Cycolac LS-natural ABS (natural resin); face side; 23°C
—·—··	GE Cycolac LS ABS (painted); face side; 23°C
— — — —	GE Cycolac LS ABS (white; white Korad laminated); face side; 23°C
Reference No.	177

GRAPH 21: West Virginia Outdoor Weathering Exposure Time vs. Flexural Modulus Retained of ABS

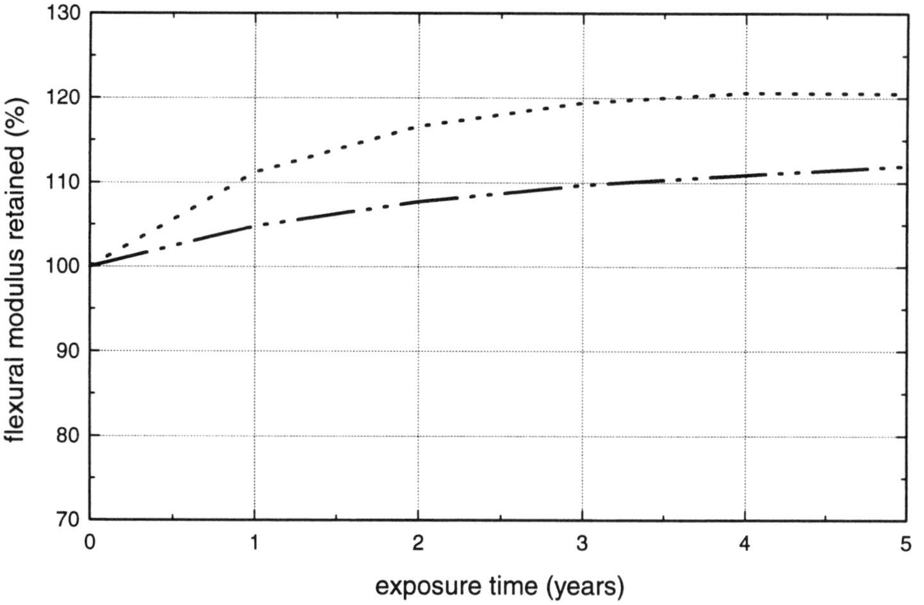

···············	GE Cycolac L-black ABS (black)
—·—··	GE Cycolac T-red ABS (red)
—··—··	GE Cycolac T-natural ABS (natural resin)
—··—··	GE Cycolac L-natural ABS (natural resin)
Reference No.	177

GRAPH 22: West Virginia Outdoor Weathering Exposure Time vs. Flexural Strength of ABS

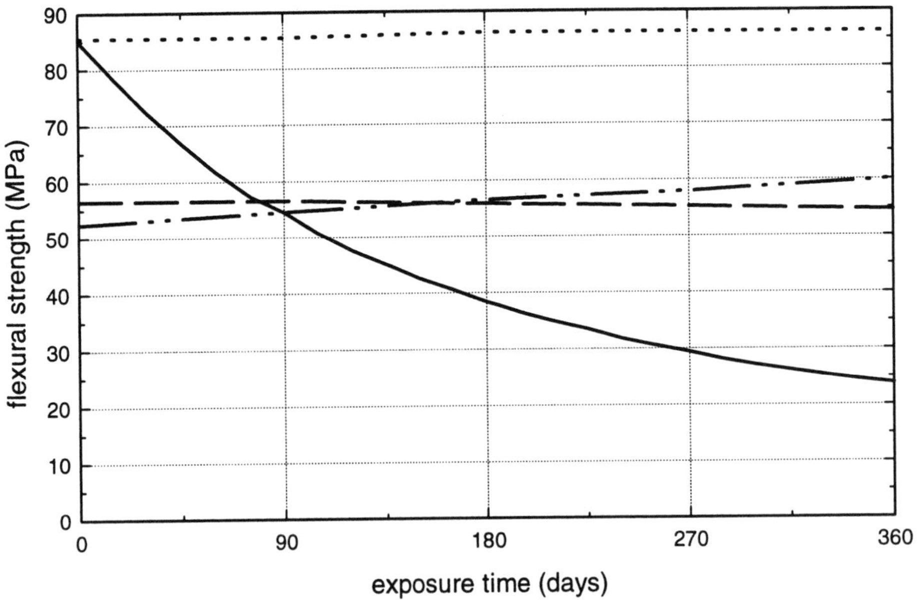

....	GE Cycolac LS-black ABS (black); -40 °C
—··—··	GE Cycolac LS-black ABS (black); 23 °C
– – – –	GE Cycolac LS-white ABS (white); 23 °C
———	GE Cycolac LS-white ABS (white); -40 °C
Reference No.	177

GRAPH 23: West Virginia Outdoor Weathering Exposure Time vs. Flexural Strength Retained of ABS

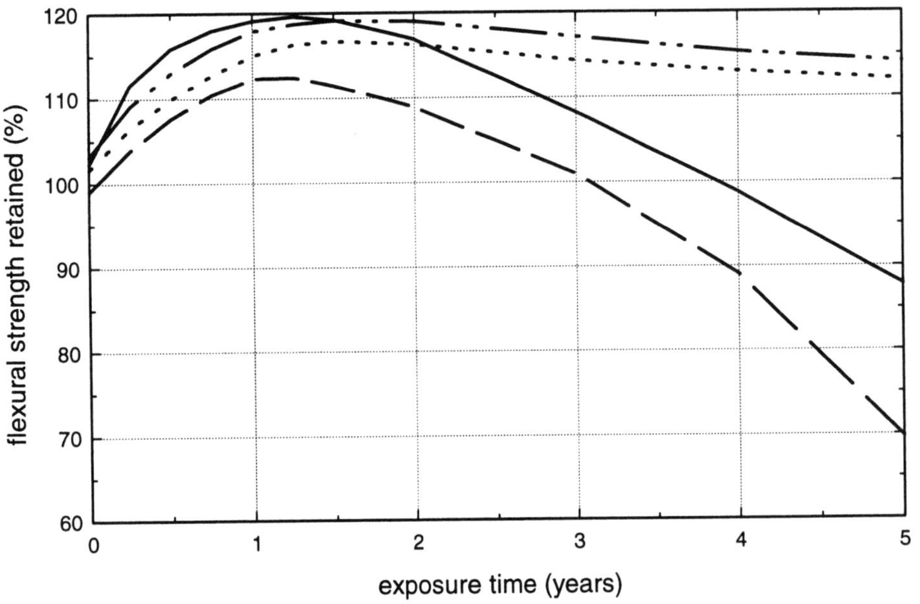

....	GE Cycolac L-black ABS (black)
—··—··	GE Cycolac T-red ABS (red)
– – – –	GE Cycolac T-natural ABS (natural resin)
———	GE Cycolac L-natural ABS (natural resin)
Reference No.	177

GRAPH 24: West Virginia Outdoor Weathering Exposure Time vs. Izod Impact Strength Retained of ABS

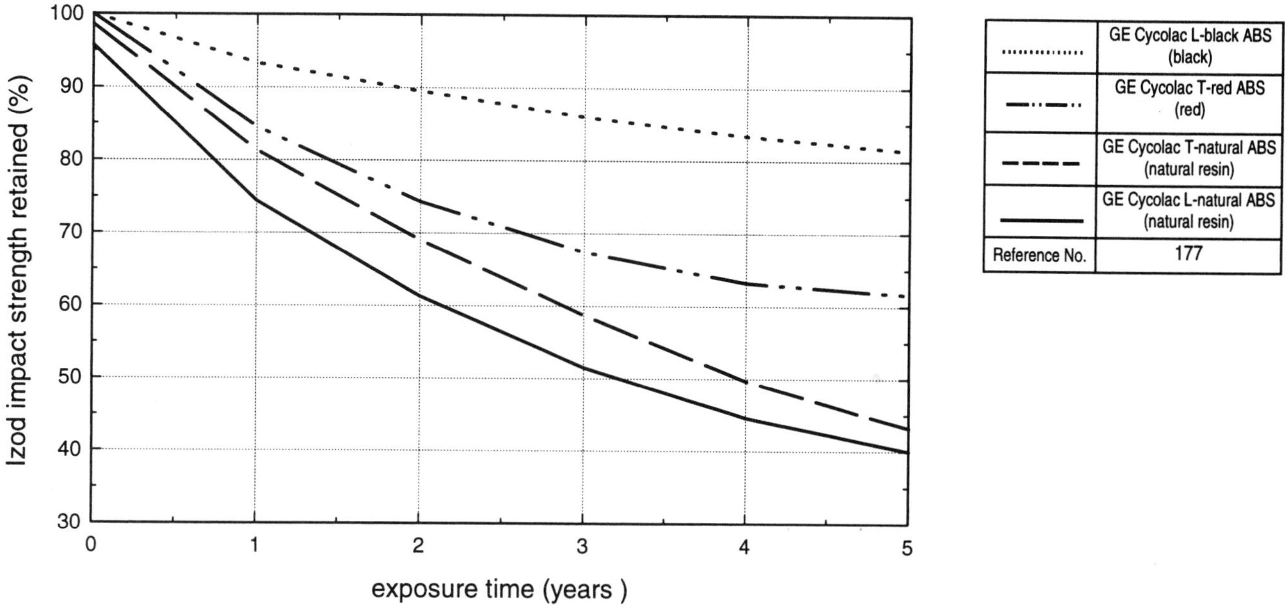

GRAPH 25: West Virginia Outdoor Weathering Exposure Time vs. Tensile Strength Retained of ABS

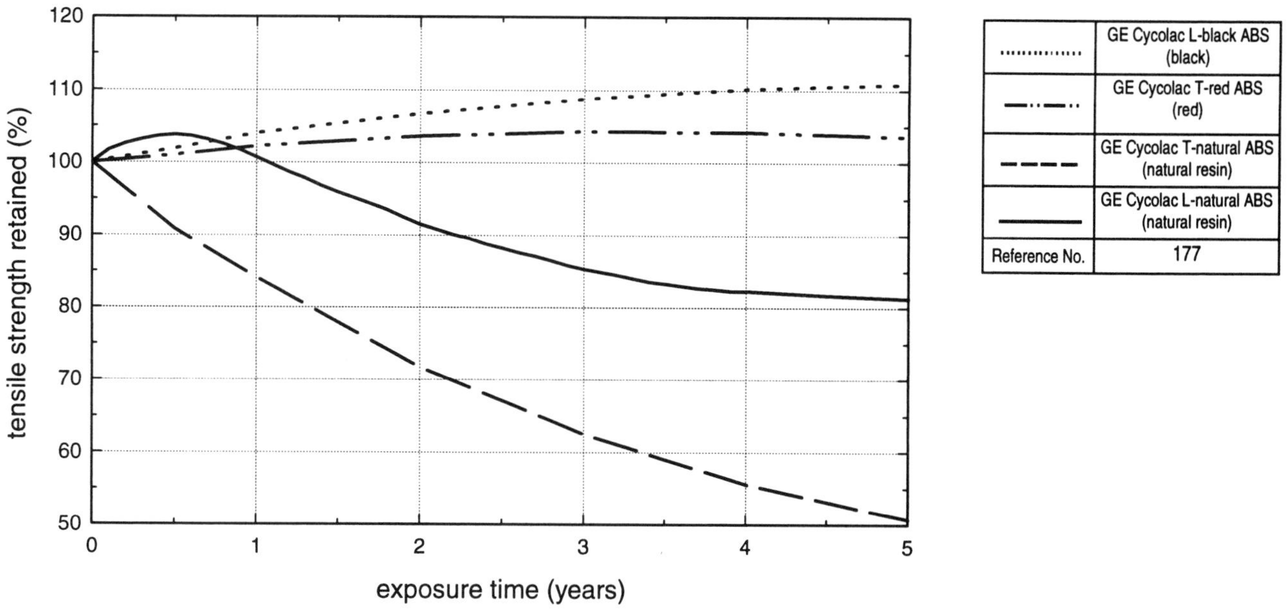

GRAPH 26: **Sunshine Weatherometer Exposure Time vs. Dynstat Impact Strength Retained of ABS**

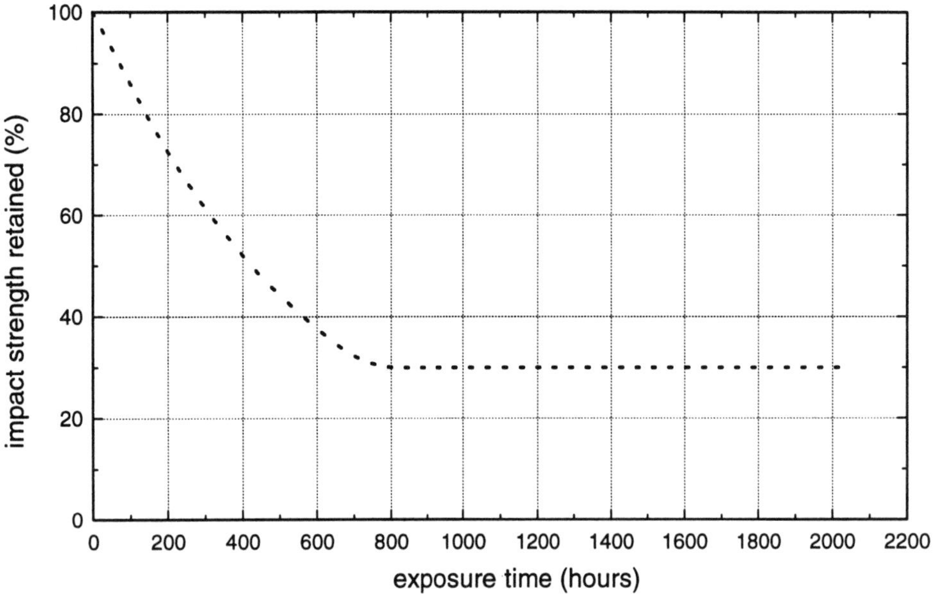

···············	ABS (weatherable, black); 63°C
Reference No.	211

GRAPH 27: **Sunshine Weatherometer Exposure Time vs. Elongation at Break Retained of ABS**

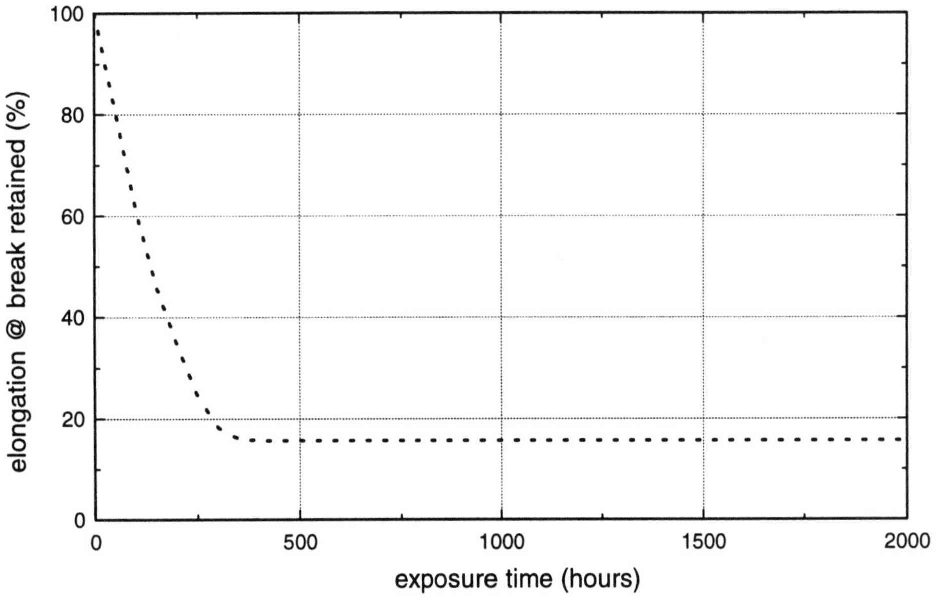

···············	ABS (weatherable, black); 63°C
Reference No.	212

GRAPH 28: **Sunshine Weatherometer Exposure Time vs. Gloss Retained of ABS**

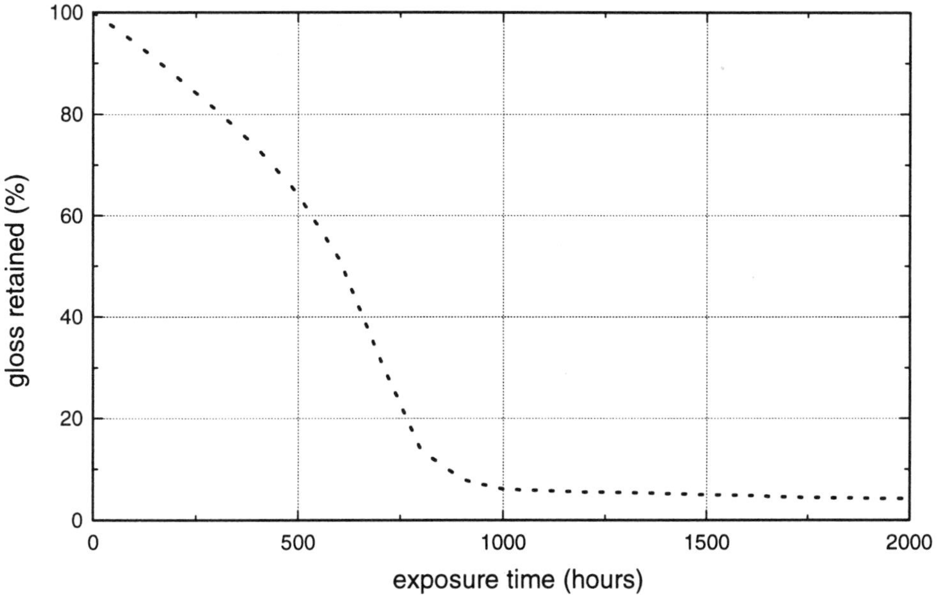

....	ABS (weatherable, black); 63°C
Reference No.	211

GRAPH 29: **Weatherometer Exposure Time vs. Impact Strength of ABS**

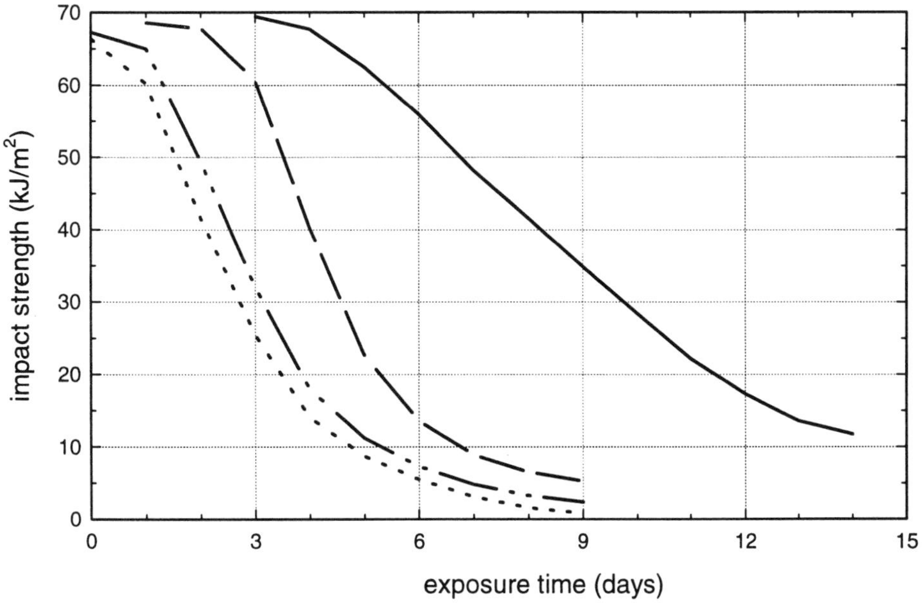

....	ABS (natural resin); tested at -40 °C DIN 53 453, blow to unexposed side; impact strength > 60kJ/m² denotes that specimen did not break
—·—·—··	ABS (natural resin); tested at -20 °C DIN 53 453, blow to unexposed side; impact strength > 60kJ/m² denotes that specimen did not break
– – – –	ABS (natural resin); tested at 0 °C DIN 53 453, blow to unexposed side; impact strength > 60kJ/m² denotes that specimen did not break
———	ABS (natural resin); tested at 20 °C DIN 53 453, blow to unexposed side; impact strength > 60kJ/m² denotes that specimen did not break
Reference No.	143

GRAPH 30: Xenotest 1200 Exposure Time vs. Impact Strength of ABS

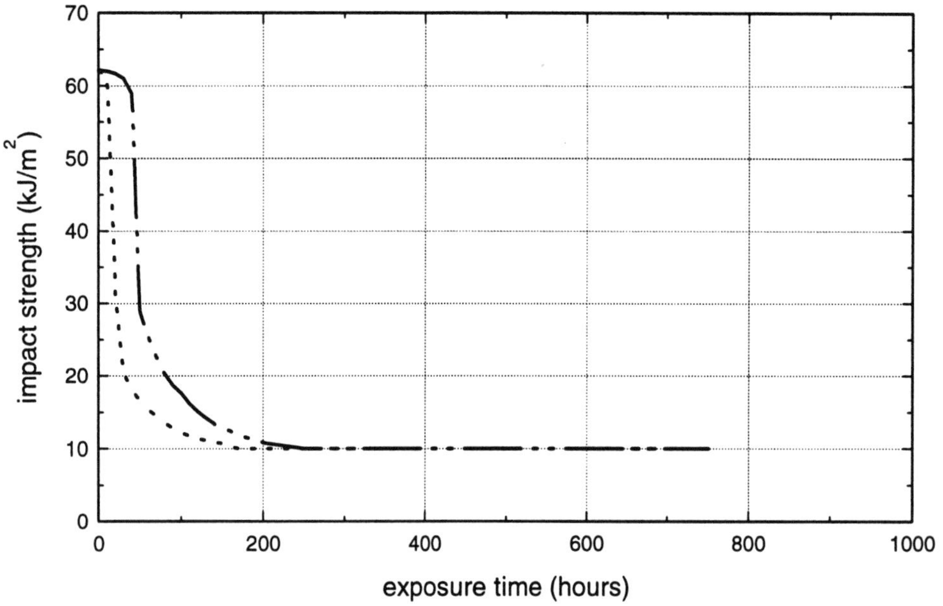

....	ABS; impact strength > 60 kJ/m² denotes that specimen did not break
—··—··	ABS (UV stabilized); impact strength > 60 kJ/m² denotes that specimen did not break
Reference No.	142

GRAPH 31: Accelerated Indoor UV Exposure Time vs. Delta E Color Change of ABS

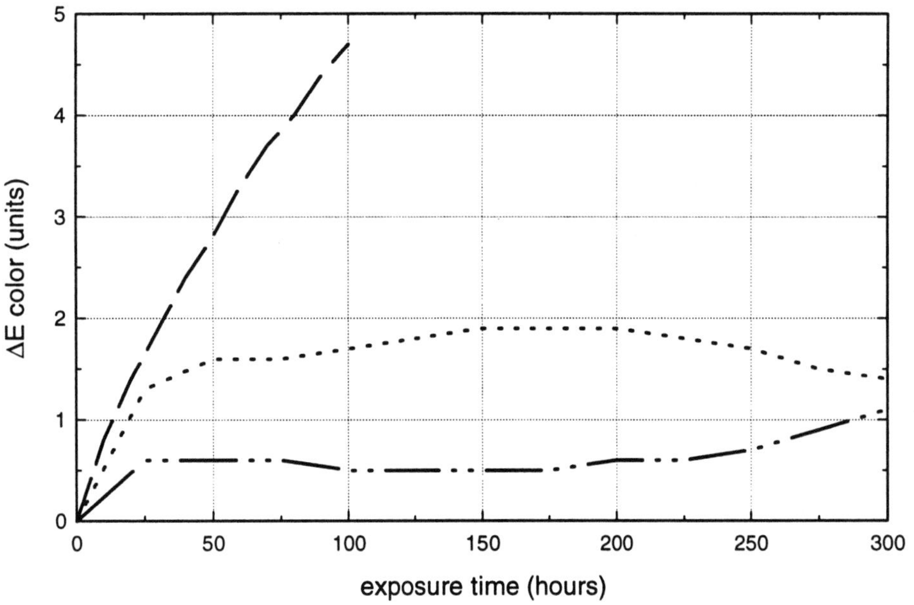

....	ABS (dibromostyrene ABS; flame retardant)
—··—··	ABS (UV stabilized, flame retardant)
————	ABS (flame retardant)
Reference No.	232

Chapter 2

Acetal Resin

UV Resistance

Acetal

Prolonged exposure to UV light will induce surface chalking and reduce the molecular weight leading to gradual embrittlement. As with the polylolefins it is found that the incorporation of a small amount of well dispersed carbon black increases resistance to UV degradation.

Reference: *Unknown Name,* reference book.

Outdoor Weather Resistance

DuPont: Delrin 507 BK-601 (additives: carbon black; features: UV stabilized); **Delrin**

Over time, exposure to UV light adversely affects the tensile strength of plastics. Weather resistant compositions of Delrin acetal resin have been developed to withstand such exposure.

For outdoor applications involving either intermittent exposure or a service life of 1 to 2 years, colors of Delrin acetal resin are generally suitable based on the property retention. For increased resistance to surface dulling and chalking and better tensile property retention, specially formulated colors containing an ultraviolet stabilizer provide significantly better performance. However, even with UV stabilized color compositions, surface dulling and chalking begin in about 6 to 8 months of exposure in Florida. The chalk may be removed by hand polishing in the early stages of development. If removal is delayed, the chalk layer hardens with time and becomes more difficult to remove.

For applications which require outdoor exposure to direct sunlight and much more than 2 years of useful life, the carbon black filled resin Delrin 507 BK-601 compositions have shown excellent retention of strength properties after 20 years of outdoor exposure in Arizona, Florida and Michigan. Over this period, essentially no loss of tensile strength occurred, but elongation was reduced to about 40% of the initial test value, with the greatest change in elongation occurring during the first 6 months of exposure.

A special more weatherable composition has the same property retention as Delrin 507 BK-601 but offers much improved gloss retention. It is used for outdoor applications where appearance is critical.

Reference: *Delrin Design Handbook For Du Pont Engineering Plastics,* supplier design guide (E-62619) - Du Pont Company, 1987.

Microbiologic Attack

Acetal

Acetal resins do not apear to be attacked by fungi, rodents and insects.

Reference: *Unknown Name,* reference book.

TABLE 06: Outdoor Weathering in Arizona of UV Stabilized DuPont Delrin Acetal Resin.

Material Family	Acetal Resin											
Material Supplier/ Grade	DuPont Delrin 507 BK-601											
Features	UV stabilized											
Reference Number	201	201	201	201	201	201	201	201	201	201	201	201
MATERIAL CHARACTERISTICS												
Sample Thickness (mm)	3.2	3.2	3.2	3.2	3.2	3.2	3.2	3.2	3.2	3.2	3.2	3.2
Sample Length (mm)	216	216	216	216	216	216	216	216	216	216	216	216
Sample Width (mm)	13	13	13	13	13	13	13	13	13	13	13	13
Sample Type	ASTM tensile specimen	ASTM tensile specimen	ASTM tensile specimen	ASTM tensile specimen	ASTM tensile specimen	ASTM tensile specimen	ASTM tensile specimen	ASTM tensile specimen	ASTM tensile specimen	ASTM tensile specimen	ASTM tensile specimen	ASTM tensile specimen
MATERIAL COMPOSITION												
additives	carbon black	carbon black	carbon black	carbon black	carbon black	carbon black	carbon black	carbon black	carbon black	carbon black	carbon black	carbon black
EXPOSURE CONDITIONS												
Exposure Type	outdoor weathering						outdoor weathering					
Exposure Location	Arizona						Arizona					
Exposure Country	USA						USA					
Exposure Time (days)	365	730	1095	1460	3650	7300	365	730	1095	1460	3650	7300
PROPERTIES RETAINED (%)												
Tensile Strength	101	102	101	104	99	99.9	100	100	100	103	99	91.6
Elongation	60	55	45	55	50	40	35	65	60	70	50	55

Chapter 3

Acetal Copolymer

Weather Resistance

Mitsubishi Gas Chemical: Iupital F20-02 (features: natural resin, general purpose grade); **Iupital** (features: black color, general purpose grade); **Iupital F20-52** (features: natural resin, weatherable); **Iupital** (features: weatherable, black color)

Iupital provides favorable weatherability (light resistance). Iupital "standard grades" can be used without any problem for weatherproof applications except for outdoor use under direct sunlight. "Weatherproof grades" are recommended for applications such as parts used for outdoor interior and exterior parts of vehicles, etc. which require high weatherability (light resistance).

Reference: *Engineering Plastics Acetal Copolymer - Iupital,* supplier design guide (M.G.C.91042000P.A.) - Mitsubishi Gas Chemical Company, Inc., 1991.

UV Resistance

BASF AG: Ultraform N 2325 U (features: UV stabilized); **Ultraform**

Acetal resins are attacked by ultraviolet radiation. If articles produced from them are exposed for long periods to direct sunlight, they lose their gloss and become brittle. Their outdoor service life can be roughly doubled by adding UV stabilizers. Even more decided improvements can be achieved by use of the special grade Ultraform N 2325 U.

Reference: *Ultraform Polyacetal (POM) Product Line, Properties, Processing,* supplier design guide (B 563/1e - (888) 4.91) - BASF Aktiengesellschaft, 1991.

Outdoor Weather Resistance

BASF AG: Ultraform N 2320 XU015 (features: weatherable, UV stabilized, developmental material); **Ultraform N 2325 U** (features: UV stabilized)

Ultraform N 2320 XU015 has been developed to meet the increasing demand for weather resistant materials. Although the properties typical of polyacetals are retained, the new product exhibits increased weathering resistance compared to the former Ultraform N2325 U.

Reference: *Topics In Chemistry - BASF Plastics Research And Development,* supplier technical report - BASF Aktiengesellschaft, 1992.

Hoechst Celanese: Celcon GC25A-CD3501 black (features: black color; material compostion: 25% glass fiber reinforcement); **Celcon M90** (features: natural resin, black color, general purpose grade); **Celcon M90-CD3068 black** (features: general purpose grade); **Celcon UV25** (features: natural resin, UV stabilized); **Celcon UV90** (features: natural resin, UV stabilized); **Celcon WR25 black** (features: weatherable, UV stabilized, black color); **Celcon WR90 black** (features: weatherable, UV stabilized, black color)

Many applications require that Celcon acetal copolymer parts withstand exposure to sunlight through use in naturally lighted areas or in outdoor applications. All plastic materials are affected by ultraviolet light exposure and suffer a certain amount of degradation. Degradation is usually noticed by fading, chalking and embrittlement.

Natural and pigmented standard Celcon grades are not recommended for applications where prolonged ultraviolet exposure is encountered. Special UV stable packages are available, such as UV90 and UV25 grades. These materials are available in natural and colored versions (precompounded or color concentrate let down). UV90 colors and M90-CD3068 black are especially recommended for applications involving UV exposure. Special black weather resistant grades of M90 and M25 designated WR90 and WR25 have been developed for maximum weathering resistance for outdoor applications. Color retention testing indicates the WR grades will have excellent long-term property retention in outdoor applications.

UV stabilized UV90 colors and M90-CD3068 black are recommended for use in UV resistant applications such as automotive interiors, lawn sprinklers, toys, and boating accessories. For maximum weathering resistance, WR25 and WR90 are recommended for outdoor applications such as irrigation gates, permanent watering systems, exterior door handles and hardware for automobiles, RV's, campers, boats and snowmobiles.

Reference: *Celcon Acetal Copolymer,* supplier design guide (90-350 7.5M/490) - Hoechst Celanese Corporation, 1990.

GRAPH 32: **Outdoor Exposure Time vs. Impact Strength Retained of Acetal Copolymer**

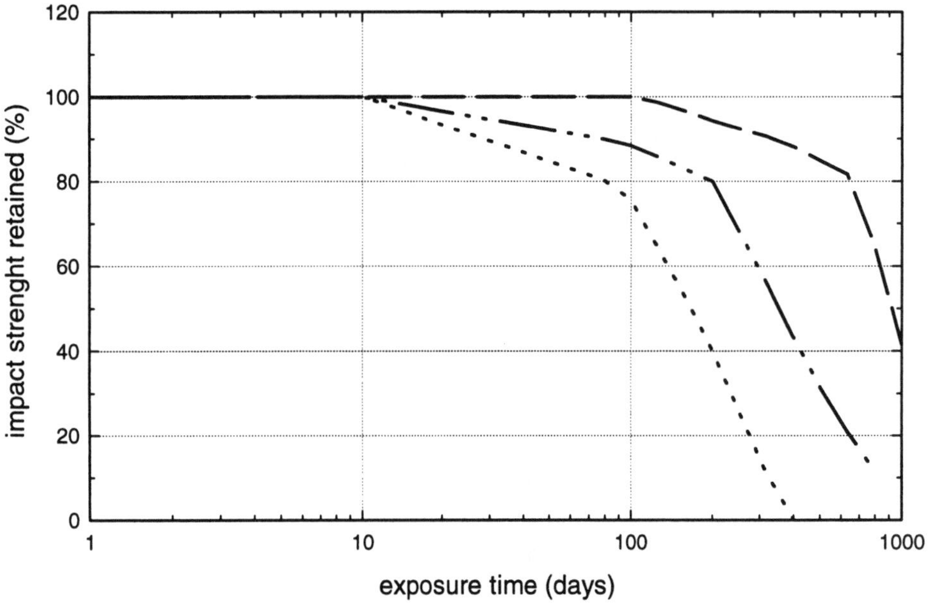

···············	BASF AG Ultraform N 2320 Natural Acetal Copol. (natural resin)
—··—··	BASF AG Ultraform N 2320 U Natural Acetal Copol. (natural resin, UV stabilized)
– – – –	BASF AG Ultraform N 2325 U Acetal Copol. (UV stabilized)
Reference No.	183

GRAPH 33: New Jersey and Arizona Outdoor Exposure Time vs. Melt Index of Acetal Copolymer

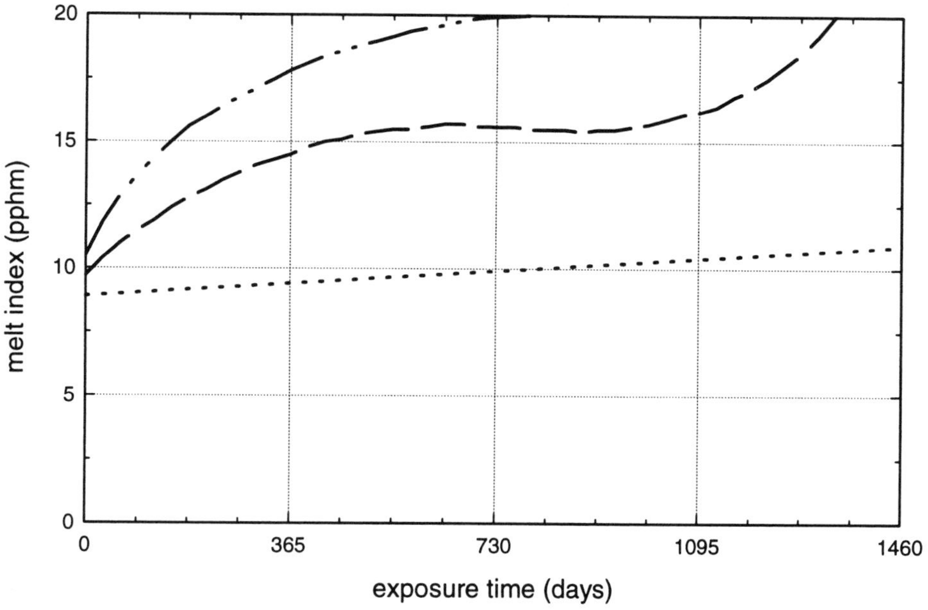

....	Hoechst Cel. Celcon M90-CD3068 black Acetal Copol. (gen. purp. grade)
—·–·—··	Hoechst Cel. Celcon M90 Acetal Copol. (natural resin, black, gen. purp. grade)
– – – –	Hoechst Cel. Celcon UV90 Acetal Copol. (natural resin, UV stabilized)
Reference No.	210

GRAPH 34: New Jersey and Arizona Outdoor Exposure Time vs. Tensile Impact of Acetal Copolymer

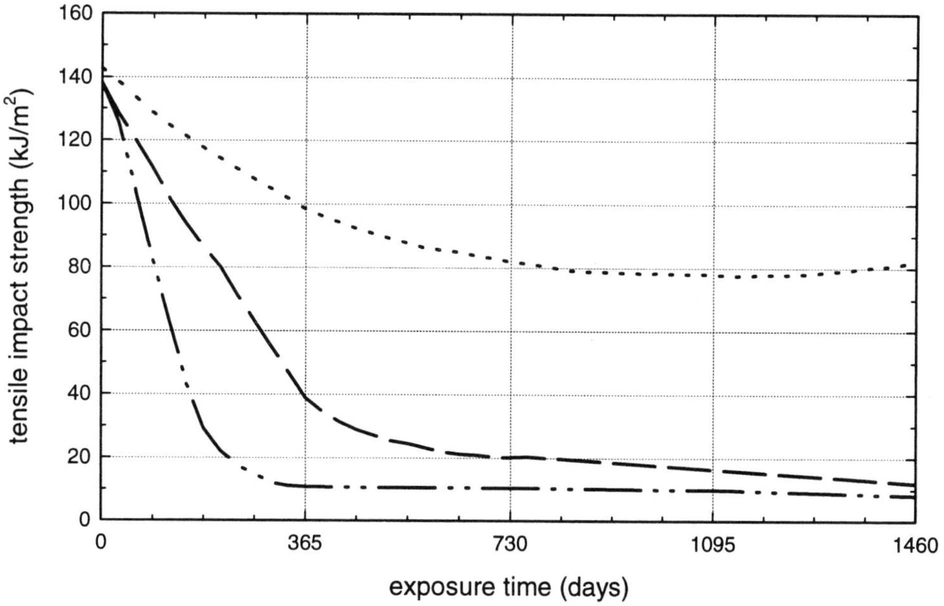

....	Hoechst Cel. Celcon M90-CD3068 black Acetal Copol. (gen. purp. grade)
—·–·—··	Hoechst Cel. Celcon M90 Acetal Copol. (natural resin, black, gen. purp. grade)
– – – –	Hoechst Cel. Celcon UV90 Acetal Copol. (natural resin, UV stabilized)
Reference No.	210

GRAPH 35: New Jersey and Arizona Outdoor Exposure Time vs. Tensile Strength at Yield of Acetal Copolymer

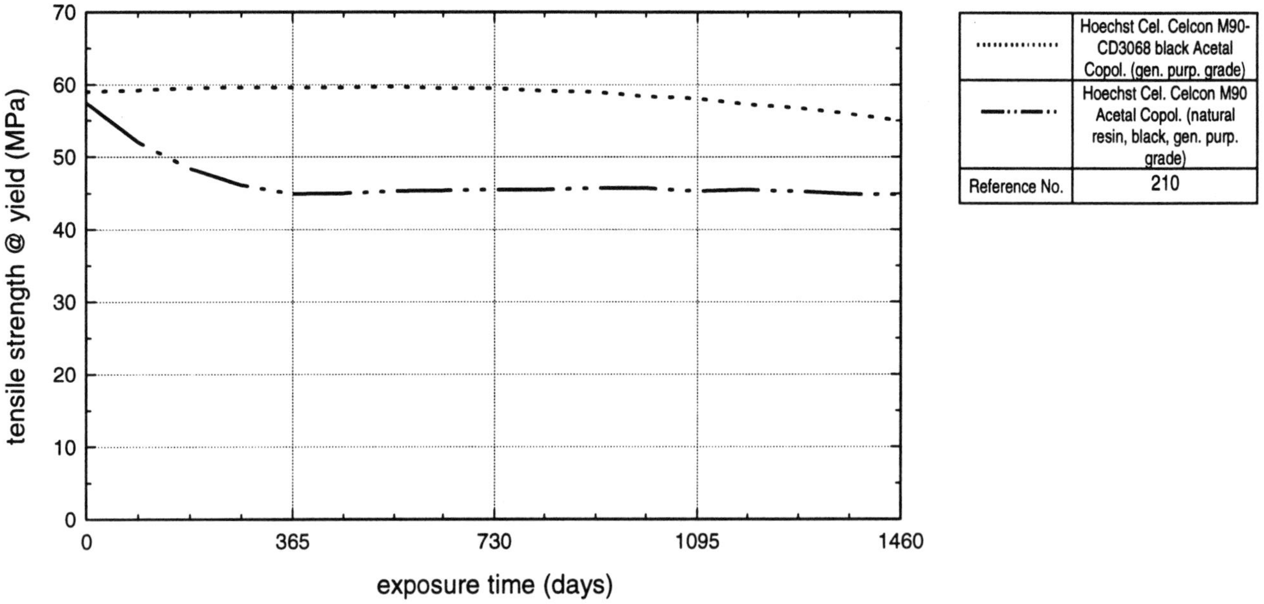

GRAPH 36: New Jersey Outdoor Exposure Time vs. Tensile Strength at Yield of Acetal Copolymer

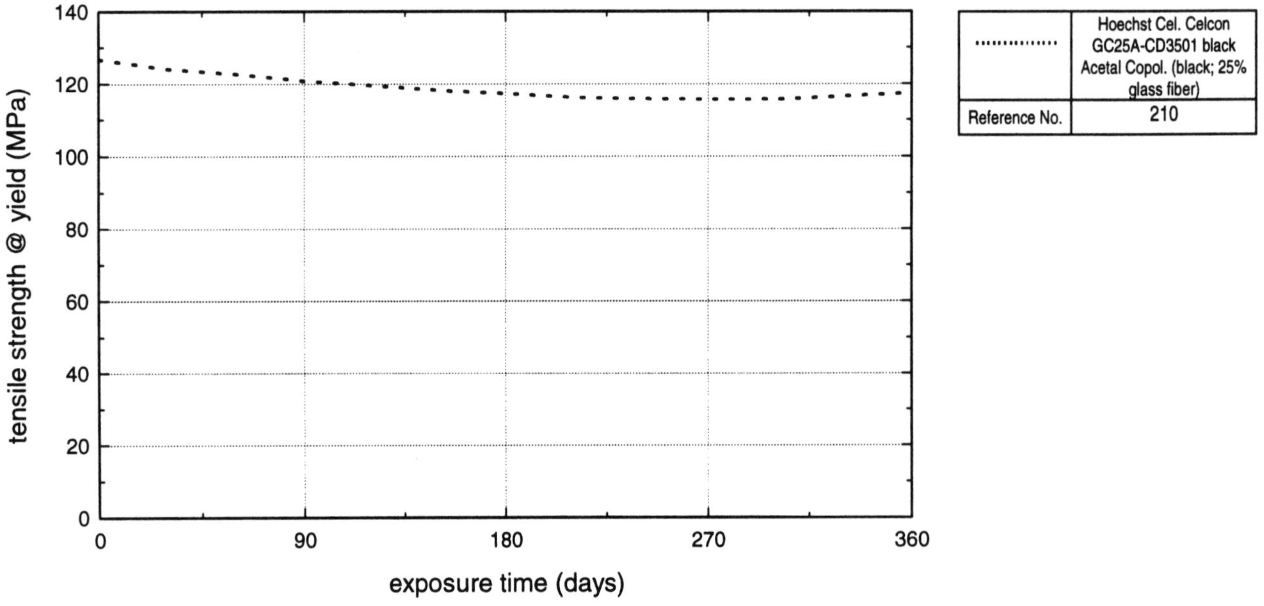

GRAPH 37: QUV Exposure Time vs. Delta E Color Change of Acetal Copolymer

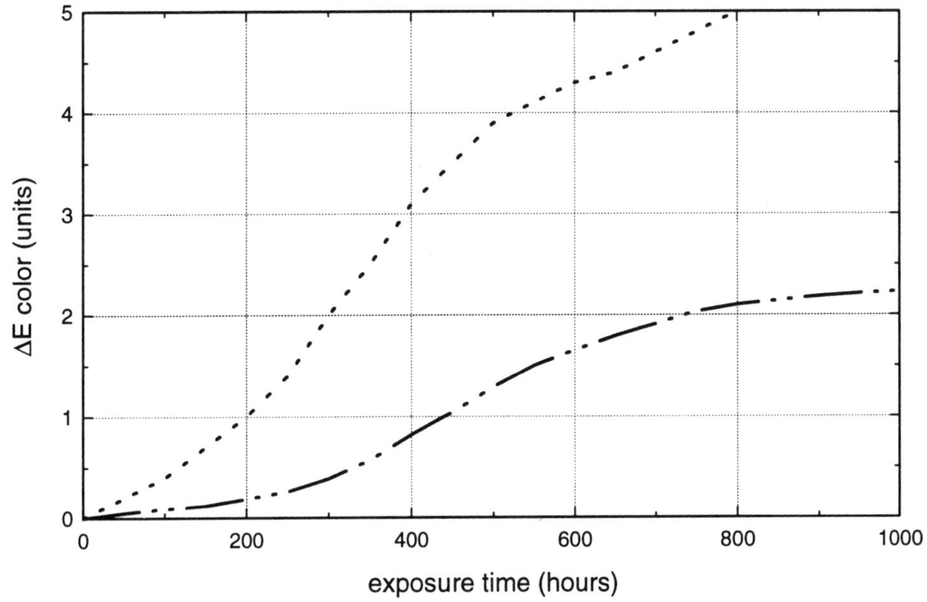

··············	Hoechst Cel. Celcon M90-CD3068 black Acetal Copol. (gen. purp. grade)
—··—··	Hoechst Cel. Celcon WR90 black Acetal Copol. (weatherable, UV stabilized, black)
—·—··	Hoechst Cel. Celcon WR25 black Acetal Copol. (weatherable, UV stabilized, black)
Reference No.	210

GRAPH 38: Sunshine Weatherometer Exposure Time vs. Discoloration of Acetal Copolymer

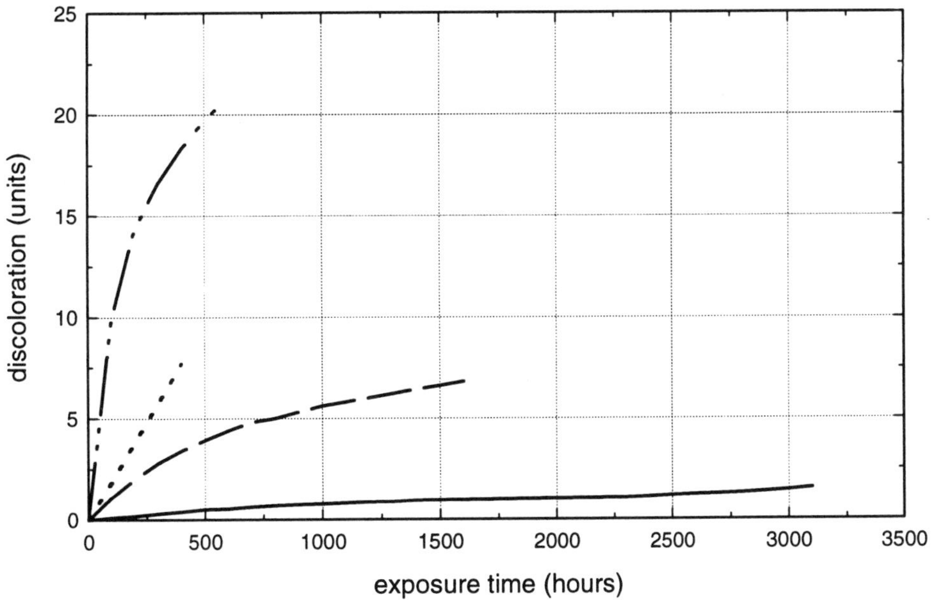

—·—··	Mitsubishi Iupital F20-02 Acetal Copol. (natural resin, gen. purp. grade); 83°C without rain
··············	Mitsubishi Iupital F20-02 Acetal Copol. (black, gen. purp. grade); 83°C without rain
— — —	Mitsubishi Iupital F20-52 Acetal Copol. (natural resin, weatherable); 83°C without rain
———	Mitsubishi Iupital F20-52 Acetal Copol. (weatherable, black); 83°C without rain
Reference No.	237

GRAPH 39: **Sunshine Weatherometer Exposure Time vs. Elongation Retained of Acetal Copolymer**

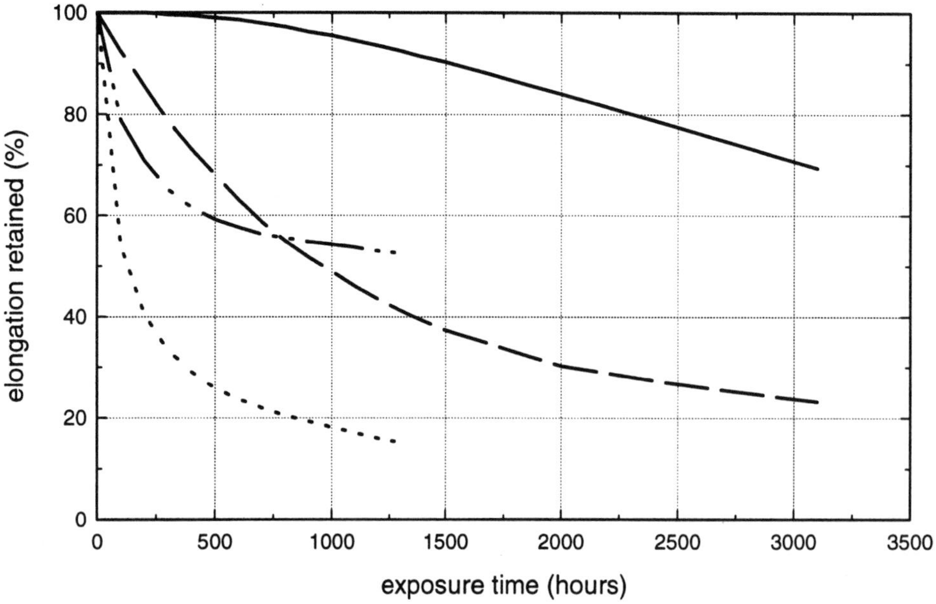

··············	Mitsubishi Iupital F20-02 Acetal Copol. (natural resin, gen. purp. grade); 83°C without rain
—··—··	Mitsubishi Iupital F20-02 Acetal Copol. (black, gen. purp. grade); 83°C without rain
— — — —	Mitsubishi Iupital F20-52 Acetal Copol. (natural resin, weatherable); 83°C without rain
————	Mitsubishi Iupital F20-52 Acetal Copol. (weatherable, black); 83°C without rain
Reference No.	237

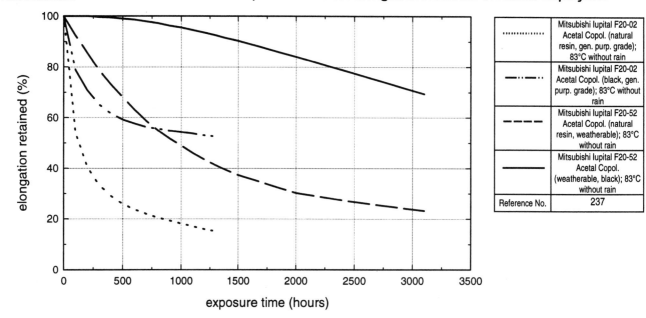

GRAPH 40: **Sunshine Weatherometer Exposure Time vs. Tensile Strength Retained of Acetal Copolymer**

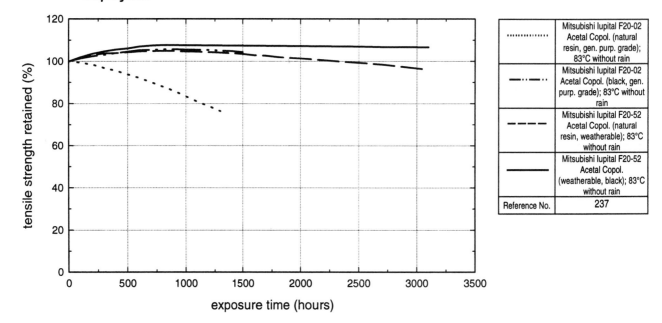

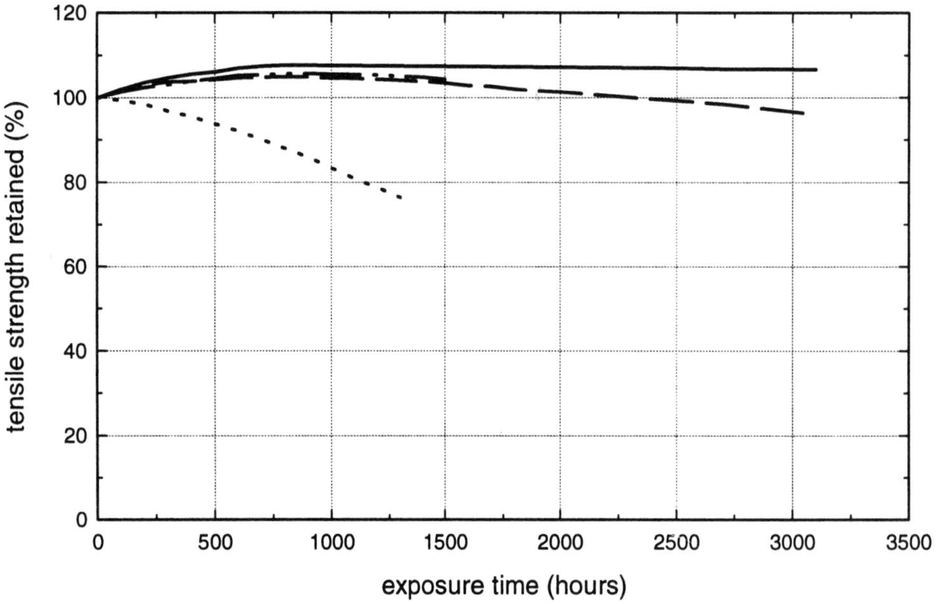

··············	Mitsubishi Iupital F20-02 Acetal Copol. (natural resin, gen. purp. grade); 83°C without rain
—··—··	Mitsubishi Iupital F20-02 Acetal Copol. (black, gen. purp. grade); 83°C without rain
— — — —	Mitsubishi Iupital F20-52 Acetal Copol. (natural resin, weatherable); 83°C without rain
————	Mitsubishi Iupital F20-52 Acetal Copol. (weatherable, black); 83°C without rain
Reference No.	237

GRAPH 41: Xenon Arc Weatherometer Exposure Time vs. Relative Gloss of Acetal Copolymer

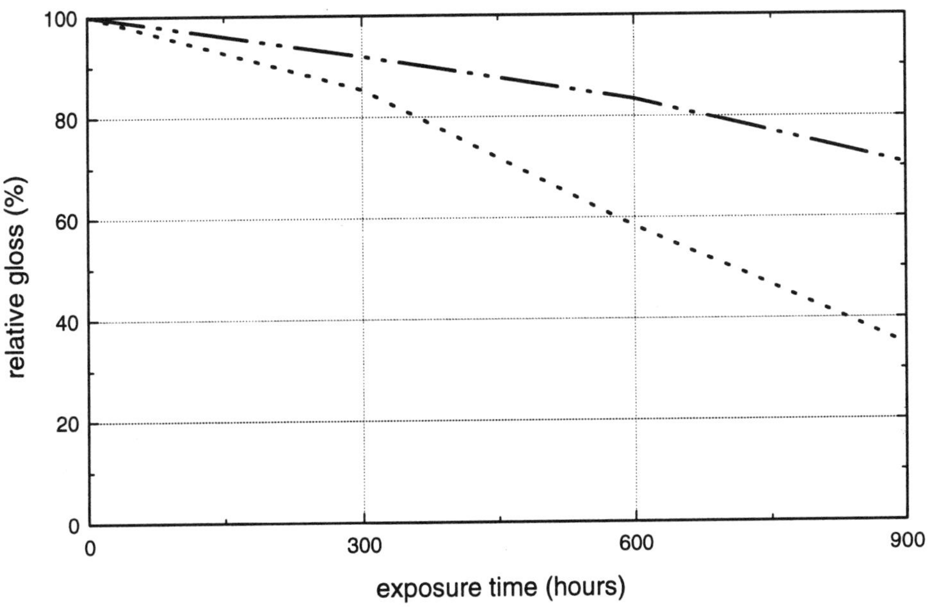

···············	BASF AG Ultraform N 2325 U Acetal Copol. (UV stabilized); test method SAE J1960
—·· —··	BASF AG Ultraform N 2320 XU015 Acetal Copol. (weatherable, UV stabilized, developmental material); test method SAE J1960
Reference No.	182

Chapter 4

Acrylate Styrene Acrylonitrile Polymer

Outdoor Weather Resistance

BASF AG: Luran S

Since their constituent acrylic elastomer is free from butadiene, Luran S resins offer much higher resistance to ultraviolet radiation and atmospheric oxygen than high impact polystyrene or ABS. However, with prolonged exposure to direct sunlight and ultraviolet radiation, even Luran S articles suffer an impairment of their mechanical properties.

In outdoor exposure tests, only one surface of the specimen usually faces the sun. (In Ludwigs-hafen, the specimens are exposed at an angle of 45° facing south). In the DIN 53453 impact test, either the exposed or the unexposed side may be struck. If it is the unexposed, the exposed side will undergo sudden tensile strain. Thus an excellent indication of the outdoor performance will be obtained, because the slightest degradation of the exposed surface immediately leads to a decrease in the original value measured. As opposed to this, if it is the exposed side that is struck, which is the most frequent case in practice, the decrease in impact strength will not become apparent until the specimen has been exposed for a very much longer period.

The decrease in impact strength that occurs after long periods of exposure outdoors can be ascribed to damage at the surface or just below it. Consequently, the values determined for the impact strength lie roughly at the same level as those for the notched impact strength.

The mechanical properties of colored articles usually commence to deteriorate later than those of articles produced from uncolored material. Particularly good outdoor performance is obtained with material colored to black shades.

Obviously, results obtained on test specimens can serve only as a means for comparing various products of colorations. The actual life of an article exposed outdoors depends not only on the properties of the raw material but also on the production conditions and other parameters, e.g. the stress pattern and the nature of the exposure site.

Change in Shade By virtue of their configuration, Luran S resins display extremely good resistance to yellowing. The period of time that elapses before the first signs of yellowing become visible is very much longer than that for comparable ABS products. In addition, the rate of yellowing on further exposure is very slow, because acrylic rubber is much more resistant than butadiene. A further improvement can be achieved by adding UV stabilizers.

Since the resins themselves have such high resistance to yellowing and great care is exercised in selecting pigments for coloration, it is obvious that the various shades in which Luran S is supplied have lightfastness ratings of at least 7 on the DIN 54003/DIN 53388 scale.

The great resistance to yellowing is particularly advantageous for automobile exterior trim. In this case, Luran S in dark shades displays very little tendency towards graying when it is exposed to ultraviolet radiation or outdoors and subsequently brought into contact with hot water and detergent solutions. These conditions correspond to those suffered by motor vehicles on leaving the product line, i.e. they are kept outdoors in a yard and subsequently dewaxed and cleaned.

Reference: *Luran S Acrylonitrile Styrene Acrylate Product Line, Properties, Processing,* supplier design guide (B 566 e / 11.90) - BASF Aktiengesellschaft, 1990.

Accelerated Artificial Weathering Resistance

BASF AG: Luran S

The effect of exposure to ultraviolet radiation on the impact strength of Luran S and ABS was plotted from the results obtained on exposing one side of 50 x 6 x 4 mm standard bar specimens in the Xenotest 1200 and subsequently subjecting the specimens to the DIN 53453 impact test with the blow struck on the unexposed side. Luran S retains its impact strength under these conditions over a long period (i.e. about seven times as long as ABS).

Reference: *Luran S Acrylonitrile Styrene Acrylate Product Line, Properties, Processing,* supplier design guide (B 566 e / 11.90) - BASF Aktiengesellschaft, 1990.

Mitsubishi Rayon: Shinko-Lac T110 (features: black color, heat stabilized); **Shinko-Lac T115** (features: black color, heat stabilized); **Shinko-Lac T120** (features: black color, heat stabilized)

Shinko-Lac ASA "T series" appearance and physical properties hardly change after accelerated weathering tests in a sunshine weatherometer.

Reference: *Shinko-Lac ASA T Weatherable And Heat Resistant ASA Resin,* supplier design guide - Mitsubishi Rayon Company.

Effect of UV Stabilizers on Weatherability

BASF AG: Luran S

The beneficial effect exerted by ultraviolet stabilizers on the outdoor performance of Luran has been shown. The impact resistance is increased by a multiple; and the excellent lightfastness and high resistance to yellowing of colored Luran S are increased even more.

Reference: *Luran S Acrylonitrile Styrene Acrylate Product Line, Properties, Processing,* supplier design guide (B 566 e / 11.90) - BASF Aktiengesellschaft, 1990.

TABLE 07: Outdoor Weathering in Ludwigshafen, Germany of BASF AG Luran Acrylate Styrene Acrylonitrile Polymer.

Material Family	Acrylate Styrene Acrylonitrile Polymer			
Material Supplier/ Grade	BASF AG Luran S 776 S			
Features	high impact, moderate flow			
Reference Number	143	143	143	143
EXPOSURE CONDITIONS				
Exposure Type	outdoor weathering			
Exposure Location	Ludwigshafen			
Exposure Country	Germany			
Exposure Note	45° angle south			
Exposure Time (days)	15	30	45	60
PROPERTIES RETAINED (%)				
Notched Impact Strength	98.6 (notched side exposed to light)	97.2 (notched side exposed to light)	101.4 (notched side exposed to light)	99.3 (notched side exposed to light)
Impact Strength	100 (no break; struck on unexposed side, thus subjecting exposed side to dynamic loading)	100 (no break; struck on unexposed side, thus subjecting exposed side to dynamic loading)	100 (no break; struck on unexposed side, thus subjecting exposed side to dynamic loading)	100 (no break; struck on unexposed side, thus subjecting exposed side to dynamic loading)

GRAPH 42: **Outdoor Weathering Exposure Time vs. Yellowness Index of ASA**

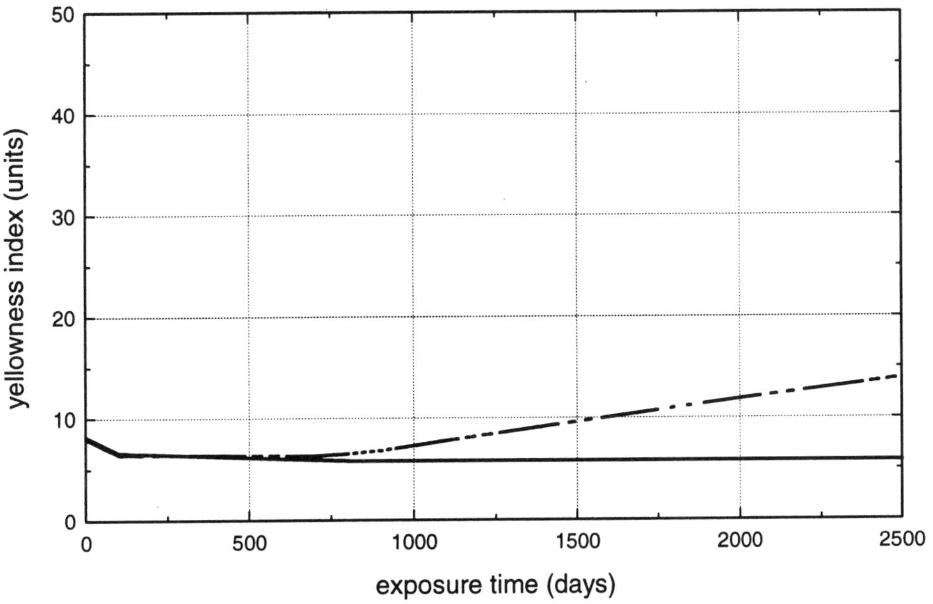

···············	BASF AG Luran S 797 SE ASA (moderate flow, high impact; extrus.); exposure to sunshine
—··—··	BASF AG Luran S 776 SE ASA (moderate flow, high impact; extrus.); exposure to sunshine
– – – –	BASF AG Luran S 797 SE UV ASA (moderate flow, high impact, UV stabilized; extrus.); exposure to sunshine
———	BASF AG Luran S 776 SE UV ASA (moderate flow, high impact, UV stabilized; extrus.); exposure to sunshine
Reference No.	142

GRAPH 43: **Okinawa, Japan Outdoor Weathering Exposure Time vs. Delta E Color Change of ASA**

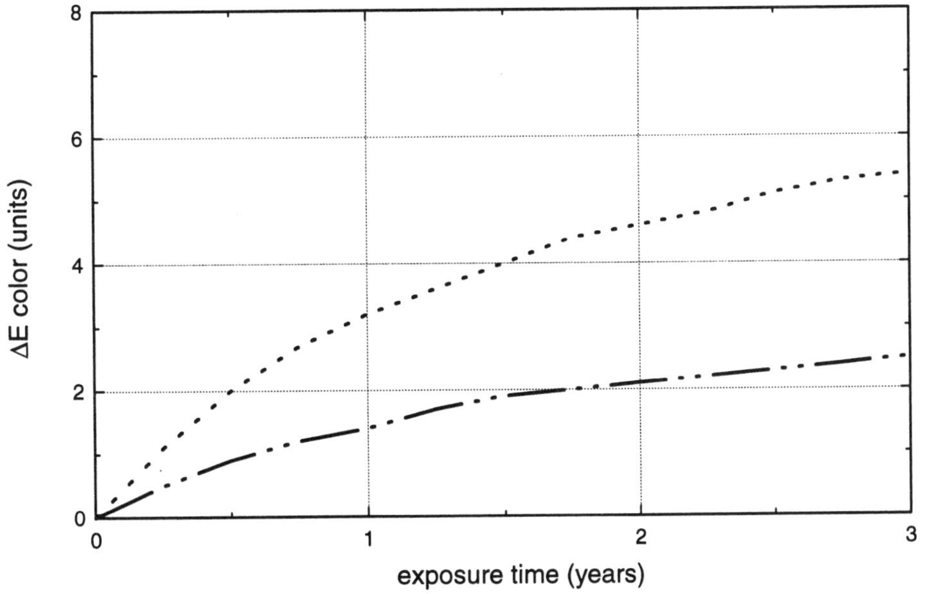

···············	Mitsub. Ray. Shinko-Lac ASA (red, appearance grade)
—··—··	Mitsub. Ray. Shinko-Lac ASA (white, appearance grade)
Reference No.	212

GRAPH 44: Okinawa, Japan Outdoor Weathering Exposure Time vs. Dynstat Impact Strength Retained of ASA

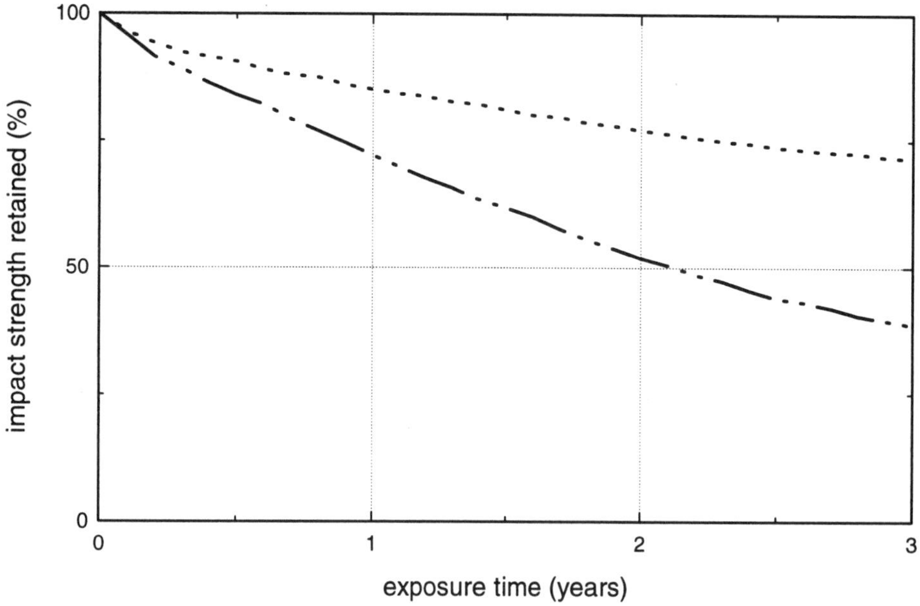

···············	Mitsub. Ray. Shinko-Lac ASA (black, gen. purp. grade)
—··—··	Mitsub. Ray. Shinko-Lac ASA (black, heat stabilized)
Reference No.	212

GRAPH 45: Okinawa, Japan Outdoor Weathering Exposure Time vs. Elongation at Break Retained of ASA

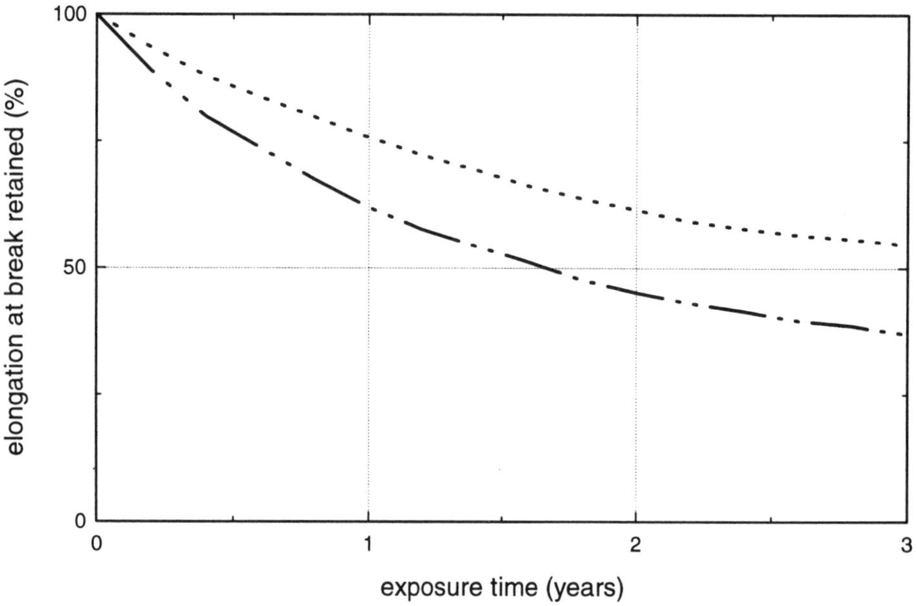

···············	Mitsub. Ray. Shinko-Lac ASA (black, gen. purp. grade)
—··—··	Mitsub. Ray. Shinko-Lac ASA (black, heat stabilized)
Reference No.	212

GRAPH 46: **Okinawa, Japan Outdoor Weathering Exposure Time vs. Gloss Retained of ASA**

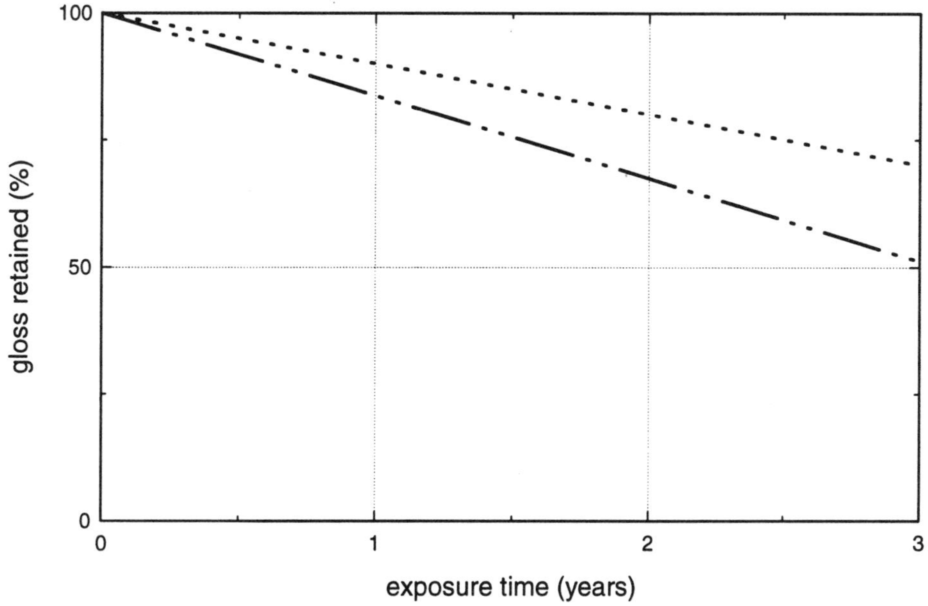

···············	Mitsub. Ray. Shinko-Lac ASA (black, gen. purp. grade)
—··—··	Mitsub. Ray. Shinko-Lac ASA (black, heat stabilized)
Reference No.	212

GRAPH 47: **Sunshine Weatherometer Exposure Time vs. Dynastat Impact Strength Retained of ASA**

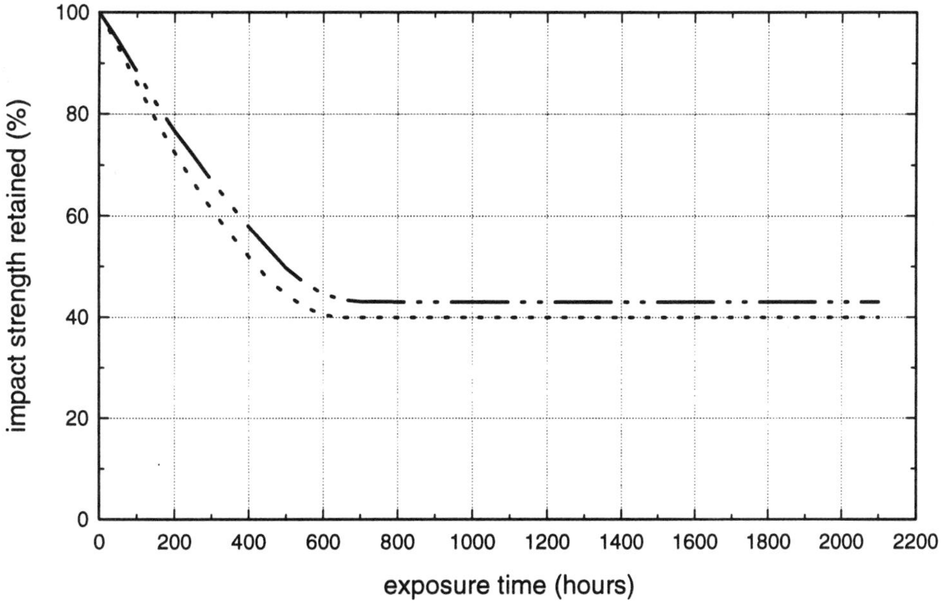

···············	Mitsub. Ray. Shinko-Lac T110 ASA (black, heat stabilized); 63°C
—··—··	Mitsub. Ray. Shinko-Lac T120 ASA (black, heat stabilized); 63°C
Reference No.	211

GRAPH 48: Sunshine Weatherometer Exposure Time vs. Dynstat Impact Strength Retained of ASA

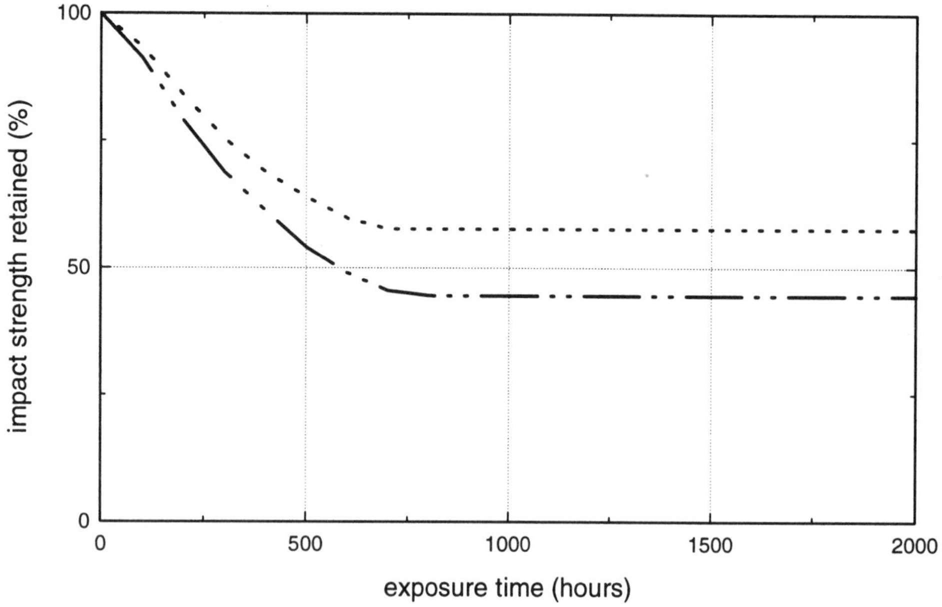

···············	Mitsub. Ray. Shinko-Lac ASA (black, gen. purp. grade); 63°C
—··—··	Mitsub. Ray. Shinko-Lac ASA (black, heat stabilized); 63°C
Reference No.	212

GRAPH 49: Sunshine Weatherometer Exposure Time vs. Elongation at Break Retained of ASA

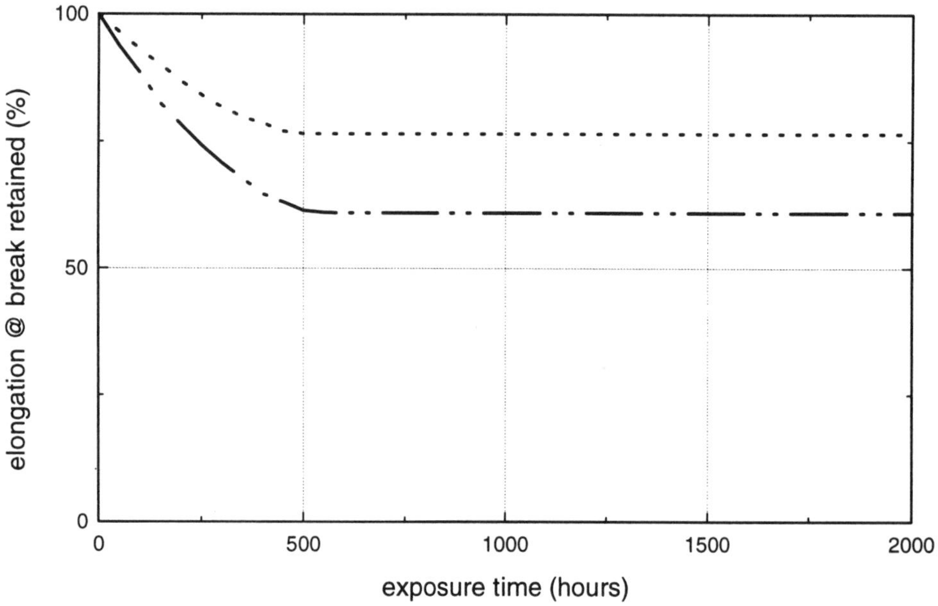

···············	Mitsub. Ray. Shinko-Lac ASA (black, gen. purp. grade); 63°C
—··—··	Mitsub. Ray. Shinko-Lac ASA (black, heat stabilized); 63°C
Reference No.	212

GRAPH 50: **Sunshine Weatherometer Exposure Time vs. Gloss Retained of ASA**

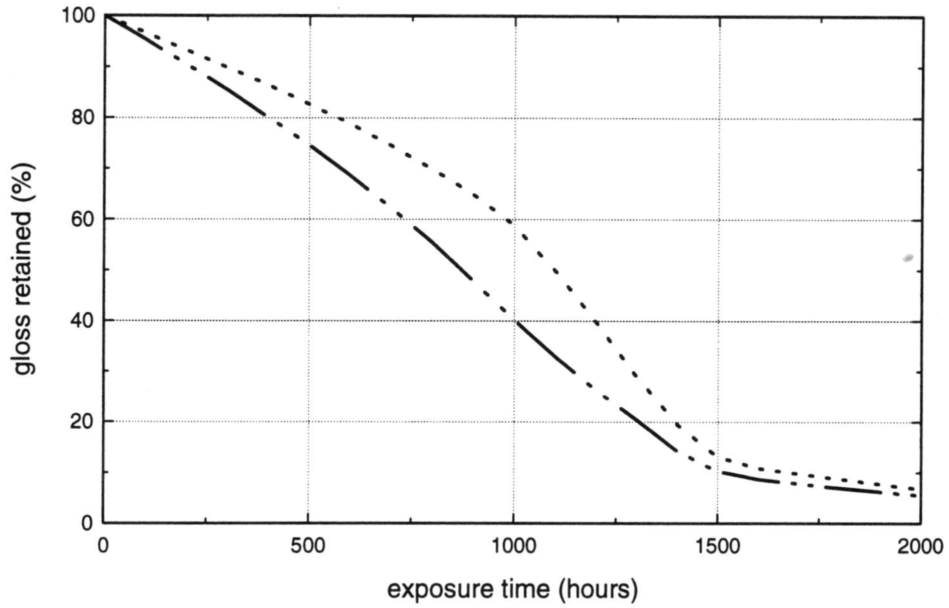

···············	Mitsub. Ray. Shinko-Lac T115 ASA (black, heat stabilized); 63°C
—··—··	Mitsub. Ray. Shinko-Lac T110 ASA (black, heat stabilized); 63°C
Reference No.	211

GRAPH 51: **Sunshine Weatherometer Exposure Time vs. Gloss Retained of ASA**

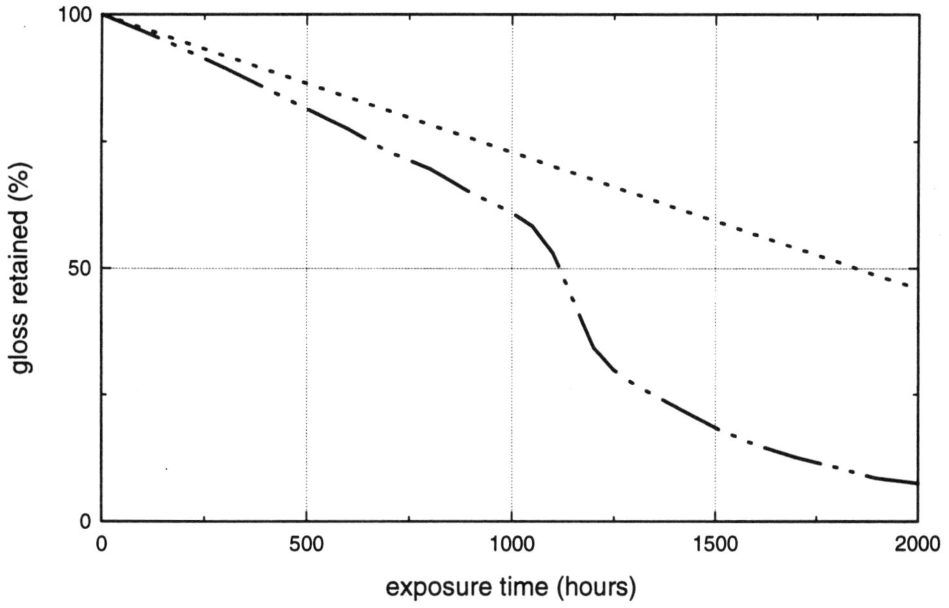

···············	Mitsub. Ray. Shinko-Lac ASA (black, gen. purp. grade); 63°C
—··—··	Mitsub. Ray. Shinko-Lac ASA (black, heat stabilized); 63°C
Reference No.	212

GRAPH 52: Weatherometer Exposure Time vs. Impact Strength of ASA

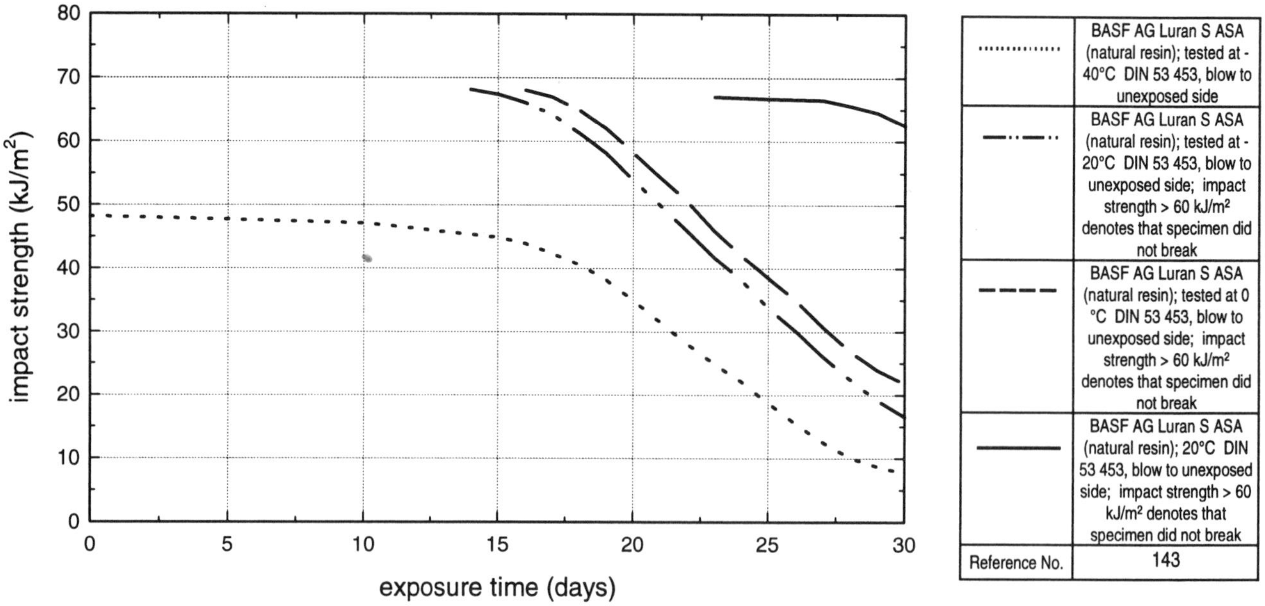

80
70
60
50
40
30
20
10
0
impact strength (kJ/m^2)
0
5
10
15
20
25
30
exposure time (days)
BASF AG Luran S ASA (natural resin); tested at -40°C DIN 53 453, blow to unexposed side
BASF AG Luran S ASA (natural resin); tested at -20°C DIN 53 453, blow to unexposed side; impact strength > 60 kJ/m^2 denotes that specimen did not break
BASF AG Luran S ASA (natural resin); tested at 0 °C DIN 53 453, blow to unexposed side; impact strength > 60 kJ/m^2 denotes that specimen did not break
BASF AG Luran S ASA (natural resin); 20°C DIN 53 453, blow to unexposed side; impact strength > 60 kJ/m^2 denotes that specimen did not break
Reference No.
143

GRAPH 53: Xenotest 1200 Exposure Time vs. Impact Strength of ASA

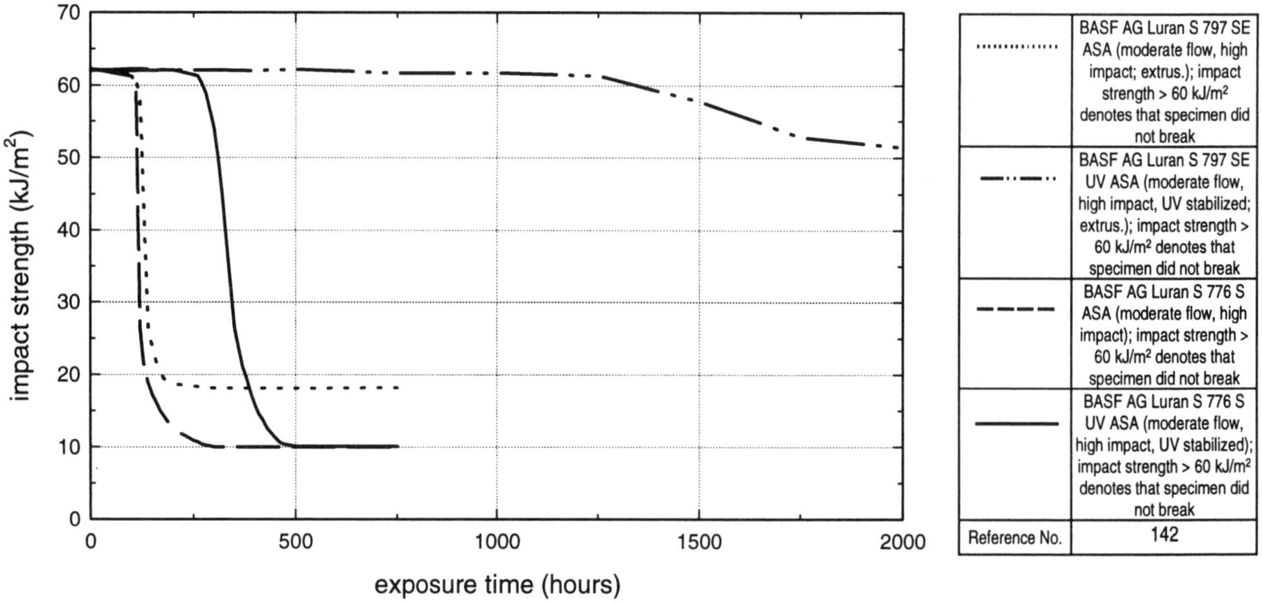

70
60
50
40
30
20
10
0
impact strength (kJ/m^2)
0
500
1000
1500
2000
exposure time (hours)
BASF AG Luran S 797 SE ASA (moderate flow, high impact; extrus.); impact strength > 60 kJ/m^2 denotes that specimen did not break
BASF AG Luran S 797 SE UV ASA (moderate flow, high impact, UV stabilized; extrus.); impact strength > 60 kJ/m^2 denotes that specimen did not break
BASF AG Luran S 776 S ASA (moderate flow, high impact); impact strength > 60 kJ/m^2 denotes that specimen did not break
BASF AG Luran S 776 S UV ASA (moderate flow, high impact, UV stabilized); impact strength > 60 kJ/m^2 denotes that specimen did not break
Reference No.
142

Chapter 5

Acrylic Resin

Weather Resistance

Aristech: Acrylic 300 (features: transparent, 3.0 mm thick; manufacturing method: continuous cast; product form: sheet); **Acrylic GPA** (features: transparent, 5.59 mm thick; manufacturing method: continuous cast; product form: sheet); **Acrylic** (features: transparent, 3.02 mm thick; manufacturing method: continuous cast; product form: sheet)

In March, 1977, it was decided to evaluate plastics currently used in glazing, signs, solar collectors, skylights and various other applications, and their relative physical properties before and after weathering exposure tests. Samples of the various materials to be tested were purchased from Industrial and Sign Distribution Companies located in Cincinnati. Extreme care was taken to insure that the materials were well labeled (masking paper, etc.) so that there could be no mistakes in product identification. The samples of Aristech Acrylic 300 and Aristech Acrylic GPA were colorless (clear) and had a nominal thicknesses of 3.2 mm and 5.6 mm.

Both these materials showed no significant change in YI (less than 1 unit) after exposure to three separate weathering tests: twin carbon arc weatherometer, outside EMMAQUA test in Arizona, and outside exposure at a 45° angle facing south. The samples exposed in the twin carbon arc weatherometer for 3,000 hours were tested for tensile strength, light transmission, and both Charpy and Izod impact (notched). They showed no significant change in tensile strength; a 1% change in light transmittance, and little or no change with impact testing.

Reference: *Comparison Of Plastics Used In Glazing, Signs, Skylights And Solar Collector Applications - Technical Bulletin 143,* competitor's technical report (ADARIS 50-1037-01) - Aristech Chemical Corporation, 1989.

Lucite L (features: transparent, 6.2 mm thick; manufacturing method: continuous cast; product form: sheet)

In March, 1977, it was decided to evaluate plastics currently used in glazing, signs, solar collectors, skylights and various other applications, and their relative physical properties before and after weathering exposure tests. Samples of the various materials to be tested were purchased from Industrial and Sign Distribution Companies located in Cincinnati. Extreme care was taken to insure that the materials were well labeled (masking paper, etc.) so that there could be no mistakes in product identification. The sample of Lucite L was colorless (clear) and had a nominal thickness of 6.20 mm.

The material showed no significant change in yellowness index (less than 1 unit) after exposure to three separate weathering tests: twin carbon arc weatherometer, outside EMMAQUA test in Arizona, and outside exposure at a 45° angle facing south. The samples exposed in the twin carbon arc weatherometer for 3,000 hours were tested for tensile strength, light transmission, and both Charpy and Izod impact (notched). They showed no significant change in tensile strength; a 1% change in light transmittance, and little or no change with impact testing.

Reference: *Comparison Of Plastics Used In Glazing, Signs, Skylights And Solar Collector Applications - Technical Bulletin 143,* competitor's technical report (ADARIS 50-1037-01) - Aristech Chemical Corporation, 1989.

Polycast (features: transparent, 3.1 mm thick; manufacturing method: cell cast process; product form: sheet)

In March, 1977, it was decided to evaluate plastics currently used in glazing, signs, solar collectors, skylights and various other applications, and their relative physical properties before and after weathering exposure tests. Samples of the various materials to be tested were purchased from Industrial and Sign Distribution Companies located in Cincinnati. The sample of Polycast was colorless (clear) and had a nominal thickness of 3.175 mm.

The material showed a YI change of two units, which is abnormal for cast acrylics, after exposure to three separate weathering tests: twin carbon arc weatherometer, outside EMMAQUA test in Arizona, and outside exposure at a 45° angle facing south. The change (YI), although significant, is not particularly objectionable. The samples exposed in the twin carbon arc weatherometer for 3,000 hours were tested for tensile strength, light transmission, and both Charpy and Izod impact (notched). Polycast showed a loss in tensile strength from 68.21 MPa to 57.88 MPa; a 1% change in light transmittance; and an unusually high Charpy value, indicating that it may be a copolymer or contain an impact additive.

Reference: *Comparison Of Plastics Used In Glazing, Signs, Skylights And Solar Collector Applications - Technical Bulletin 143,* competitor's technical report (ADARIS 50-1037-01) - Aristech Chemical Corporation, 1989.

Outdoor Weather Resistance

Cyro: Acrylite Plus (features: transparent)

Acrylite Plus compounds resist the adverse effects of outdoor weathering and will retain both their physical properties and appearance after long periods of outdoor exposure.

Reference: *Acrylite Plus Acrylic Based Molding and Extrusion Compounds,* supplier marketing literature (1511A-293-5CG) - Cyro Industries, 1993.

Rohm & Haas: Plexiglas G (features: transparent; manufacturing method: cell cast process; product form: sheet); **Plexiglas MC** (features: transparent; manufacturing method: melt calendering; product form: sheet)

Samples of colorless Plexiglas sheet exposed outdoors in Arizona, Florida, and Pennsylvania for 20 years or more show no significant discoloration, crazing, surface dulling, loss of light transmission, or development of haze or turbidity. Although these samples were Plexiglas G sheet, we would expect Plexiglas MC sheet, which has not been on the market for 20 years at this writing, to behave in the same way.

In these tests the samples were mounted on outdoor racks at a 45° angle facing south. Angling the racks in this manner increases the rigors of exposure, significantly. Actual outdoor applications ordinarily involve less severe conditions.

Reference: *Plexiglas Acrylic Sheet General Information And Physical Properties,* supplier design guide (PLA-22a) - AtoHaas North America Inc., 1992.

Rohm & Haas: Plexiglas (features: transparent; product form: sheet)

Plexiglas plastic has time-proved ability to withstand weather, sun and a wide range of temperatures in outdoor use. This permanence is due to the inherent stability of the resins from which the plastic is made. Outdoor exposure tests on a large number of clear samples, over a ten-year period in Pennsylvania show an average of more than 91% light transmission, which is a loss of only 1%. Inspection reveals no readily visible effects in nearly all test samples.

Reference: *Plexiglas Acrylic Sheet General Information,* supplier technical report (PL-1p) - Rohm and Haas Company, 1985.

Acrylic (features: transparent, crosslinked; manufacturing method: cast)

One of the prime contenders for solar reflectors or concentrators is acrylic plastic. Other than the basic requirement of being a highly efficient reflector or transmitter of the solar spectrum, it must offer excellent long-term weatherability with minimum cost. The weatherability of acrylic polymers, although known to be good compared to other polymers, has been an unknown with regard to actual exposure data in the southwestern United States where the potential is high for solar energy installations.

On January 19, 1956 a 4"x12"x1/8" piece of cast crosslinked acrylic, was mounted in the static test frame at Sandia Laboratories exposure site. This semi-arid location is some five miles southeast of Albuquerque on a sandy plain (elevation - 5,200'). The panels as mounted on the exposure racks, faced due south at 45° to the horizontal. The weathered panel was removed from the rack on September 24, 1973 for an elapsed exposure time of 17 years and 8 months.

In general, chemical changes are not detectable while mechanical response shows some evidence of embrittlement. In spite of these, the decrease in optical transmission is surprisingly low. The as-recovered material (with dust eroded surface) had an integrated or total transmittance which was 10% less than the control material, based on a solar radiation spectrum. In order to isolate the degradation loss due to change in the basic polymer, a polished specimen of aged material was tested and showed only a three percent drop in transmission across the solar spectrum. It appears reasonable that this low loss could be duplicated in a solar reflector.

Reference: Rainhart, L. G., Schimmel, Jr., W. P., *Effect Of Outdoor Aging On Acrylic Sheet,* 1974 International Solar Energy Society, U. S. Section Annual Meeting, conference proceedings (SAND 74-0241) - Sandia Laboratories, 1974.

Accelerated Outdoor Weathering Resistance

Cyro: Acrylite AR (features: transparent, abrasion resistant coating, 3.2 mm thick; manufacturing method: continuous cast; product form: sheet)

Acrylite AR sheet withstands the adverse effects of outdoor weathering. The abrasion resistant properties are maintained during prolonged periods of outdoor use. Accelerated weathering studies indicate that there is no significant loss in light transmittance or any appreciable increase in yellowing, as measured by the yellowness index after an accelerated weathering period of three years.

Reference: *Physical Properties Acrylite AR Acrylic Sheet And Cyrolon AR Polycarbonate Sheet,* supplier design guide (1632B-0193-10BP) - Cyro Industries, 1993.

TABLE 08: Outdoor Weathering in Arizona of Aristech Acrylic Resin.

Material Family	Acrylic Resin	
Material Supplier/ Grade	Aristech GPA	Aristech Acrysteel IGP
Features	transparent	transparent
Manufacturing Method	continuous cast	continuous cast
Product Form	sheet	sheet
Reference Number	117	117
EXPOSURE CONDITIONS		
Exposure Type	outdoor weathering	outdoor weathering
Exposure Location	Arizona	Arizona
Exposure Country	USA	USA
Exposure Note	45° angle south	45° angle south
Exposure Time (days)	2069	2069
SURFACE AND APPEARANCE		
Δ Yellowness Index	0.58	0.08
Haze Retained (%)	508	468
Luminous Transmittance Retained (%)	99.8	98.1

TABLE 09: Outdoor Weathering in Arizona of Cyro Acrylite Plus Acrylic Resin.

Material Family	Acrylic Resin		
Material Supplier/ Grade	Cyro Acrylite Plus		
Features	transparent	transparent	transparent
Reference Number	113	113	113
EXPOSURE CONDITIONS			
Exposure Type	outdoor weathering		
Exposure Location	Arizona	Arizona	Arizona
Exposure Country	USA	USA	USA
Exposure Time (days)	365	730	1096
SURFACE AND APPEARANCE			
Luminous Transmittance (%)	90 {j}	90 {j}	90 {j}

TABLE 10: Outdoor Weathering in Florida of ICI Perspex Acrylic Resin.

Material Family	Acrylic Resin									
Material Supplier/ Grade	ICI Perspex CP-61	ICI Perspex CP-61	ICI Perspex CP-82	ICI Perspex CP-82	ICI Perspex CP-82 UVA	ICI Perspex CP-82 UVA	ICI Perspex CP-82 Red 190	ICI Perspex CP-82 Red 190	ICI Perspex CP-82 Red 192	ICI Perspex CP-82 Red 192
Features	transparent	transparent	transparent	transparent	transparent	transparent	red color	red color	red color	red color
Reference Number	218	218	218	218	218	218	218	218	218	218
MATERIAL CHARACTERISTICS										
SampleThickness	1.6 mm	3.2 mm	1.6 mm	3.2 mm	1.6 mm	3.2 mm		3.2 mm	1.6 mm	3.2 mm
EXPOSURE CONDITIONS										
Exposure Type	outdoor weathering									
Exposure Location	Miami, Florida									
Exposure Country	USA									
Exposure Test Lab	DSET Laboratories									
Exposure Test Method	ASTM G7-89									
Exposure Note	45° south									
Exposure Time (days)	1095	1095	1095	1095	1095	1095	1095	1095	1095	1095
Total Radiation (MJ/m²)	19,135	19,135	19,135	19,135	19,135	19,135	19,135	19,135	19,135	19,135
Total Radiation (Langleys)	457,347	457,347	457,347	457,347	457,347	457,347	457,347	457,347	457,347	457,347
SURFACE AND APPEARANCE										
Haze (%)	2.2 {o}	1.8 {o}	1.8 {o}	1.9 {o}	1.8 {o}	2.2 {o}	0 {o}	0 {o}	0.6 {o}	0 {o}
Luminous Transmittance (%)	91.7 {p}	92.1 {p}	92.6 {p}	92.1 {p}	91.9 {p}	91.5 {p}	25.7 {p}	19.9 {p}	58 {p}	41.7 {p}
Luminous Transmittance Change (%)	-0.8 {p}	0 {p}	0.5 {p}	0.1 {p}	-0.1 {p}	-0.7 {p}	7.5 {p}	8.2 {p}	17.6 {p}	4.5 {p}

TABLE 11: Outdoor Weathering in Florida of ICI Perspex Acrylic Resin.

Material Family	Acrylic Resin									
Material Supplier/ Grade	ICI Perspex CP-924	ICI Perspex CP-927	ICI Perspex CP-927	ICI Perspex CP-927	ICI Perspex CP-1000E	ICI Perspex CP-1000E	ICI Perspex CP-1000E	ICI Perspex CP-1000I	ICI Perspex CP-1000I	ICI Perspex CP-1000I
Features	transparent	transparent	transparent	transparent	transparent	transparent	transparent	transparent	transparent	transparent
Reference Number	218	218	218	218	218	218	218	218	218	218
MATERIAL CHARACTERISTICS										
SampleThickness	3.2 mm	1.6 mm	3.2 mm	6.4 mm	1.6 mm	3.2 mm	6.4 mm	1.6 mm	3.2 mm	6.4 mm
EXPOSURE CONDITIONS										
Exposure Type	outdoor weathering									
Exposure Location	Miami, Florida									
Exposure Country	USA									
Exposure Test Lab	DSET Laboratories									
Exposure Test Method	ASTM G7-89									
Exposure Note	45° south									
Exposure Time (days)	1095	1095	1095	1095	1095	1095	1095	1095	1095	1095
Total Radiation (MJ/m^2)	19,135	19,135	19,135	19,135	19,135	19,135	19,135	19,135	19,135	19,135
Total Radiation (Langleys)	457,347	457,347	457,347	457,347	457,347	457,347	457,347	457,347	457,347	457,347
SURFACE AND APPEARANCE										
Haze (%)	5 {o}	5.3 {o}	5.7 {o}	12.5 {o}	4.4 {o}	6.6 {o}	18.2 {o}	5 {o}	7.4 {o}	24.2 {o}
Luminous Transmittance (%)	89.2 {p}	90.5 {p}	88 {p}	81.8 {p}	90.3 {p}	89.2 {p}	78.6 {p}	89.6 {p}	89.1 {p}	74.9 {p}
Luminous Transmittance Change (%)	-1.7 {p}	-1.3 {p}	-2.1 {p}	-8.2 {p}	-0.8 {p}	-0.6 {p}	-9.1 {p}	-1.2 {p}	-0.4 {p}	-13.2 {p}

TABLE 12: Outdoor Weathering in Florida of ICI Perspex Acrylic Resin.

Material Family	Acrylic Resin									
Material Supplier/ Grade	ICI Perspex CP-82 YL-99	ICI Perspex CP-82 YL-99	ICI Perspex CP-82 YL-130	ICI Perspex CP-82 YL-130	ICI Perspex CP-82 YL-131	ICI Perspex CP-82 YL-131	ICI Perspex CP-82 YL-198	ICI Perspex CP-82 YL-198	ICI Perspex CP-86 UVA	ICI Perspex CP-924
Features	yellow color	yellow color	yellow color	yellow color	yellow color	yellow color	yellow color	yellow color	transparent	transparent
Reference Number	218	218	218	218	218	218	218	218	218	218

MATERIAL CHARACTERISTICS

SampleThickness	1.6 mm	3.2 mm	1.6 mm	3.2 mm	1.6 mm	3.2 mm	1.6 mm	3.2 mm	1.6 mm	1.6 mm

EXPOSURE CONDITIONS

Exposure Type	outdoor weathering									
Exposure Location	Miami, Florida									
Exposure Country	USA									
Exposure Test Lab	DSET Laboratories									
Exposure Test Method	ASTM G7-89									
Exposure Note	45° south									
Exposure Time (days)	1095	1095	1095	1095	1095	1095	1095	1095	1095	1095
Total Radiation (MJ/m^2)	19,135	19,135	19,135	19,135	19,135	19,135	19,135	19,135	19,135	19,135
Total Radiation (Langleys)	457,347	457,347	457,347	457,347	457,347	457,347	457,347	457,347	457,347	457,347

SURFACE AND APPEARANCE

Haze (%)	2.5 {o}	3.8 {o}	1.1 {o}	1.4 {o}	2.3 {o}	1.9 {o}	3.2 {o}	6.4 {o}	1.7 {o}	5 {o}
Luminous Transmittance (%)	90.2 {p}	81.8 {p}	56.5 {p}	51.6 {p}	91.5 {p}	89.1 {p}	84 {p}	65 {p}	92 {p}	90 {p}
Luminous Transmittance Change (%)	6.5 {p}	4.9 {p}	-7.8 {p}	-8.8 {p}	4.2 {p}	4.1 {p}	7.7 {p}	1.4 {p}	0.4 {p}	-1.5 {p}

TABLE 13: Outdoor Weathering in Kentucky and Accelerated Outdoor Weathering by EMMAQUA of Acrylic Resin.

Material Family	Acrylic Resin									
Material Supplier/ Grade	Aristech Acrylic 300		Aristech Acrylic GPA		Du Pont Lucite L		Polycast		Aristech Acrylic GPA	
Features	transparent		transparent		transparent		transparent		transparent	
Manufacturing Method	continuous cast		continuous cast		continuous cast		cell cast process		continuous cast	
Product Form	sheet		sheet		sheet		sheet		sheet	
Reference Number	152	152	152	152	152	152	152	152	152	152
MATERIAL CHARACTERISTICS										
SampleThickness (mm)	3.0	3.0	3.02	3.02	6.2	6.2	3.1	3.1	5.59	5.59
EXPOSURE CONDITIONS										
Exposure Type	accelerated outdoor weatheing	outdoor weathering	accelerated outdoor weatheing	outdoor weathering	accelerated outdoor weatheing	outdoor weathering	accelerated outdoor weatheing	outdoor weathering	accelerated outdoor weatheing	outdoor weathering
Exposure Location	Arizona	Florence, Kentucky	Arizona	Florence, Kentucky	Arizona	Florence, Kentucky	Arizona	Florence, Kentucky	Arizona	Florence, Kentucky
Exposure Country	USA	USA	USA	USA	USA	USA	USA	USA	USA	USA
Exposure Apparatus	EMMAQUA		EMMAQUA		EMMAQUA		EMMAQUA		EMMAQUA	
Exposure Note		45° angle south		45° angle south		45° angle south		45° angle south		45° angle south
Exposure Time (days)	365	730	365	730	365	730	365	730	365	730
SURFACE AND APPEARANCE										
Δ Yellowness Index	0.42	0.29	0.67	0.41	0.7	0.54	2.9	2.2	0.72	0.21

TABLE 14: Accelerated Weathering in a Xenon Arc Weatherometer of Aristech Acrylic Resin.

Material Family	Acrylic Resin							
Material Supplier/ Grade	Aristech GPA (#2662)	Aristech GPA (#2577)	Aristech GPA (#7328)	Aristech Altair I300 (#6370)	Aristech Altair I300 (#6345)	Aristech Altair I300 (#6391)	Aristech Altair I300 (#6385)	Aristech Altair I300 (#6064)
Features	red color	red color	white color	silver color	bone color	almond color	white color	white color
Manufacturing Method	continuous cast	continuous cast	continuous cast	continuous cast	continuous cast	continuous cast	continuous cast	continuous cast
Product Form	sheet	sheet	sheet	sheet	sheet	sheet	sheet	sheet
Reference Number	117	117	117	117	117	117	117	117

EXPOSURE CONDITIONS

Exposure Type	accelerated weathering							
Exposure Apparatus	xenon arc weatherometer							
Exposure Note	equivalent to 6-7 years of conventional weathering	equivalent to 6-7 years of conventional weathering	equivalent to 13 years of conventional weathering	equivalent to 6-7 years of conventional weathering	equivalent to 6-7 years of conventional weathering	equivalent to 6-7 years of conventional weathering	equivalent to 6-7 years of conventional weathering	equivalent to 6-7 years of conventional weathering
Exposure Time (days)	83.3	83.3	160.8	83.2	83.2	83.2	83.2	83.2

SURFACE AND APPEARANCE

Δa Color Change	-0.35	-0.13	-0.41	-0.02	-0.05	-0.06	0.08	-0.09
Δb Color Change	-0.23	0.17	0.36	-0.05	-0.8	-0.13	-0.63	-0.09
ΔL Color Change	-0.33	-0.26	0.28	-0.09	0.17	-9	-0.14	-0.12
ΔE Color Change	0.54	0.34	0.61	0.1	0.82	0.17	0.66	0.18

TABLE 15: Accelerated Weathering in a Carbon Arc Weatherometer of Aristech Acrylic and DuPont Lucite Acrylic Resin.

Material Family	Acrylic Resin				
Material Supplier/ Grade	Aristech Acrylic 300	Aristech Acrylic GPA	Du Pont Lucite L	Polycast	Aristech Acrylic GPA
Features	transparent	transparent	transparent	transparent	transparent
Manufacturing Method	continuous cast	continuous cast	continuous cast	cell cast process	continuous cast
Product Form	sheet	sheet	sheet	sheet	sheet
Reference Number	152	152	152	152	152
MATERIAL CHARACTERISTICS					
SampleThickness (mm)	3.0	3.02	6.2	3.1	5.59
EXPOSURE CONDITIONS					
Exposure Type	accelerated weathering				
Exposure Apparatus	carbon arc weatherometer				
Exposure Time (days)	125	125	125	125	125
PROPERTIES RETAINED (%)					
Tensile Strength	106	100	103	89	101
Notched Izod Impact Strength	100	97.6	71.7	74.4	102.9
Charpy Notched Impact Strength	90.5	118.8	79.6	86.7	88.4
SURFACE AND APPEARANCE					
Δ Yellowness Index	0.45	0.52	0.45	2.45	1.0
Luminous Transmittance Retained (%)	98.9	98.9	98.9	98.9	98.9

GRAPH 54: Florence, Kentucky Outside Weathering Exposure Time vs. Yellowness Index of Acrylic Resin

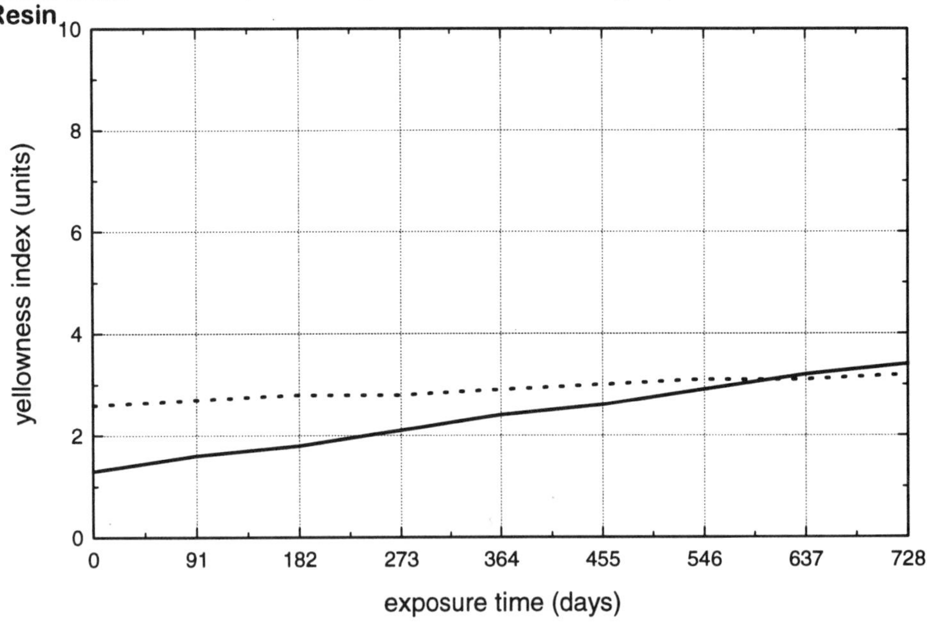

··············	Aristech Acrylic 300 Acrylic (transparent, 3.0 mm thick; continuous cast; sheet); test lab: Aristech
··············	Aristech Acrylic GPA Acrylic (transparent, 3.02 mm thick; continuous cast; sheet); test lab: Aristech
··············	Lucite L Acrylic (transparent, 6.2 mm thick; continuous cast; sheet); test lab: Aristech
———	Polycast Acrylic (transparent, 3.1 mm thick; cell cast; sheet); test lab: Aristech
··············	Aristech Acrylic GPA Acrylic (transparent, 5.59 mm thick; continuous cast; sheet); test lab: Aristech
Reference No.	152

GRAPH 55: EMMAQUA Accelerated Weathering Exposure Time vs. Light Transmission of Acrylic Resin

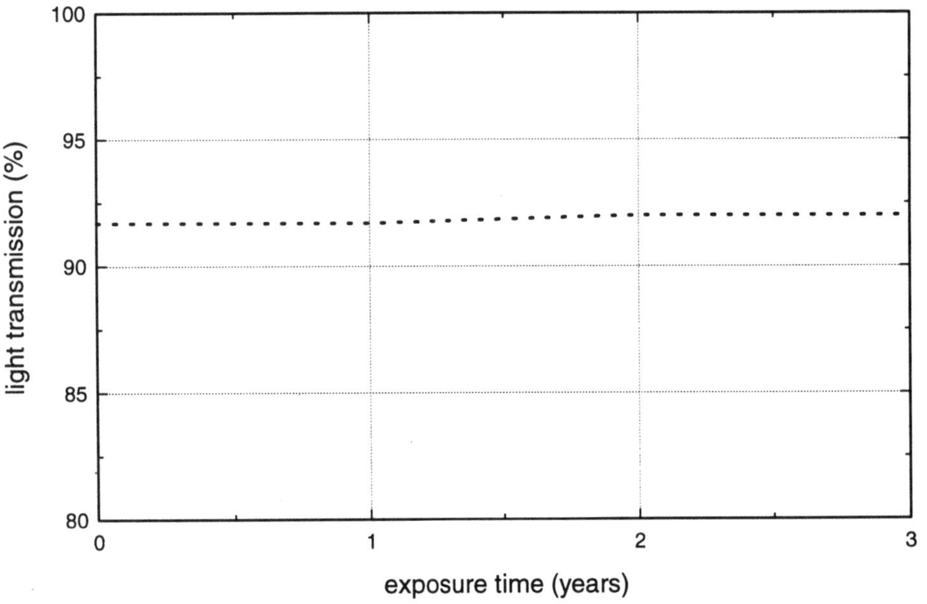

··············	Cyro Acrylite AR Acrylic (transparent, 3.2 mm thick; continuous cast; abras. resist. coating; sheet); ASTM D1003
Reference No.	151

GRAPH 56: EMMAQUA Accelerated Weathering Exposure Time vs. Yellowness Index of Acrylic Resin

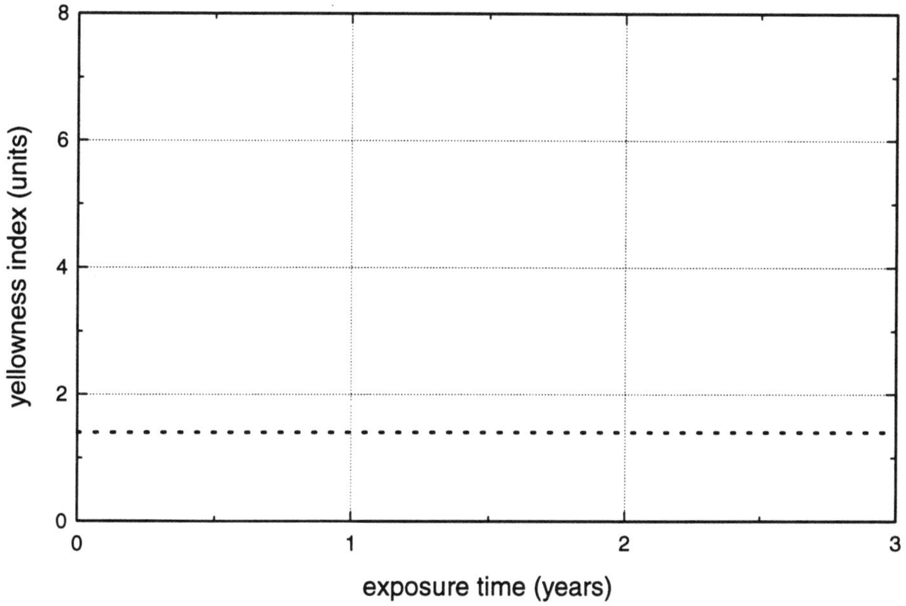

··············	Cyro Acrylite AR Acrylic (transparent, 3.2 mm thick; continuous cast; abras. resist. coating; sheet); ASTM D1003
Reference No.	151

GRAPH 57: EMMAQUA Arizona Accelerated Weathering Exposure Time vs. Yellowness Index of Acrylic Resin

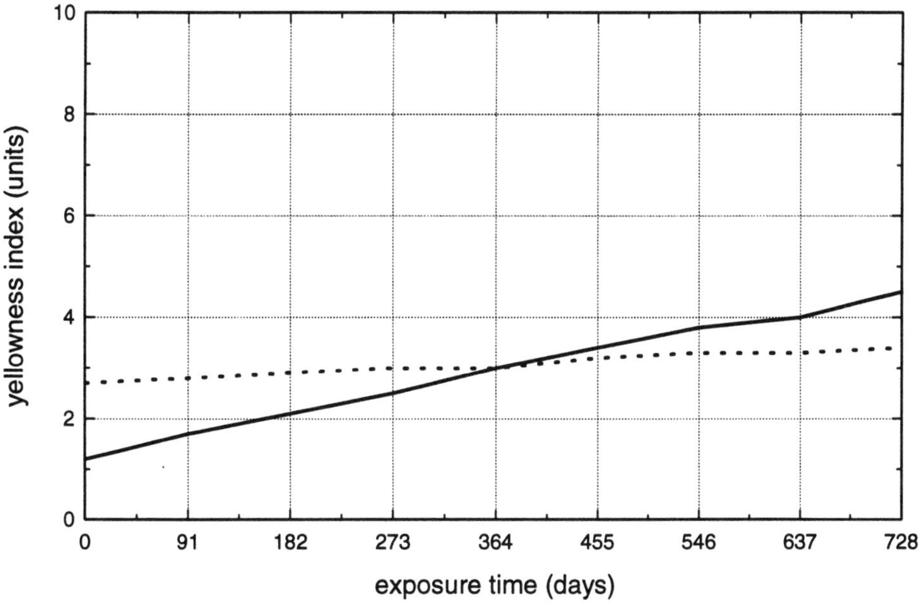

··············	Aristech Acrylic 300 Acrylic (transparent, 3.0 mm thick; continuous cast; sheet); test lab: DSET
··············	Aristech Acrylic GPA Acrylic (transparent, 3.02 mm thick; continuous cast; sheet); test lab: DSET
··············	Lucite L Acrylic (transparent, 6.2 mm thick; continuous cast; sheet); test lab: DSET
———	Polycast Acrylic (transparent, 3.1 mm thick; cell cast; sheet); test lab: DSET
··············	Aristech Acrylic GPA Acrylic (transparent, 5.59 mm thick; continuous cast; sheet); test lab: DSET
Reference No.	152

GRAPH 58: Twin Carbon Arc Weatherometer Exposure Time vs. Yellowness Index of Acrylic Resin

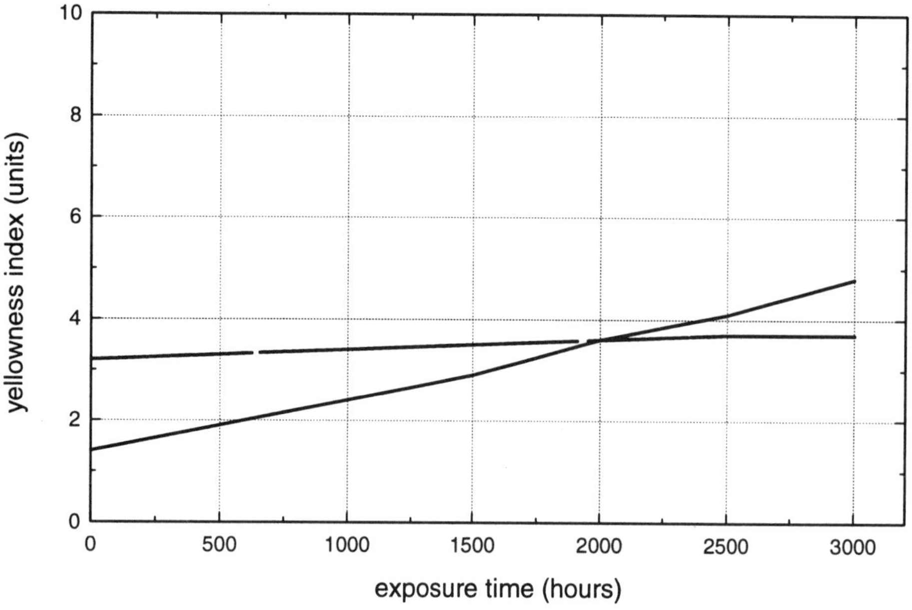

···············	Aristech Acrylic 300 Acrylic (transparent, 3.0 mm thick; continuous cast; sheet)
···············	Aristech Acrylic GPA Acrylic (transparent, 3.02 mm thick; continuous cast; sheet)
···············	Lucite L Acrylic (transparent, 6.2 mm thick; continuous cast; sheet)
————	Polycast Acrylic (transparent, 3.1 mm thick; cell cast; sheet)
···············	Aristech Acrylic GPA Acrylic (transparent, 5.59 mm thick; continuous cast; sheet)
Reference No.	152

GRAPH 59: Fadeometer Exposure Time vs. Yellowness Index of Acrylic Resin

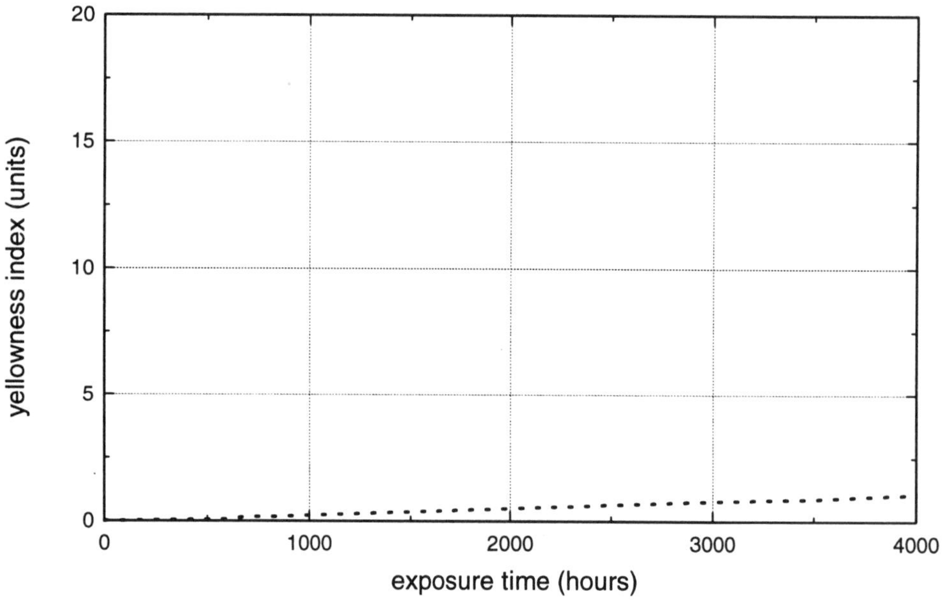

···············	Acrylic (transparent); Atlas fadeometer
Reference No.	230

GRAPH 60: **UV-CON Accelerated Weathering Exposure Time vs. Yellowness Index of Acrylic Resin**

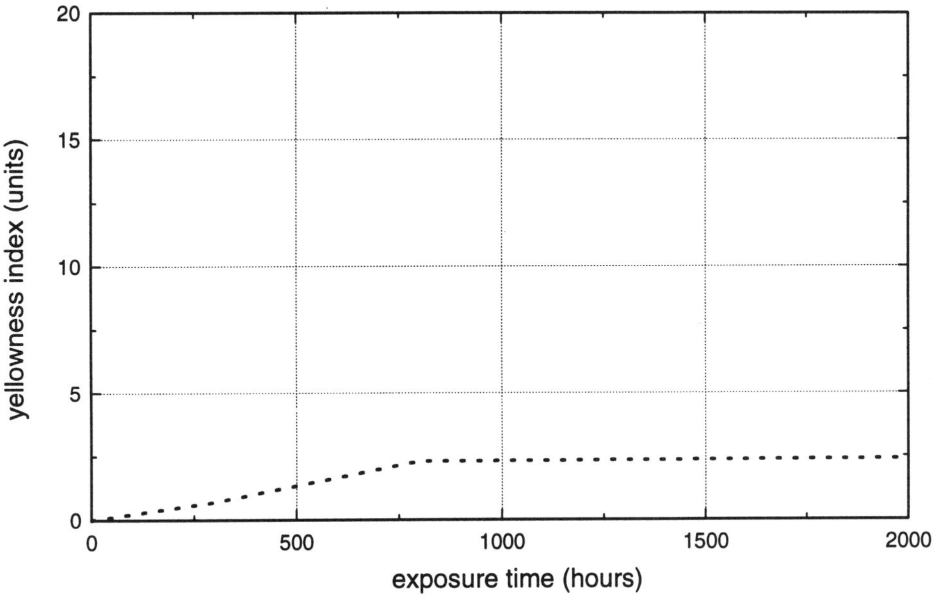

	Acrylic (transparent)
Reference No.	231

Chapter 6

Acrylic Copolymer

GRAPH 61: Fadeometer Exposure Time vs. Yellowness Index of Acrylic Copolymer

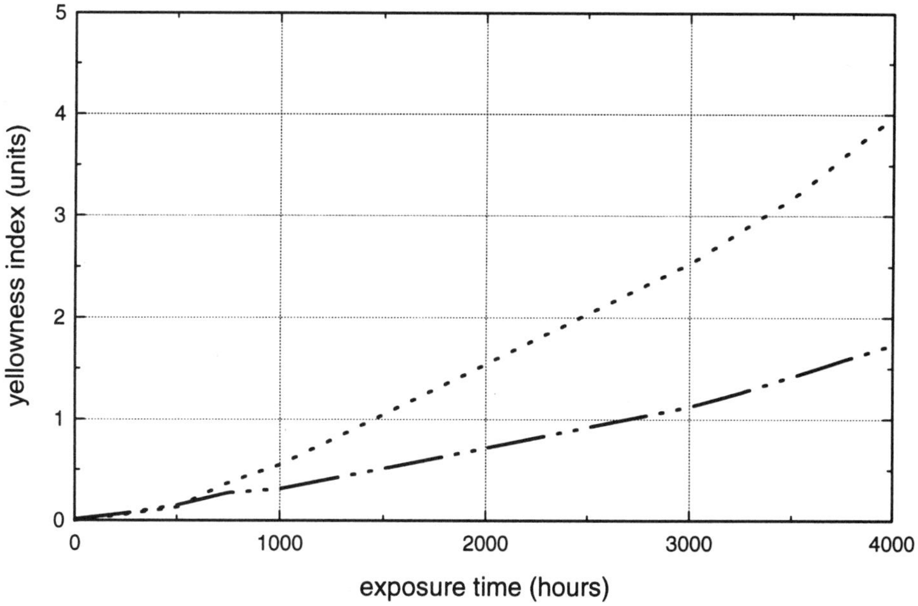

....	Novacor NAS 50 Acrylic Copol. (transparent); Atlas fadeometer
—·· —··	Novacor NAS 55 Acrylic Copol. (transparent); Atlas fadeometer
Reference No.	230

GRAPH 62: UV-CON Accelerated Weathering Exposure Time vs. Yellowness Index of Acrylic Copolymer

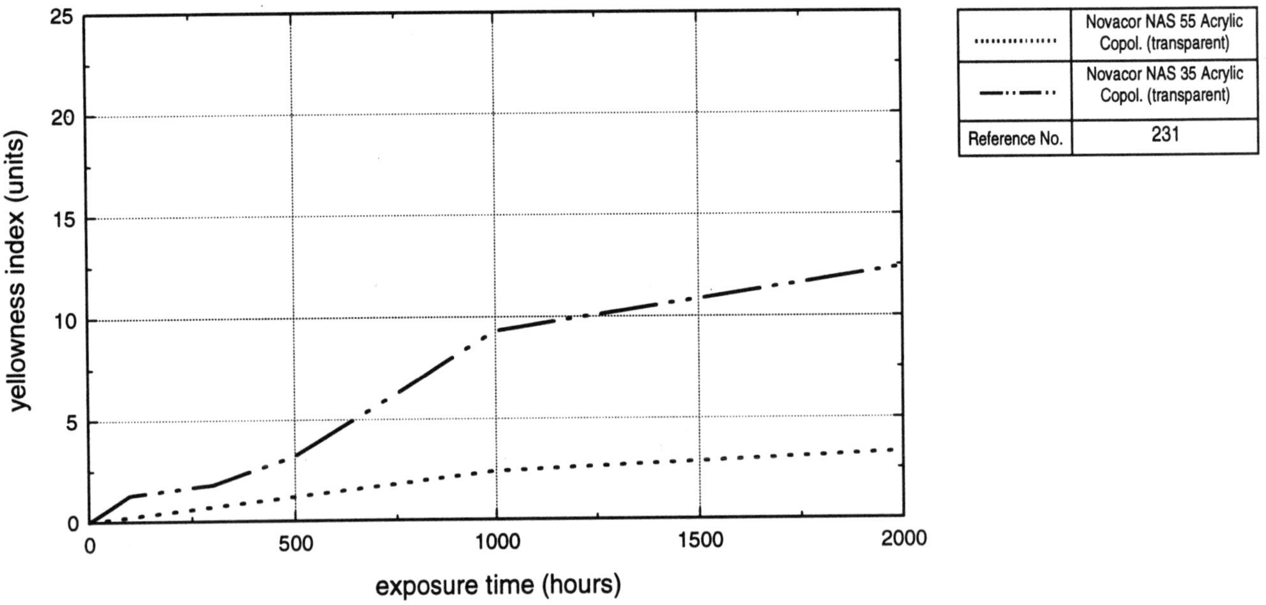

Chapter 7

Cellulose Acetate Butyrate

Weather Resistance

Eastman: Tenite Butyrate

Most of the effects of weathering result from the ultraviolet portion of the sun's radiation. The energy contained in ultraviolet light is capable of rupturing the long chains of a cellulose ester directly; and in addition, in the presence of oxygen, ultraviolet radiation causes oxidation of the plastic. The infrared portion of sunlight warms the plastic and accelerates the harmful effects of ultraviolet light.

Cellulose esters can be partially protected from both the direct chain scission and the photocatalyzed oxidation discussed in the previous paragraph. Protection from chain scission is obtained with special stabilizers known as ultraviolet inhibitors; protection from oxidation, with antioxidants. Ultraviolet inhibitors, commonly referred to as UVI's, are chemical compounds that absorb ultraviolet light and disperse the energy contained in the UV radiation in a form that is less harmful to the plastic. Most materials synthesized for the purpose of being used as UVI's are transparent and essentially colorless, but there are also some pigments and dyes that function as ultraviolet inhibitors.

Plastics in certain colors may have a significantly longer useful life outdoors than clear-transparent formulations that are otherwise identical to the colored materials. Specimens of a heavily pigmented black have weathered in Tennessee for 16 years without significant changes in physical properties.

The deterioration caused in a cellulose ester plastic by weathering will depend upon the particular cellulose ester, the plasticizer, the stabilizer system, the wavelength of the incident radiation, the total amount of radiation absorbed, the temperature of the plastic, atmospheric humidity, industrial contaminants in the atmosphere, and possibly other factors.

The development of weather-resistant formulations of Tenite butyrate has been based on information accumulated in a test program carried on continuously by Eastman since the early 1940's. The total program has involved outdoor weathering, accelerated weathering tests of many types, preparation and testing of hundreds of different plastic formulations, and synthesis of many organic compounds that might be a useful UVI.

The degree by which the useful life of a normal application can exceed that of test specimens is illustrated by the history of a butyrate sign installed over a shop in New York City several years ago. The sign was made from a butyrate formulation containing a UVI of only moderate effectiveness (a useful life of only about two years in Arizona), yet it remained in service for about nine years.

Deterioration of cellulose ester plastics caused by weathering appears first as a dulling of the surface. As the deterioration proceeds into advanced stages, the surface crazes and checks, the formation of each fissure exposing underlying plastic to the action of the weather.

The onset of surface crazing does not mean the end of usefulness for Tenite butyrate. It will still have good tensile strength, elongation, and impact strength.

Reference: *Weathering Of Tenite Butyrate,* supplier technical report (TR-25C) - Eastman Plastics, 1984.

Uvex (features: transparent, 3.07 mm thick; manufacturing method: extrusion; product form: sheet)

In March, 1977, it was decided to evaluate plastics currently used in glazing, signs, solar collectors, skylights and various other applications, and their relative physical properties before and after weathering exposure tests. Samples of the various materials to be tested were purchased from Industrial and Sign Distribution Companies located in Cincinnati. The sample of Uvex butyrate was colorless (clear) and had a nominal thickness of 3.175 mm.

The material showed visually objectionable changes and had high yellowness indices after exposure to three separate weathering tests: twin carbon arc weatherometer, outside EMMAQUA test in Arizona, and outside exposure at a 45° angle facing south. The samples exposed in the twin carbon arc weatherometer for 3,000 hours were tested for tensile strength, light transmission, and both Charpy and Izod impact (notched). Uvex showed a loss in tensile strength from 28.94 MPa to 18.6 MPa. The material had a significant loss in light transmittence of 16%. Uvex also showed an 86% loss by Charpy and a 92% loss by Izod.

Reference: *Comparison Of Plastics Used In Glazing, Signs, Skylights And Solar Collector Applications - Technical Bulletin 143,* competitor's technical report (ADARIS 50-1037-01) - Aristech Chemical Corporation, 1989.

Outdoor Weather Resistance

Eastman: Tenite Butyrate 205 (features: moderate weather resistance); **Tenite Butyrate 460** (features: weatherable); **Tenite Butyrate 461** (features: weatherable); **Tenite Butyrate 465** (features: weatherable); **Tenite Butyrate 485** (features: weatherable); **Tenite Butyrate 513** (features: weatherable); **Tenite Butyrate 527** (features: moderate weather resistance); **Tenite Butyrate 554** (features: moderate weather resistance)

Various formulations of Tenite butyrate have different degrees of resistance to solar radiation. Formulations unprotected by UVI or heavy pigmentation, although ordinarily suitable for indoor use indefinitely, generally give only a few months of satisfactory outdoor service.

One group of formulations, typified by Tenite butyrate 554, may be expected to remain serviceable for two years or more outdoors in Arizona, and can be classified as having moderate weatherability. The most widely used formulas in this group are 205, 527, and 554. The formulas have excellent color retention and for many years have served admirably in applications that require some resistance to weathering but where continuous, longtime outdoor exposure is not anticipated. A typical application of this type is handles for small garden tools.

The most weather-resistant butyrate formulations are typified by Tenite butyrate 460. The most used formulas in this group are 460, 461, 465, 485, and 513. These special outdoor materials ordinarily remain useful for five years or more when exposed continuously in Arizona. These formulas are used for such applications as covers for electric meters and housings for fire alarms, where continuous, longtime outdoor exposure is intended. The weathering inhibitor in these formulations has a slight tendency to discolor after long exposure where heavy industrial atmospheric contamination is prevalent; discoloration sometimes becomes noticeable if light colors are used.

Like cellulose esters, dyes and pigments may be affected adversely by sunlight. In many instances, they change color or lose their color under the influence of light and oxygen. An investigation of many years' duration involving exposure of samples under a carbon arc to accelerate the effects of weathering, as well as actual outdoor weathering tests, has resulted in the selection of colorants for butyrate outdoor formulations as stable as the plastic materials themselves.

Articles made of outdoor types of Tenite butyrate in suggested colors, should therefore give at least five years' service under even the most adverse weather conditions found in the continental United States. These most adverse conditions represent exposure to solar radiation that measures 185,000 to 200,000 langleys per year on a horizontal surface, and are found, in general, south of about 35°N latitude and between about 100° and 115°W longitude. Comparable exposure in other parts of the world should have similar effects.

Reference: *Weathering Of Tenite Butyrate,* supplier technical report (TR-25C) - Eastman Plastics, 1984.

TABLE 16: **Outdoor Weathering in Kentucky, Accelerated Outdoor Weathering by EMMAQUA and Accelerated Weathering in a Carbon Arc Weatherometer of Uvex Cellulose Acetate Butyrate.**

Material Family	Cellulose Acetate Butyrate		
Trade Name	Uvex		
Features	transparent		
Manufacturing Method	extrusion		
Product Form	sheet		
Reference Number	152	152	152
MATERIAL CHARACTERISTICS			
Sample Thickness (mm)	3.07	3.07	3.07
EXPOSURE CONDITIONS			
Exposure Type	accelerated outdoor weatheing	accelerated weathering	outdoor weathering
Exposure Location	Arizona		Florence, Kentucky
Exposure Country	USA		USA
Exposure Apparatus	EMMAQUA	carbon arc weatherometer	
Exposure Note			45° angle south
Exposure Time (days)	365	125	730
PROPERTIES RETAINED (%)			
Tensile Strength		64.3	
Notched Izod Impact Strength		8.2	
Charpy Impact Strength		13.9	
SURFACE AND APPEARANCE			
Δ Yellowness Index	15.7	13.9	9.09
Luminous Transmittance Retained (%)		82	

GRAPH 63: **Florence, Kentucky Outside Weathering Exposure Time vs. Yellowness Index of Cellulose Acetate Butyrate**

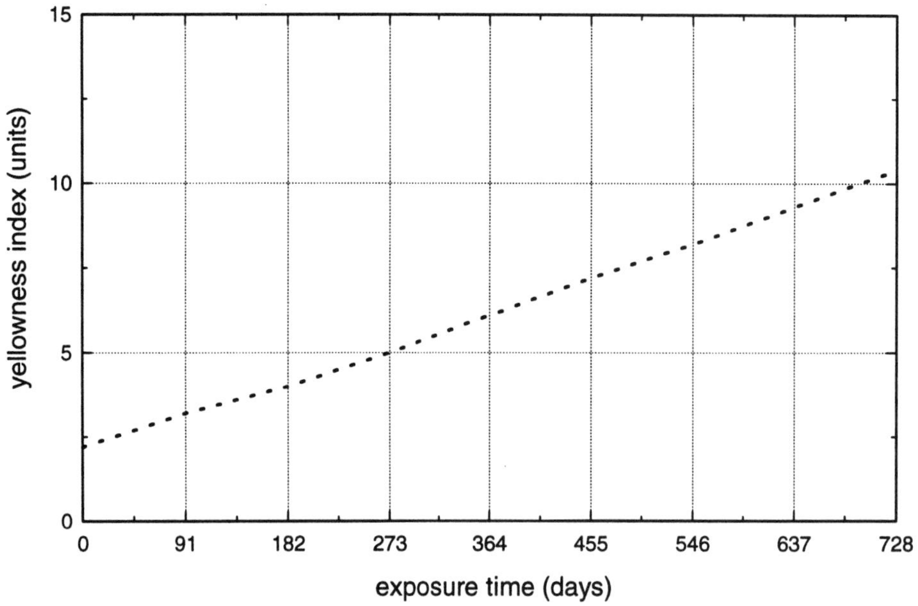

....	Uvex CAB (transparent, 3.07 mm thick; extrusion; sheet); test lab: Aristech Chemical Corp
Reference No.	152

GRAPH 64: **Arizona Outdoor Weathering Exposure Time vs. Elongation at Break of Cellulose Acetate Butyrate**

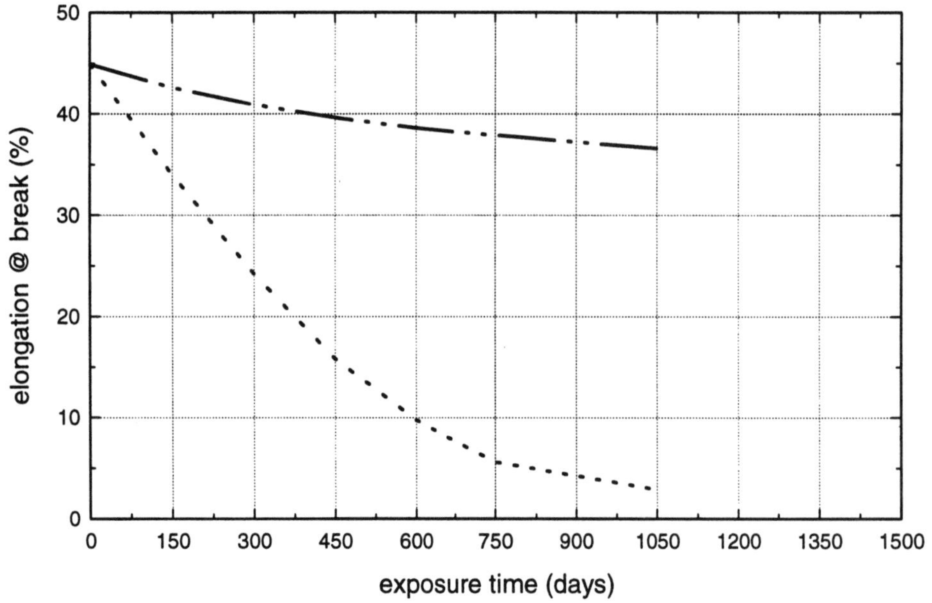

....	Eastman Tenite Butyrate CAB (weatherable, transparent, colored, 3.2 mm thick); Pheonix 45° south
—··—··	Eastman Tenite Butyrate CAB (weatherable, black, 3.2 mm thick); Pheonix 45° south
Reference No.	167

GRAPH 65: Arizona Outdoor Weathering Exposure Time vs. Tensile Strength at Break of Cellulose Acetate Butyrate

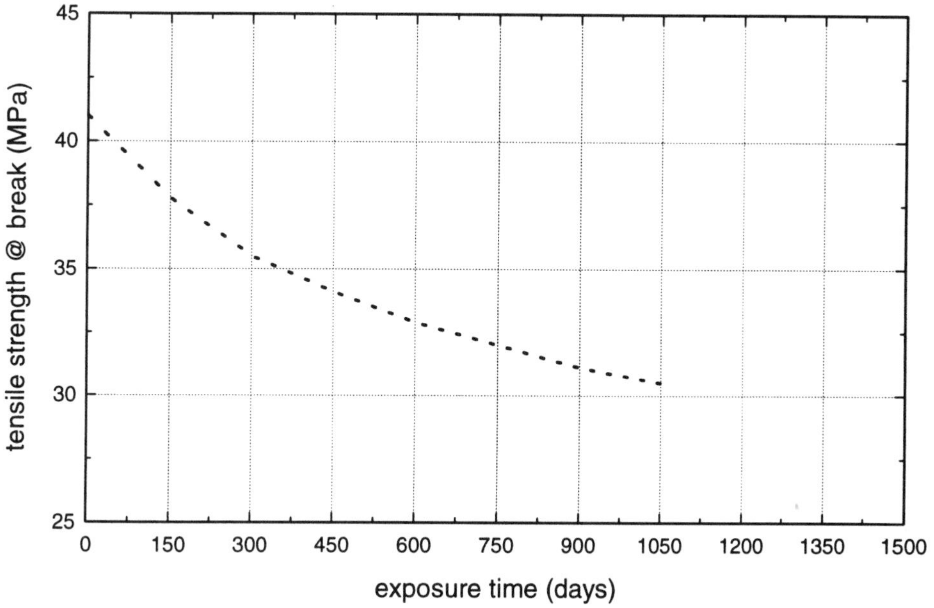

....	Eastman Tenite Butyrate CAB (weatherable, transparent, colored, 3.2 mm thick)
....	Eastman Tenite Butyrate CAB (weatherable, black, 3.2 mm thick)
Reference No.	167

GRAPH 66: Kingsport, Tennessee Outdoor Weathering Exposure Time vs. Plaque Impact Strength of Cellulose Acetate Butyrate

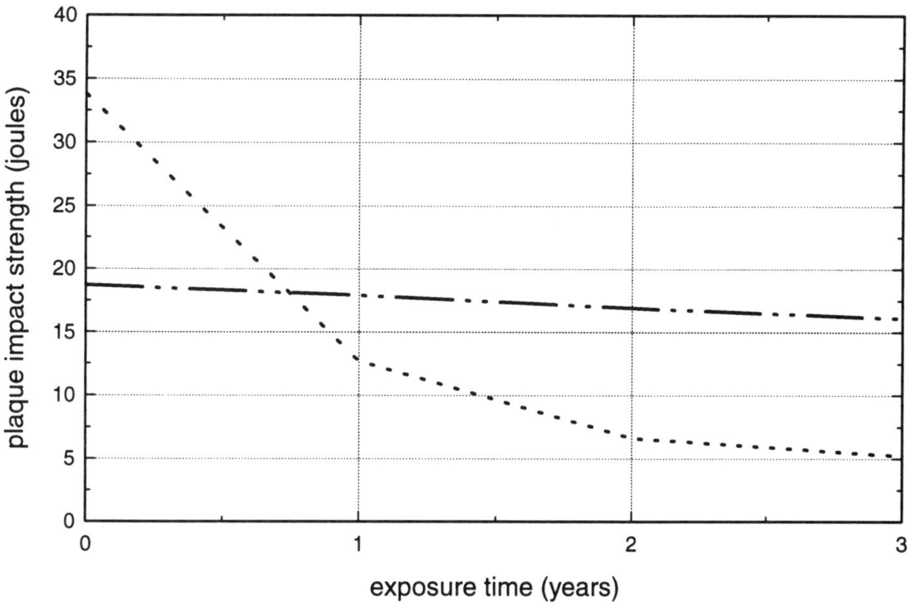

....	Eastman Tenite Butyrate CAB (weatherable, transparent, colored, 3.2 mm thick); test at 23 °C; vertical position due south, struck on weathered side
—··—··	Eastman Tenite Butyrate CAB (weatherable, black, 3.2 mm thick); test at 23 °C; vertical position due south, struck on weathered side
Reference No.	167

GRAPH 67: **EMMAQUA Arizona Accelerated Weathering Exposure Time vs. Yellowness Index of Cellulose Acetate Butyrate**

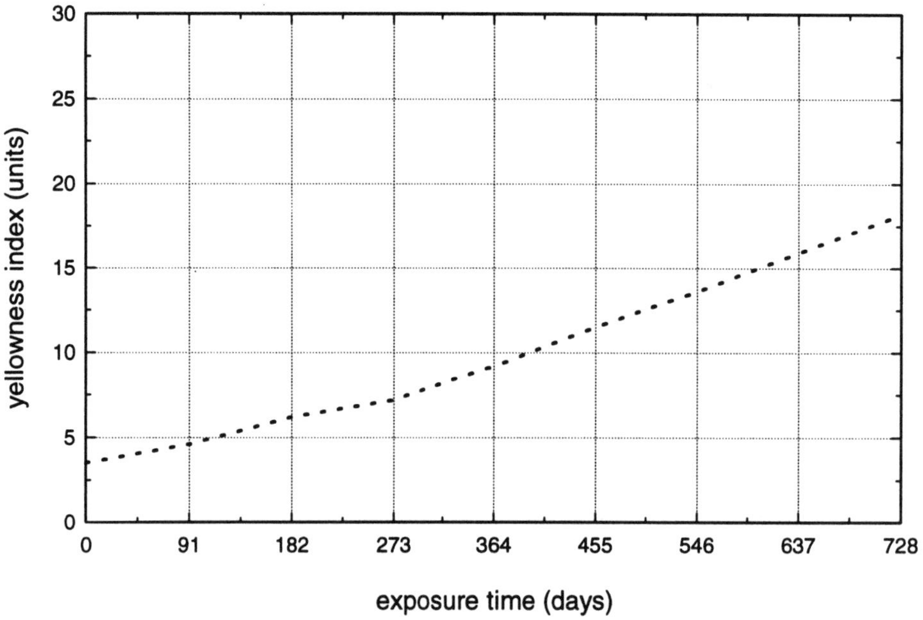

···············	Uvex CAB (transparent, 3.07 mm thick; extrusion; sheet); test lab: DSET
Reference No.	152

GRAPH 68: **Twin Carbon Arc Weatherometer Exposure Time vs. Yellowness Index of Cellulose Acetate Butyrate**

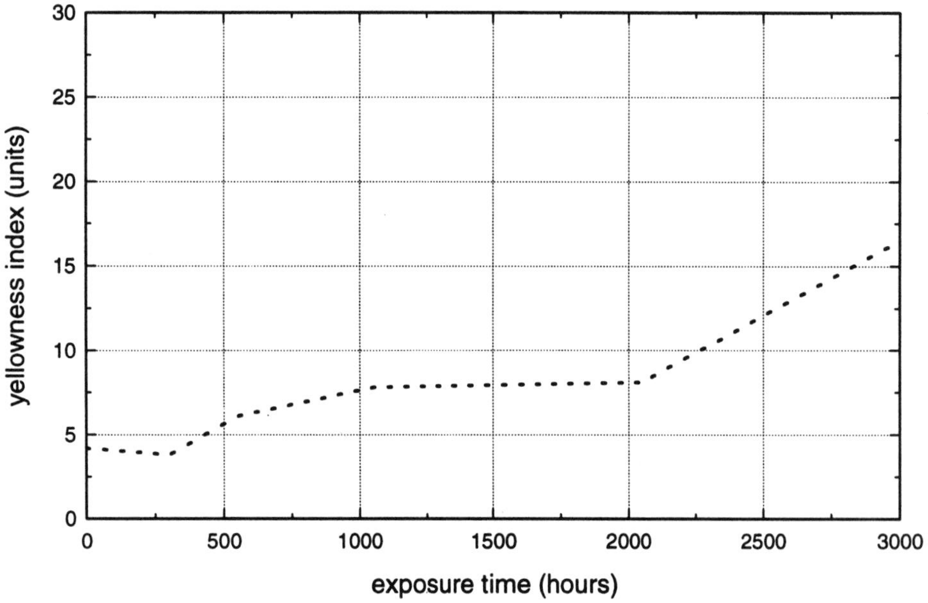

···············	Uvex CAB (transparent, 3.07 mm thick; extrusion; sheet); test lab Aristech Chemical
Reference No.	152

Chapter 8

Ethylene Chlorotrifluoroethylene Copolymer

Outdoor Weather Resistance

Ausimont: Halar

Glass Halar fluoropolymer undergoes very little change in properties or appearance on outdoor exposure to sunlight.

Reference: *Thermal And Other Properties Of Halar Fluoropolymer,* supplier technical report (GHG) - Ausimont.

Accelerated Artificial Weathering Resistance

Ausimont: Halar

Accelerated weathering studies demonstrate the remarkable stability of Halar. This is particularly evident in the elongation at break which is a good indication of polymer degradation. After an exposure of 1,000 hours in a xenon arc weatherometer, this critical property is barely affected.

Reference: *Thermal And Other Properties Of Halar Fluoropolymer,* supplier technical report (GHG) - Ausimont.

TABLE 17: Accelerated Weathering in a Xenon Arc Weatherometer of Ausimont Halar Ethylene Chlorotrifluoroethylene Copolymer.

Material Family	Ethylene Chlorotrifluoroethylene Copolymer	
Material Supplier/ Grade	Ausimont Halar	Ausimont Halar
Reference Number	131	131
MATERIAL CHARACTERISTICS		
Sample Thickness	2.03 mm	2.03 mm
EXPOSURE CONDITIONS		
Exposure Type	accelerated weathering	accelerated weathering
Exposure Apparatus	xenon arc weatherometer	xenon arc weatherometer
Exposure Time (days)	20.8	41.6
PROPERTIES RETAINED (%)		
Relative Tensile Modulus	89 {z}	95 {z}
Tensile Strength @ Yield	103 {y}	101 {y}
Tensile Strength @ Break	97 {y}	94 {y}
Elongation @ Break	94 {y}	102 {y}

Chapter 9

Ethylene Tetrafluoroethylene Copolymer

Outdoor Weather Resistance

DuPont: Tefzel 200 (features: general purpose grade)

Long term outdoor testing is in progress, and exposure for more than a year in Florida and Michigan has had no effect on Tefzel 200.

Reference: *Tefzel Fluoropolymer Design Handbook,* supplier design guide (E-31301-1) - Du Pont Company, 1973.

Accelerated Artificial Weathering Resistance

Ausimont: Hyflon 700 (features: high molecular weight, 0.012 mm thick; product form: film); **Hyflon 800** (features: low molecular weight, 0.012 mm thick; product form: film)

The influence of UV light or 2000 hours of weatherometer exposure do not have any measurable or adverse effect on the tensile properties of Hyflon ETFE.

Reference: *Hyflon ETFE 700/800 Properties and Application Guide,* supplier design guide - Ausimont USA, Inc..

DuPont: Tefzel 200 (features: general purpose grade); **Tefzel HT-2004** (material compostion: 25% glass fiber reinforcement)

Based on accelerated laboratory tests, Tefzel 200 has excellent weather resistance, whereas Tefzel HT-2004 is affected by weather conditions. No correlation exists for Tefzel relating weatherometer exposure time to outdoor exposure.

Reference: *Tefzel Fluoropolymer Design Handbook,* supplier design guide (E-31301-1) - Du Pont Company, 1973.

TABLE 18: Accelerated Weathering in a Weatherometer of Ausimont Hyflon Ethylene Tetrafluoroethylene Copolymer.

Material Family	Ethylene Tetrafluoroethylene Copolymer			
Material Supplier/ Grade	Ausimont Hyflon 700	Ausimont Hyflon 700	Ausimont Hyflon 800	Ausimont Hyflon 800
Features	high molecular weight	high molecular weight	low molecular weight	low molecular weight
Product Form	film	film	film	film
Reference Number	114	114	114	114
MATERIAL CHARACTERISTICS				
SampleThickness	0.012 mm	0.012 mm	0.012 mm	0.012 mm
EXPOSURE CONDITIONS				
Exposure Type	accelerated weathering			
Exposure Apparatus	weatherometer	weatherometer	weatherometer	weatherometer
Exposure Time (days)	41.6	83.3	41.6	83.3
PROPERTIES RETAINED (%)				
Elastic Modulus	105	105	105	105
Tensile Strength	102	102	102	102
Elongation @ Break	115	115	115	115

TABLE 19: Accelerated Weathering in a Weatherometer of Ausimont Hyflon Ethylene Tetrafluoroethylene Copolymer.

Material Family	Ethylene Tetrafluoroethylene Copolymer			
Material Supplier/ Grade	Ausimont Hyflon 700	Ausimont Hyflon 700	Ausimont Hyflon 800	Ausimont Hyflon 800
Features	high molecular weight	high molecular weight	low molecular weight	low molecular weight
Product Form	film	film	film	film
Reference Number	114	114	114	114
MATERIAL CHARACTERISTICS				
SampleThickness	0.016 mm	0.016 mm	0.016 mm	0.016 mm
EXPOSURE CONDITIONS				
Exposure Type	accelerated weathering			
Exposure Apparatus	weatherometer	weatherometer	weatherometer	weatherometer
Exposure Time (days)	41.6	83.3	41.6	83.3
PROPERTIES RETAINED (%)				
Elastic Modulus	98	98	98	98
Tensile Strength	100	100	100	100
Elongation @ Break	100	100	100	100

TABLE 20: Accelerated Weathering in a Weatherometer of DuPont Tefzel Ethylene Tetrafluoroethylene Copolymer.

Material Family	Ethylene Tetrafluoroethylene Copolymer			
Material Supplier/ Grade	DuPont Tefzel 200		DuPont Tefzel HT-2004	
Features	general purpose grade			
Reference Number	205	205	205	205
MATERIAL CHARACTERISTICS				
sample type	injection molded tensile bar	injection molded tensile bar	injection molded tensile bar	injection molded tensile bar
MATERIAL COMPOSITION				
glass fiber reinforcement			25%	25%
EXPOSURE CONDITIONS				
Exposure Type	accelerated weathering		accelerated weathering	
Exposure Apparatus	weatherometer		weatherometer	
Exposure Note	two hour cycle: 102 minutes of sunshine plus 18 minutes of sunshine and rain (rain is distilled and deionized water)		two hour cycle: 102 minutes of sunshine plus 18 minutes of sunshine and rain (rain is distilled and deionized water)	
Exposure Temperature (°C)	63-66	63-66	63-66	63-66
Exposure Time (days)	41.7	83.3	41.7	83.3
PROPERTIES RETAINED (%)				
Tensile Strength	97.6	99.8	60.7	53.8
Elongation	87.6	100.5	40	20

Chapter 10

Polychlorotrifluoroethylene

Accelerated Artificial Weathering Resistance

Allied Signal: Aclar 22A (features: transparent; product form: film); **Aclar 22C** (features: transparent; product form: film); **Aclar 33C** (features: transparent; product form: film); **Aclar 88A** (features: transparent; product form: film)

Accelerated tests in a weatherometer show that Aclar is extremely resistant to UV radiation and water spray.

Reference: *Aclar Performance Films,* supplier technical report (SFI-14 Rev. 9-89) - Allied-Signal Enineered Plastics, 1989.

GRAPH 69: Weatherometer Exposure Time vs. Elongation Retained of Polychlorotrifluoroethylene

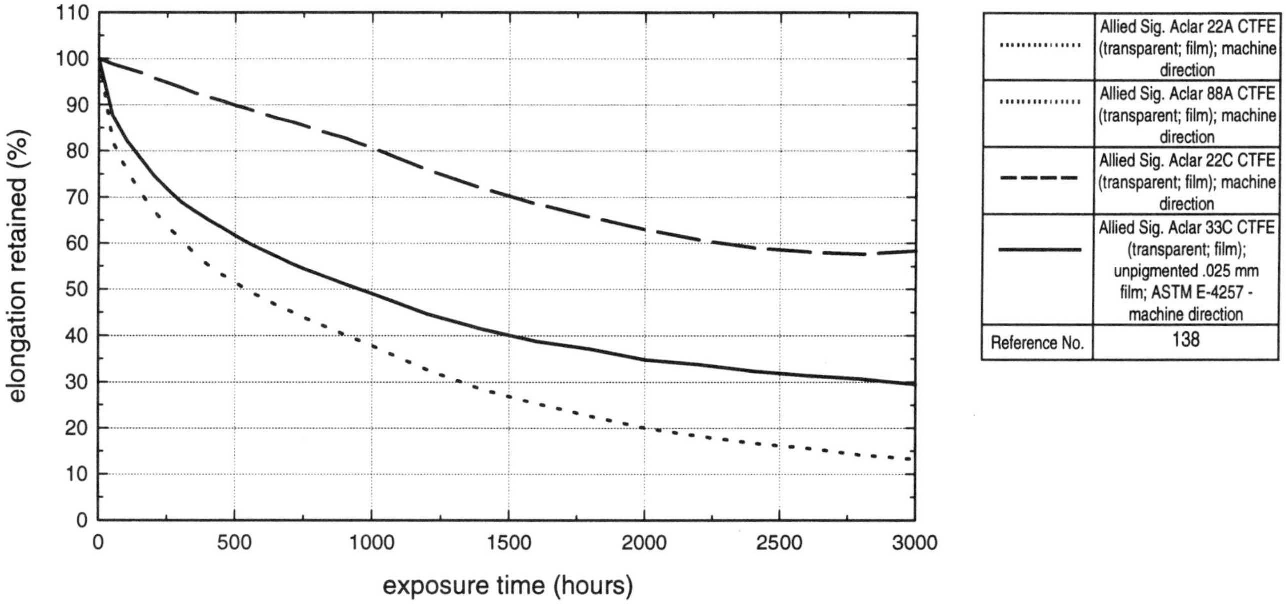

GRAPH 70: Weatherometer Exposure Time vs. Elongation Retained of Polychlorotrifluoroethylene

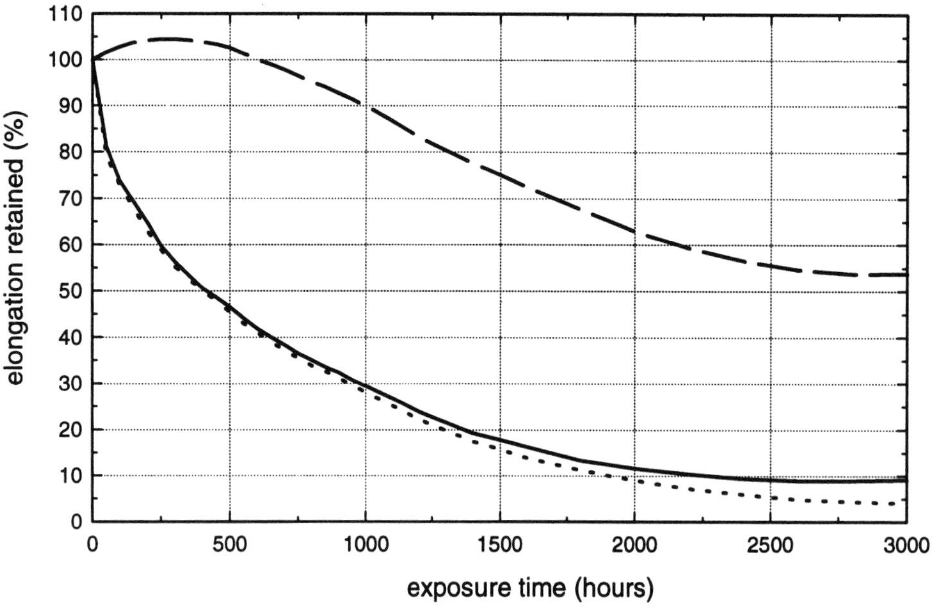

···············	Allied Sig. Aclar 22A CTFE (transparent; film); transverse direction
···············	Allied Sig. Aclar 88A CTFE (transparent; film); transverse direction
– – – –	Allied Sig. Aclar 22C CTFE (transparent; film); transverse direction
———	Allied Sig. Aclar 33C CTFE (transparent; film); transverse direction
Reference No.	138

GRAPH 71: Weatherometer Exposure Time vs. Tensile Strength Retained of Polychlorotrifluoroethylene

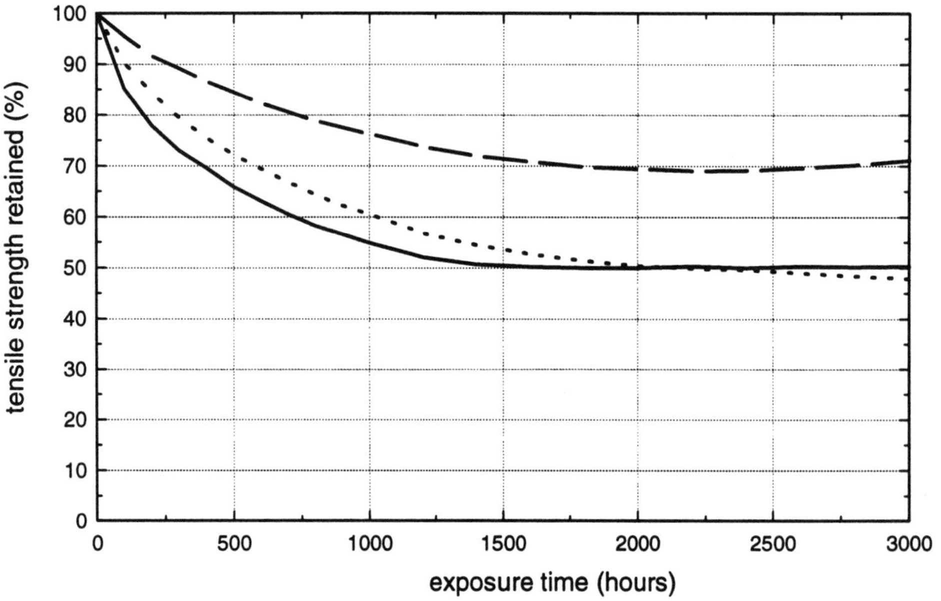

···············	Allied Sig. Aclar 22A CTFE (transparent; film); machine direction
···············	Allied Sig. Aclar 88A CTFE (transparent; film); machine direction
– – – –	Allied Sig. Aclar 22C CTFE (transparent; film); machine direction
———	Allied Sig. Aclar 33C CTFE (transparent; film); machine direction
Reference No.	138

GRAPH 72: Weatherometer Exposure Time vs. Tensile Strength Retained of Polychlorotrifluoroethylene

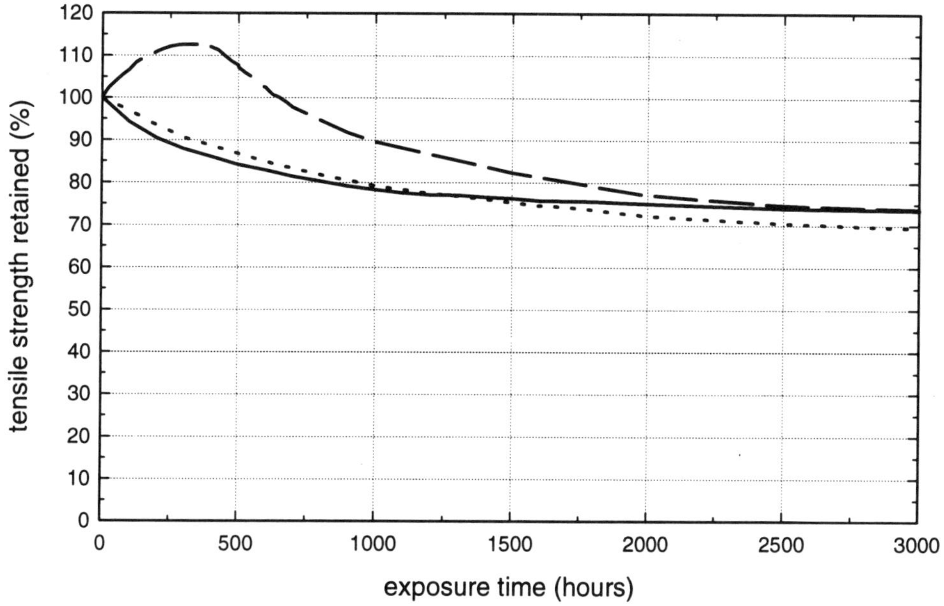

....	Allied Sig. Aclar 22A CTFE (transparent; film); unpigmented .025 mm film; ASTM E-4257 - transverse direction
....	Allied Sig. Aclar 88A CTFE (transparent; film); unpigmented .025 mm film; ASTM E-4257 - transverse direction
– – – –	Allied Sig. Aclar 22C CTFE (transparent; film); unpigmented .025 mm film; ASTM E-4257 - transverse direction
———	Allied Sig. Aclar 33C CTFE (transparent; film); unpigmented .025 mm film; ASTM E-4257 - transverse direction
Reference No.	138

Chapter 11

Polyvinylidene Fluoride

Outdoor Weather Resistance

Atochem: Kynar 500 (product form: coating)

Two decades of continuous exposure testing have repeatedly demonstrated the superiority of Kynar 500 over anodized metals, PVC plastisols, porcelain enamels, acrylics, and silicone polyester-based materials. In exposure tests conducted in the southeastern United States (Florida), six commercial finishes were exposed to the tropical sun and salt air for 12 years. Only the Kynar 500-based finish resisted significant color change.

When exposed to various climates of hot sun, salt spray and atmospheric pollution, (e.g. wintering in Scandinavia and Japan, wind and sand in Saudi Arabia and the extreme weather changes in the American Middle West) Kynar 500 based finish was significantly resistant.

Durability of coatings is measured by their ability to resist erosion when exposed to the falling sand test method. Kynar 500-based finishes are able to withstand sand abrasion because of their toughness.

After 160 months of continuous exposure to a subtropical environment, at a 45° angle facing south, a metal panel with a Kynar 500 finish shows very little change even when examined at 1,000 times magnification with a scanning electron microscope.

Reference: *The Enduring Beauty Of Architectural Finishes Based On Kynar 500,* supplier marketing literature (PL500-TR-10M 12-90) - Atochem North America, 1990.

Atochem: Kynar (features: transparent, 0.204 mm thick; product form: film)

The mechanical properties of Kynar film are maintained throughout many years of outdoor exposure. Clear films, exposed to the sun at a 45° angle South, retained their tensile strengths over a 17-year period. During the first few months of exposure when normal crystallization takes place, the percent of elongation at break decreases to a level that then remains essentially constant with time. In addition, the weathered films remain flexible and are capable of being bent 180° without cracking.

Reference: *Kynar Polyvinylidene Fluoride,* supplier technical report (PL705-Rev4-1-91) - Atochem North America, Inc., 1991.

Solvay: Solef

Natural aging tests have been carried out for more than ten years with no alteration in observed properties..

After 25,000 hours of aging at 165°C, the tensile stress at yield and tensile strength varied by only approximately 10%.

Reference: *Solvay Polyvinylidene Fluoride,* supplier design guide (B-1292c-B-2.5-0390) - Solvay, 1992.

Accelerated Artificial Weathering Resistance

Atochem: Foraflon

An exposure of 5,000 hours in the Xenotest 450 does not alter the mechanical properties of Foraflon. Its surface gloss is not affected and no yellowing is observed.

Reference: *Foraflon PVDF,* supplier design guide (694.E/07.87/20) - Atochem S. A., 1987.

Solvay: Solef

The resistance to aging of Solef PVDF is outstanding; after 7,200 hours exposure to the weatherometer, no alteration of its properties was observed.

Reference: *Solvay Polyvinylidene Fluoride,* supplier design guide (B-1292c-B-2.5-0390) - Solvay, 1992.

PVDF

After an exposure of 1000 hours in a xenon arc weatherometer, PVDF undergoes a decrease of 32% in elongation at break.

Reference: *Thermal And Other Properties Of Halar Fluoropolymer,* supplier technical report (GHG) - Ausimont.

Indoor UV Light Resistance

Atochem: Kynar (product form: film)

The stability of Kynar resin to ultraviolet radiation (200-400 nm) is demonstrated by the fold endurance of thin films after accelerated aging. When tested by the MIT Fold Endurance Test at 0.5 kg load, a 0.05 mm (0.002-in.) film withstood 500,000 cycles before accelerated exposure and 450,000 cycles after one year exposure to a GE S-1 lamp.

Reference: *Kynar Polyvinylidene Fluoride,* supplier design guide (15M-8-88-TR PL705-REV-2) - Penwalt Corporation, 1988.

Solvay: Solef

Solef PVDF transmits and is inert to ultraviolet radiation having a wavelength above 300 nm. An ultraviolet screen effect in multilayer films may be obtained in combination with other polymers, using a carefully selected additive.

Reference: *Solvay Polyvinylidene Fluoride,* supplier design guide (B-1292c-B-2.5-0390) - Solvay, 1992.

TABLE 21: Outdoor Weathering of Atochem Kynar Polyvinylidene Fluoride.

Material Family	Polyvinylidene Fluoride
Material Supplier/ Grade	Atochem Kynar
Features	transparent
Product Form	film
Reference Number	2
MATERIAL CHARACTERISTICS	
Sample Thickness (mm)	0.204
EXPOSURE CONDITIONS	
Exposure Type	outdoor weathering
Exposure Time (days)	6209
Exposure Note	45° angle south
PROPERTIES RETAINED (%)	
Tensile Strength	124 {d}
Elongation @ Break	22 {d}

TABLE 22: Accelerated Weathering in a Weatherometer of Polyvinylidene Fluoride.

Material Family	Polyvinylidene Fluoride	
Reference Number	131	131

MATERIAL CHARACTERISTICS

Sample Thickness	2.03 mm	2.03 mm

EXPOSURE CONDITIONS

Exposure Type	accelerated weathering	accelerated weathering
Exposure Apparatus	xenon arc weatherometer	xenon arc weatherometer
Exposure Time (days)	20.8	41.6

PROPERTIES RETAINED (%)

Relative Tensile Modulus	109 {z}	93 {z}
Tensile Strength @ Yield	117 {y}	104 {y}
Tensile Strength @ Break	109 {y}	101 {y}
Elongation @ Break	77 {y}	78 {y}

Chapter 12

Ionomer

Outdoor Weather Resistance

DuPont: Surlyn

Under continuous exposure to sunlight and weather, it is essential that UV stabilizers be used.

The most traditional and positive method of stabilizing ionomers for long-term usage in all-weather environments requires the addition of 0.2% antioxidant and 5% (by weight) of well dispersed micron-sized carbon black. With this modification, products made from Surlyn have been in continuous service for over ten years.

The development of technology for stabilizing clear or color-pigmented ionomers is a dynamic process. Earlier recommendations, based on the incorporation of antioxidant and UV absorbers, have produced pigmented products that still retain their physical integrity and appearance after five years of normal exposure to an Arizona environment.

Subsequent development of newer stabilizers and "energy quencher" additives has led to broader recommendations for both clear and pigmented systems. There is no complete, comprehensive system for UV protection. However, based on continued research with accelerated testing and evaluation of long-term Florida exposure, it is possible to present a series of basic rules that provide an opportunity to customize the use of polymer modifiers.

Six basic rules for UV protection in ionomers:

1. For a more stable base and long-term performance use zinc type ionomers.
2. It is essential to use antioxidant with all stabilizer systems.
3. Both sodium and zinc type ionomers may be modified for protection from occasional exposure to sunlight (less than 200 hrs./year).
4. For maximum retention of tensile and impact properties, a combination of antioxidant, UV absorber, and energy quencher must be used. (In pigmented parts, this should not present any limitations to product appearance. However, in clear, transparent applications the presence of currently recommended UV absorbers may create unacceptable levels of yellowness, depending on part thickness.)
5. When maximum retention of clarity, surface brilliance, and absence of color formation are primary end-use considerations, then the combination of antioxidant with energy quencher are recommended. (In this system, tensile and impact character will decline to one-third the level of natural grade properties.)
6. In either of the above cases (4 & 5), addition of 2-10 ppm of Monastral blue or violet ("transparent" pigment) will act to neutralize the observation of slightly yellow tints.

In applications where retention of "water white" clarity is necessary, elimination of the UV absorber component will reduce the yellow coloration. However, tensile properties will degrade to approximately 30% of original. Use of a masking agent neutralizes the slight color due to the energy quencher.

Reference: *Resistance to Ultraviolet Irradiation for Surlyn Ionomer Resins,* supplier technical report (E-78693-103520/A) - Du Pont Company, 1986.

DuPont: Surlyn

Outdoor weathering experience has confirmed the outstanding performance of UV stabilized Surlyn. Parts containing carbon black have been in service and exposed to all types of weather for over 10 years with no significant change in physical integrity or appearance. Other pigmented parts have retained their physical integrity and appearance after 5 years of exposure to an Arizona environment.

Reference: *Weatherability,* supplier marketing literature (E-53525) - Du Pont Company, 1983.

Accelerated Artificial Weathering Resistance

DuPont: Surlyn

In up to 5,000 hours accelerated weathering tests, production samples of automotive exterior trim extrusions faced with clear, UV stabilized Surlyn and clear, UV stabilized PVC were exposed side by side. Besides the obvious edge in UV stability, Surlyn ionomer resin requires no liquid plasticizer, so there can be no migration problem in the finished part.

Reference: *Weatherability,* supplier marketing literature (E-53525) - Du Pont Company, 1983.

TABLE 23: Outdoor Weathering in Florida and Arizona of DuPont Surlyn Ionomer.

Material Family	Ionomer						
Material Supplier/ Grade	DuPont Surlyn 9520	DuPont Surlyn 9520	DuPont Surlyn 9520	DuPont Surlyn 9520	DuPont Surlyn 9910	DuPont Surlyn 8528	DuPont Surlyn 8920
Reference Number	134	134	134	134	134	134	134
MATERIAL COMPOSITION							
Irganox 1010 (antioxidant - Ciba Geigy)				0.3 wt.%	0.3 wt.%		0.3 wt.%
Santonox R (antioxidant)	0.2 wt.%	0.2 wt.%	0.2 wt.%			0.2 wt.%	
Cyasorb 531 (UV absorber - American Cyanamid)		1.0 wt.%	1.0 wt.%				
Tinuvin 770 (hindered amine light stabilizer - Ciba Geigy)				0.5 wt.%	0.5 wt.%		0.5 wt.%
argent pigment			0.2 wt.%				
black pigment	5.0 wt.%					5.0 wt.%	
bronze pigment		0.5 wt.%					
ion type	zinc	zinc	zinc	zinc	zinc	sodium	sodium
EXPOSURE CONDITIONS							
Exposure Type	outdoor weathering						
Exposure Location	Florida	Florida	Florida	Florida	Arizona	Florida	Florida
Exposure Country	USA	USA	USA	USA	USA	USA	USA
Exposure Time (days)	1095	1825	913	913	365	1095	365
PROPERTIES RETAINED (%)							
Physical Properties	>90	no change apparent, but no quantitative test data	87	no change apparent, but no quantitative test data	no change apparent, but no quantitative test data	no change apparent, but no quantitative test data	50
SURFACE AND APPEARANCE							
Visual Appearance	slightly dull	slightly dull	slightly dull	slight haze	no visible change	slightly dull	slight haze

TABLE 24: Effect of Pigments, UV Stabilizers and Antioxidants on the Accelerated Weathering in an Atlas Weatherometer of Zinc Ion Type DuPont Surlyn Ionomer.

Material Family	Ionomer								
Material Supplier/ Grade	DuPont Surlyn 9910	DuPont Surlyn 9910	DuPont Surlyn 9910	DuPont Surlyn 9910	DuPont Surlyn 9910	DuPont Surlyn 9720	DuPont Surlyn 9020	DuPont Surlyn 9020	DuPont Surlyn 9020
Features								unstabilized	unstabilized
Reference Number	134	134	134	134	134	134	134	134	134
MATERIAL COMPOSITION									
Irganox 1010 (antioxidant - Ciba Geigy)	0.1 wt.%	0.1 wt.%	0.1 wt.%	0.1 wt.%	0.1 wt.%		0.1 wt.%		
Santonox R (antioxidant)						0.2 wt.%			
Cyasorb 531 (UV absorber - American Cyanamid)						0.4 wt.%	0.2 wt.%		
Tinuvin 328 (UV absorber - Ciba Geigy)				0.3 wt.%	0.3 wt.%				
Tinuvin 770 (hindered amine light stabilizer - Ciba Geigy)	0.3 wt.%	0.6 wt.%	0.6 wt.%	0.3 wt.%	0.3 wt.%		0.2 wt.%		
orange pigment							0.2 wt.%		
sulfur pigment						2.0 wt.%			
ion type	zinc	zinc	zinc	zinc	zinc	zinc	zinc	zinc	zinc
EXPOSURE CONDITIONS									
Exposure Type	accelerated weathering								
Exposure Apparatus	Atlas weatherometer								
Exposure Apparatus Note	filtered carbon arc								
Exposure Note	60° dry, 50° wet								
Exposure Time (days)	125	208	125	208	125	340	42	4	42
PROPERTIES RETAINED (%)									
Physical Properties	22	25	29	33	38	46	no change apparent, but no quantitative test data	no change apparent, but no quantitative test data	poor
SURFACE AND APPEARANCE									
Visual Appearance	moderately yellow	slightly yellow, slightly crazed	slightly yellow	slightly yellow, slightly crazed	slightly yellow	no visible change	slightly dull	no visible change	yellow, crazed

TABLE 25: Effect of Pigments, UV Stabilizers and Antioxidants on the Accelerated Weathering in an Atlas Weatherometer of Zinc Ion Type DuPont Surlyn Ionomer.

Material Family	Ionomer							
Material Supplier/ Grade	DuPont Surlyn 9520							
Features						unstabilized		
Reference Number	134	134	134	134	134	134	134	134
MATERIAL COMPOSITION								
Irganox 1010 (antioxidant - Ciba Geigy)			0.2 wt.%	0.21 wt.%	0.2 wt.%			
Santonox R (antioxidant)	0.2 wt.%	0.2 wt.%					0.2 wt.%	0.2 wt.%
Cyasorb 531 (UV absorber - American Cyanamid)			1.0 wt.%		0.1 wt.%		1.0 wt.%	1.0 wt.%
Tinuvin 770 (hindered amine light stabilizer - Ciba Geigy)				0.2 wt.%	0.1 wt.%			
argent pigment								0.2 wt.%
black pigment	2.7 wt.%	5.0 wt.%						
bronze pigment							0.5 wt.%	
ion type	zinc	zinc	zinc	zinc	zinc	zinc	zinc	zinc
EXPOSURE CONDITIONS								
Exposure Type	accelerated weathering							
Exposure Apparatus	Atlas weatherometer							
Exposure Apparatus Note	filtered carbon arc							
Exposure Note	60° dry, 50° wet							
Exposure Time (days)	292	67	58	58	58	58	67	67
PROPERTIES RETAINED (%)								
Physical Properties	89	>90	100	80	100	50	no change apparent, but no quantitative test data	87
SURFACE AND APPEARANCE								
Visual Appearance	slightly dull	slightly dull	good	good	good	slightly crazed	slightly dull	slightly dull

TABLE 26: **Effect of Pigments, UV Stabilizers and Antioxidants on the Accelerated Weathering in an Atlas Weatherometer of Sodium Ion Type DuPont Surlyn Ionomer.**

Material Family	Ionomer											
Material Supplier/ Grade	DuPont Surlyn 8528	DuPont Surlyn 8528	DuPont Surlyn 8528	DuPont Surlyn 8528	DuPont Surlyn 8528	DuPont Surlyn 8528	DuPont Surlyn 8528	DuPont Surlyn 8920	DuPont Surlyn 8920	DuPont Surlyn 8020	DuPont Surlyn 8020	DuPont Surlyn 8020
Features							unsta-bilized				unsta-bilized	unsta-bilized
Reference Number	134	134	134	134	134	134	134	134	134	134	134	134

MATERIAL COMPOSITION

Irganox 1010 (antioxidant - Ciba Geigy)				0.2 wt.%	0.2 wt.%	0.1 wt.%				0.1 wt.%		
Santonox R (antioxidant)	0.2 wt.%	0.2 wt.%	0.2 wt.%									
Cyasorb 531 (UV absorber - American Cyanamid)			1.0 wt.%		0.1 wt.%	0.1 wt.%				0.2 wt.%		
Tinuvin 770 (hindered amine light stabilizer - Ciba Geigy)				0.2 wt.%	0.1 wt.%	0.1 wt.%		1.0 wt.%	1.0 wt.%	0.2 wt.%		
black pigment	5.0 wt.%	2.7 wt.%										
orange pigment						0.1 wt.%				0.2 wt.%		
ion type	sodium	sodium	sodium	sodium	sodium	sodium	sodium	sodium	sodium	sodium	sodium	sodium

EXPOSURE CONDITIONS

Exposure Type	accelerated weathering											
Exposure Apparatus	Atlas weatherometer											
Exposure Apparatus Note	filtered carbon arc											
Exposure Note	60° dry, 50° wet											
Exposure Time (days)	67	292	58	58	58	58	58	100	67	42	4	42

PROPERTIES RETAINED (%)

Physical Properties	no change apparent, but no quantitative test data	100	100	60	65	75	25	no change apparent, but no quantitative test data	no change apparent, but no quantitative test data	no change apparent, but no quantitative test data	no change apparent, but no quantitative test data	0

SURFACE AND APPEARANCE

Visual Appearance	slightly dull	slightly dull	yellow	good	yellow	no visible change	yellow, crazed	slight haze	no visible change	slightly dull	no visible change	yellow, crazed

TABLE 27: Accelerated Weathering in a QUV of Zinc Ion Type DuPont Surlyn Ionomer.

Material Family	Ionomer								
Material Supplier/ Grade	DuPont Surlyn 9910								
Features				unstabilized	unstabilized				
Reference Number	134	134	134	134	134	134	134	134	134
MATERIAL COMPOSITION									
Irganox 1010 (antioxidant - Ciba Geigy)	0.1 wt.%	0.1 wt.%	0.1 wt.%			0.2 wt.%	0.2 wt.%	0.1 wt.%	0.1 wt.%
Cyasorb 531 (UV absorber - American Cyanamid)		0.2 wt.%	0.5 wt.%						
Tinuvin 328 (UV absorber - Ciba Geigy)									0.3 wt.%
Tinuvin 770 (hindered amine light stabilizer - Ciba Geigy)	0.3 wt.%	0.2 wt.%	0.2 wt.%			0.2 wt.%	0.2 wt.%	0.6 wt.%	0.3 wt.%
ion type	zinc	zinc	zinc	zinc	zinc	zinc	zinc	zinc	zinc
EXPOSURE CONDITIONS									
Exposure Type	accelerated weathering								
Exposure Apparatus	QUV								
Exposure Note	8 hrs at 71 °C dry, 4 hrs at 48°C wet	8 hrs at 71 °C dry, 4 hrs at 48°C wet	8 hrs at 71 °C dry, 4 hrs at 48°C wet	8 hrs at 71 °C dry, 4 hrs at 48°C wet	8 hrs at 60 °C dry, 4 hrs at 50°C wet	8 hrs at 71 °C dry, 4 hrs at 48°C wet	8 hrs at 60 °C dry, 4 hrs at 50°C wet	8 hrs at 71 °C dry, 4 hrs at 48°C wet	8 hrs at 71 °C dry, 4 hrs at 48°C wet
Exposure Time (days)	125	125	125	83	84	83	84	125	125
PROPERTIES RETAINED (%)									
Physical Properties	22	46	66	0	29	18	36	15	70
SURFACE AND APPEARANCE									
Visual Appearance	moderately yellow	slightly yellow	slight haze	crazed	crazed	good	good	slightly yellow, slightly crazed	slightly yellow, slightly crazed

Chapter 13

Modified Polyphenylene Oxide

Outdoor Weather Resistance

GE Plastics: Noryl

Noryl resins should not degrade, decompose, chalk, craze, or crack on exposure to outdoor weathering.

Noryl resins will:

1. Lose some impact and elongation (20-40% depending on grade).
2. Gain tensile and flexural strength (5-15% on long term exposure).
3. Lose any surface gloss within a few months and become dull.
4. Change shades of color which are more yellow or darker on exposure.

Only surface discoloration will occur. However, very thin sections (under 12.7 mm) may become more brittle. This brittleness occurs because the surface layers which are losing impact and becoming stiffer make up a proportionately larger volume of a thin section versus a thicker section.

Reference: *Weatherability Of Noryl Resins,* supplier technical report - General Electric Company, 1992.

GE Plastics: Noryl

When exposed to outdoor light, parts of Noryl resin undergo a color change with a tendency to darken slightly and drift toward yellow. When selecting Noryl resins for outdoor use, dark colors - black and brown - are recommended, as well as reds, yellows and oranges, which show excellent color stability where the tendency to yellow is masked.

Reference: *Noryl Extrusion Resins,* supplier design guide (CDX-265) - General Electric Company.

TABLE 28: Outdoor Weathering in Arizona, Florida and New York of General Electric Noryl Modified Polyphenylene Oxide.

Material Family	Modified Polyphenylene Oxide								
Material Supplier/ Grade	GE Plastics Noryl SE100-8078								
Features	flame retardant								
Reference Number	173	173	173	173	173	173	173	173	173
EXPOSURE CONDITIONS									
Exposure Type	outdoor weathering			outdoor weathering			outdoor weathering		
Exposure Location	Arizona			Florida			Selkirk, New York		
Exposure Country	USA			USA			USA		
Exposure Time (days)	365	730	1096	365	730	1096	365	730	1096
PROPERTIES RETAINED (%)									
Flexural Modulus	115.9	111.4	116.5	115.3	111.4	115	112.9	109.9	113.8
Ultimate Tensile Strength	91.3	87	85.5	87	85.5	84.1	89.9	88.4	85.5
Tensile Strength @ Yield	88.1	89.6	91	86.6	88.1	89.6	85.1	86.6	86.6
Flexural Strength	106.1	106.1	105.3	99.1	100.9	107.9	101.8	98.2	101.8
Elongation	33.3	22.2	13.9	33.3	25	16.7	55.6	41.7	30.6
Notched Izod Impact Strength	72.9	72.9	64.6	70.8	64.6	58.3	89.6	87.5	83.3
SURFACE AND APPEARANCE									
Δ Yellowness Index	31.7	30	37.9	28.2	31.1	31.7	26.3	30.8	39
Gloss Retained (%)	7.3	7.1	6.5	5.8	3.7	3.7	6	4.3	4.7

GRAPH 73: **Accelerated Indoor UV Exposure Time vs. Delta E Color Change of Modified Polyphenylene Oxide**

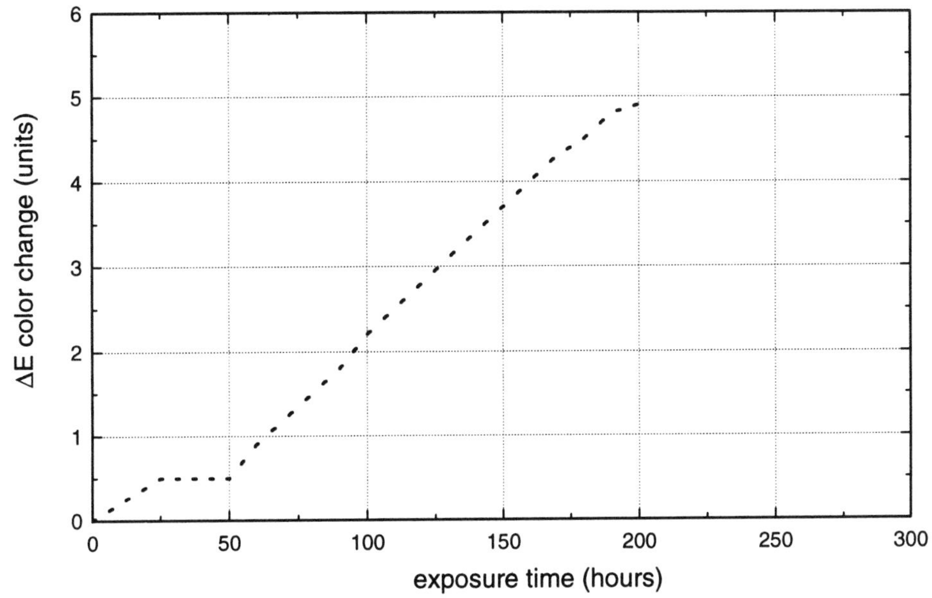

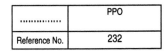

....	PPO
Reference No.	232

GRAPH 74: **Arizona Outdoor Weathering Exposure Time vs. Drop Dart Impact Strength of Modified Polyphenylene Oxide**

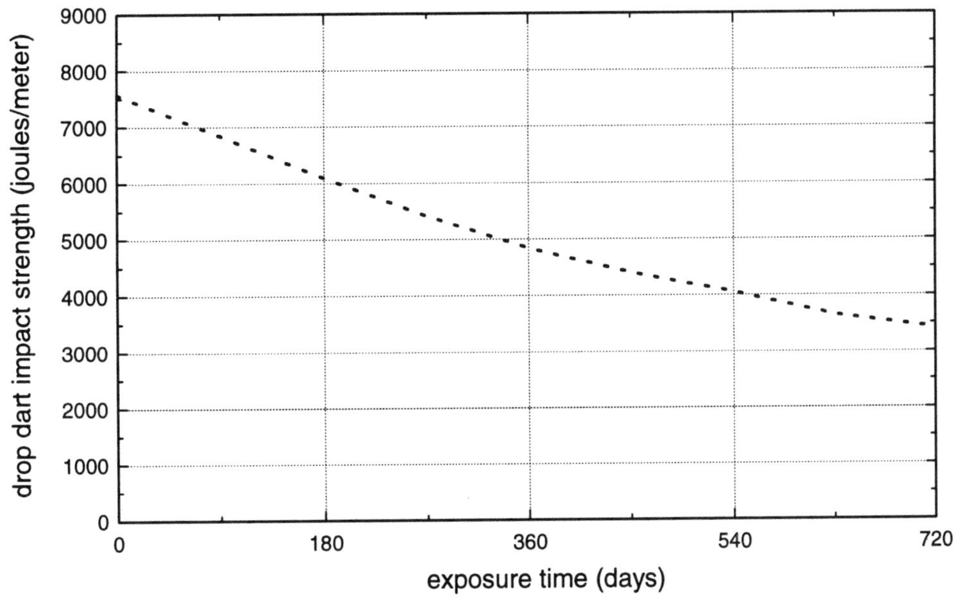

....	PPO (grey, 3.2 mm thick)
Reference No.	189

GRAPH 75: **Arizona Outdoor Weathering Exposure Time vs. Elongation of Modified Polyphenylene Oxide**

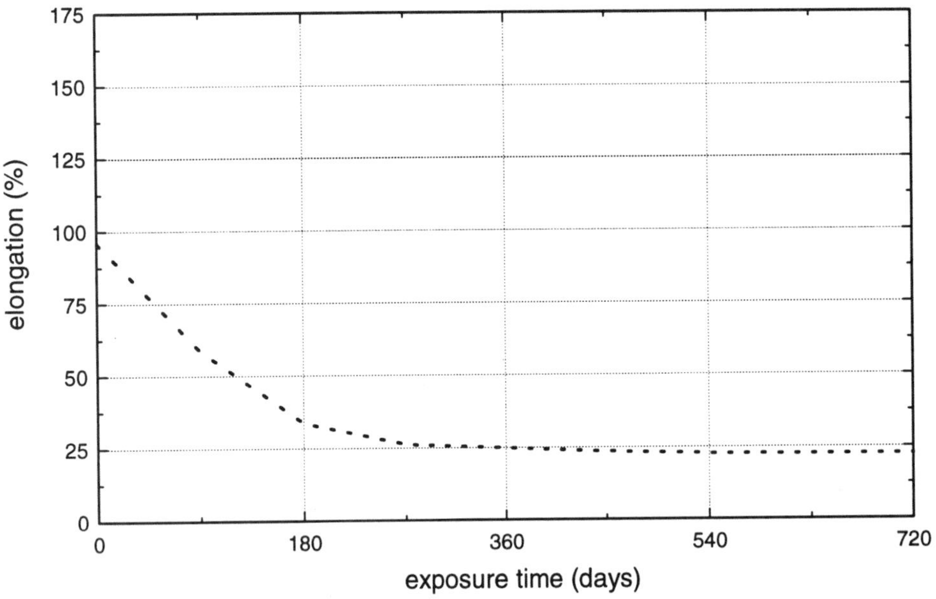

....	PPO (grey)
Reference No.	189

GRAPH 76: **Arizona Outdoor Weathering Exposure Time vs. Tensile Strength at Yield of Modified Polyphenylene Oxide**

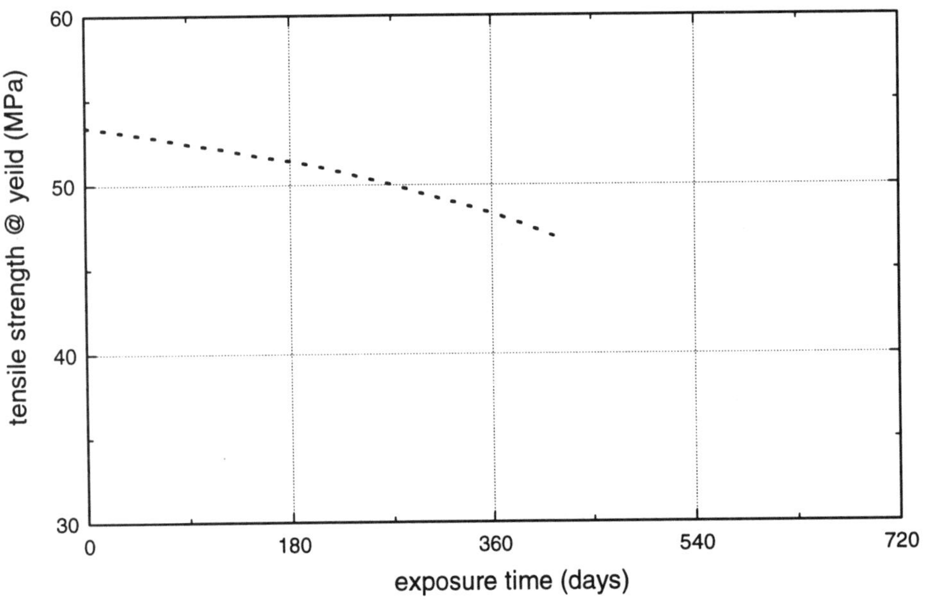

....	PPO
Reference No.	189

GRAPH 77: **Arizona Outdoor Weathering Exposure vs. Delta E Color Change of Modified Polyphenylene Oxide**

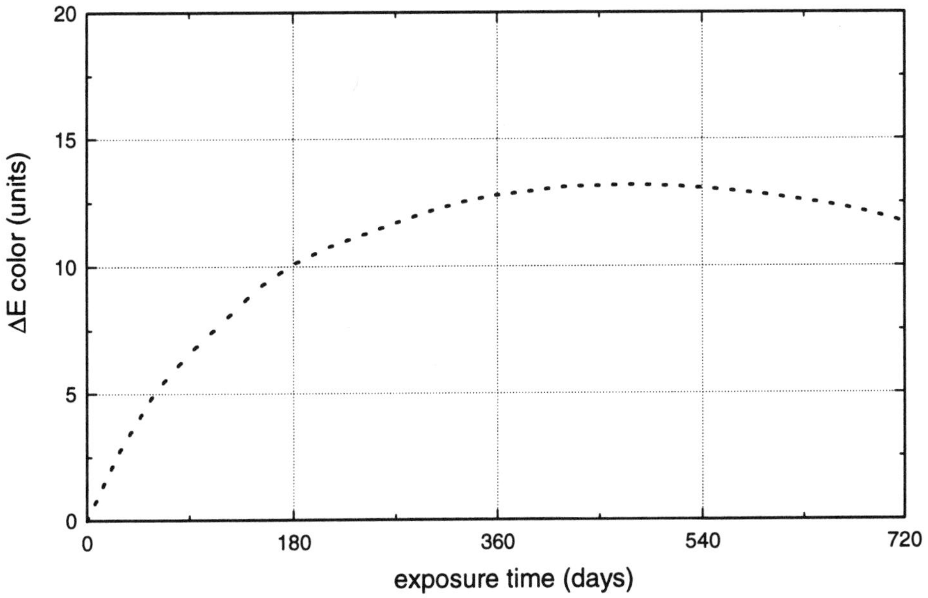

....	PPO (grey); CIE lab color scale
Reference No.	189

GRAPH 78: **Ohio Outdoor Weathering Exposure vs. Delta E Color Change of Modified Polyphenylene Oxide**

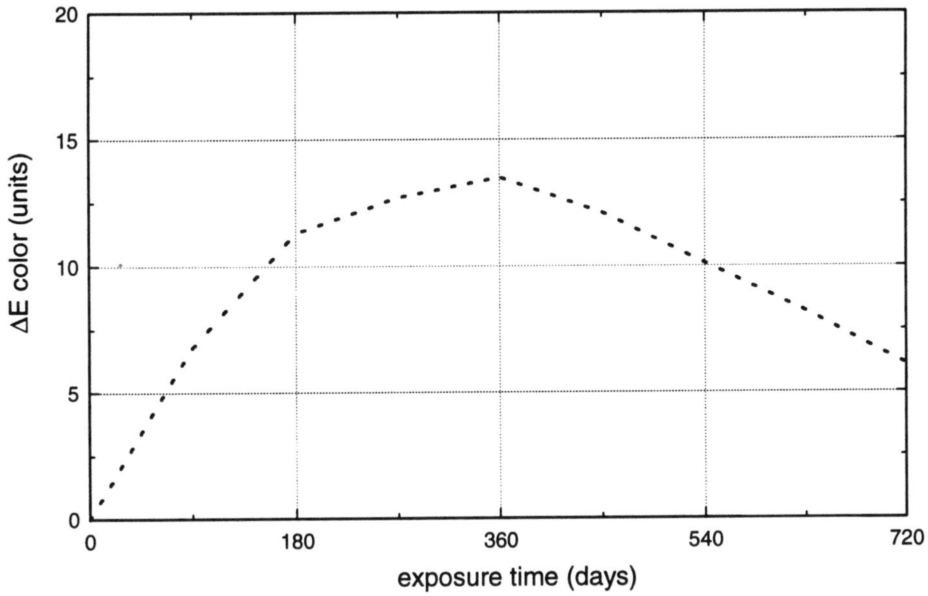

....	PPO (grey); CIE lab color scale
Reference No.	189

GRAPH 79: Ohio Outdoor Weathering Exposure vs. Drop Dart Impact Strength of Modified Polyphenylene Oxide

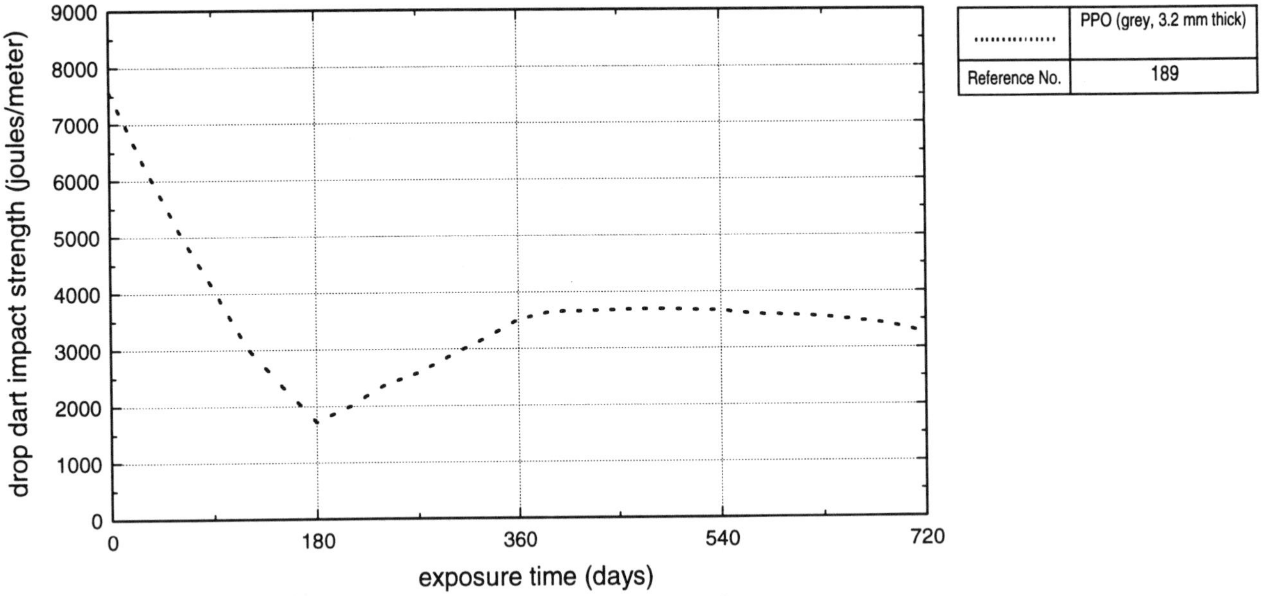

Chapter 14

Nylon 12

Outdoor Weather Resistance

EMS-Chemie: Grilamid TR55 (features: transparent); **Grilamid TR55UV** (features: UV stabilized, transparent)

Grilamid TR55 and TR55UV test plaques have been exposed for 40 months in Southeast Switzerland at 45°, facing south. Unstabilized Grilamid TR55 retained good transparency, but with some discoloration after 4 months exposure and some brittleness after 25 months.

In contrast, Grilamid TR55UV maintained its transparency with no brittleness, surface crazing or degradation, and no effect on relative viscosity. A small increase in yellowness occurred for up to 11 months exposure, and then showed no further increase. This compares very favorably with a typical stabilized polycarbonate in the same test, which showed significant loss of molecular weight and an increase in yellowness that seriously impaired transparency.

Reference: *Grilamid TR55 Transparent Nylons,* supplier design guide (GR1-104) - EMS-Chemie.

Accelerated Artificial Weathering Resistance

EMS-Chemie: Grilamid TR55 (features: transparent); **Grilamid TR55UV** (features: UV stabilized, transparent)

Samples of Grilamid TR55 and TR55UV were tested in an Atlas weatherometer, cycling for 20 minutes (17 minutes of UV exposure followed by 3 minutes of UV exposure plus water spray). After 2,000 hours, Grilamid TR55UV showed no measurable change of color or surface appearance. In contrast, the unstabilized Grilamid TR55, while retaining transparency, showed an increase in yellowness and a slightly matte surface - a behavior similar to unstabilized polycarbonate.

Reference: *Grilamid TR55 Transparent Nylons,* supplier design guide (GR1-104) - EMS-Chemie.

GRAPH 80: Weatherometer Exposure Time vs. Delta E Color Change of Nylon 12

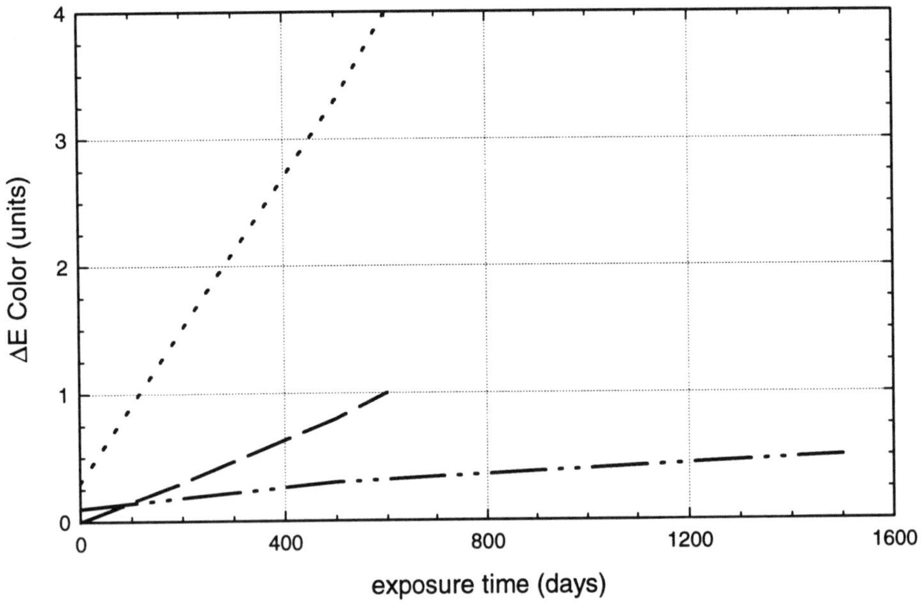

···············	EMS Grilamid TR55 Nylon 12 (transparent); tested to ASTM D2565; 0.35 W/m^2 @ 340 nm, 63°C BP Temp; 102/18 min. rain cycle; linear least square fit
—·—··	EMS Grilamid TR55LX Nylon 12 (high impact, UV stabilized, transparent); tested to ASTM D2565; 0.35 W/m^2 @ 340 nm, 63°C BP Temp; 102/18 min. rain cycle; linear least square fit
– – – –	EMS Grilamid TR55LY Nylon 12 (high impact, UV stabilized, transparent); tested to ASTM D2565; 0.35 W/m^2 @ 340 nm, 63°C BP Temp; 102/18 min. rain cycle; linear least square fit
Reference No.	171

GRAPH 81: Weatherometer Exposure Time vs. Delta E Color Change of Nylon 12

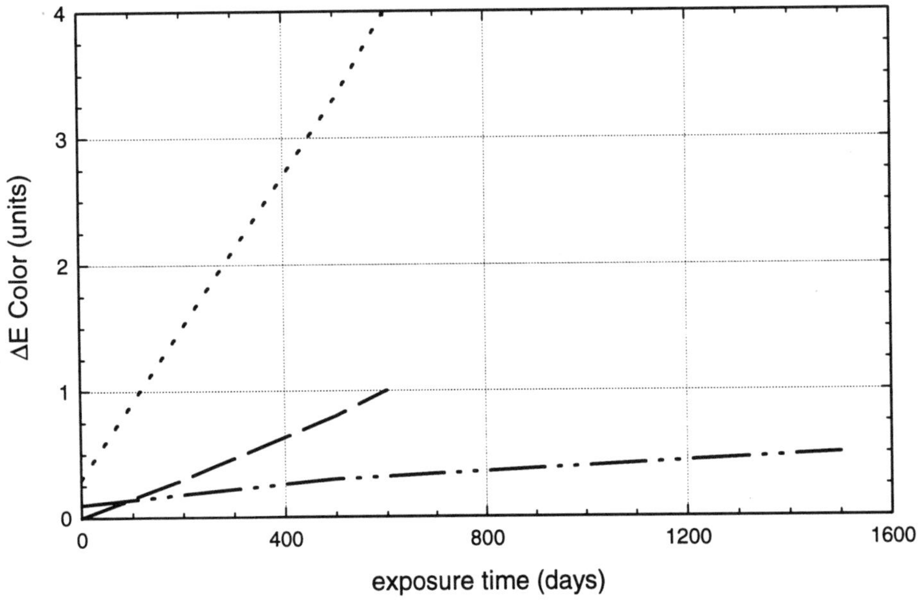

···············	EMS Grilamid TR55 Nylon 12 (transparent); tested to ASTM D2565; 0.35 W/m^2 @ 340 nm, 63°C BP Temp; 102/18 min. rain cycle; linear least square fit
—·—··	EMS Grilamid TR55LX Nylon 12 (high impact, UV stabilized, transparent); tested to ASTM D2565; 0.35 W/m^2 @ 340 nm, 63°C BP Temp; 102/18 min. rain cycle; linear least square fit
– – – –	EMS Grilamid TR55LY Nylon 12 (high impact, UV stabilized, transparent); tested to ASTM D2565; 0.35 W/m^2 @ 340 nm, 63°C BP Temp; 102/18 min. rain cycle; linear least square fit
Reference No.	171

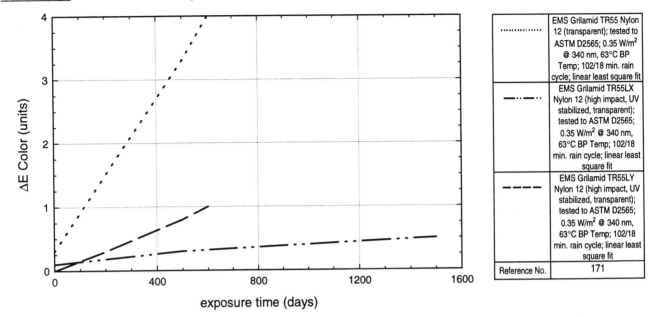

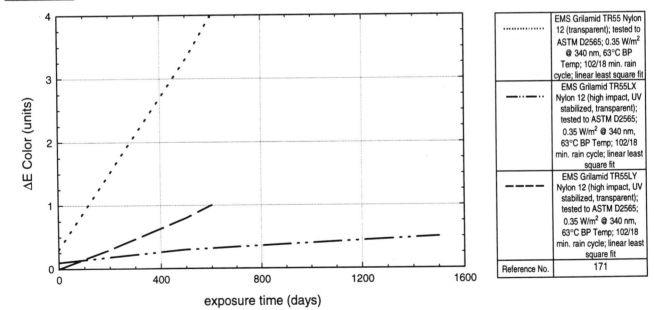

GRAPH 82: **Weatherometer Exposure Time vs. Tensile Impact Strength of Nylon 12**

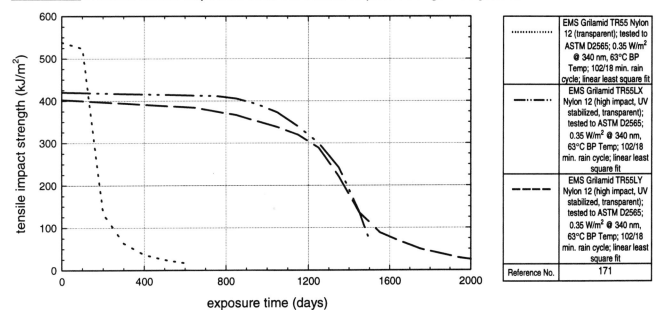

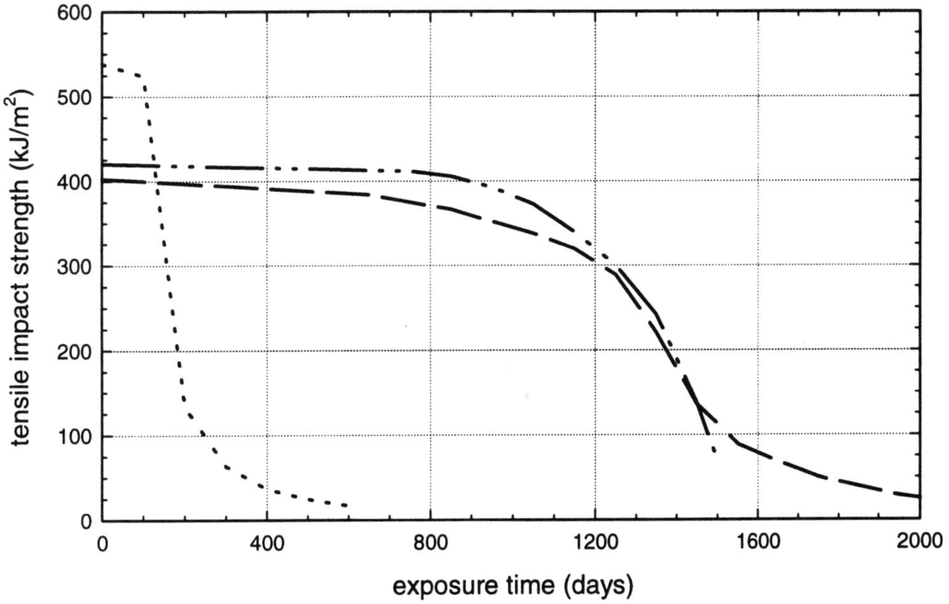

··············	EMS Grilamid TR55 Nylon 12 (transparent); tested to ASTM D2565; 0.35 W/m^2 @ 340 nm, 63°C BP Temp; 102/18 min. rain cycle; linear least square fit
—··—··	EMS Grilamid TR55LX Nylon 12 (high impact, UV stabilized, transparent); tested to ASTM D2565; 0.35 W/m^2 @ 340 nm, 63°C BP Temp; 102/18 min. rain cycle; linear least square fit
— — — —	EMS Grilamid TR55LY Nylon 12 (high impact, UV stabilized, transparent); tested to ASTM D2565; 0.35 W/m^2 @ 340 nm, 63°C BP Temp; 102/18 min. rain cycle; linear least square fit
Reference No.	171

Chapter 15

Nylon 6

UV Resistance

BIP Chemicals: Beetle

Jonylon

Nylons tend to become embrittled on long-term exposure to high intensity UV radiation, such as sunlight, and this may also be accompanied by slight discoloration.

The incorporation of heat stabilizers results in some improvement in weatherability and UV resistance, and special light stabilized grades are also available. (e.g. Beetle A41 and Jonylon L and HL grades). Heavily pigmented grades, particularly black, will give further improvements, and are recommended for extensive exposure to strong sunlight.

Reference: *Beetle & Jonylon Engineering Thermoplastics,* supplier marketing literature (BJ2/1291/SP/5) - BIP Chemicals Limited, 1991.

Outdoor Weather Resistance

BASF: Ultramid B

Ultramid B35EG3 Black 20 590 (features: natural resin, stabilized, black color; material compostion: 15% glass fiber reinforcement); **Ultramid B35K** (features: natural resin, stabilized); **Ultramid B3EG5** (features: natural resin, stabilized; material compostion: 25% glass fiber reinforcement); **Ultramid B3K** (features: natural resin, stabilized)

Many Ultramid resins are suitable for outdoor applications. The unreinforced stabilized Ultramid resins, i.e. those with the letters K and H in the type nomenclature, are extremely resistant to weathering, even if they are uncolored. Their outdoor performance can be further improved by suitable pigments, best effects being achieved with carbon black. For instance, seats that have been produced from Ultramid B3K and B35K containing special UV stabilizers and have been exposed for more than ten years in an open-air stadium have remained unbreakable, and their appearance has undergone hardly any change.

Thin articles for outdoor use should be produced from Ultramid resins with a high carbon black content, e.g. the Black 20 590 and 20 592 types, in order to ensure that their strength remains undiminished. Moldings with a high proportion of carbon black can also withstand several years' exposure to tropical conditions.

The reinforced Ultramid resins also give good outdoor performance, and the stabilized types (e.g. Ultramid B3EG5) can be relied upon to withstand exposure for periods of much more than five years. Nevertheless, the constituent glass fibers cause the surface to be attacked more severely than that of unreinforced Ultramid articles. As a consequence, the texture and the hue may undergo a change after comparatively brief exposure periods. If the glass-reinforced moldings remain exposed for a number of years, erosion to a depth of a few tenths of 1 mm (0.04 in) can generally be expected, but experience has shown that this does not exert any significant effect on the mechanical properties.

Housings for automobile rear-view mirrors are examples of articles that must retain an attractive appearance for many years. In applications of this nature, good results have been obtained by products with special UV stabilizers and products with a high carbon black content, e.g. Ultramid B35EG3 Black 20590.

Reference: *Ultramid Nylon Resins Product Line, Properties, Processing,* supplier design guide (B 568/1e/4.91) - BASF Corporation, 1991.

Ube: Ube 1013B (features: general purpose grade); **Ube 1013NU2** (features: high heat grade)

Nylon plastics are relatively stable when exposed to ultraviolet light and heat, and are particularly resistant to ozone. Thus, they are well suited to outdoor applications.

Reference: *Ube Nylon Technical Brochure,* supplier design guide (1989.8.1000) - Ube Industries, Ltd., 1989.

TABLE 29: Outdoor Weathering in Florida of Allied Signal Capron Nylon 6.

Material Family	Nylon 6										
Material Supplier/ Grade			Allied Signal Capron BK 102			Allied Signal Capron BK104			Allied Signal Capron BK106		
Features	natural resin, unstabilized		black color, unstabilized			black color, UV stabilized			black color, UV stabilized		
Product Form	monofilament		monofilament			monofilament			monofilament		
Reference Number	149	149	149	149	149	149	149	149	149	149	149
MATERIAL CHARACTERISTICS											
Sample Diameter (mm)	0.38		0.38			0.38			0.38		
EXPOSURE CONDITIONS											
Exposure Type	outdoor weathering										
Exposure Location	Florida (southern)		Florida (southern)			Florida (southern)			Florida (southern)		
Exposure Country	USA		USA			USA			USA		
Exposure Note	45° angle south		45° angle south			45° angle south			45° angle south		
Exposure Time (days)	152	304	152	304	1217	152	304	1217	152	304	1522
PROPERTIES RETAINED (%)											
Tensile Strength	80	68	87	77	70	97	94	92	100	99	98
Elongation	65	45	74	57	47	100	99	93	100	99	98

TABLE 30: Outdoor Weathering in California and Pennsylvania of Nylon 6.

Material Family	Nylon 6							
Material Supplier	LNP Engineering Plastics							
Reference Number	215	215	215	215	215	215	215	215
MATERIAL COMPOSITION								
carbon black	1%	1%	1%	1%	1%	1%	1%	1%
glass fiber reinforcement	30%	30%	30%	30%	30%	30%	30%	30%
EXPOSURE CONDITIONS								
Exposure Type	outdoor weathering				outdoor weathering			
Exposure Location	Los Angeles, California	Los Angeles, California	Los Angeles, California	Los Angeles, California	Philadelphia, Pennsylvania	Philadelphia, Pennsylvania	Philadelphia, Pennsylvania	Philadelphia, Pennsylvania
Exposure Country	USA	USA	USA	USA	USA	USA	USA	USA
Exposure Test Method	ASTM D1435	ASTM D1435	ASTM D1435	ASTM D1435	ASTM D1435	ASTM D1435	ASTM D1435	ASTM D1435
Exposure Time (days)	91	182	365	730	91	182	365	730
PROPERTIES RETAINED (%)								
Tensile Strength	105	102	101	102	103	102	101	100
Notched Izod Impact Strength	106	100			100	106		
Unnotched Izod Impact Strength	99	120			92	115		

GRAPH 83: Outdoor Exposure Time vs. Elongation at Break of Nylon 6

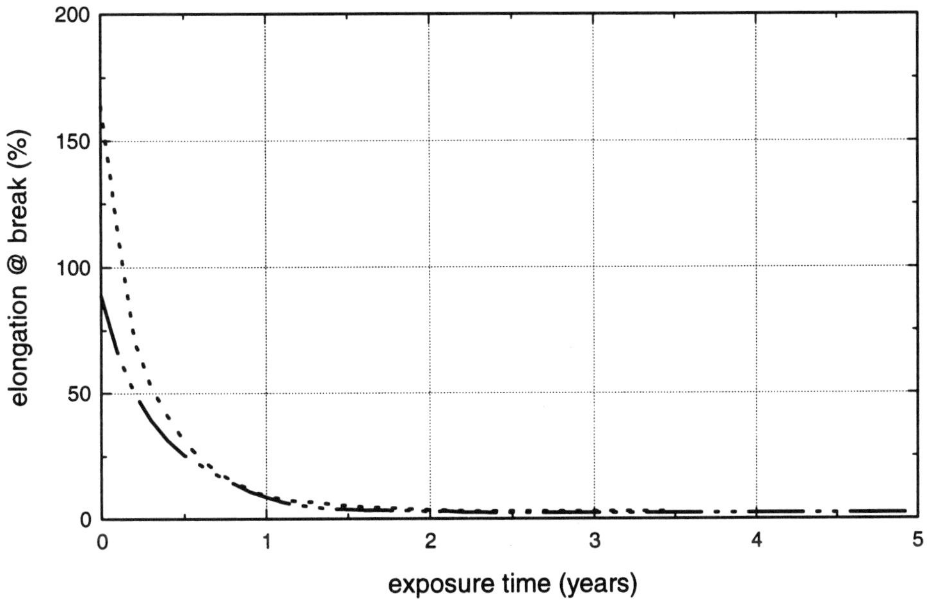

···············	Ube Ube 1013B Nylon 6 (gen. purp. grade)
—··—··	Ube Ube 1013NU2 Nylon 6 (high heat grade)
Reference No.	238

GRAPH 84: Outdoor Exposure Time vs. Flexural Modulus of Nylon 6

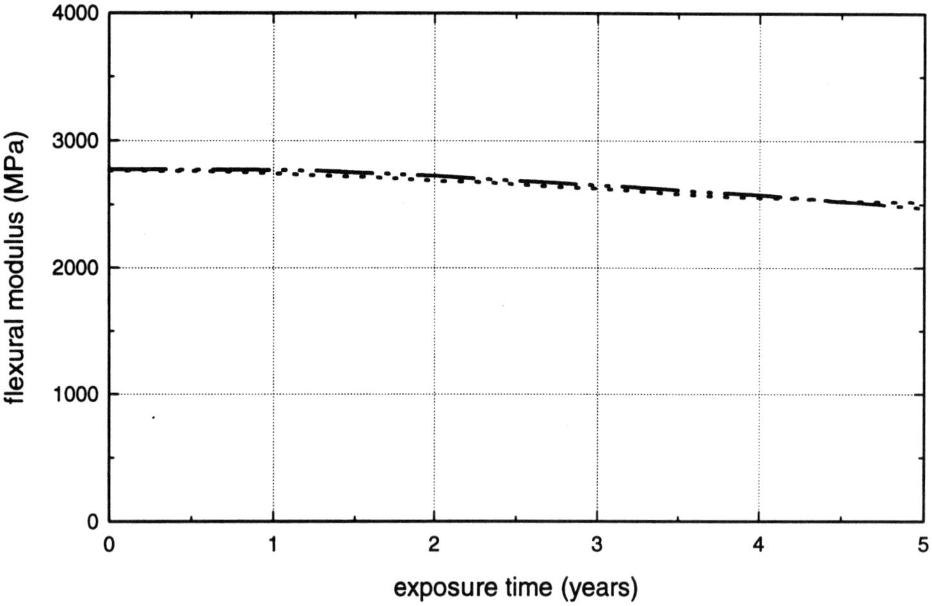

....	Ube Ube 1013B Nylon 6 (gen. purp. grade)
—··—··	Ube Ube 1013NU2 Nylon 6 (high heat grade)
Reference No.	238

GRAPH 85: Outdoor Exposure Time vs. Notched Izod Impact Strength of Nylon 6

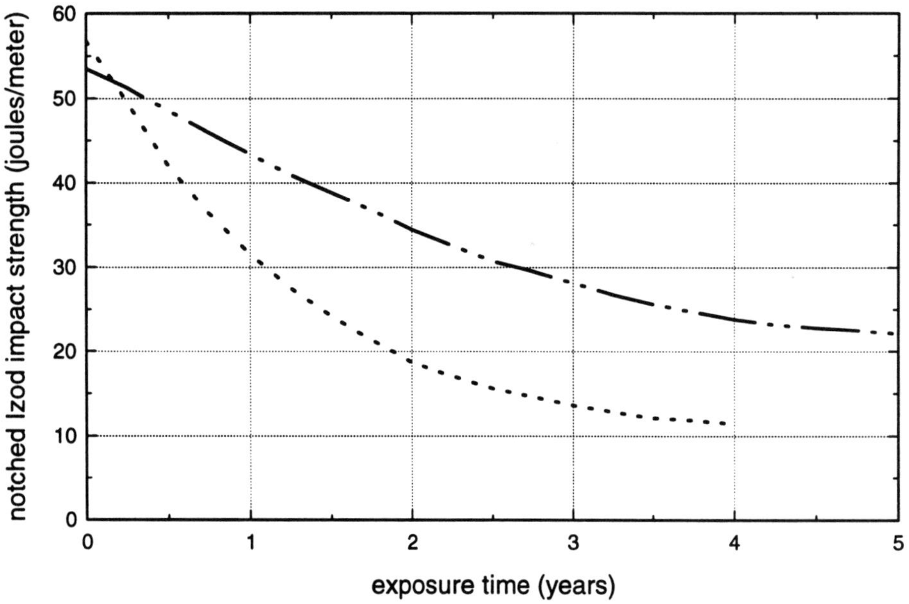

....	Ube Ube 1013B Nylon 6 (gen. purp. grade)
—··—··	Ube Ube 1013NU2 Nylon 6 (high heat grade)
Reference No.	238

GRAPH 86: Outdoor Exposure Time vs. Tensile Strength of Nylon 6

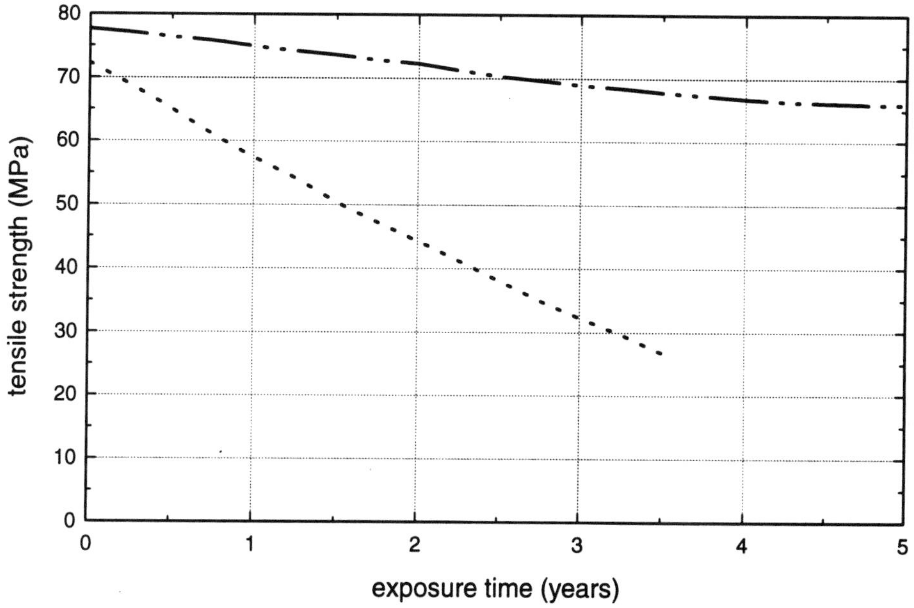

....	Ube Ube 1013B Nylon 6 (gen. purp. grade)
—·—··	Ube Ube 1013NU2 Nylon 6 (high heat grade)
Reference No.	238

GRAPH 87: Hiratsuka, Japan Outdoor Exposure Time vs. Breaking Stress in Flexure of Nylon 6

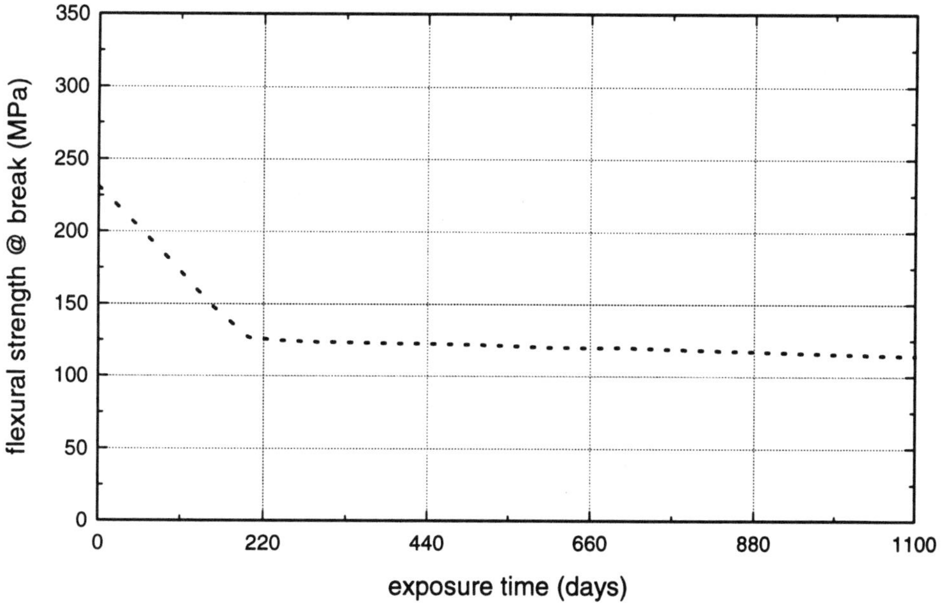

....	Nylon 6 (30% glass fiber); natural aging
Reference No.	135

GRAPH 88: Hiratsuka, Japan Outdoor Exposure Time vs. Flexural Modulus of Nylon 6

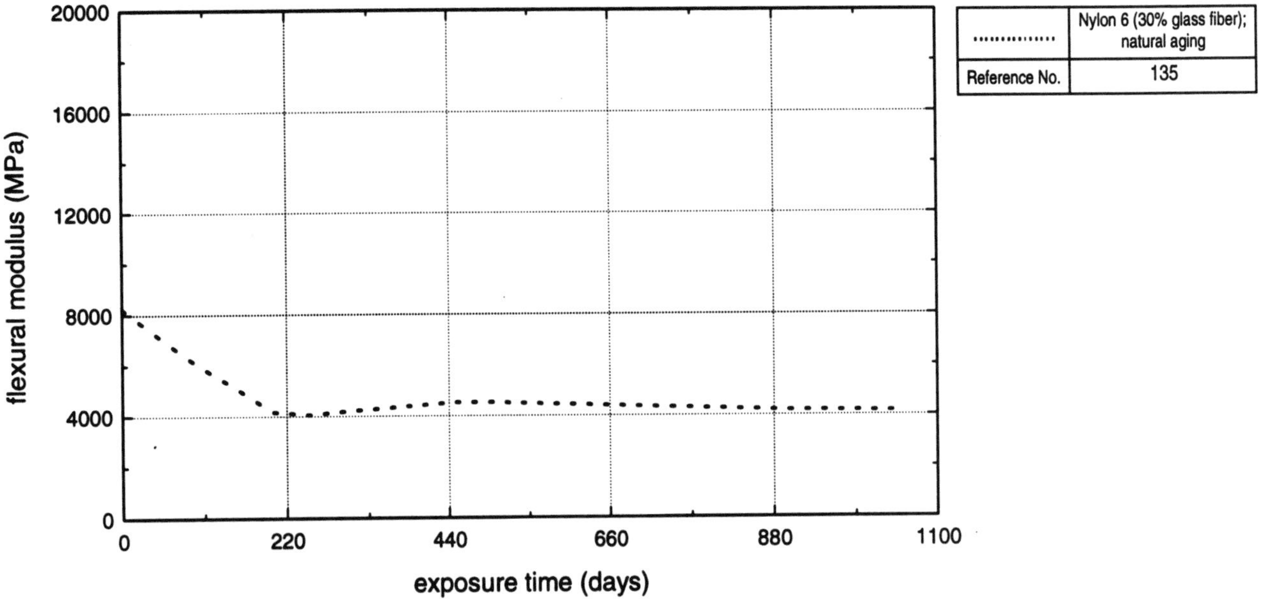

GRAPH 89: Hiratsuka, Japan Outdoor Exposure Time vs. Notched Izod Impact Strength of Nylon 6

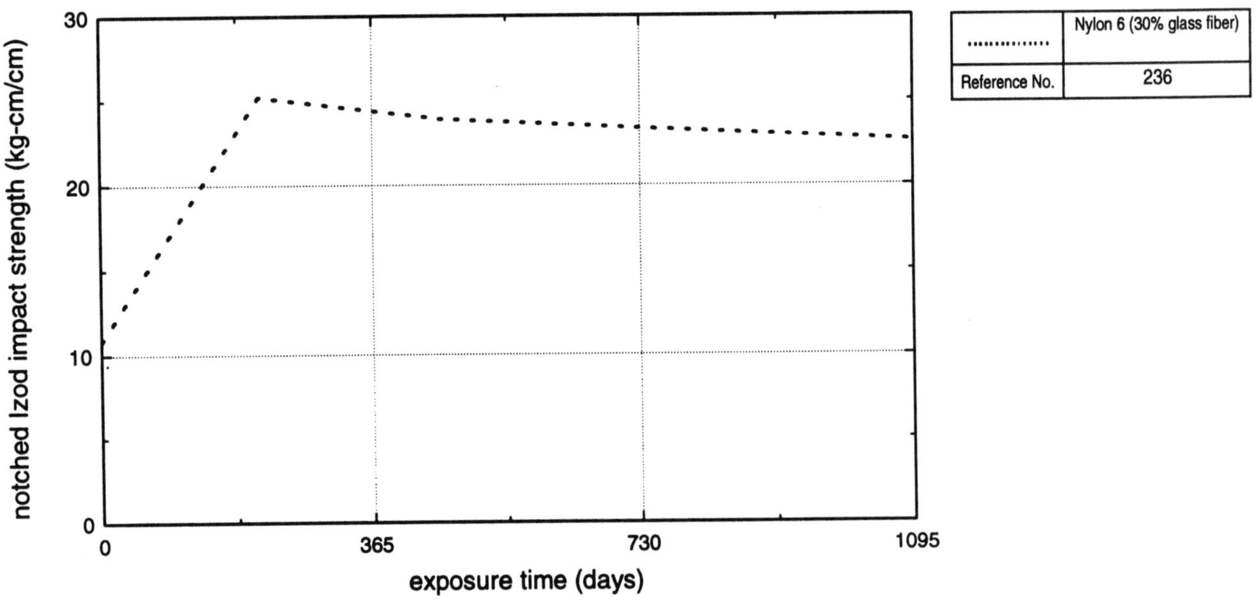

GRAPH 90: Hiratsuka, Japan Outdoor Exposure Time vs. Percent Weight Change of Nylon 6

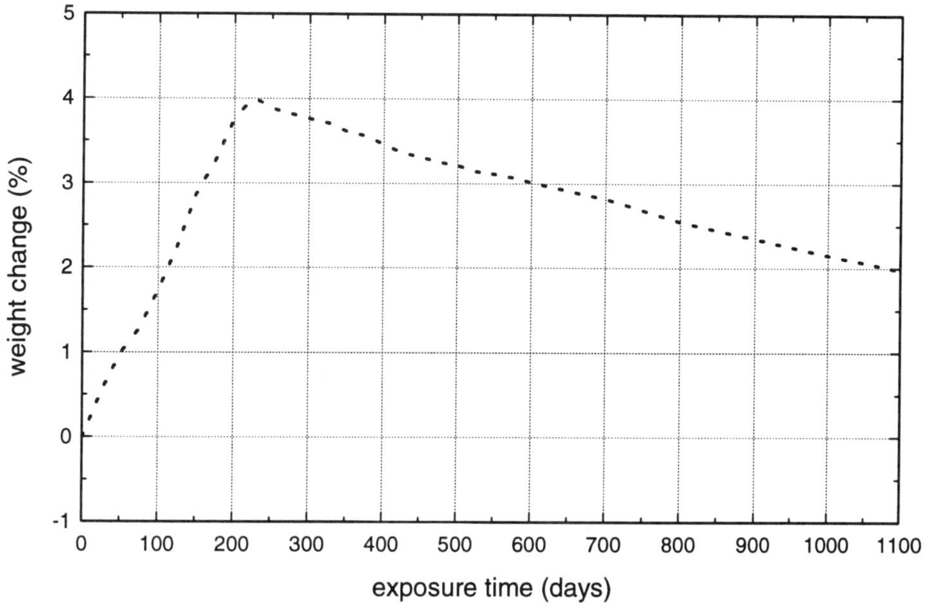

....	Nylon 6 (30% glass fiber); specimen thickness 3.2mm (1/8") under tensile stress - ASTM D638
Reference No.	135

GRAPH 91: Hiratsuka, Japan Outdoor Weathering Exposure Time vs. Flexural Strength of Nylon 6

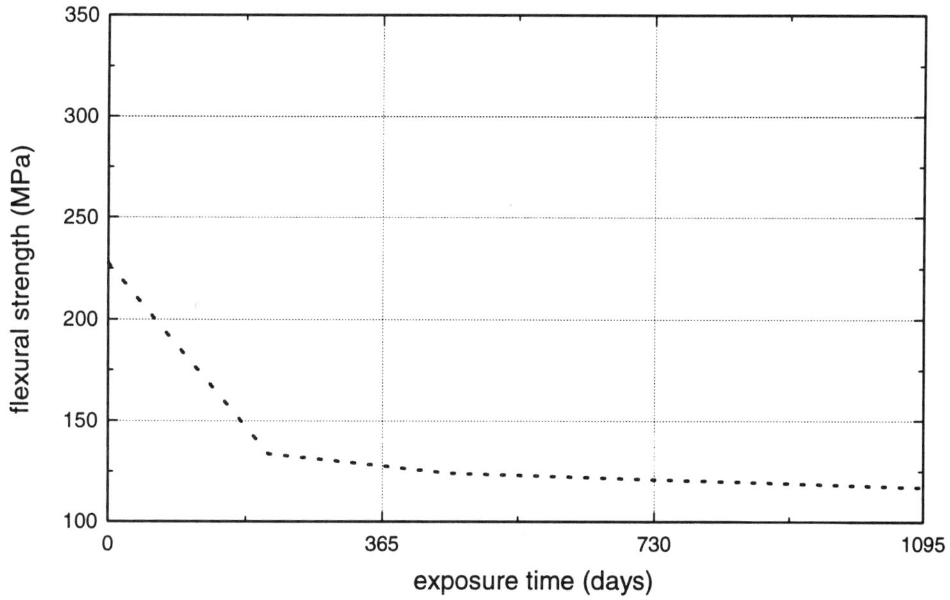

....	Nylon 6 (30% glass fiber)
Reference No.	236

GRAPH 92: Hiratsuka, Japan Outdoor Weathering Exposure Time vs. Tensile Strength of Nylon 6

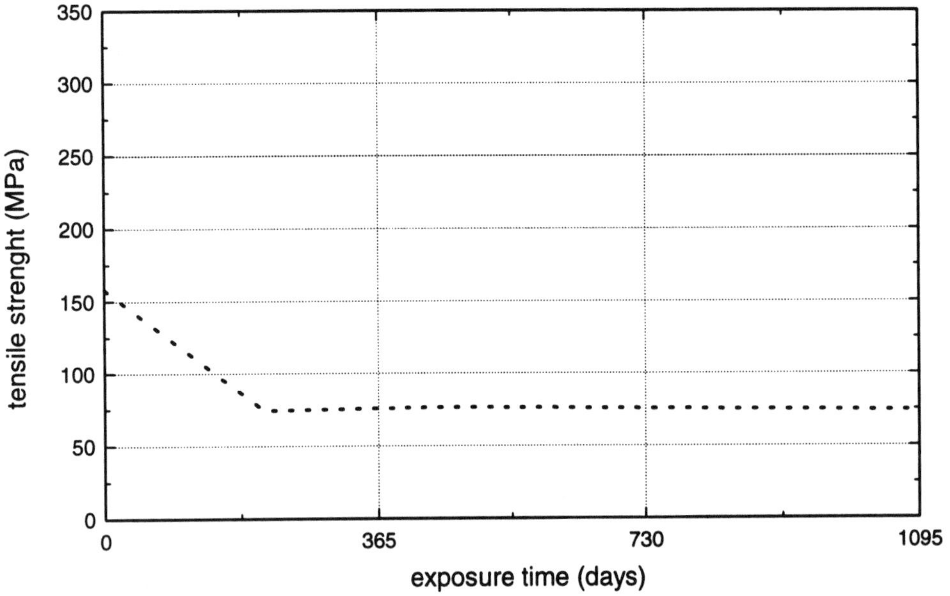

....	Nylon 6 (30% glass fiber)
Reference No.	236

GRAPH 93: Sunshine Weatherometer Exposure Time vs. Elongation of Nylon 6

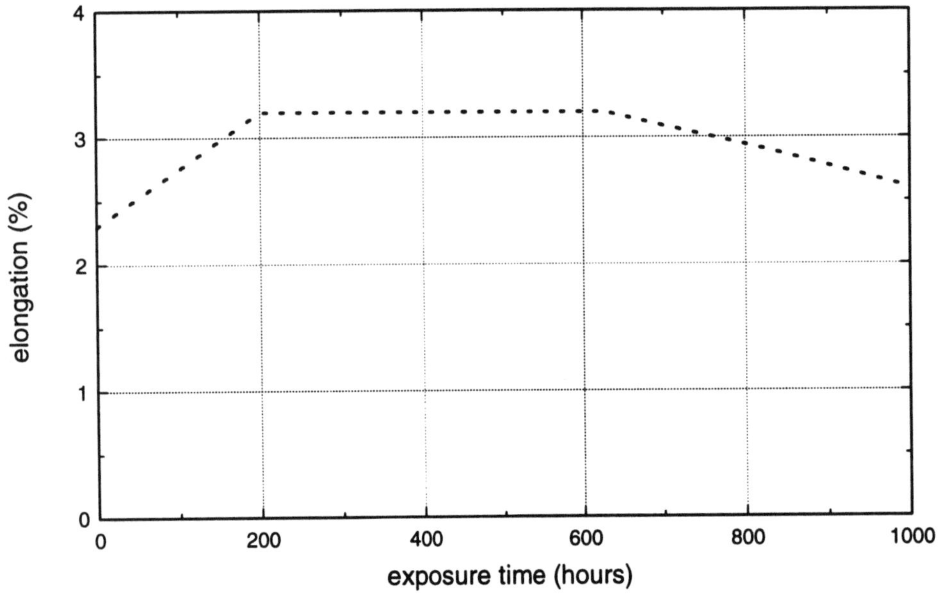

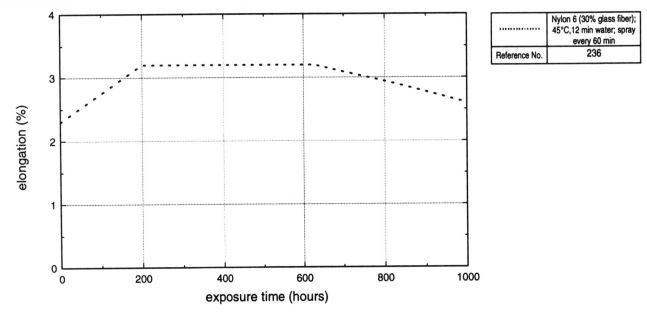

....	Nylon 6 (30% glass fiber); 45°C,12 min water; spray every 60 min
Reference No.	236

GRAPH 94: Sunshine Weatherometer Exposure Time vs. Tensile Strength of Nylon 6

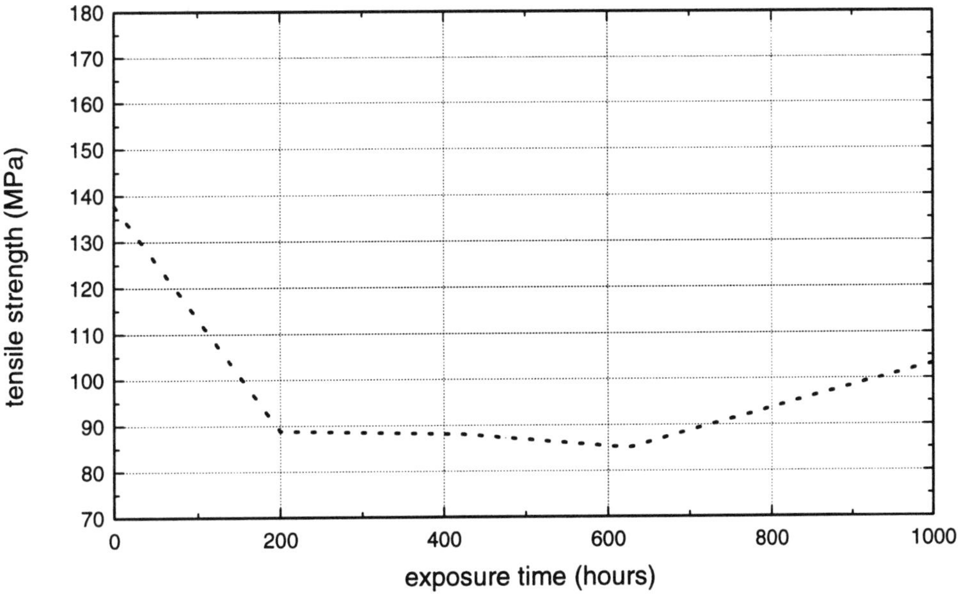

....	Nylon 6 (30% glass fiber); 45°C,12 min water spray every 60 min
Reference No.	236

Chapter 16

Nylon 610

Outdoor Weather Resistance

BASF: Ultramid S; Ultramid S3K (features: natural resin, stabilized)

Many Ultramid resins are suitable for outdoor applications. The unreinforced stabilized Ultramid resins, i.e. those with the letters K and H in the type nomenclature, are extremely resistant to weathering, even if they are uncolored. Their outdoor performance can be further improved by suitable pigments, best effects being achieved with carbon black.

Thin articles for outdoor use should be produced from Ultramid resins with a high carbon black content, e.g. the Black 20 590 and 20 592 types, in order to ensure that their strength remains undiminished. Moldings with a high proportion of carbon black can also withstand several years' exposure to tropical conditions.

The gloss retention on outdoor exposure of nylon 610, e.g. Ultramid S3K, is very high.

Reinforced Ultramid resins give good outdoor performance, and the stabilized types can be relied upon to withstand exposure for periods of much more than five years. Nevertheless, the constituent glass fibers cause the surface to be attacked more severely than that of unreinforced Ultramid articles. As a consequence, the texture and the hue may undergo a change after comparatively brief exposure periods. If the glass-reinforced moldings remain exposed for a number of years, erosion to a depth of a few tenths of 1 mm (0.04 in) can generally be expected, but experience has shown that this does not exert any significant effect on the mechanical properties.

Housings for automobile rear-view mirrors are examples of articles that must retain an attractive appearance for many years. In applications of this nature, good results have been obtained by products with special UV stabilizers and products with a high carbon black content.

Reference: *Ultramid Nylon Resins Product Line, Properties, Processing,* supplier design guide (B 568/1e/4.91) - BASF Corporation, 1991.

TABLE 31: Outdoor Weathering in California and Pennsylvania of Glass Reinforced Nylon 610.

Material Family	Nylon 6/6							
Material Supplier	LNP Engineering Plastics							
Reference Number	215	215	215	215	215	215	215	215
MATERIAL COMPOSITION								
carbon black	1%	1%	1%	1%	1%	1%	1%	1%
glass fiber reinforcement	30%	30%	30%	30%	30%	30%	30%	30%
EXPOSURE CONDITIONS								
Exposure Type	outdoor weathering				outdoor weathering			
Exposure Location	Los Angeles, California	Los Angeles, California	Los Angeles, California	Los Angeles, California	Philadelphia, Pennsylvania	Philadelphia, Pennsylvania	Philadelphia, Pennsylvania	Philadelphia, Pennsylvania
Exposure Country	USA	USA	USA	USA	USA	USA	USA	USA
Exposure Test Method	ASTM D1435	ASTM D1435	ASTM D1435	ASTM D1435	ASTM D1435	ASTM D1435	ASTM D1435	ASTM D1435
Exposure Time (days)	91	182	365	730	91	182	365	730
PROPERTIES RETAINED (%)								
Tensile Strength	90	90.3	91	96	86	81.5	90	93
Notched Izod Impact Strength	89	83.3			94	94.4		
Unnotched Izod Impact Strength	104	107			86	97.2		

Chapter 17

Nylon 66

UV Resistance

BIP Chemicals: Beetle

Jonylon

Nylons tend to become embrittled on long-term exposure to high intensity UV radiation, such as sunlight, and this may also be accompanied by slight discoloration.

The incorporation of heat stabilizers results in some improvement in weatherability and UV resistance, and special light stabilized grades are also available. (e.g. Beetle A41 and Jonylon L and HL grades). Heavily pigmented grades, particularly black, will give further improvements, and are recommended for extensive exposure to strong sunlight.

Reference: *Beetle & Jonylon Engineering Thermoplastics,* supplier marketing literature (BJ2/1291/SP/5) - BIP Chemicals Limited, 1991.

DuPont: Zytel 101; Zytel ST 801 (features: high impact)

Exposure of nylon that is inadequately stabilized against UV light results in surface degradation with a corresponding drop in relative viscosity or molecular weight. Serious loss in this property is related to a comparable loss in toughness.

Reference: *Design Handbook For Du Pont Engineering Plastics - Module II,* supplier design guide (E-42267) - Du Pont Engineering Polymers.

Outdoor Weather Resistance

BASF: Ultramid A; Ultramid A3EG6 Black 20 591 (features: natural resin, stabilized, black color; material compostion: 30% glass fiber reinforcement)

Many Ultramid resins are suitable for outdoor applications.

The unreinforced stabilized Ultramid resins, i.e. those with the letters K and H in the type nomenclature, are extremely resistant to weathering, even if they are uncolored. Their outdoor performance can be further improved by suitable pigments, best effects being achieved with carbon black.

Thin articles for outdoor use should be produced from Ultramid resins with a high carbon black content, e.g. the Black 20 590 and 20 592 types, in order to ensure that their strength remains undiminished. Moldings with a high proportion of carbon black can also withstand several years' exposure to tropical conditions.

The reinforced Ultramid resins also give good outdoor performance, and the stabilized types can be relied upon to withstand exposure for periods of much more than five years. Nevertheless, the constituent glass fibers cause the surface to be attacked more severely than that of unreinforced Ultramid articles. As a consequence, the texture and the hue may undergo a change after comparatively brief exposure periods. If the glass-reinforced moldings remain exposed for a number of years, erosion to a depth of a few tenths of 1mm (0.04 in) can generally be expected, but experience has shown that this does not exert any significant effect on the mechanical properties.

Housings for automobile rear-view mirrors are examples of articles that must retain an attractive appearance for many years. In applications of this nature, good results have been obtained by products with special UV stabilizers and products with a high carbon black content, e.g. Ultramid A3EG6 Black 20591.

Reference: *Ultramid Nylon Resins Product Line, Properties, Processing,* supplier design guide (B 568/1e/4.91) - BASF Corporation, 1991.

DuPont: Minlon 10B NC-10 (features: natural resin; filler: mineral)

Minlon retains much of its original tensile strength and elongation after 24 months of exposure in Florida. Minlon engineering thermoplastic resins are more resistant to UV light than are the unreinforced Zytel nylon resins. For maximum resistance to outdoor weathering, black compositions containing uniformly dispersed carbon as a UV screen are available.

Reference: *Design Handbook For Du Pont Engineering Plastics - Module II,* supplier design guide (E-42267) - Du Pont Engineering Polymers.

DuPont: Zytel 101; Zytel 101 WT-07 (pigment: TiO_2)

Experience with Arizona exposure tests show this climate to be more severe on Zytel 101 WT-07 and Zytel 101 NC-10 than on Zytel 105 BK-10A.

Reference: *Design Handbook For Du Pont Engineering Plastics - Module II,* supplier design guide (E-42267) - Du Pont Engineering Polymers.

DuPont: Zytel 101 NC-10

Zytel 101 NC-10 shows substantial loss of toughness at six months of Florida weathering. The tensile strength, however, remains at 24 MPa (3,480 psi) after 180 months exposure.

Reference: *Design Handbook For Du Pont Engineering Plastics - Module II,* supplier design guide (E-42267) - Du Pont Engineering Polymers.

DuPont: Zytel 105 BK-10A (additives: carbon black; features: weatherable, UV stabilized)

Zytel 105 BK-10A is still tough and strong after 180 months of Florida weathering. Experience with Arizona exposure tests show this climate to be more severe on Zytel 101 WT-07 and Zytel 101NC-10 than on Zytel 105 BK-10A. For Arizona or similar climates, black stabilized compositions such as Zytel 105 BK-10A should be used.

Reference: *Design Handbook For Du Pont Engineering Plastics - Module II,* supplier design guide (E-42267) - Du Pont Engineering Polymers.

DuPont: Zytel 70G 13L (features: lubricated, impact modified; material compostion: 13% glass fiber reinforcement); **Zytel 70G 33L** (features: lubricated; material compostion: 33% glass fiber reinforcement); **Zytel 71G 13L** (features: lubricated, impact modified; material compostion: 13% glass fiber reinforcement); **Zytel 71G 33L** (features: lubricated; material compostion: 33% glass fiber reinforcement)

Glass reinforcement improves the outdoor weatherability of nylon. Actual weathering studies in Florida with GRZ resins, have shown the tensile strength values to be reduced only slightly after seven years exposure.

Reference: *Design Handbook For Du Pont Engineering Plastics - Module II,* supplier design guide (E-42267) - Du Pont Engineering Polymers.

DuPont: Zytel ST 801 (features: high impact)

The natural grade of Zytel ST Super Tough Nylon (NC-10) will provide limited service in outdoor applications and is not recommended for extensive UV exposure.

Reference: *Design Handbook For Du Pont Engineering Plastics - Module II,* supplier design guide (E-42267) - Du Pont Engineering Polymers.

Ube: Ube 2020B (features: general purpose grade); **Ube 2020UW1** (features: high heat grade)

Nylon plastics are relatively stable when exposed to ultraviolet light and heat, and are particularly resistant to ozone. Thus, they are well suited to outdoor applications.

Reference: *Ube Nylon Technical Brochure,* supplier design guide (1989.8.1000) - Ube Industries, Ltd., 1989.

Accelerated Artificial Weathering Resistance

DuPont: Zytel 70G 13L (features: lubricated, impact modified; material compostion: 13% glass fiber reinforcement); **Zytel 70G 33L** (features: lubricated; material compostion: 33% glass fiber reinforcement); **Zytel 71G 13L** (features: lubricated, impact modified; material compostion: 13% glass fiber reinforcement); **Zytel 71G 33L** (features: lubricated; material compostion: 33% glass fiber reinforcement)

Glass reinforcement improves the outdoor weatherability of nylon. Laboratory X-W weatherometer tests show Zytel 70G33L experiences only a slight decrease in strength after 5,000 hours of exposure in accordance with ASTM D 1499.

Reference: *Design Handbook For Du Pont Engineering Plastics - Module II,* supplier design guide (E-42267) - Du Pont Engineering Polymers.

TABLE 32: Outdoor Weathering in Arizona of DuPont Zytel Nylon 66.

Material Family	Nylon 66								
Material Supplier/ Grade	DuPont Zytel 101			DuPont Zytel 105 BK-10A			DuPont Zytel 101 WT-07		
Features				UV stabilized, weatherable					
Reference Number	68	68	68	68	68	68	68	68	68
MATERIAL COMPOSITION									
additives				carbon black	carbon black	carbon black			
pigment							TiO_2	TiO_2	TiO_2
table-id	177	177	177	177	177	177	177	177	177
EXPOSURE CONDITIONS									
Exposure Type	outdoor weathering			outdoor weathering			outdoor weathering		
Exposure Location	Arizona			Arizona			Arizona		
Exposure Country	USA			USA			USA		
Exposure Time (days)	183	365	730	183	365	730	183	365	730
PROPERTY VALUES AFTER EXPOSURE[1]									
Tensile Strength @ Yield (MPa)	no yield	no yield	no yield				no yield	no yield	no yield
PROPERTIES RETAINED (%)[1]									
Tensile Strength	39.2	31.6 (broad range of values)	57 (broad range of values)	97.8	90.2	95.6	51.8	32.1 (broad range of values)	53.1 (broad range of values)
Yield Stress				97.8	90.2	95.6			
Elongation	9.1	9.1 (broad range of values)	9.1 (broad range of values)	80	100	100	11.1	11.1 (broad range of values)	11.1 (broad range of values)
SURFACE AND APPEARANCE									
Visual Appearance		surface cracking	surface cracking					surface cracking	surface cracking

[1]All test bars exposed in dry-as-molded condition.

TABLE 33: Outdoor Weathering in Florida and Accelerated Weathering in XW Weatherometer of Mineral Filled DuPont Minlon Nylon 66.

Material Family	Nylon 66				
Material Supplier/ Grade	DuPont Minlon 10B NC-10				
Features	natural resin				
Reference Number	68	68	68	68	68
MATERIAL COMPOSITION					
filler	mineral	mineral	mineral	mineral	mineral
EXPOSURE CONDITIONS					
Exposure Type	outdoor weathering		accelerated weathering		
Exposure Location	Florida				
Exposure Country	USA				
Exposure Apparatus			X-W weatherometer		
Exposure Note			2 hour wet-dry cycle; 400 to 1000 hours X-W weatheromether exposure is equivalent to one year of outdoor weathering in Florida		
Exposure Time (days)	365	730	41.7	125	208.3
PROPERTIES RETAINED (%)[1]					
Tensile Strength	78.9	74.4	81.3	78.9	61.2
Elongation	85.7	85.7	100	100	133.3

[1]Values are based on moisture contents as removed from equipment and range from 0.8 to 1.2%.

TABLE 34: Outdoor Weathering in Florida of DuPont Zytel Nylon 66.

Material Family	Nylon 66													
Material Supplier/ Grade	DuPont Zytel 101					DuPont Zytel 105 BK-10A					DuPont Zytel 101 WT-07			
Features	unstabilized					UV stabilized, weatherable								
Reference Number	68	68	68	68	68	68	68	68	68	68	68	68	68	68
MATERIAL COMPOSITION														
additives						carbon black	carbon black	carbon black	carbon black	carbon black				
pigment											TiO_2	TiO_2	TiO_2	TiO_2
EXPOSURE CONDITIONS														
Exposure Type	outdoor weathering					outdoor weathering					outdoor weathering			
Exposure Location	Florida					Florida					Florida			
Exposure Country	USA					USA					USA			
Exposure Time (days)	183	365	730	2557	5479	183	365	730	2557	5479	183	365	730	1095
PROPERTY VALUES AFTER EXPOSURE														
Tensile Strength @ Yield (MPa)	no yield	no yield	no yield	no yield	no yield									
PROPERTIES RETAINED (%)[1]														
Tensile Strength	50.7	48	42.5	21.9	32.9	98.4	104.8	87.3	74.6	65.1	85	64	64	56.9
Tensile Strength @ Yield						124	132	110	94	82	80	83	85	75.9
Elongation	3.3	2	2	1.7		37.5	25.6	20	25.6	20 (can still be bent 180° around 3.2 mm mandrel)	141	112	32	14.6

[1]Tensile bars tested as received; moisture contents ranged from 2-3%.

TABLE 35: Outdoor Weathering in California and Pennsylvania of Glass Reinforced Nylon 66.

Material Family	Nylon 6/6							
Material Supplier	LNP Engineering Plastics							
Reference Number	215	215	215	215	215	215	215	215
MATERIAL COMPOSITION								
carbon black	1%	1%	1%	1%	1%	1%	1%	1%
glass fiber reinforcement	30%	30%	30%	30%	30%	30%	30%	30%
EXPOSURE CONDITIONS								
Exposure Type	outdoor weathering				outdoor weathering			
Exposure Location	Los Angeles, California	Los Angeles, California	Los Angeles, California	Los Angeles, California	Philadelphia, Pennsylvania	Philadelphia, Pennsylvania	Philadelphia, Pennsylvania	Philadelphia, Pennsylvania
Exposure Country	USA	USA	USA	USA	USA	USA	USA	USA
Exposure Test Method	ASTM D1435	ASTM D1435	ASTM D1435	ASTM D1435	ASTM D1435	ASTM D1435	ASTM D1435	ASTM D1435
Exposure Time (days)	91	182	365	730	91	182	365	730
PROPERTIES RETAINED (%)								
Tensile Strength	115	107	107	112	112	104	103	101
Notched Izod Impact Strength	112	106			106	106		
Unnotched Izod Impact Strength	98	87.9			82	138		

TABLE 36: Outdoor Weathering in Delaware of DuPont Zytel Nylon 66.

Material Family	Nylon 66					
Material Supplier/ Grade	DuPont Zytel 105 BK-10A			DuPont Zytel 101 WT-07		
Features	UV stabilized, weatherable					
Reference Number	68	68	68	68	68	68
MATERIAL COMPOSITION						
additives	carbon black	carbon black	carbon black			
pigment				TiO_2		
EXPOSURE CONDITIONS						
Exposure Type	outdoor weathering			outdoor weathering		
Exposure Location	Delaware			Delaware		
Exposure Country	USA			USA		
Exposure Time (days)	183	365	730	183	365	730
PROPERTIES RETAINED (%)[1]						
Tensile Strength	78.9	84.8	84.8	67.6	64.8	81.8
Yield Stress	78.9	84.8	84.8	76.4	83.6	63.4
Elongation	93	32.6	20.9	84.8	32.2	22

[1]Bars contained 2.5% moisture at start of test.

TABLE 37: Accelerated Weathering in an XW Weatherometer of DuPont Zytel Nylon 66.

Material Family	Nylon 66														
Material Supplier/ Grade	DuPont Zytel 101					DuPont Zytel 105 BK-10A					DuPont Zytel 101 WT-07				
Features						UV stabilized, weatherable									
Reference Number	68	68	68	68	68	68	68	68	68	68	68	68	68	68	68

MATERIAL CHARACTERISTICS

Sample Thickness (mm)	3.2	3.2	3.2	3.2	3.2	3.2	3.2	3.2	3.2	3.2	3.2	3.2	3.2	3.2	3.2
Sample Type	tensile bars	tensile bars	tensile bars	tensile bars	tensile bars	tensile bars	tensile bars	tensile bars	tensile bars	tensile bars	tensile bars	tensile bars	tensile bars	tensile bars	tensile bars

MATERIAL COMPOSITION

additives						carbon black	carbon black	carbon black	carbon black	carbon black					
pigment											TiO_2	TiO_2	TiO_2	TiO_2	TiO_2

EXPOSURE CONDITIONS

Exposure Type	accelerated weathering					accelerated weathering					accelerated weathering				
Exposure Apparatus	X-W weatherometer					X-W weatherometer					X-W weatherometer				
Exposure Note	2 hour wet-dry cycle; 400 to 1000 hours X-W weatheromether exposure is equivalent to one year of outdoor weathering in Florida					2 hour wet-dry cycle; 400 to 1000 hours X-W weatheromether exposure is equivalent to one year of outdoor weathering in Florida					2 hour wet-dry cycle; 400 to 1000 hours X-W weatheromether exposure is equivalent to one year of outdoor weathering in Florida				
Exposure Time (days)	8.3	25	41.7	83.3	250	8.3	25	41.7	83.3	250	8.3	25	41.7	83.3	250

PROPERTY VALUES AFTER EXPOSURE[1]

Tensile Strength @ Yield (MPa)		no yield	no yield	no yield	no yield				no yield	no yield				no yield	no yield

PROPERTIES RETAINED (%)[1]

Tensile Strength	88.6	75.7	60	47.1	55.7	100	103.9	98	125.5	176.5	93	78.9	64.8	84.5	91.6
Yield Stress	107.4					104.5	114.9	107.5			105.5	107.3	100		
Elongation	103.3	3.3	3.3	3.3	3.3	50	28.6	21.9	4.8	56.2	105	96.7	70	18	9.3

[1]Based on specimens conditioned to equilibrium at 50% RH.

TABLE 38: Accelerated Weathering in an XW Weatherometer of DuPont Zytel Nylon 66.

Material Family	Nylon 66							
Material Supplier/ Grade	DuPont Zytel 408 BK-10		DuPont Zytel ST 801			DuPont Zytel ST 801 BK-10		
Features	UV stabilized, weatherable, black color		high impact			high impact, UV stabilized, black color		
Reference Number	68	68	68	68	68	68	68	68
MATERIAL CHARACTERISTICS								
Sample Thickness (mm)	3.2	3.2	3.2	3.2	3.2	3.2	3.2	3.2
Sample Type	tensile bars	tensile bars	tensile bars	tensile bars	tensile bars	tensile bars	tensile bars	tensile bars
MATERIAL COMPOSITION								
additives	carbon black	carbon black				carbon black	carbon black	carbon black
EXPOSURE CONDITIONS								
Exposure Type	accelerated weathering		accelerated weathering			accelerated weathering		
Exposure Apparatus	X-W weatherometer		X-W weatherometer			X-W weatherometer		
Exposure Note	2 hour wet-dry cycle; 400 to 1000 hours X-W weatheromether exposure is equivalent to one year of outdoor weathering in Florida		2 hour wet-dry cycle; 400 to 1000 hours X-W weatheromether exposure is equivalent to one year of outdoor weathering in Florida			2 hour wet-dry cycle; 400 to 1000 hours X-W weatheromether exposure is equivalent to one year of outdoor weathering in Florida		
Exposure Time (days)	25	83.3	41.7	83.3	416.7	41.7	83.3	416.7
PROPERTIES RETAINED (%)[1]								
Tensile Strength	108.5	111.9	87.8	82.9	73.2	102.4	95.1	90.2
Yield Stress	120.8	124.5						
Elongation	115.4	64.1	27.4	26.1	28.4	100	103.3	87

[1]Based on specimens conditioned to equilibrium at 50% RH.

GRAPH 95: **Outdoor Exposure Time vs. Elongation at Break of Nylon 66**

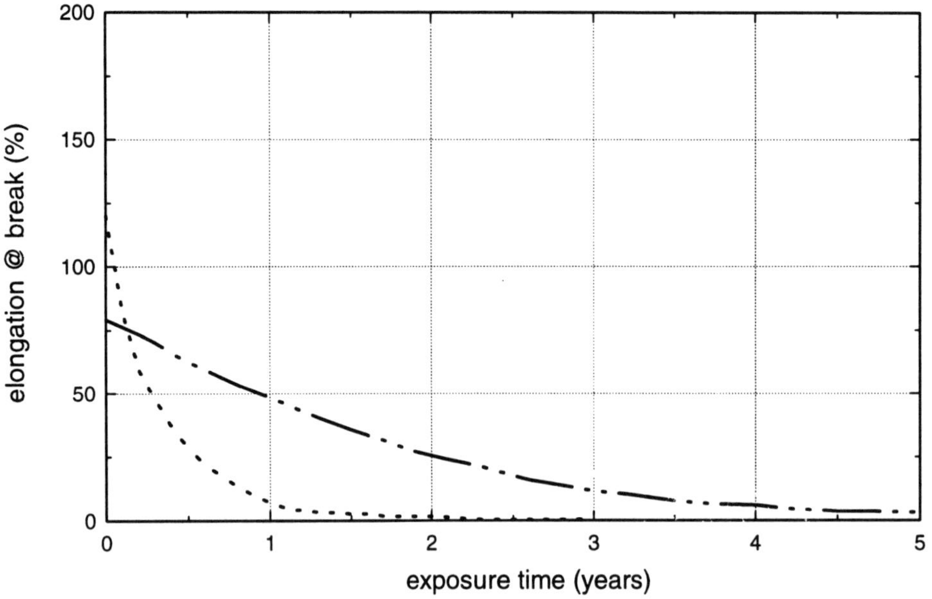

....	Ube Ube 2020B Nylon 66 (gen. purp. grade)
—··—··	Ube Ube 2020UW1 Nylon 66 (high heat grade)
Reference No.	238

GRAPH 96: **Outdoor Exposure Time vs. Flexural Modulus of Nylon 66**

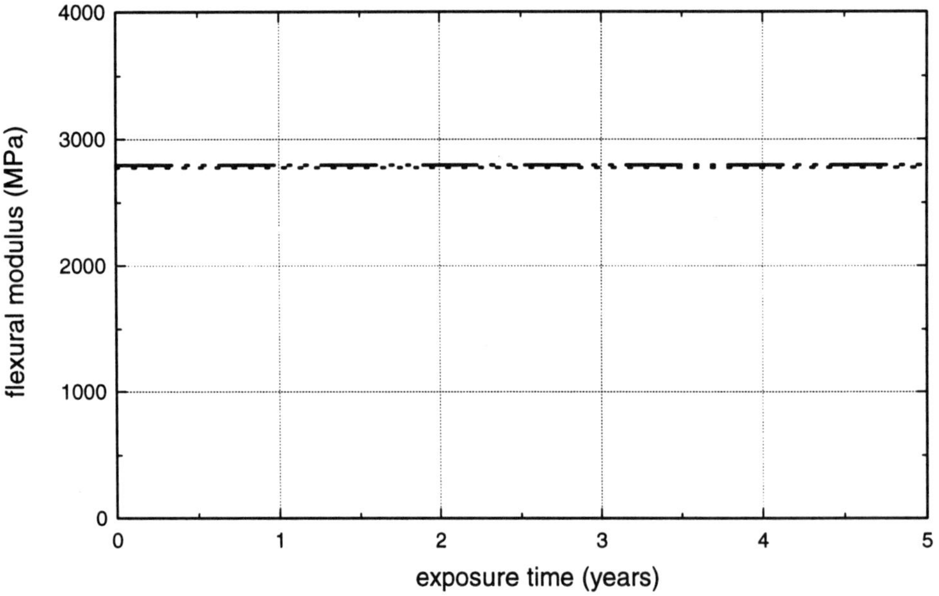

....	Ube Ube 2020B Nylon 66 (gen. purp. grade)
—··—··	Ube Ube 2020UW1 Nylon 66 (high heat grade)
Reference No.	238

GRAPH 97: Outdoor Exposure Time vs. Notched Izod Impact Strength of Nylon 66

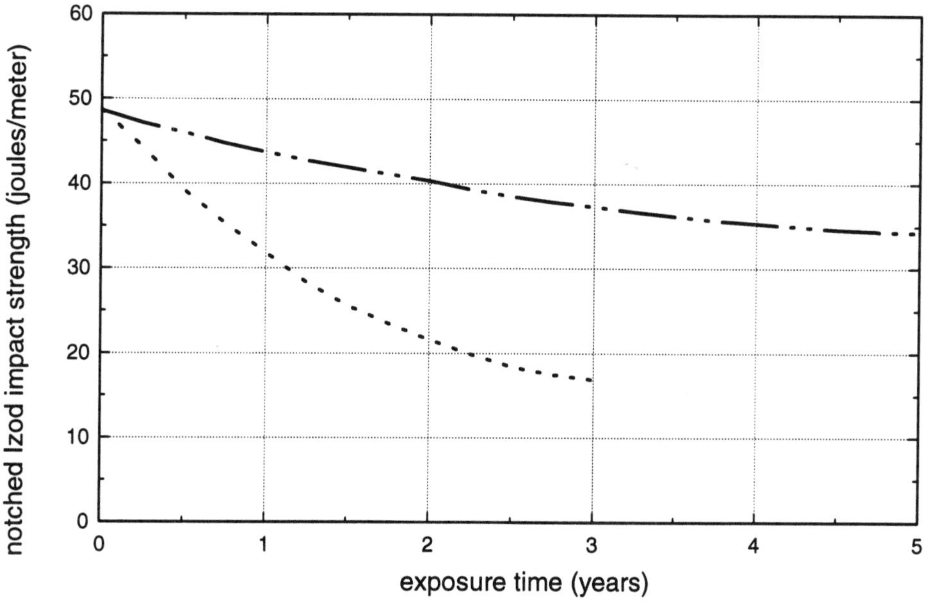

··············	Ube Ube 2020B Nylon 66 (gen. purp. grade)
—··—··	Ube Ube 2020UW1 Nylon 66 (high heat grade)
Reference No.	238

GRAPH 98: Outdoor Exposure Time vs. Tensile Strength of Nylon 66

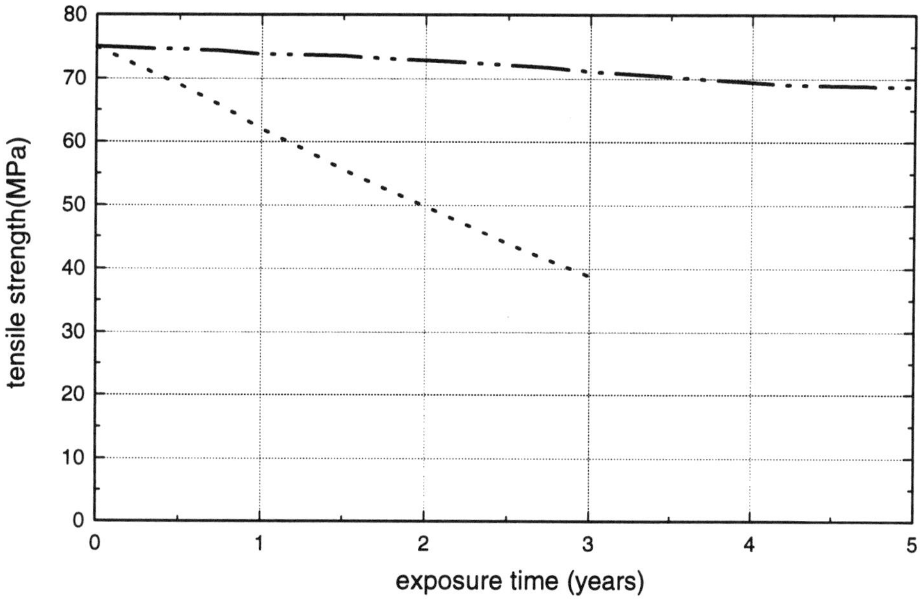

··············	Ube Ube 2020B Nylon 66 (gen. purp. grade)
—··—··	Ube Ube 2020UW1 Nylon 66 (high heat grade)
Reference No.	238

GRAPH 99: Florida Outdoor Weathering Exposure Time vs. Tensile Strength of Nylon 66

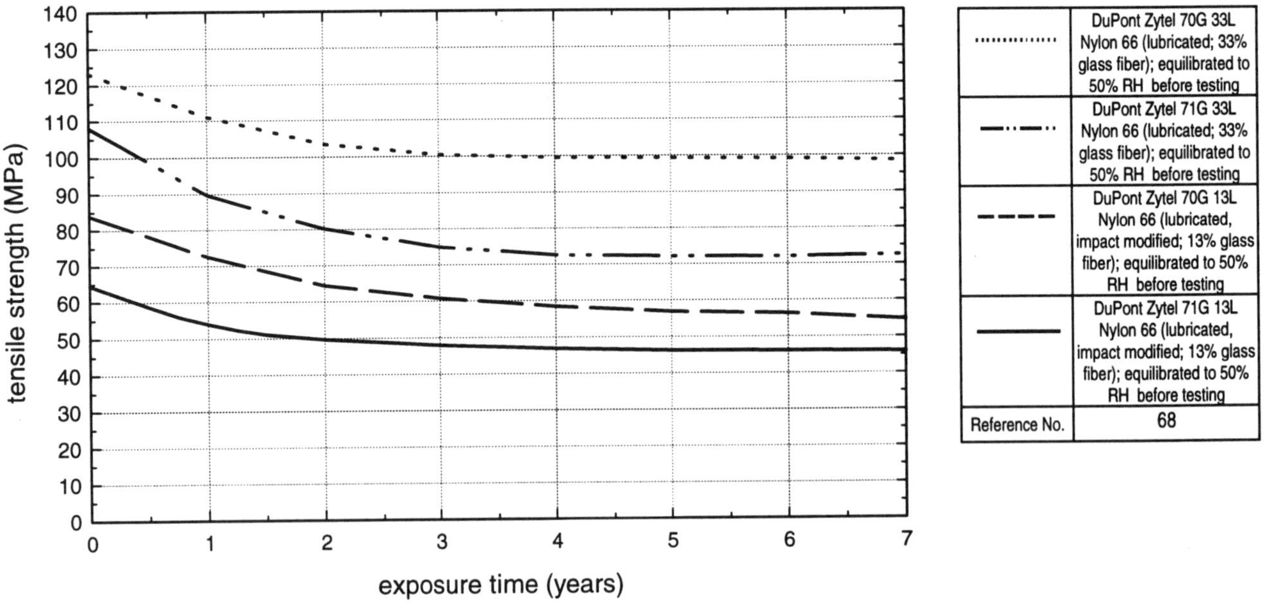

GRAPH 100: Hiratsuka, Japan Outdoor Exposure Time vs. Breaking Stress in Flexure of Nylon 66

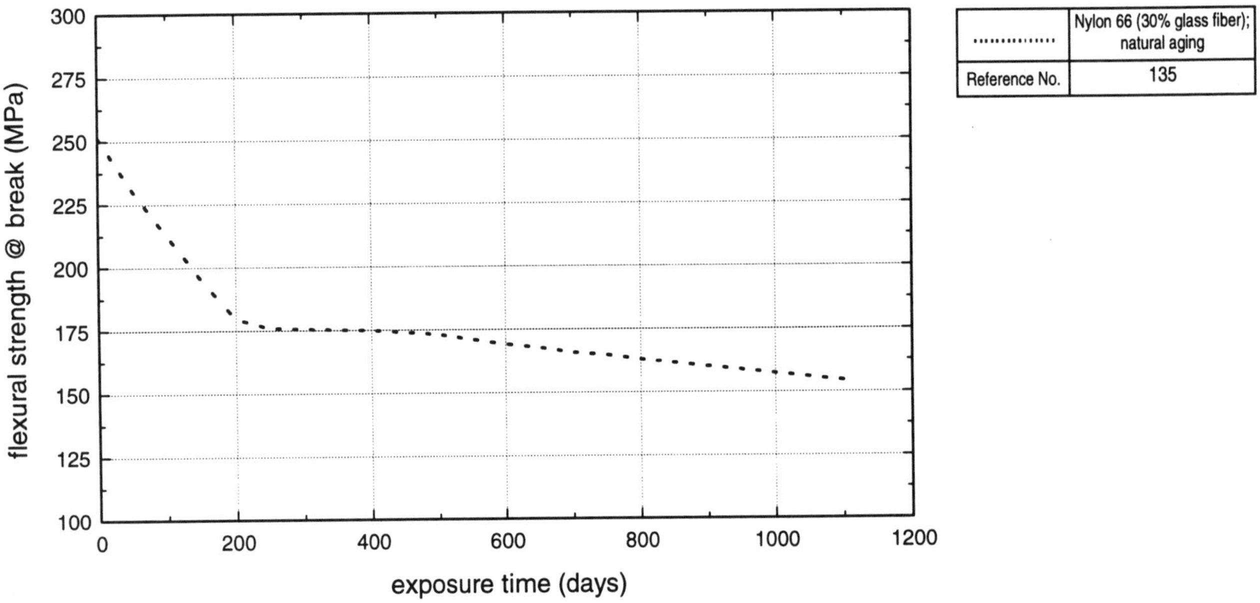

GRAPH 101: Hiratsuka, Japan Outdoor Exposure Time vs. Flexural Modulus of Nylon 66

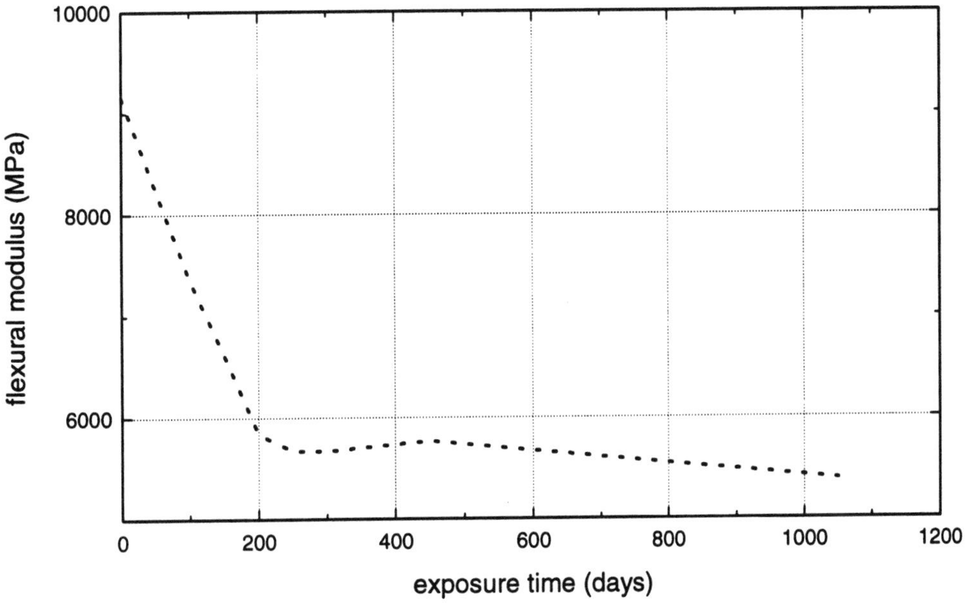

....	Nylon 66 (30% glass fiber); natural aging
Reference No.	135

GRAPH 102: Hiratsuka, Japan Outdoor Exposure Time vs. Notched Izod Impact Strength of Nylon 66

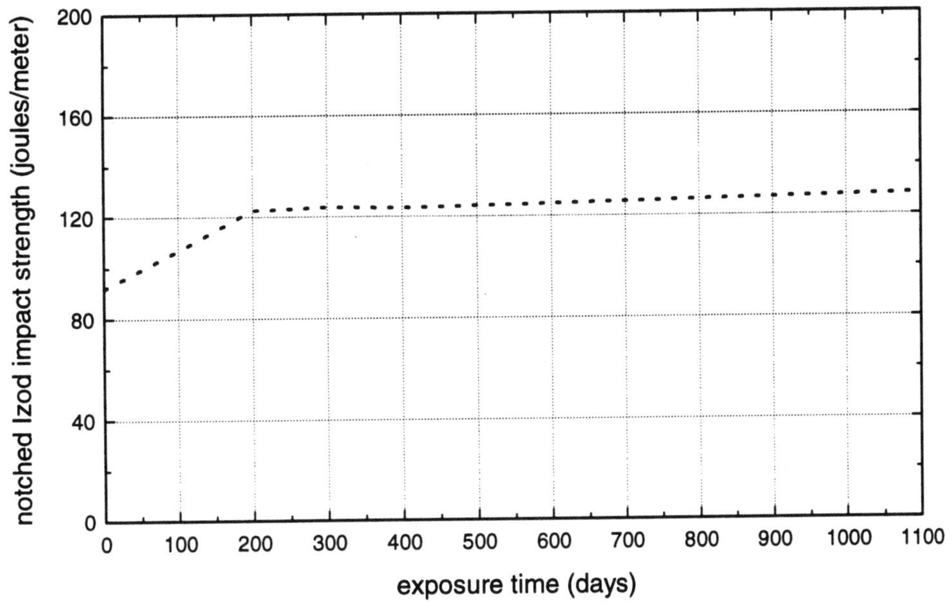

....	Nylon 66 (30% glass fiber); natural aging
Reference No.	135

GRAPH 103: Hiratsuka, Japan Outdoor Exposure Time vs. Percent Weight Change of Nylon 66

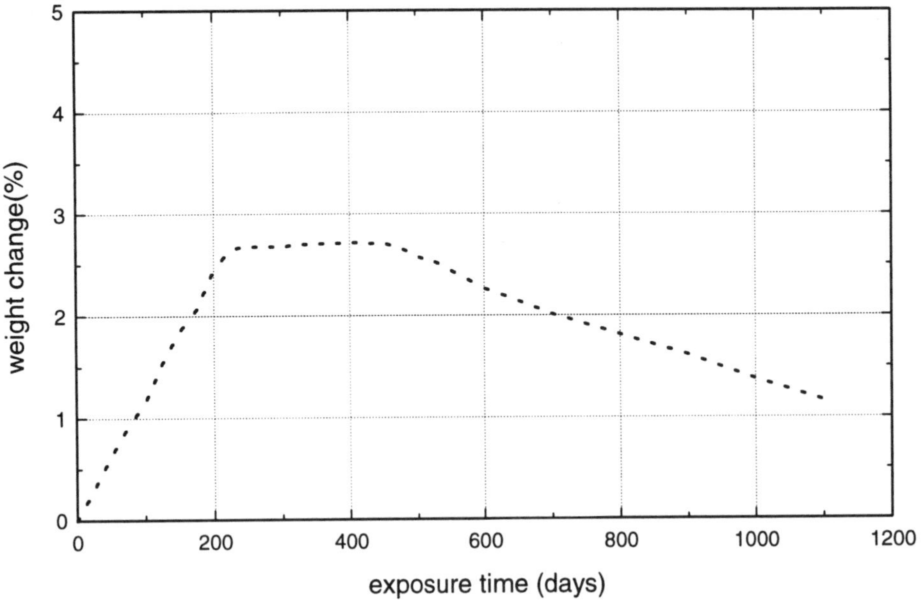

....	Nylon 66 (30% glass fiber); specimen thickness 3.2mm (1/8") under tensile stress - ASTM D638
Reference No.	135

GRAPH 104: Hiratsuka, Japan Outdoor Weathering Exposure Time vs. Flexural Strength of Nylon 66

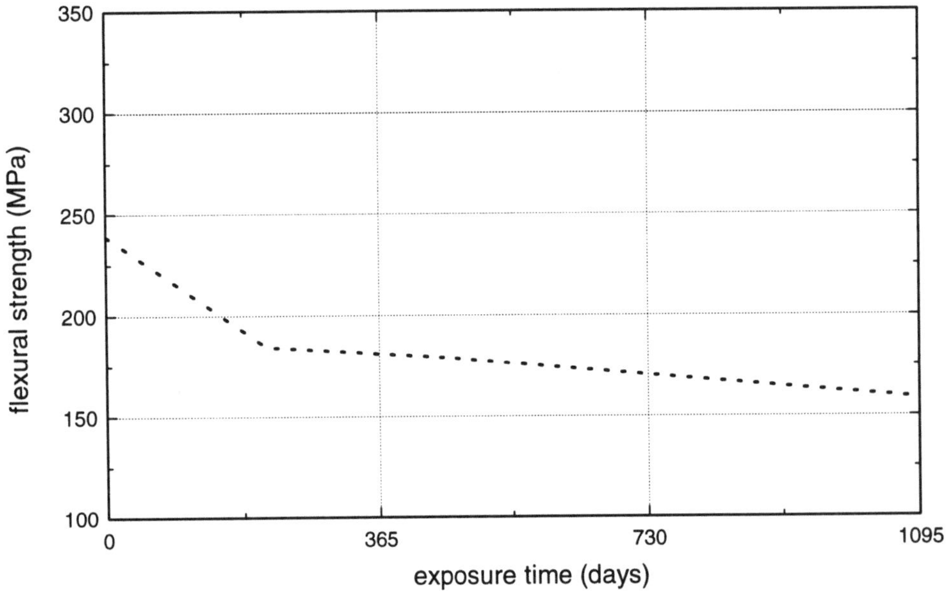

....	Nylon 66 (30% glass fiber)
Reference No.	236

GRAPH 105: Hiratsuka, Japan Outdoor Weathering Exposure Time vs. Tensile Strength of Nylon 66

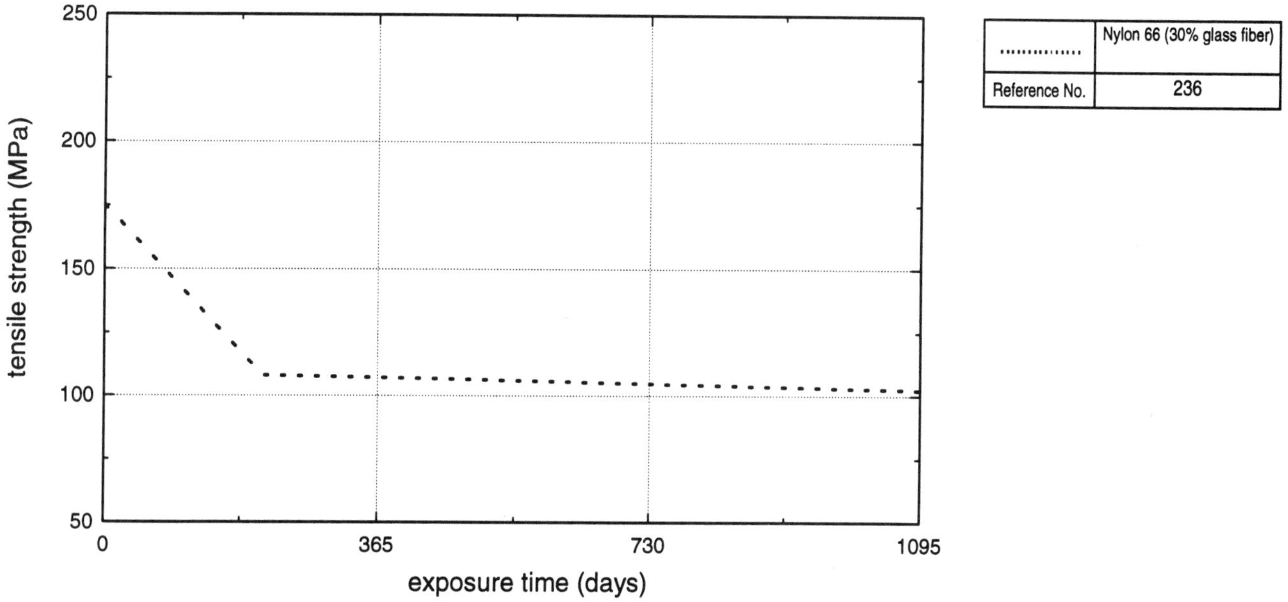

GRAPH 106: X-W Weatherometer Exposure Time vs. Tensile Strength of Nylon 66

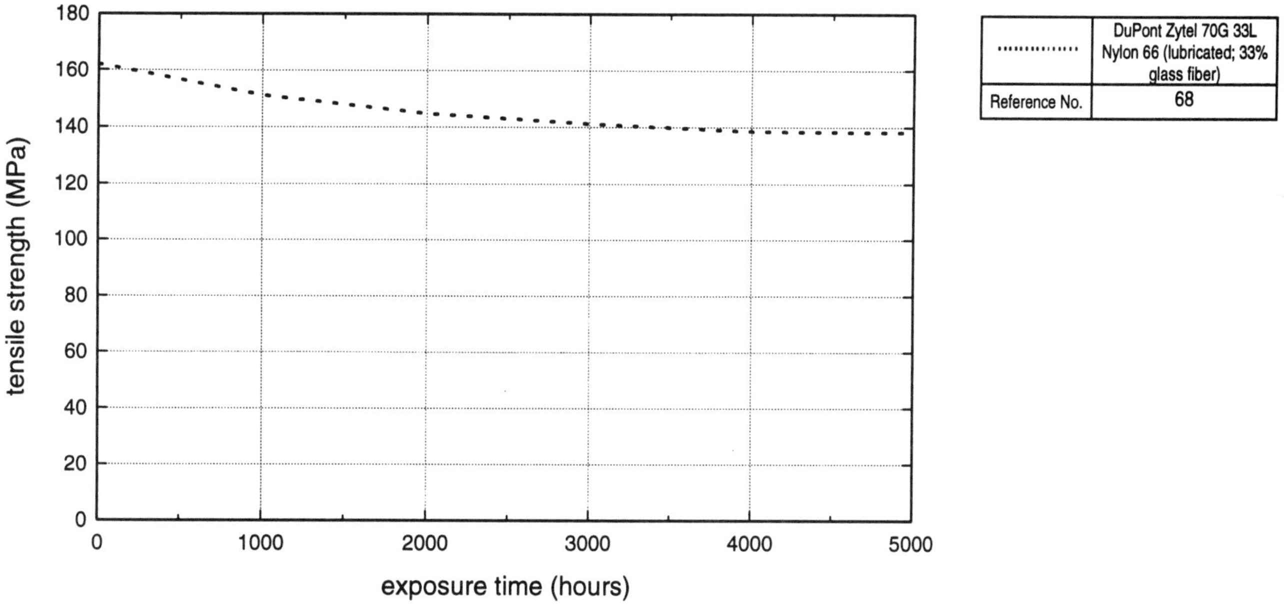

Chapter 18

Nylon 6/6T

Outdoor Weather Resistance

BASF AG: Ultramid T (chemical type: aromatic and aliphatic building blocks)

The experience gained on the outdoor performance of Ultramid A and B also applies essentially to Ultramid T. However, Ultramid T is degraded and discolored somewhat more than nylon 66 and nylon 6 when exposed to prolonged UV radiation.

Reference: *Ultramid T Polyamid 6/6T (PA) Product Line, Properties, Processing,* supplier design guide (B 605 e / 3.93) - BASF Aktiengesellschaft, 1993.

Chapter 19

Nylon MXD6

GRAPH 107: Hiratsuka, Japan Outdoor Exposure Time vs. Flexural Modulus of Nylon MXD6

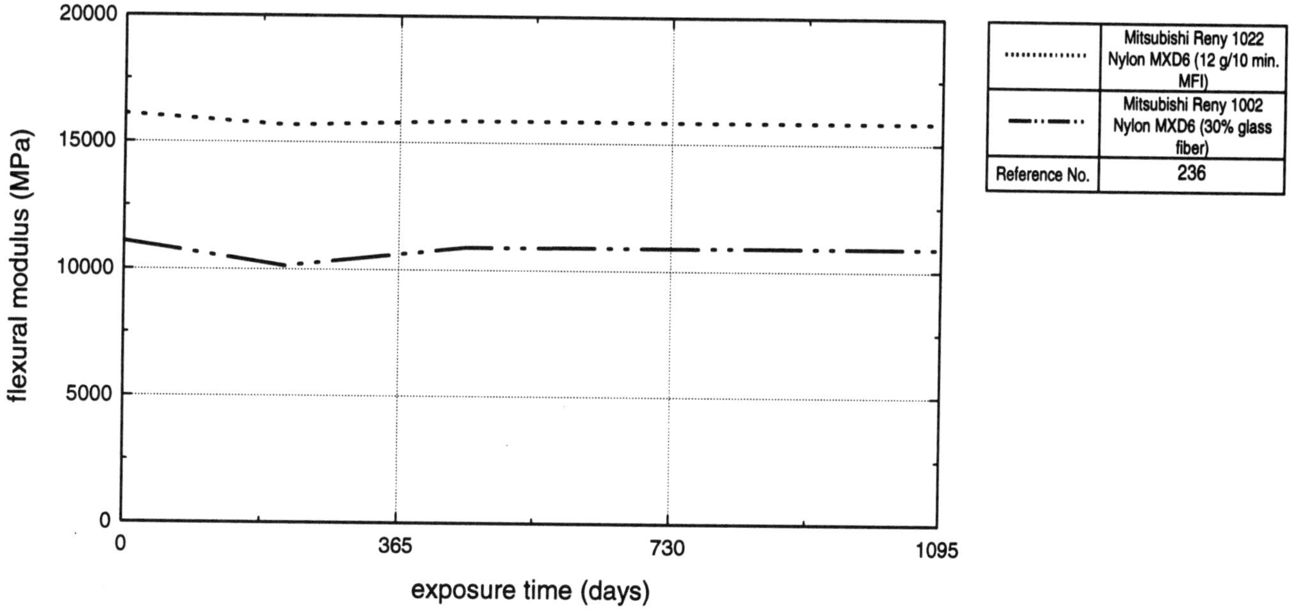

GRAPH 108: Hiratsuka, Japan Outdoor Exposure Time vs. Notched Izod Impact Strength of Nylon MXD6

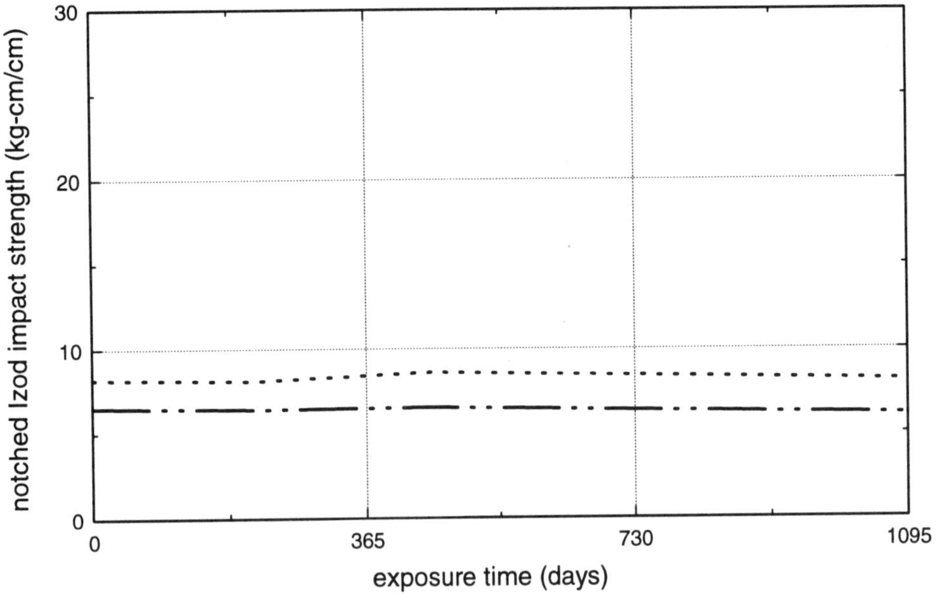

···············	Mitsubishi Reny 1022 Nylon MXD6 (12 g/10 min. MFI)
—··—··	Mitsubishi Reny 1002 Nylon MXD6 (30% glass fiber)
Reference No.	236

GRAPH 109: Hiratsuka, Japan Outdoor Weathering Exposure Time vs. Flexural Strength of Nylon MXD6

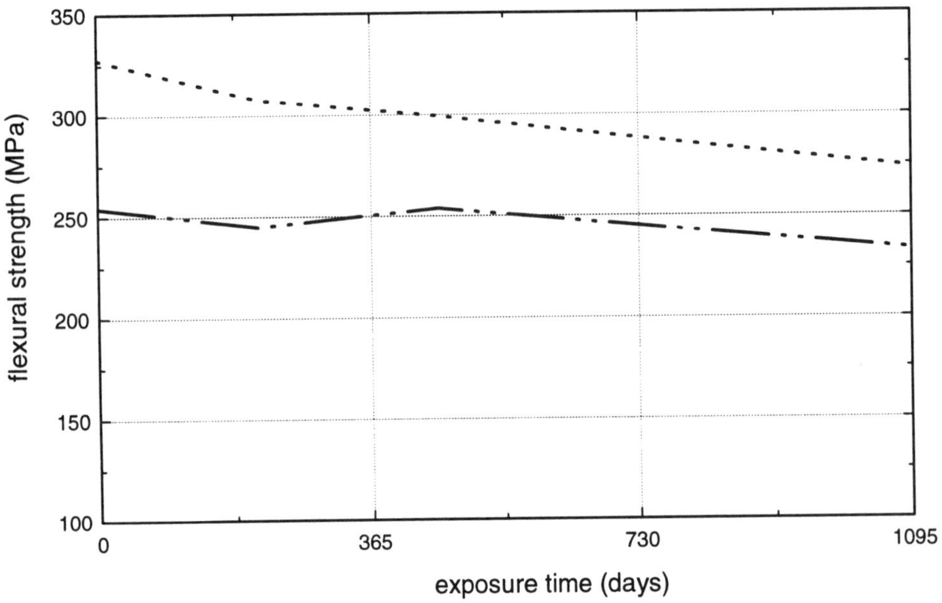

···············	Mitsubishi Reny 1022 Nylon MXD6 (12 g/10 min. MFI)
—··—··	Mitsubishi Reny 1002 Nylon MXD6 (30% glass fiber)
Reference No.	236

GRAPH 110: Hiratsuka, Japan Outdoor Weathering Exposure Time vs. Tensile Strength of Nylon MXD6

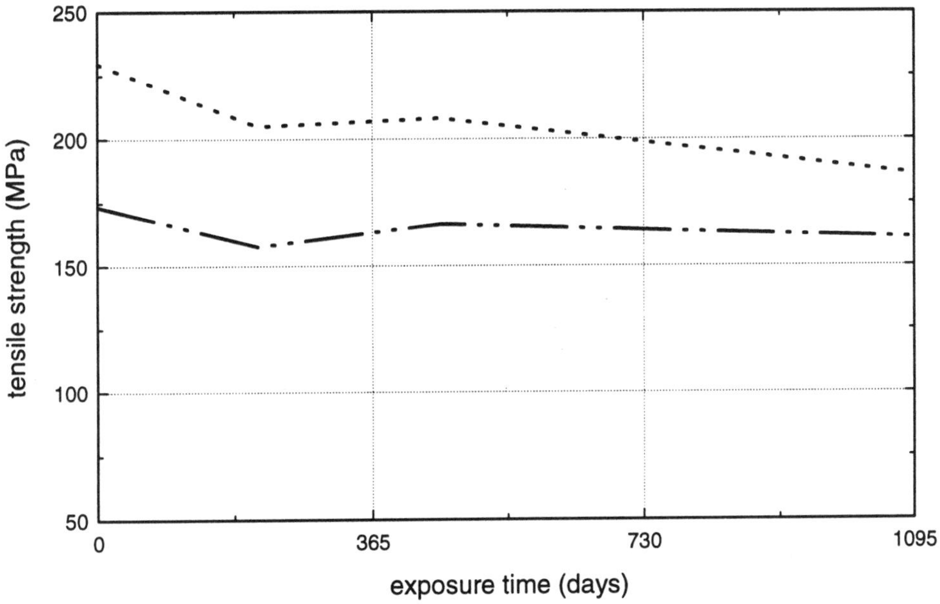

··············	Mitsubishi Reny 1022 Nylon MXD6 (12 g/10 min. MFI)
—··—··	Mitsubishi Reny 1002 Nylon MXD6 (30% glass fiber)
Reference No.	236

GRAPH 111: Sunshine Weatherometer Exposure Time vs. Elongation of Nylon MXD6

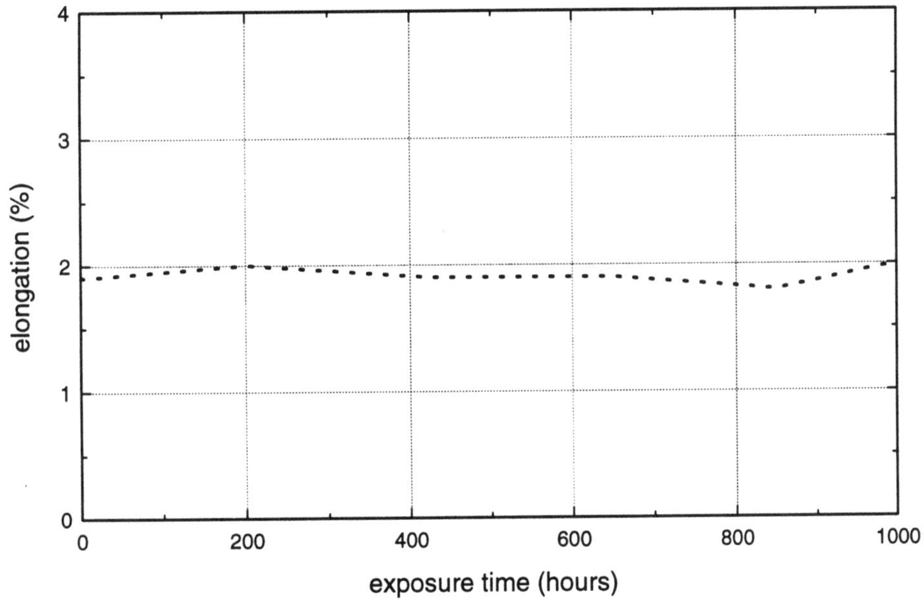

··············	Mitsubishi Reny 1002 Nylon MXD6 (30% glass fiber); 45°C; 12 min water spray/60 min
Reference No.	236

GRAPH 112: Sunshine Weatherometer Exposure Time vs. Tensile Strength of Nylon MXD6

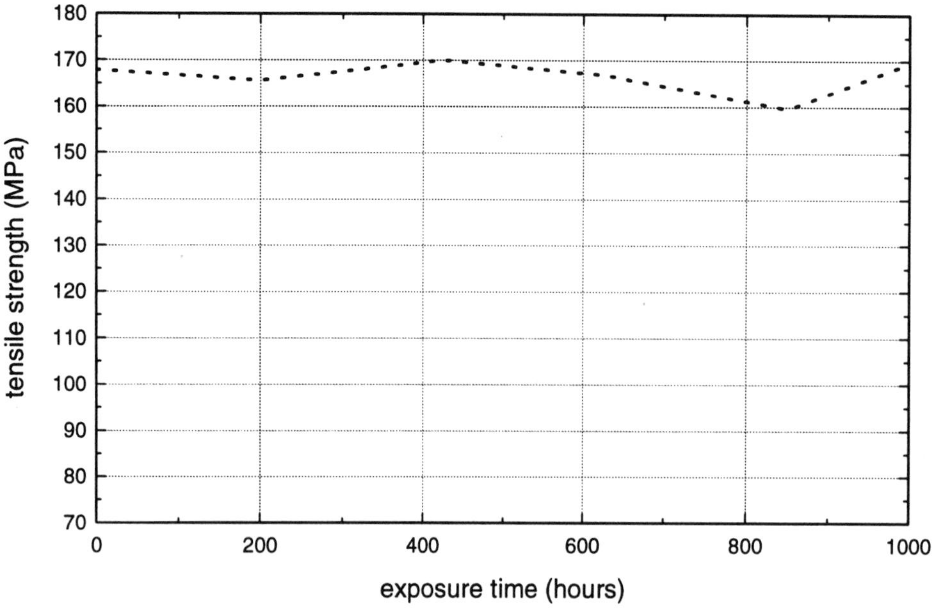

··············	Mitsubishi Reny 1002 Nylon MXD6 (30% glass fiber); 45°C; 12 min water spray/60 min
Reference No.	236

Chapter 20

Polyarylamide

Outdoor Weather Resistance

Solvay: Ixef 1002; Ixef 1022 (material compostion: 50% glass fiber reinforcement)

Specimens were exposed to the weather for 4 years at the Hiratsuka Test Station (average temperature 23°C, extremes 0°C, 30°C; average precipitation = 130 mm per month, extremes 50-200 mm per month; total solar irradiance 500 kj/cm^2 per year).

The results obtained on specimens 3.2mm thick show:

1. water absorption of approx. 0.8%.
2. an approx. 30% reduction in the maximum stress corresponding essentially to the reversible plasticization brought about by water.
3. no change in modulus.

The surface of a part made of IXEF product is a layer of pure polymer approximately 1 micron thick. This layer allows a very good gloss finish to be obtained. If photo-oxidation occurs, this layer deteriorates as a result of a change in the structure of the surface state (e.g. an increase from Ra = 0.15 microns to Ra = 2 microns). Oxidation affecting a very small quantity of material (3 mg/m^2) therefore results in a change in the appearance of the surface in this case (gloss and color) without the properties of the material being affected in any way.

When choosing the surface appearance of parts likely to be exposed to UV, it is advisable to avoid excessively low roughness levels which will be affected to a considerable degree by very superficial photo-oxidation.

To date, experience with outdoor IXEF applications, has established that the variations in shades observed are acceptable for many colors. Some particularly exacting sectors of the market have very stringent requirements; special IXEF grades may satisfy these requirements in certain cases.

The flame-resistant grades, on the other hand, exhibit variations in shade which are generally unacceptable for light colors.

Reference: *IXEF Reinforced Polyarylamide Based Thermoplastic Compounds Technical Manual,* supplier design guide (Br 1409c-B-2-1190) - Solvay, 1990.

GRAPH 113: **Hiratsuka, Japan Outdoor Exposure Time vs. Breaking Stress in Flexure of Polyarylamide**

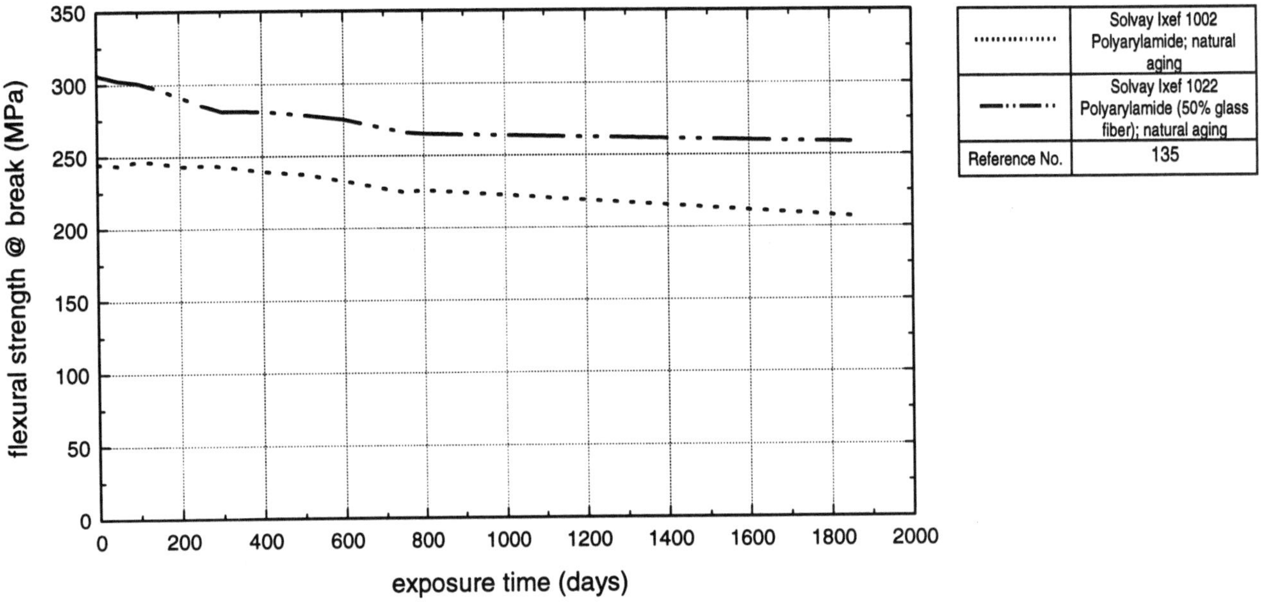

GRAPH 114: **Hiratsuka, Japan Outdoor Exposure Time vs. Flexural Modulus of Polyarylamide**

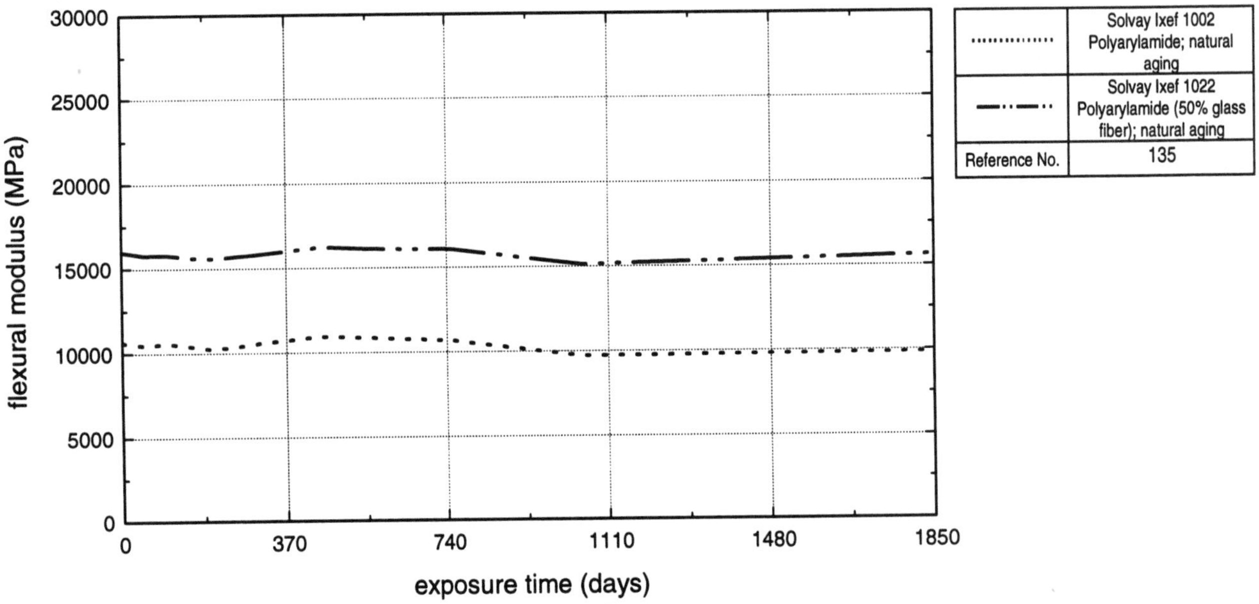

GRAPH 115: Hiratsuka, Japan Outdoor Exposure Time vs. Notched Izod Impact Strength of Polyarylamide

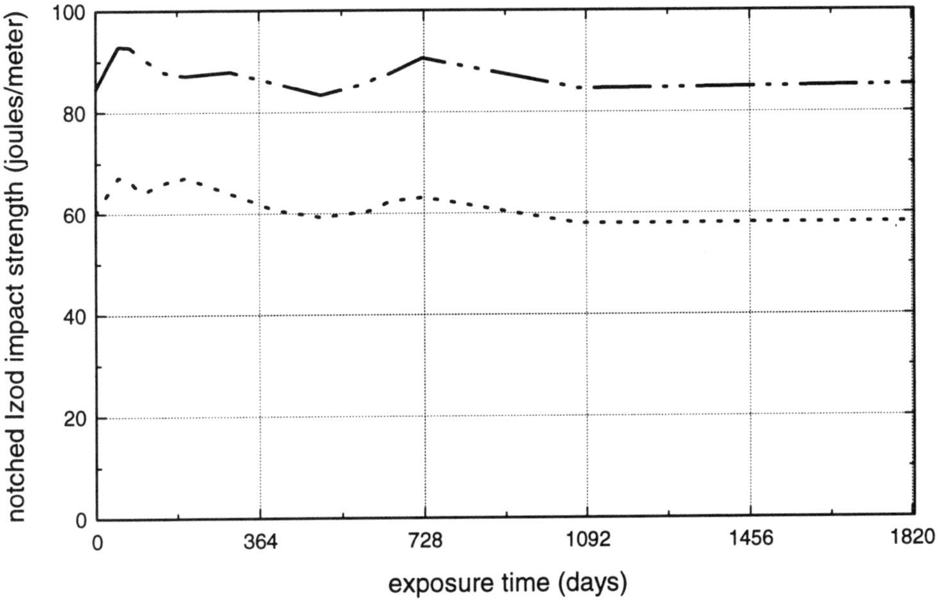

	Solvay Ixef 1002 Polyarylamide; natural aging
	Solvay Ixef 1022 Polyarylamide (50% glass fiber); natural aging
Reference No.	135

GRAPH 116: Hiratsuka, Japan Outdoor Exposure Time vs. Percent Weight Change of Polyarylamide

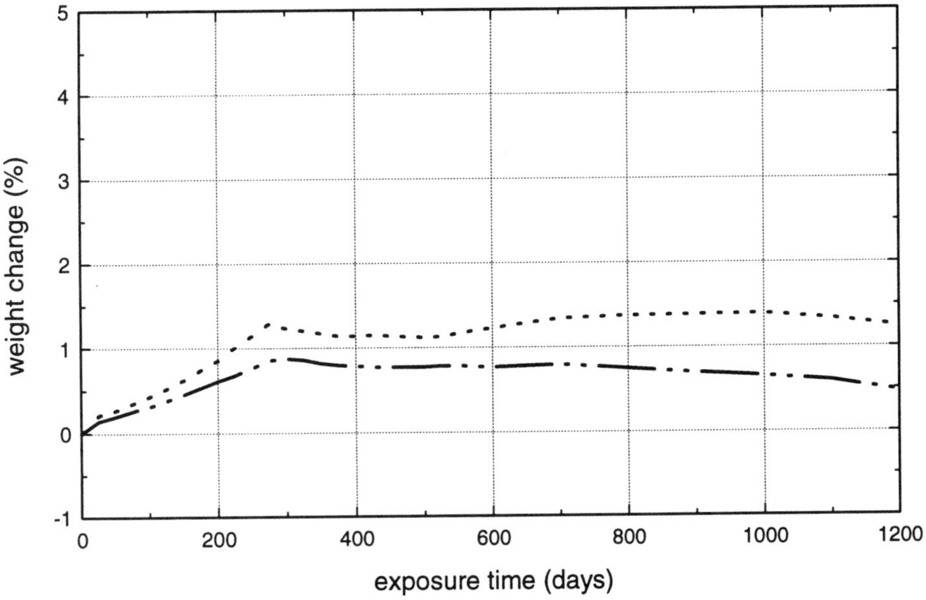

	Solvay Ixef 1002 Polyarylamide; natural aging: specimen thickness 3.2mm; under a tensile stress
	Solvay Ixef 1022 Polyarylamide (50% glass fiber); specimen thickness 3.2mm; under a tensile stress
Reference No.	135

Chapter 21

Parylene

Outdoor Weather Resistance

Union Carbide Specialty Coating Sys: Parylene (features: highly crystalline, high molecular weight, completely linear)

Although stable indoors, the Parylenes are not recommended for long term use outdoors when exposed to direct sunlight.

Reference: *Parylene Conformal Coatings Specifications and Properties,* supplier technical report - Union Carbide Specialty Coating Systems, 1992.

Chapter 22

Polycarbonate

Weather Resistance

GE Plastics: Lexan 100 (features: transparent, 2.36 mm thick; manufacturing method: extrusion; product form: sheet); **Lexan S-100** (features: transparent, 3.28 mm thick; manufacturing method: extrusion; product form: sheet)

In March, 1977, it was decided to evaluate plastics currently used in glazing, signs, solar collectors, skylights and various other applications, and their relative physical properties before and after weathering exposure tests. Samples of the various materials to be tested were purchased from Industrial and Sign Distribution Companies located in Cincinnati. Two samples of Lexan S-100 were tested; they were colorless (clear) and had a nominal thickness of 2.36mm and 3.28mm.

The material showed visually objectionable changes and had high yellowness indices after exposure to three separate weathering tests: twin carbon arc weatherometer, outside EMMAQUA test in Arizona, and outside exposure at a 45° angle facing south. The samples exposed in the twin carbon arc weatherometer for 3,000 hours were tested for tensile strength, light transmission, and both Charpy and Izod impact (notched). Lexan showed no significant change in tensile strength; 8% (2.36mm sample) and 10% (3.28mm sample) changes in light transmittance, respectively; both samples showed significant changes by both impact tests but still retained a reasonably high impact value.

Reference: *Comparison Of Plastics Used In Glazing, Signs, Skylights And Solar Collector Applications - Technical Bulletin 143,* competitor's technical report (ADARIS 50-1037-01) - Aristech Chemical Corporation, 1989.

UV Resistance

GE Plastics: Lexan 103 (features: UV stabilized, transparent); **Lexan** (features: transparent)

Like many other thermoplastics, polycarbonate may undergo a degradation process due to exposure to UV radiation, resulting in yellowing. In lighting we can find this risk in luminaries, which are provided with a mercury lamp. This sort of lamp emits a high level of UV radiation, which may influence the optical properties of Lexan. The evaluation of this effect is less predictable than in the case of heat aging because of the influence of other variables. If we consider the example of a street lantern bowl, we can distinguish the following action: 1) A weathering action. 2) A direct UV irradiation from the mercury lamp which varies according to the lamp, its distance from the transparent polycarbonate part, the possible concentration of the light flux on certain spots of the bowl because of the reflector. 3) A heat aging action. 4) An action due to the material degradation caused by incorrect processing. (Material degradation may not always be represented by visual defects.)

These are some of the possible factors which can influence the polycarbonate at the same time. In the case of a sodium lamp the low UV content of the radiation limits the risk of damage through UV exposure, but thermal degradation may occur if temperature recommendations are exceeded.

According to our experience, we advise a minimum distance between the Lexan bowl and mercury lamp of 160 mm and not to exceed a temperature of 80°C on the Lexan surface.

Reference: *Engineering Thermoplastics For Lighting,* supplier technical report (6m/0387) - General Electric Plastics Europe, 1987.

Outdoor Weather Resistance

Dow Chemical: Calibre (features: transparent)

Calibre polycarbonate resins are excellent materials for parts requiring a high degree of resistance to ultra-violet (UV) light from either indoor or outdoor sources.

For parts exposed to the higher levels of UV light, found in outdoor environments, Calibre polycarbonate resins are available in formulations that incorporate a UV stabilizer. UV stabilization also can greatly extend the retention of key physical properties such as impact strength and appearance. Data indicate the superior performance of the UV-stabilized grades of Calibre polycarbonate in outdoor applications.

Reference: *Calibre Engineering Thermoplastics Basic Design Manual,* supplier design guide (301-1040-1288) - Dow Chemical Company, 1988.

GE Plastics: Lexan 103 (features: UV stabilized, transparent); **Lexan** (features: transparent)

Exposure trials were carried out by General Electric Plastics. While they are an incomplete evaluation, they indicate a substantial retention of mechanical and optical properties in a subtropical climate as exists in Florida.

In all cases where outdoor applications are envisaged, Lexan UV stabilized grades are recommended.

Reference: *Engineering Thermoplastics For Lighting,* supplier technical report (6m/0387) - General Electric Plastics Europe, 1987.

Accelerated Outdoor Weathering Resistance

Cyro: Cyrolon AR (features: transparent, abrasion resistant coating, 3.2 mm thick; manufacturing method: continuous cast; product form: sheet)

Cyrolon AR sheet offers an improved resistance to weathering over standard grade polycarbonate sheet. No significant loss in abrasion resistance, light transmission or impact will be seen over many years of service. Minimal change in light transmittance and yellowness index occur after an accelerated weathering cycle of three years.

Reference: *Physical Properties Acrylite AR Acrylic Sheet And Cyrolon AR Polycarbonate Sheet,* supplier design guide (1632B-0193-10BP) - Cyro Industries, 1993.

Indoor UV Light Resistance

Dow Chemical: Calibre (features: transparent)

Calibre polycarbonate resins are excellent materials for parts requiring a high degree of resistance to ultraviolet (UV) light from either indoor or outdoor sources. All Calibre resins, even those without UV stabilizers, pass the accelerated indoor colorfastness tests developed by International Business Machine Corp. (IBM) and Hewlett-Packard Co. Parts made from Calibre resins are not affected by the levels of UV light normally found indoors.

Reference: *Calibre Engineering Thermoplastics Basic Design Manual,* supplier design guide (301-1040-1288) - Dow Chemical Company, 1988.

TABLE 39: Outdoor Weathering in Arizona of Dow Calibre Polycarbonate.

Material Family	Polycarbonate					
Material Supplier/ Grade	Dow Chemical Calibre 300-6			Dow Chemical Calibre 302-6		
Features	general purpose grade/unstabilized			UV stabilized		
Reference Number	78	78	78	78	78	78

MATERIAL CHARACTERISTICS

Melt Flow Rate	6 grams/10 min.	6 grams/10 min.	6 grams/10 min.	6 grams/10 min.	6 grams/10 min.	6 grams/10 min.

EXPOSURE CONDITIONS

Exposure Type	outdoor weathering			outdoor weathering		
Exposure Location	Arizona	Arizona	Arizona	Arizona	Arizona	Arizona
Exposure Country	USA	USA	USA	USA	USA	USA
Exposure Time (days)	182	365	730	182	365	730
Exposure Note	45° south	45° south	45° south	45° south	45° south	45° south

PROPERTIES RETAINED (%)

Notched Izod Impact Strength	101.7	6.8	3.4	98.4	97.2	101.6

SURFACE AND APPEARANCE

Δ Yellowness Index	12.3	15.7	20	2.3	3.3	6.1
Haze Retained (%)	482	747	1165	425	562	925
Luminous Transmittance Retained (%)	96	94.7	92.1	98.7	98.8	97.4

TABLE 40: Outdoor Weathering in California and Pennsylvania of Glass Reinforced Polycarbonate.

Material Family	Polycarbonate							
Material Supplier	LNP Engineering Plastics							
Reference Number	215	215	215	215	215	215	215	215
MATERIAL COMPOSITION								
carbon black	1%	1%	1%	1%	1%	1%	1%	1%
glass fiber reinforcement	30%	30%	30%	30%	30%	30%	30%	30%
EXPOSURE CONDITIONS								
Exposure Type	outdoor weathering				outdoor weathering			
Exposure Location	Los Angeles, California	Los Angeles, California	Los Angeles, California	Los Angeles, California	Philadelphia, Pennsylvania	Philadelphia, Pennsylvania	Philadelphia, Pennsylvania	Philadelphia, Pennsylvania
Exposure Country	USA	USA	USA	USA	USA	USA	USA	USA
Exposure Test Method	ASTM D1435	ASTM D1435	ASTM D1435	ASTM D1435	ASTM D1435	ASTM D1435	ASTM D1435	ASTM D1435
Exposure Time (days)	91	182	365	730	91	182	365	730
PROPERTIES RETAINED (%)								
Tensile Strength	90	90.4	89	87	88	86	89	87
Notched Izod Impact Strength	94	83.3			94	83.3		
Unnotched Izod Impact Strength	83	72.3			80	73.8		

TABLE 41: Outdoor Weathering in Pennsylvania of Miles Makrolon Polycarbonate.

Material Family	Polycarbonate											
Material Supplier/ Grade	Miles Makrolon 3203			Miles Makrolon 3203			Miles Merlon M-60 UV			Miles Merlon M-60 white		
Features	natural resin, UV stabilized			tinted, UV stabilized			natural resin, UV stabilized			white color		
Material Note							Merlon trade name replaced by Makrolon, equivalent Makrolon grade not known			Merlon trade name replaced by Makrolon, equivalent Makrolon grade not known		
Reference Number	219	219	219	219	219	219	219	219	219	219	219	219

EXPOSURE CONDITIONS

Exposure Type	outdoor weathering			outdoor weathering			outdoor weathering			outdoor weathering		
Exposure Location	Pittsburgh, Pennsylvania			Pittsburgh, Pennsylvania			Pittsburgh, Pennsylvania			Pittsburgh, Pennsylvania		
Exposure Country	USA			USA			USA			USA		
Exposure Time (days)	365	730	1095	365	730	1095	1095	365	730	365	730	1095

PROPERTIES RETAINED (%)

Tensile Strength @ Yield	99	105	101	99	108	103	98	106	101	101	103	103
Ultimate Elongation	100	99	98	98	96	101	99	82	88	96	94	103
Notched Izod Impact Strength	100 {x}	99 {x}	98 {x}	98 {x}	96 {x}	101 {x}	99 {x}	82 {x}	88 {x}	96 {x}	94 {x}	103 {x}

SURFACE AND APPEARANCE

Haze Retained (%)	1240	2640	2400	790	1150	1030	500	715	1000			
Gloss Retained (%)	79.8	50	48.9	80.7	53.4	56.8	74.7	49.5	38.5	12.1	7.6	7.6
Luminous Transmittance Retained (%)	98.8	95	96.9	100.6	97.6	99.9	98.8	94.7	93.4			

TABLE 42: Outdoor Weathering in Pennsylvania of Miles Makrolon Polycarbonate.

Material Family	Polycarbonate								
Material Supplier/ Grade	Miles Makrolon 2800			Miles Makrolon 3200			Miles Merlon M-60		
Features	natural resin			natural resin			natural resin		
Material Note							Merlon trade name replaced by Makrolon, equivalent Makrolon grade no known		
Reference Number	219	219	219	219	219	219	219	219	219
EXPOSURE CONDITIONS									
Exposure Type	outdoor weathering			outdoor weathering			outdoor weathering		
Exposure Location	Pittsburgh, Pennsylvania			Pittsburgh, Pennsylvania			Pittsburgh, Pennsylvania		
Exposure Country	USA			USA			USA		
Exposure Time (days)	365	730	1095	1095	1095	1095	1095	1095	1095
PROPERTIES RETAINED (%)									
Tensile Strength @ Yield	100	103	99	98	105	99	96	105	98
Ultimate Elongation	62	8	7	61	8	5	65	8	5
Notched Izod Impact Strength	100 {x}	13 {x}	3.4 {x}	98 {x}	34 {x}	3 {x}	100 {x}	101 {x}	4 {x}
SURFACE AND APPEARANCE									
Haze Retained (%)	640	2190	1880	590	1490	1090	300	1430	960
Gloss Retained (%)	78.1	35.4	43.8	80	38.9	46.7	90	38.9	47.9
Luminous Transmittance Retained (%)	96.9	92.2	93.3	97.2	91.2	95.4	98.9	95.1	94.6

TABLE 43: Outdoor Weathering in Kentucky, Accelerated Outdoor Weathering by EMMAQUA and Accelerated Weathering in a Carbon Arc Weatherometer of General Electric Lexan Polycarbonate.

Material Family	Polycarbonate					
Material Supplier/ Grade	GE Plastics Lexan S-100	GE Plastics Lexan 100	GE Plastics Lexan S-100	GE Plastics Lexan 100	GE Plastics Lexan S-100	GE Plastics Lexan 100
Features	transparent	transparent	transparent	transparent	transparent	transparent
Manufacturing Method	extrusion	extrusion	extrusion	extrusion	extrusion	extrusion
Product Form	sheet	sheet	sheet	sheet	sheet	sheet
Reference Number	152	152	152	152	152	152
MATERIAL CHARACTERISTICS						
Sample Thickness (mm)	3.28	2.36	3.28	2.36	3.28	2.36
EXPOSURE CONDITIONS						
Exposure Type	accelerated outdoor weatheing		outdoor weathering		accelerated weathering	
Exposure Location	Arizona		Florence, Kentucky		Arizona	
Exposure Country	USA		USA		USA	
Exposure Apparatus	EMMAQUA				carbon arc weatherometer	
Exposure Note			45° angle south			
Exposure Time (days)	365	365	730	730	125	125
PROPERTIES RETAINED (%)						
Tensile Strength					101	98
Notched Izod Impact Strength					84.2	64.7
Charpy Notched Impact Strength					10.9	55.7
SURFACE AND APPEARANCE						
Δ Yellowness Index	17.29	20.91	10.46	9.0	25.43	25.0
Luminous Transmittance Retained (%)					90.6	88.4

TABLE 44: Accelerated Weathering in an XW Weatherometer of General Electric Lexan Polycarbonate.

Material Family	Polycarbonate			
Material Supplier/ Grade	GE Plastics Lexan 303			
Features	transparent			
Reference Number	145	145	145	145

EXPOSURE CONDITIONS

Exposure Type	accelerated weathering			
Exposure Apparatus	XW weatherometer			
Exposure Time (days)	41.7	83.3	166.7	333.3

PROPERTIES RETAINED (%)

Notched Izod Impact Strength	42	9	9	9
Tensile Impact Strength		30	38	29
Heat Deflection Temperature	100.8	101.5	100	98.5

TABLE 45: Accelerated Indoor Exposure by HPUV of General Electric Lexan Polycarbonate.

Material Family	Polycarbonate	
Material Supplier/ Grade	GE Plastics Lexan 920	GE Plastics Lexan 920A
Features	flame retardant; Hewlett Packard parchment white color	flame retardant; Hewlett Packard parchment white color
Reference Number	176	176

EXPOSURE CONDITIONS

Exposure Type	Accelerated Indoor UV Light	
Exposure Apparatus	HPUV	HPUV
Exposure Note	3 year office exposure simulation	3 year office exposure simulation

SURFACE AND APPEARANCE

ΔE Color	0.5	0.5

GRAPH 117: **Florence, Kentucky Outside Weathering Exposure Time vs. Yellowness Index of Polycarbonate**

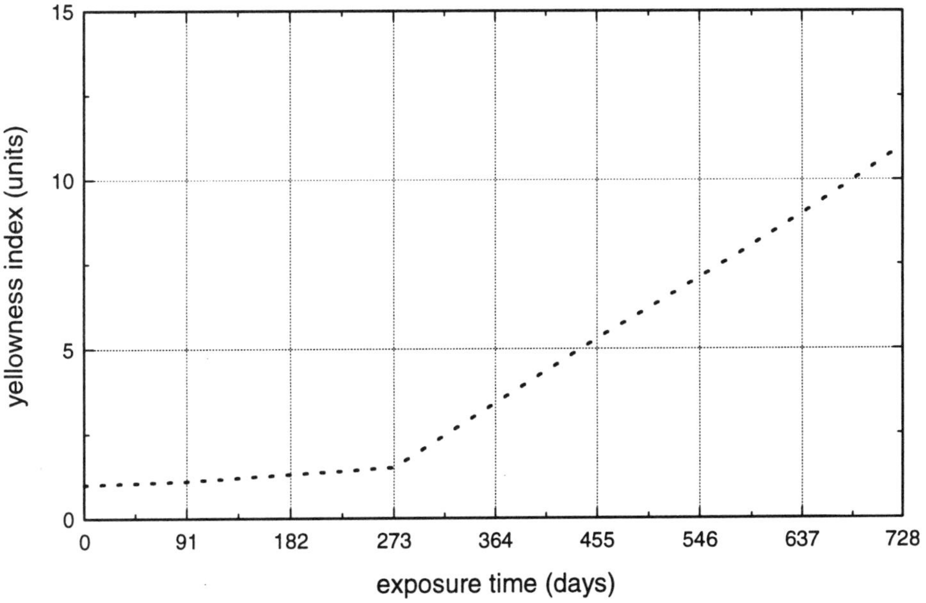

....	GE Lexan S-100 Polycarbonate (transparent, 3.28 mm thick; extrusion; sheet); test lab: Aristech Chemical.
Reference No.	152

GRAPH 118: **Arizona Outdoor Weathering Exposure Time vs. Drop Dart Impact Strength of Polycarbonate**

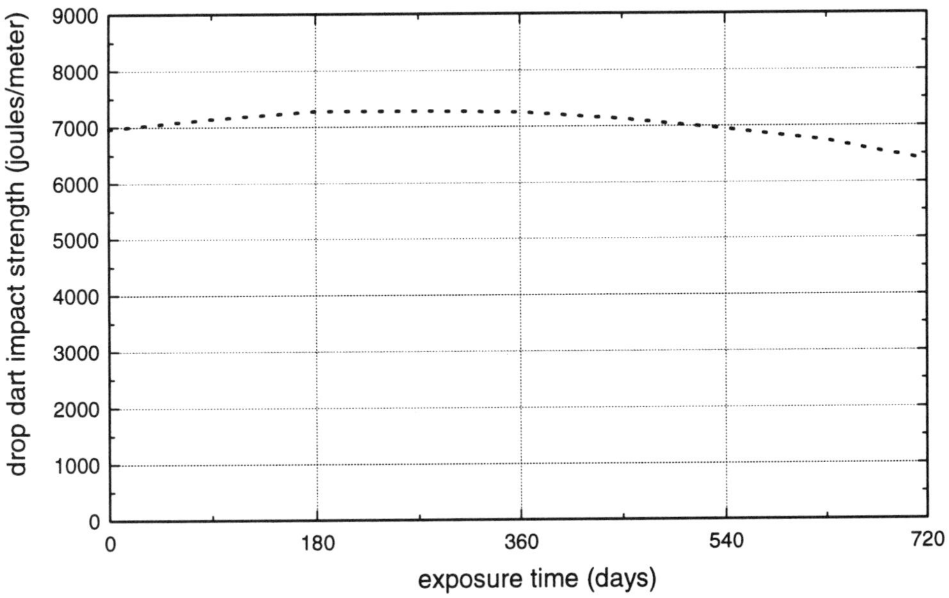

....	Polycarbonate (white, 3.2 mm thick)
Reference No.	189

GRAPH 119: Arizona Outdoor Weathering Exposure Time vs. Elongation of Polycarbonate

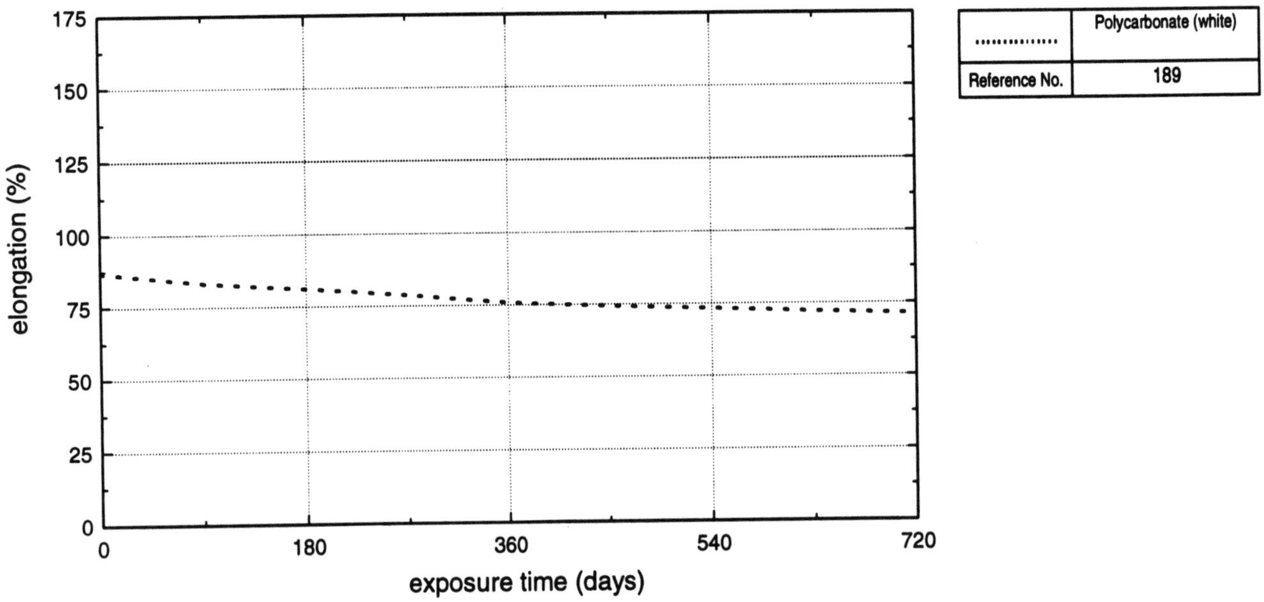

GRAPH 120: Arizona Outdoor Weathering Exposure Time vs. Tensile Strength at Yield of Polycarbonate

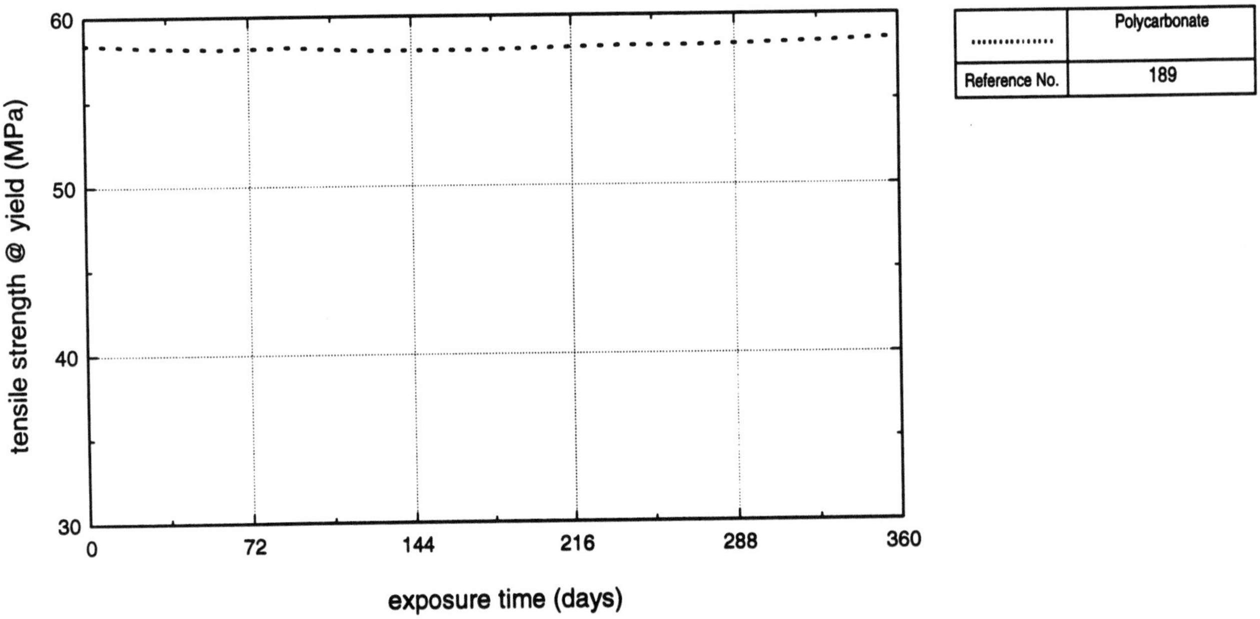

GRAPH 121: **Arizona Outdoor Weathering Exposure Time vs. Delta E Color Change of Polycarbonate**

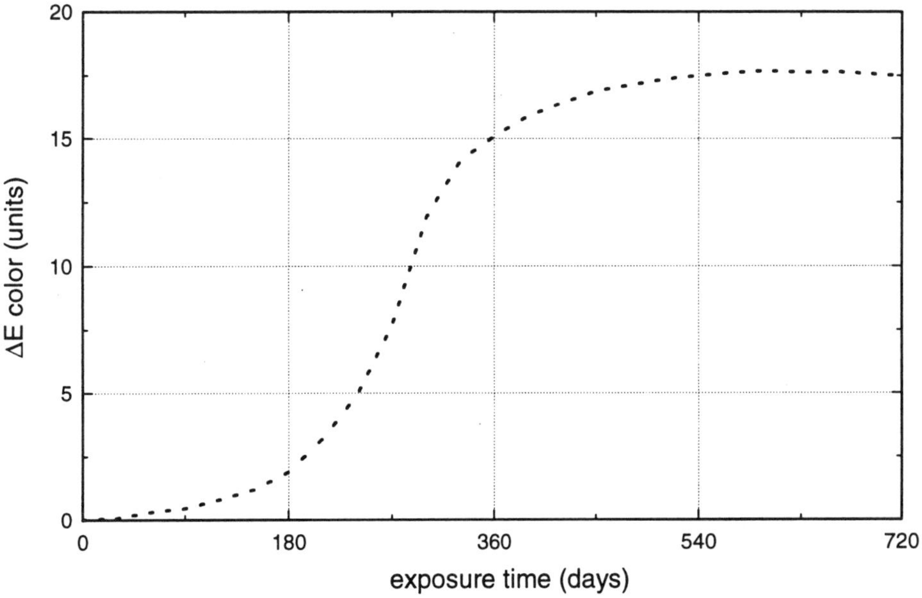

....	Polycarbonate (white); CIE lab color scale
Reference No.	189

GRAPH 122: **Florida Outdoor Weathering Exposure Time vs. Drop Dart Impact Strength of Polycarbonate**

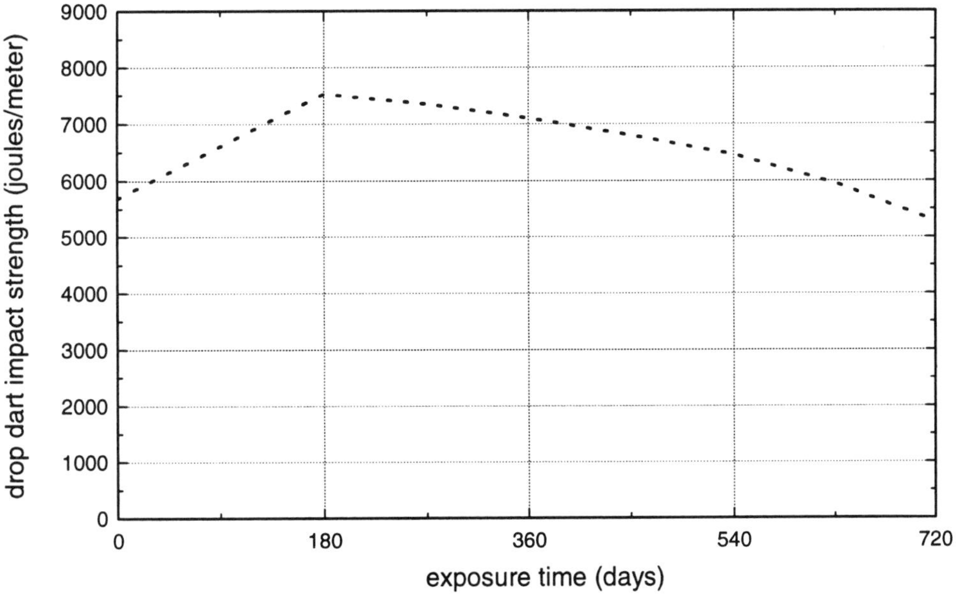

....	Polycarbonate (white, 3.2 mm thick)
Reference No.	189

GRAPH 123: **Florida Outdoor Weathering Exposure vs. Delta E Color Change of Polycarbonate**

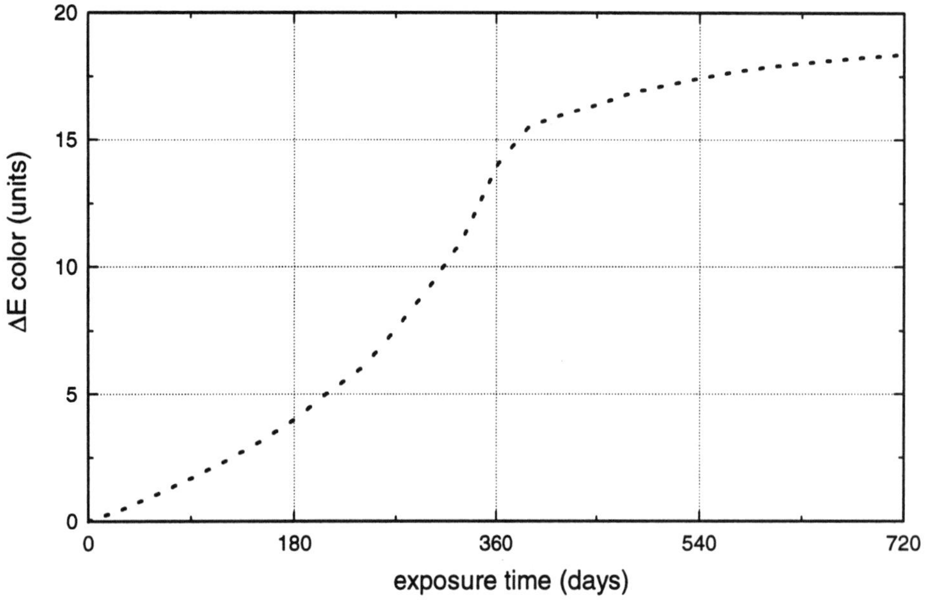

....	Polycarbonate; CIE lab color scale
Reference No.	189

GRAPH 124: **Ohio Outdoor Weathering Exposure vs. Delta E Color Change of Polycarbonate**

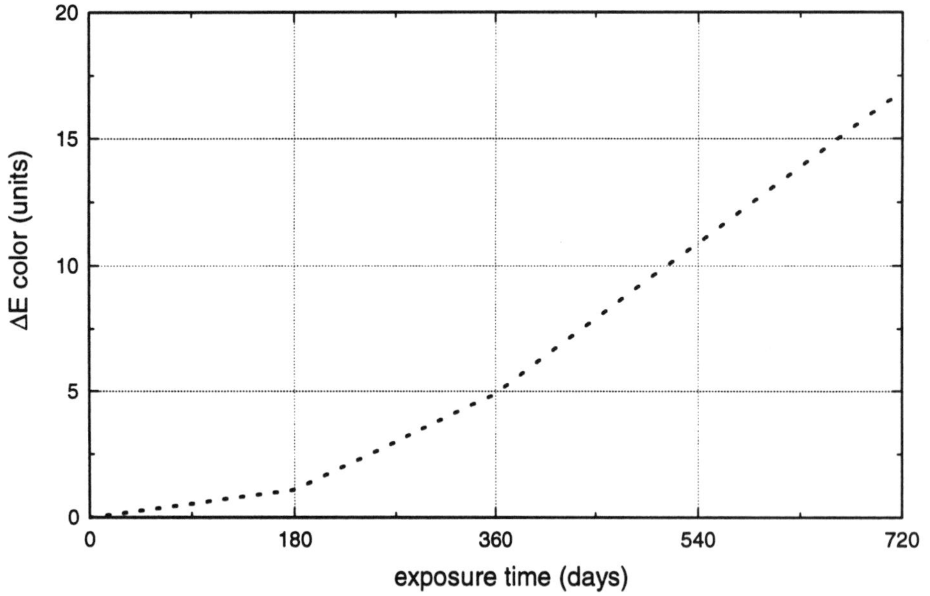

....	Polycarbonate (white); CIE lab color scale
Reference No.	189

GRAPH 125: Ohio Outdoor Weathering Exposure vs. Drop Dart Impact Strength of Polycarbonate

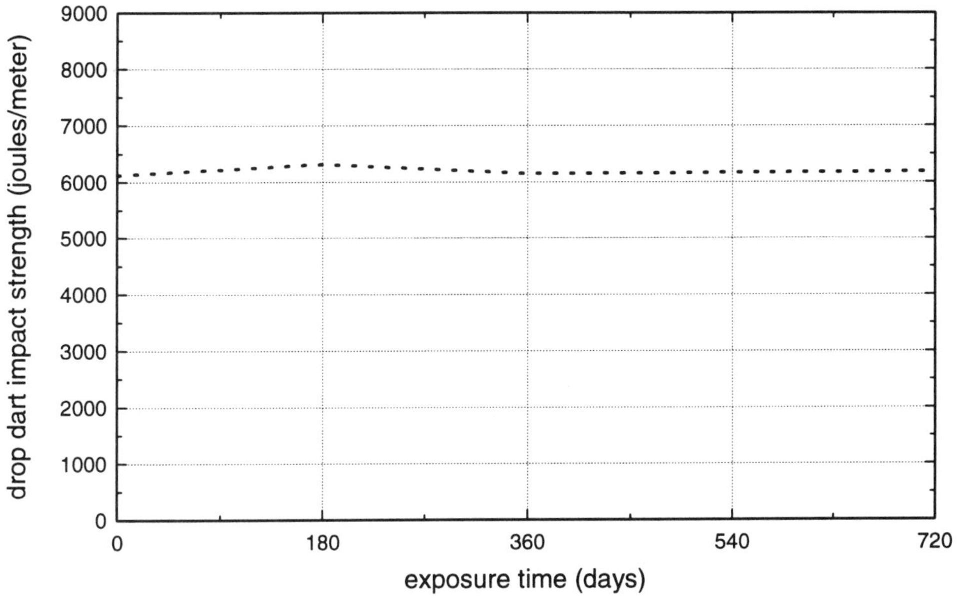

....	Polycarbonate (white, 3.2 mm thick)
Reference No.	189

GRAPH 126: EMMAQUA Accelerated Weathering Exposure Time vs. Light Transmission of Polycarbonate

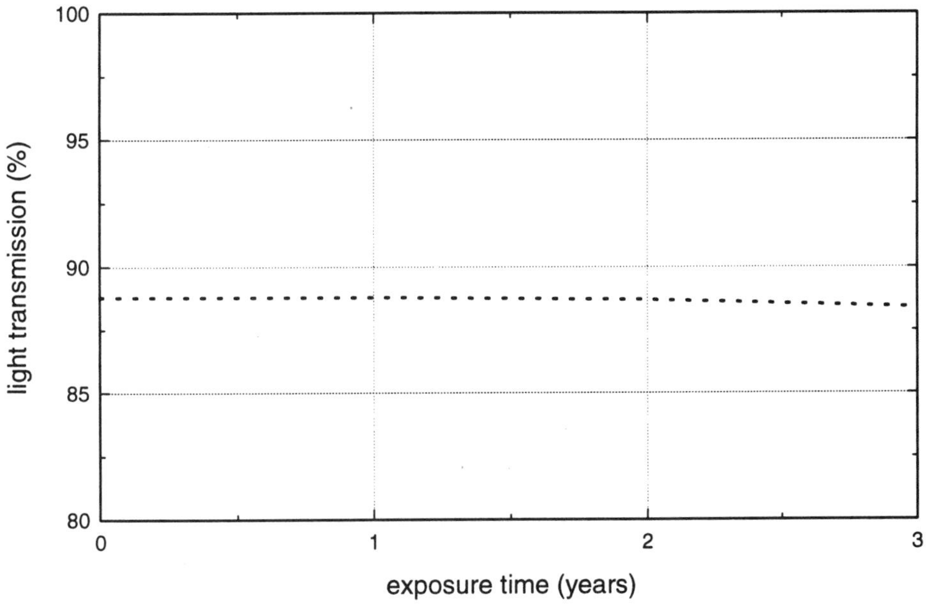

....	Cyro Cyrolon AR Polycarbonate (transparent, 3.2 mm thick; continuous cast; abras. resist. coating; sheet); ASTM D1003
Reference No.	151

GRAPH 127: **EMMAQUA Accelerated Weathering Exposure Time vs. Yellowness Index of Polycarbonate**

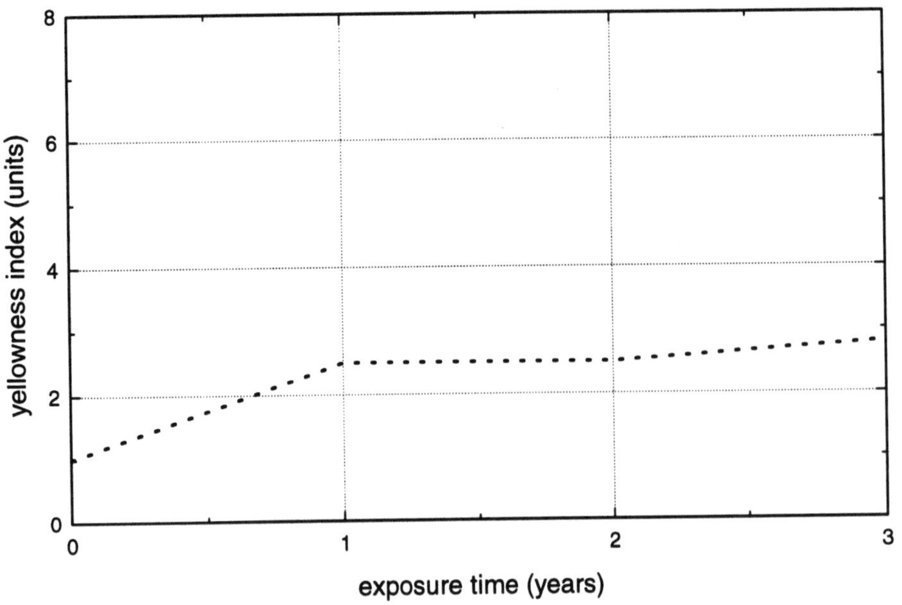

··············	Cyro Cyrolon AR Polycarbonate (transparent, 3.2 mm thick; continuous cast; abras. resist. coating; sheet)
Reference No.	151

GRAPH 128: **EMMAQUA Arizona Accelerated Weathering Exposure Time vs. Yellowness Index of Polycarbonate**

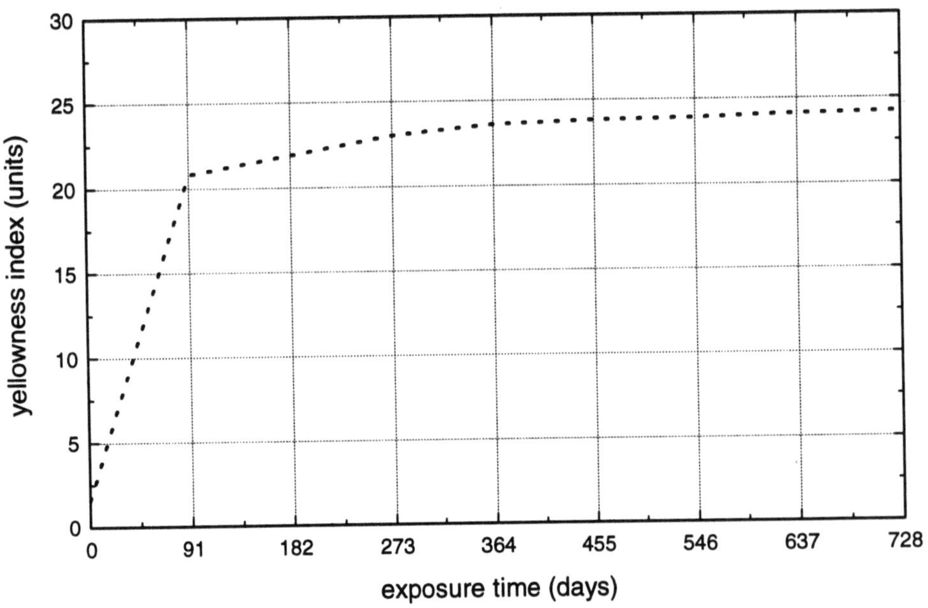

··············	GE Lexan S-100 Polycarbonate (transparent, 3.28 mm thick; extrusion; sheet); test lab: DSET
Reference No.	152

GRAPH 129: Carbon Arc XW Weatherometer Exposure Time vs. Haze of Polycarbonate

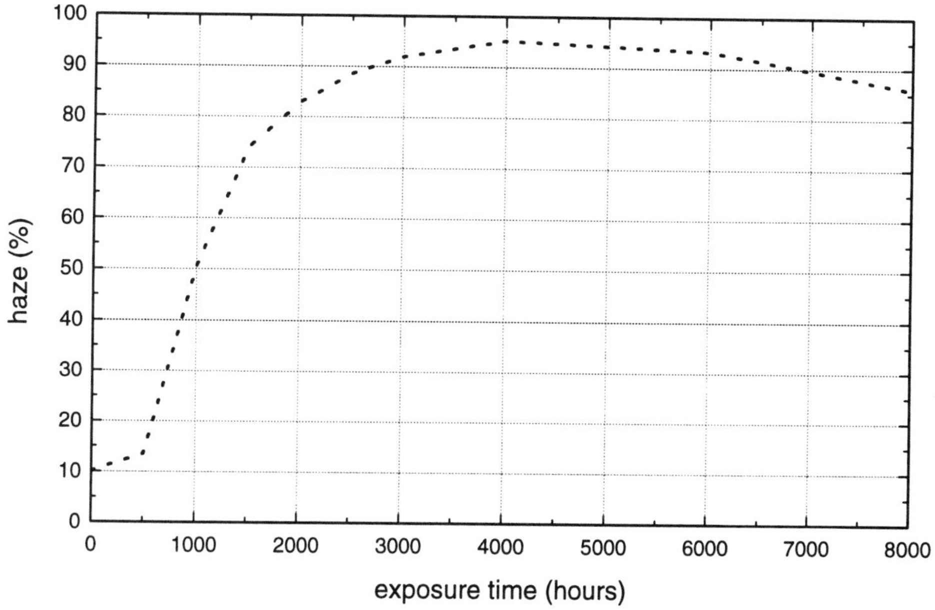

....	GE Lexan 153 Polycarbonate (transparent)
Reference No.	145

GRAPH 130: Twin Carbon Arc Weatherometer Exposure Time vs. Yellowness Index of Polycarbonate

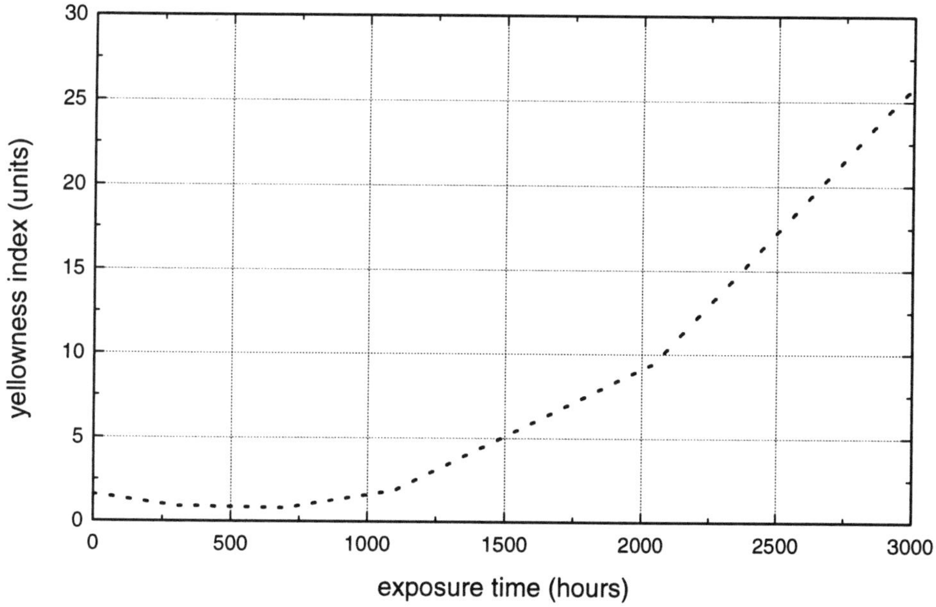

....	GE Lexan S-100 Polycarbonate (transparent, 3.28 mm thick; extrusion; sheet); Aristech test lab Florence, KY
Reference No.	152

GRAPH 131: Accelerated Indoor UV Exposure Time vs. Delta E Color Change of Polycarbonate

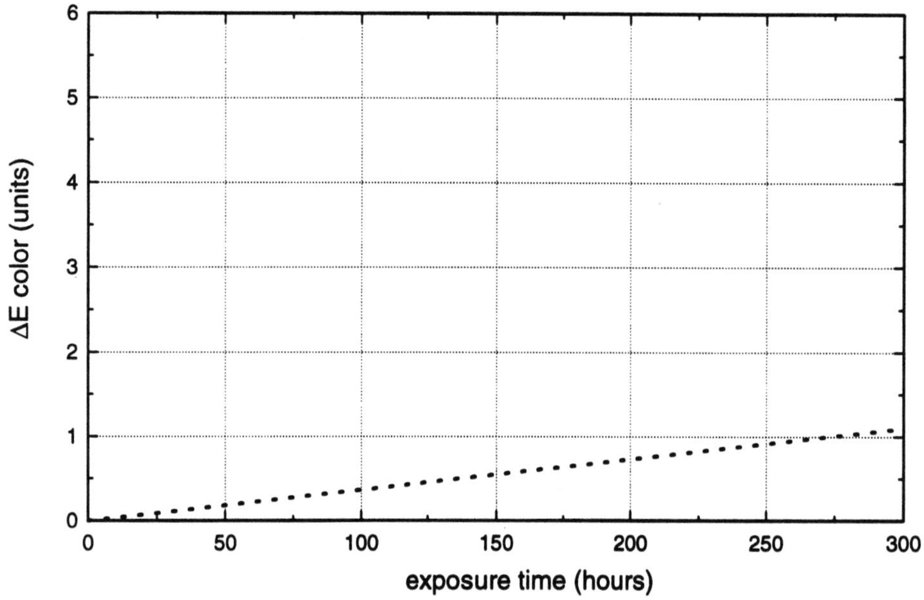

....	Polycarbonate
Reference No.	232

Chapter 23

Polybutylene Terephthalate

Weather Resistance

BASF AG: Ultradur B

Ultradur B moldings exposed outdoors for three years in Central Europe showed hardly any signs of yellowing, and change in their appearance. The effect of the outdoor exposure on the mechanical properties, e.g. the rigidity and tensile strength, was also very slight. After exposure for 3,600 hours in the Xenotest 1200, the tensile strength and ultimate elongation were still 90% of the original value.

If the experience gained on other engineering thermoplastics, e.g. nylon 6 and nylon 66, can be adopted as a guideline, 3,600 hours exposure in the Xenotest 1200 correspond to an outdoor exposure period of about five to six years.

Parts intended for outdoor use should be colored black, in order to avoid any risk that the exposed surfaces will suffer an impairment of strength.

Reference: *Ultradur Polybutylene Terephthalate (PBT) Product Line, Properties, Processing,* supplier design guide (B 575/1e - (819) 4.91) - BASF Aktiengesellschaft, 1991.

Hoechst Celanese: Celanex 3210 (features: natural resin, flame retardant; material compostion: 18% glass fiber reinforcement); **Celanex 3300** (features: natural resin, general purpose grade; material compostion: 30% glass fiber reinforcement); **Celanex 3310** (features: natural resin, flame retardant; material compostion: 30% glass fiber reinforcement)

Results taken after three years outdoor exposure indicate that there has been no fundamental change in physical properties. Predictably, black Celanex 3300 polyester resin exhibits better property retention than natural and therefore should be considered where long term outdoor exposure is required.

Reference: *Celanex Thermoplastic Polyester Properties and Proecessing (CX-1A),* supplier design guide (HCER 91-343/10M/692) - Hoechst Celanese Corporation, 1992.

Accelerated Artificial Weathering Resistance

Hoechst Celanese: Celanex 3210 (features: natural resin, flame retardant; material compostion: 18% glass fiber reinforcement); **Celanex 3300** (features: black color, general purpose grade; material compostion: 30% glass fiber reinforcement); **Celanex** (features: natural resin, general purpose grade; material compostion: 30% glass fiber reinforcement); **Celanex 3310** (features: natural resin, flame retardant; material compostion: 30% glass fiber reinforcement)

Laboratory weatherometer testing of moldings of unpigmented and pigmented Celanex has shown relatively little loss of tensile properties.

Reference: *Celanex Thermoplastic Polyester Properties and Proecessing (CX-1A),* supplier design guide (HCER 91-343/10M/692) - Hoechst Celanese Corporation, 1992.

GRAPH 134: Florida and Arizona Outdoor Weathering Exposure Time vs. Notched Izod Impact Strength of Polybutylene Terephthalate

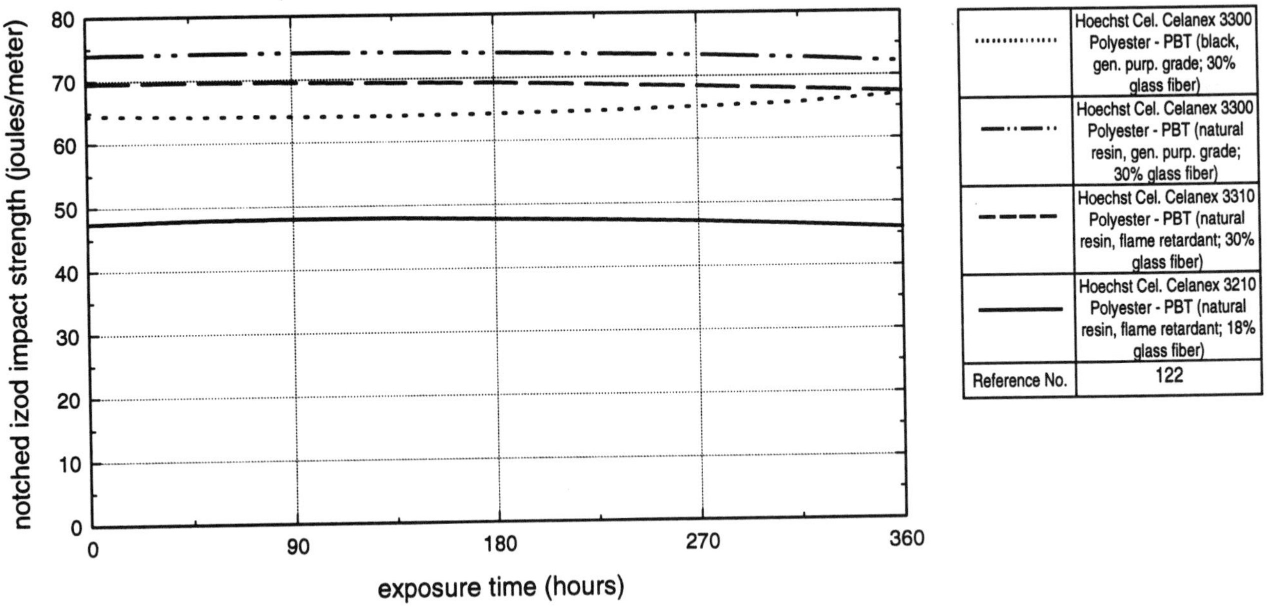

GRAPH 135: Florida and Arizona Outdoor Weathering Exposure Time vs. Tensile Strength of Polybutylene Terephthalate

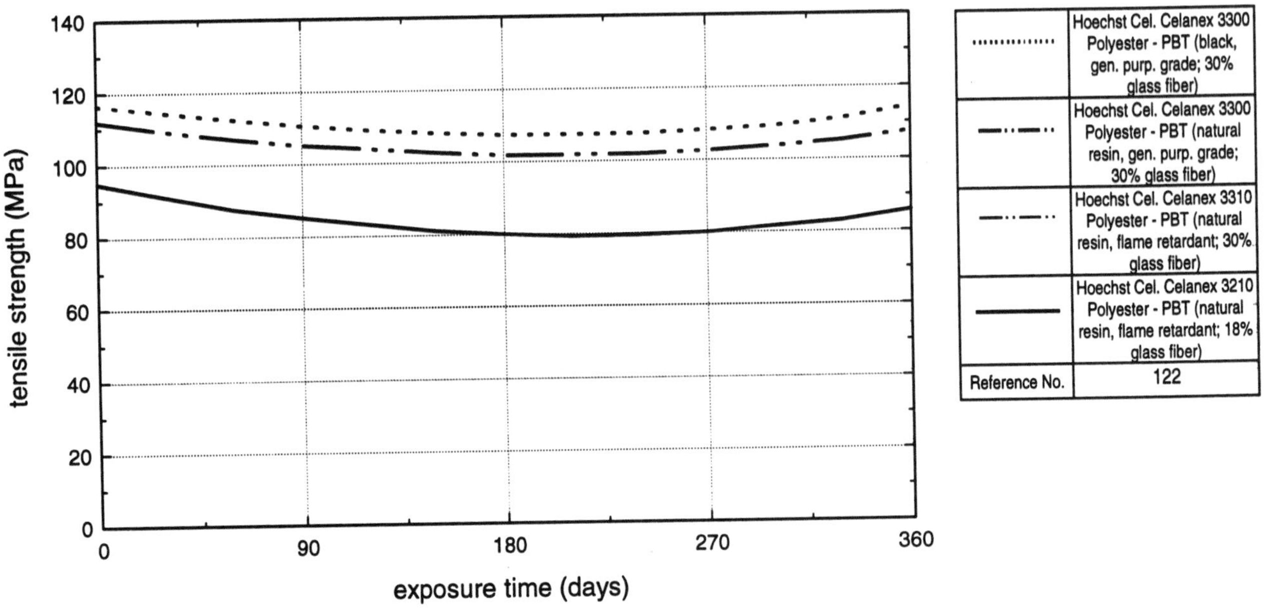

GRAPH 136: **Hiratsuka, Japan Outdoor Exposure Time vs. Breaking Stress in Flexure of Polybutylene Terephthalate**

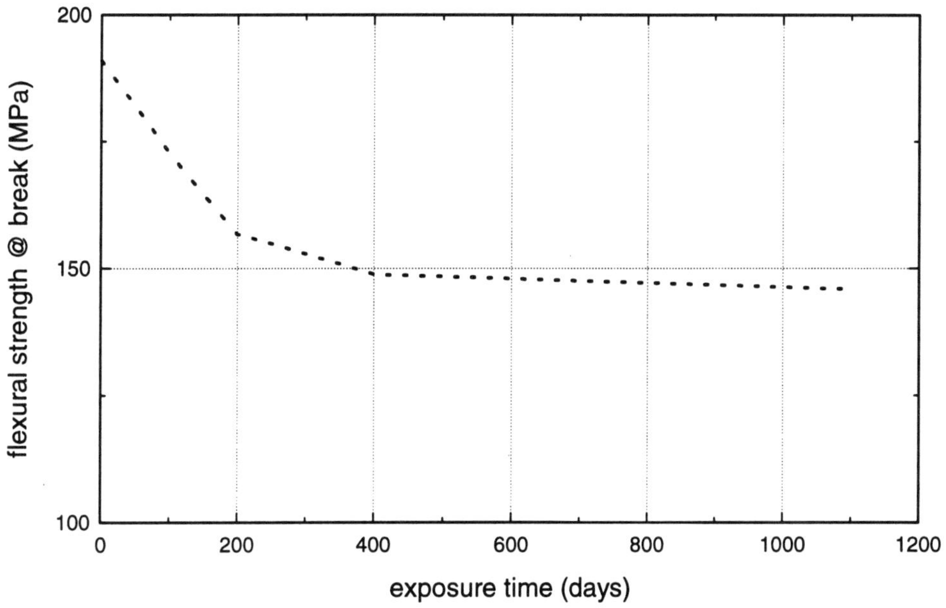

....	Polyester - PBT (30% glass fiber); natural aging
Reference No.	135

GRAPH 137: **Hiratsuka, Japan Outdoor Exposure Time vs. Flexural Modulus of Polybutylene Terephthalate**

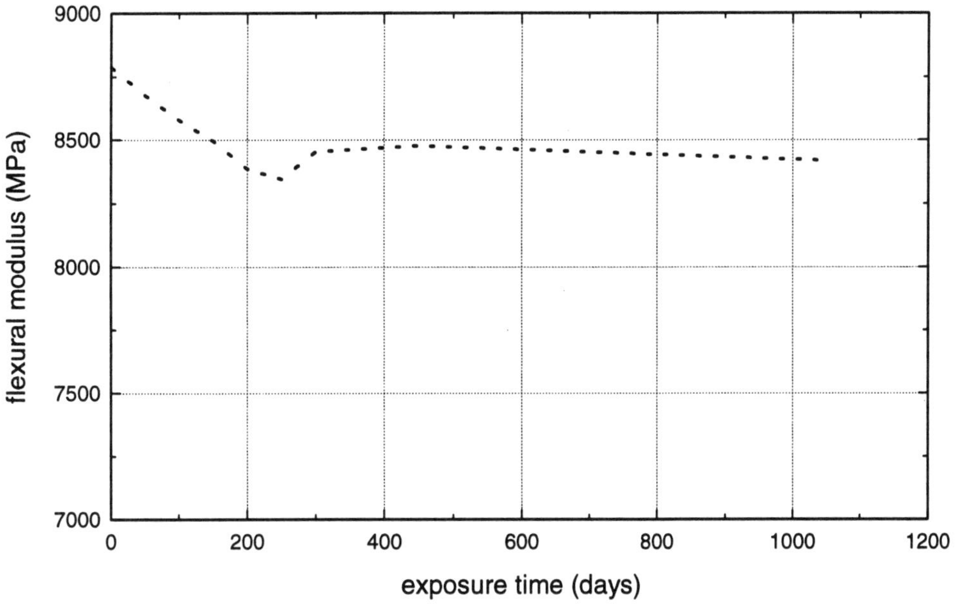

....	Polyester - PBT (30% glass fiber); natural aging
Reference No.	135

GRAPH 138: **Hiratsuka, Japan Outdoor Exposure Time vs. Notched Izod Impact Strength of Polybutylene Terephthalate**

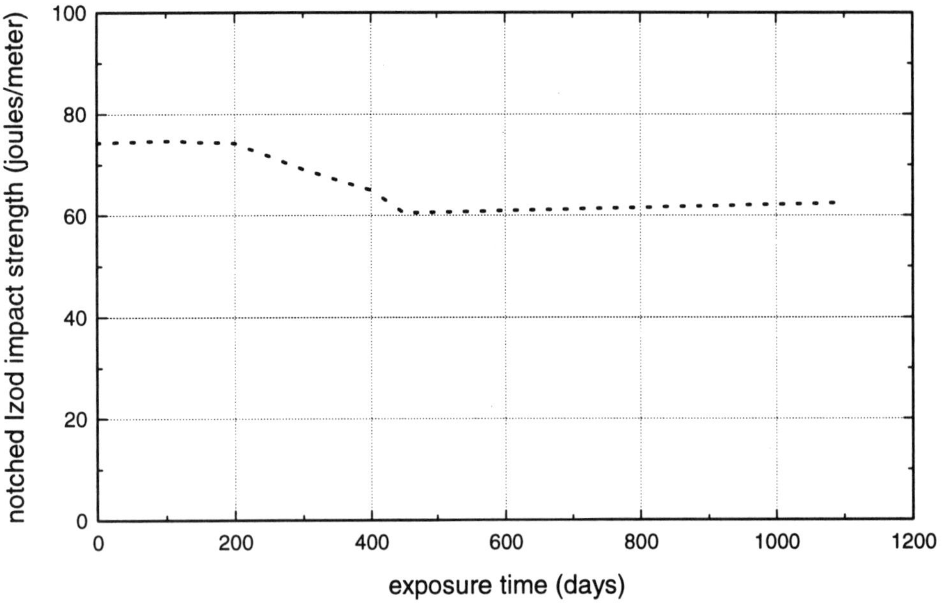

....	Polyester - PBT (30% glass fiber); natural aging
Reference No.	135

GRAPH 139: **Hiratsuka, Japan Outdoor Exposure Time vs. Percent Weight Change of Polybutylene Terephthalate**

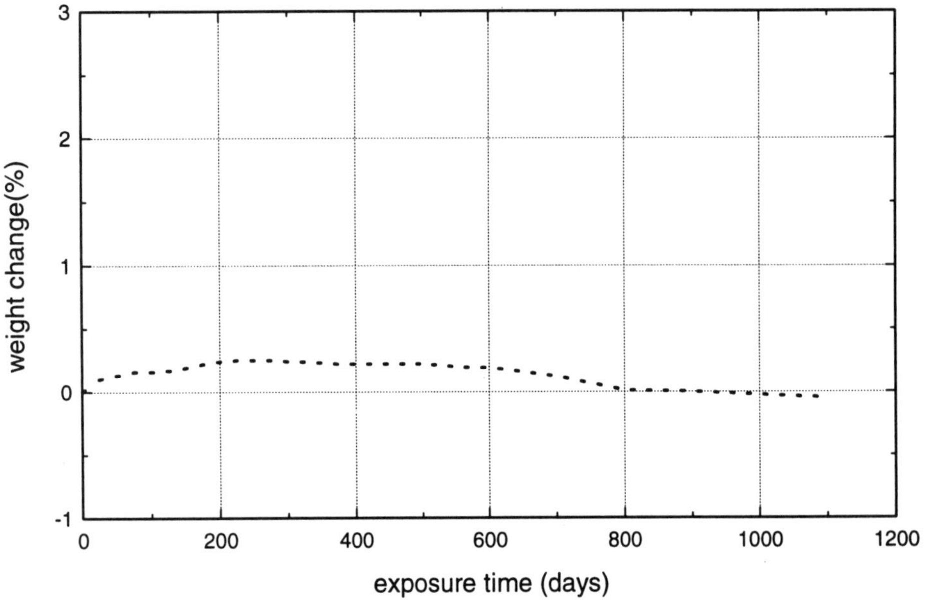

....	Polyester - PBT (30% glass fiber); specimen thickness 3.2mm(1/8") under tensile stress - ASTM D638
Reference No.	135

GRAPH 140: Sunshine Carbon Arc Weatherometer Exposure Time vs. Impact Strength of Polybutylene Terephthalate

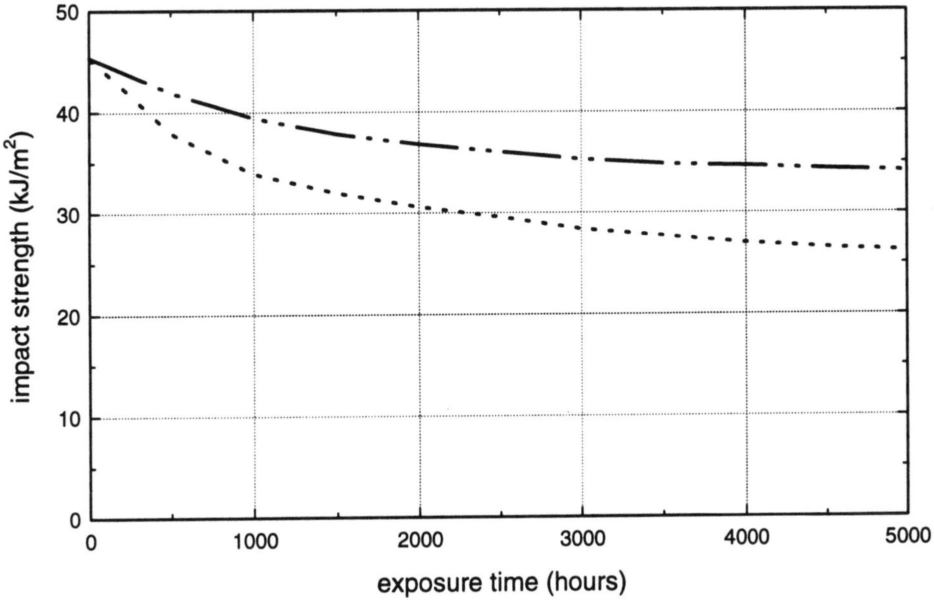

···············	Bayer Pocan B3235 Polyester - PBT (natural resin; 30% glass fiber)
—··—··	Bayer Pocan B3235 Polyester - PBT (black; 30% glass fiber)
Reference No.	227

GRAPH 141: Sunshine Weatherometer Exposure Time vs. Elongation of Polybutylene Terephthalate

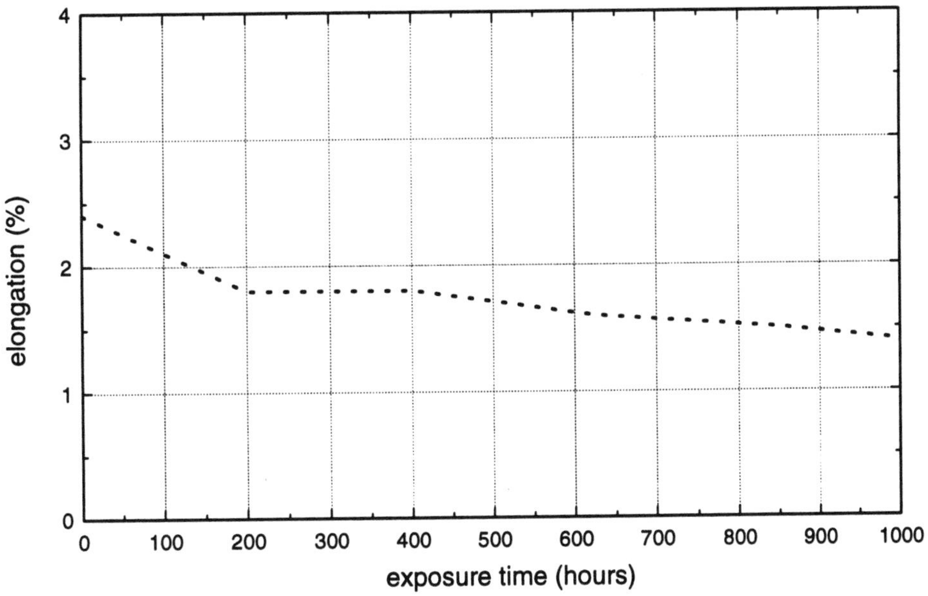

···············	Polyester - PBT (30% glass fiber); 45°C, 12 min water spray/ 60 min
Reference No.	236

GRAPH 142: Sunshine Weatherometer Exposure Time vs. Tensile Strength of Polybutylene Terephthalate

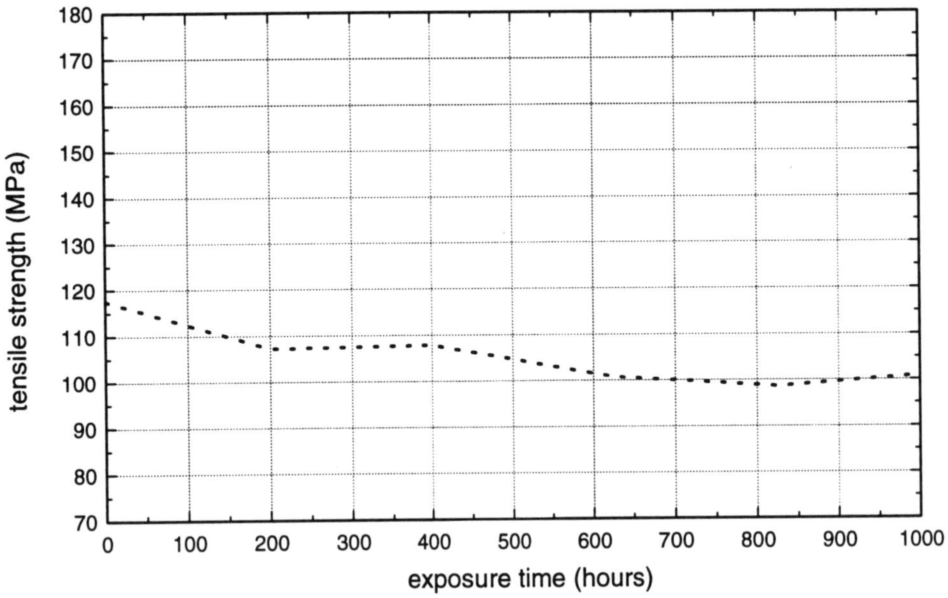

....	Polyester - PBT (30% glass fiber); 45°C,12 min water spray/ 60 min
Reference No.	236

GRAPH 143: Weatherometer Exposure Time vs. Tensile Strength Retained of Polybutylene Terephthalate

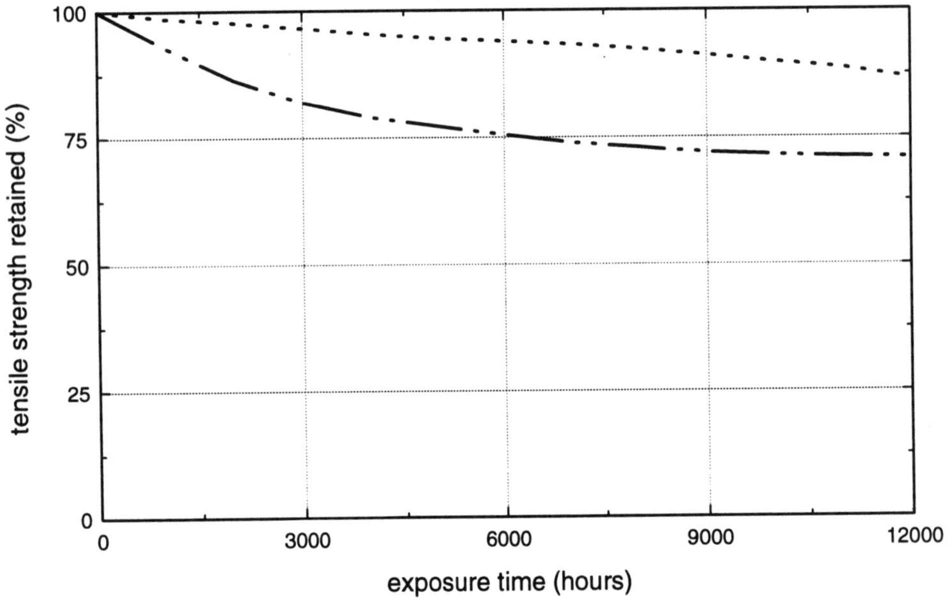

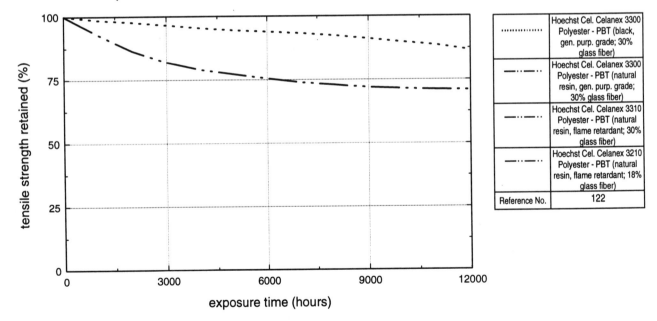

....	Hoechst Cel. Celanex 3300 Polyester - PBT (black, gen. purp. grade; 30% glass fiber)
—·—··	Hoechst Cel. Celanex 3300 Polyester - PBT (natural resin, gen. purp. grade; 30% glass fiber)
—··—··	Hoechst Cel. Celanex 3310 Polyester - PBT (natural resin, flame retardant; 30% glass fiber)
—··—··	Hoechst Cel. Celanex 3210 Polyester - PBT (natural resin, flame retardant; 18% glass fiber)
Reference No.	122

Chapter 24

Polyethylene Terephthalate

Outdoor Weather Resistance

DuPont: Rynite 530 BK503 (features: black color; material compostion: 30% glass fiber reinforcement); **Rynite 530 NC10** (features: natural resin; material compostion: 30% glass fiber reinforcement); **Rynite 545 BK504** (features: black color; material compostion: 45% glass fiber reinforcement); **Rynite 545 NC10** (features: natural resin; material compostion: 45% glass fiber reinforcement); **Rynite 935 BK505** (features: low warp grade, black color; material compostion: 35% mica/ glass fiber reinforcement)

Rynite 530 NC10 and BK503 and Rynite 545 NC10 and BK504 resins have been exposed outdoors in Florida and Arizona facing 45° South for three years. The data on these exposed samples indicate that the resins have retained over 72% of their initial tensile strength and over 50% of their initial elongation. As expected, the compositions containing carbon black have a higher property retention. After three years, all the test samples were slightly "etched."

After three years of exposure in Arizona, Rynite 935 BK505 retains over 97% of original tensile strength and 76% of original elongation.

Reference: *Rynite Design Handbook For Du Pont Engineering Plastics,* supplier design guide (E-62620) - Du Pont Company, 1987.

Accelerated Outdoor Weathering Resistance

DuPont: Rynite 530 BK503 (features: black color; material compostion: 30% glass fiber reinforcement); **Rynite 530 NC10** (features: natural resin; material compostion: 30% glass fiber reinforcement); **Rynite 545 BK504** (features: black color; material compostion: 45% glass fiber reinforcement); **Rynite 545 NC10** (features: natural resin; material compostion: 45% glass fiber reinforcement)

After 500,000 Langleys of exposure in the equitorial mount with mirrors (EMMA) and EMMAQUA (EMMA with water spray) environments, Rynite 530 NC10 and BK503 and Rynite 545 NC10 and BK504 resins retain over 90% of their original tensile strength and 73% of their original elongation properties. The EMMA and EMMAQUA environments have similar effects on the properties of the Rynite 530 and Rynite 545 resins. All test specimens had reduced gloss levels after exposure. On the average, samples exposed in Arizona received approximately 150,000 Langleys of sunlight per year. These tests correspond to about 3.3 years of natural weathering in Arizona.

Reference: *Rynite Design Handbook For Du Pont Engineering Plastics,* supplier design guide (E-62620) - Du Pont Company, 1987.

TABLE 46: Outdoor Weathering in Arizona of DuPont Rynite Polyethylene Terephthalate.

Material Family	Polyethylene Terephthalate											
Material Supplier/ Grade	DuPont Rynite 545 NC10				DuPont Rynite 545 BK504				DuPont Rynite 935 BK505			
Features	natural resin				black color				black color, low warp grade			
Reference Number	200	200	200	200	200	200	200	200	200	200	200	200
MATERIAL CHARACTERISTICS												
glass fiber reinforcement	45%	45%	45%	45%	45%	45%	45%	45%				
mica/ glass fiber reinforcement									35%	35%	35%	35%
EXPOSURE CONDITIONS												
Exposure Type	outdoor weathering				outdoor weathering				outdoor weathering			
Exposure Location	Arizona				Arizona				Arizona			
Exposure Country	USA				USA				USA			
Exposure Note	45° south				45° south				45° south			
Exposure Time (days)	182	365	730	1095	182	365	730	1095	182	365	730	1095
PROPERTIES RETAINED (%)												
Tensile Strength	98	88	87	82	94	97	94	90	100	100	100	97.5
Elongation	82	77	73	68	83	94	89	78	100	94	94	76

TABLE 47: Outdoor Weathering in Arizona of DuPont Rynite Polyethylene Terephthalate.

Material Family	Polyethylene Terephthalate							
Material Supplier/ Grade	DuPont Rynite 530 NC10				DuPont Rynite 530 BK503			
Features	natural resin				black color			
Reference Number	200	200	200	200	200	200	200	200
MATERIAL COMPOSITION								
glass fiber reinforcement	30%	30%	30%	30%	30%	30%	30%	30%
EXPOSURE CONDITIONS								
Exposure Type	outdoor weathering				outdoor weathering			
Exposure Location	Arizona				Arizona			
Exposure Country	USA				USA			
Exposure Note	45° south				45° south			
Exposure Time (days)	182	365	730	1095	182	365	730	1095
PROPERTIES RETAINED (%)								
Tensile Strength	100	98	90	87	98	100	98	98
Elongation	85	88	77	73	91	96	96	83

TABLE 48: Outdoor Weathering in Florida of DuPont Rynite Polyethylene Terephthalate.

Material Family	Polyethylene Terephthalate							
Material Supplier/ Grade	DuPont Rynite 530 NC10				DuPont Rynite 530 BK503			
Features	natural resin				black color			
Reference Number	200	200	200	200	200	200	200	200
MATERIAL COMPOSITION								
glass fiber reinforcement	30%	30%	30%	30%	30%	30%	30%	30%
EXPOSURE CONDITIONS								
Exposure Type	outdoor weathering				outdoor weathering			
Exposure Location	Florida				Florida			
Exposure Country	USA				USA			
Exposure Note	45° south				45° south			
Exposure Time (days)	182	365	730	1095	182	365	730	1095
PROPERTIES RETAINED (%)								
Tensile Strength	98	92	82	76	100	100	93	98
Elongation	85	77	69	58	87	91	91	87

TABLE 49: Outdoor Weathering in Florida of DuPont Rynite Polyethylene Terephthalate.

Material Family	Polyethylene Terephthalate							
Material Supplier/ Grade	DuPont Rynite 545 NC10				DuPont Rynite 545 BK504			
Features	natural resin				black color			
Reference Number	200	200	200	200	200	200	200	200
MATERIAL COMPOSITION								
glass fiber reinforcement	45%	45%	45%	45%	45%	45%	45%	45%
EXPOSURE CONDITIONS								
Exposure Type	outdoor weathering				outdoor weathering			
Exposure Location	Florida				Florida			
Exposure Country	USA				USA			
Exposure Note	45° south				45° south			
Exposure Time (days)	182	365	730	1095	182	365	730	1095
PROPERTIES RETAINED (%)								
Tensile Strength	89	84	75	72	88	90	91	91
Elongation	77	68	73	50	67	78	89	78

TABLE 50: Accelerated Outdoor Weathering in Arizona by EMMA and EMMAQUA of DuPont Rynite Polyethylene Terephthalate.

Material Family	Polyethylene Terephthalate							
Material Supplier/ Grade	DuPont Rynite 530 NC10	DuPont Rynite 530 BK503	DuPont Rynite 545 NC10	DuPont Rynite 545 BK504	DuPont Rynite 530 NC10	DuPont Rynite 530 BK503	DuPont Rynite 545 NC10	DuPont Rynite 545 BK504
Features	natural resin	black color	natural resin	black color	natural resin	black color	natural resin	black color
Reference Number	200	200	200	200	200	200	200	200
MATERIAL COMPOSITION								
glass fiber reinforcement	30%	30%	45%	45%	30%	30%	45%	45%
EXPOSURE CONDITIONS								
Exposure Type	accelerated outdoor weathering				accelerated outdoor weathering			
Exposure Location	Arizona				Arizona			
Exposure Country	USA				USA			
Exposure Apparatus	EMMA				EMMAQUA			
Total Radiation (Langleys)	500,000				500,000			
Exposure Note	150,000 Langleys is approximately equal to one year							
PROPERTIES RETAINED (%)								
Tensile Strength	100	100	92	93	100	100	92	93
Elongation	85	87	73	89	81	87	73	94

GRAPH 144: Sunshine Weatherometer Exposure Time vs. Elongation of Polyethylene Terephthalate

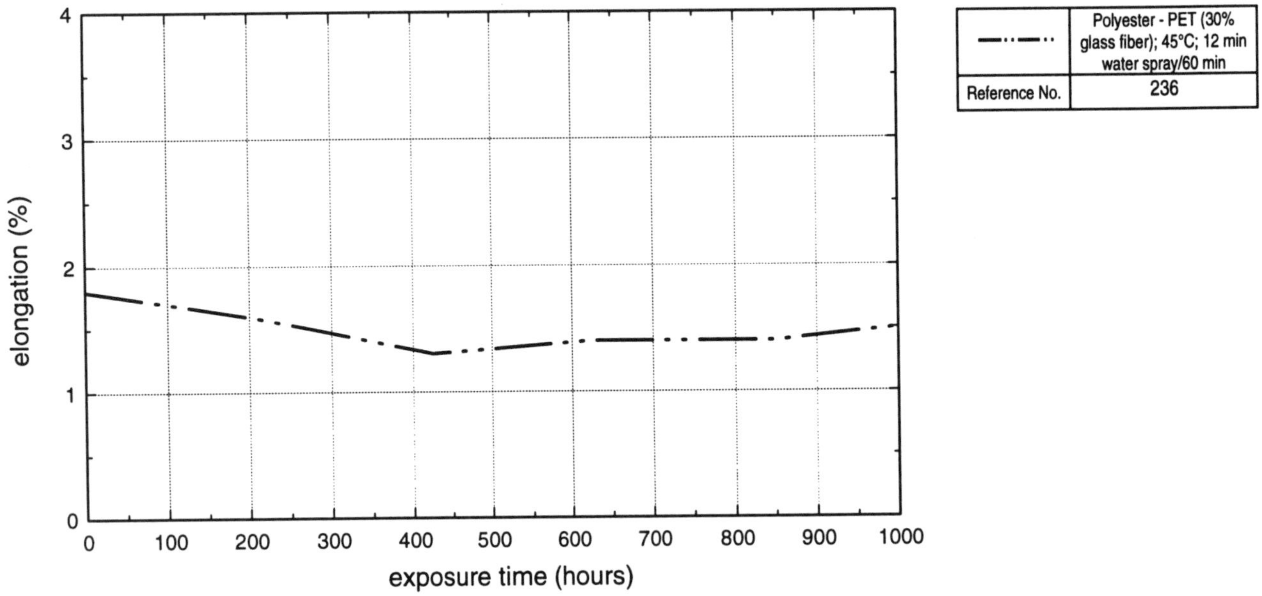

GRAPH 145: Sunshine Weatherometer Exposure Time vs. Tensile Strength of Polyethylene Terephthalate

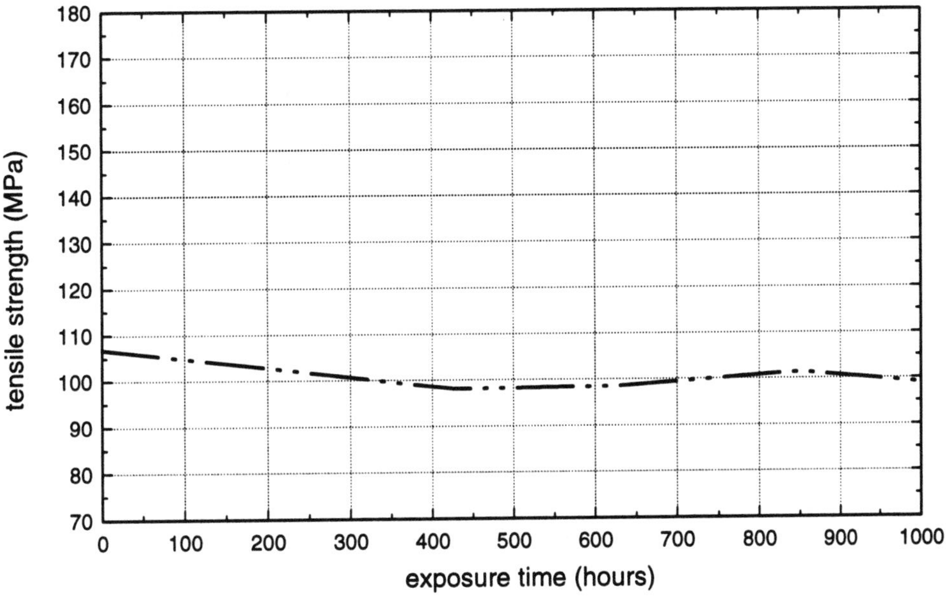

	Polyester - PET (30% glass fiber); 45°C; 12 min water spray/60 min
Reference No.	236

Chapter 25

Glycol Modified Polycyclohexylenedimethylene Terephthalate

Outdoor Weather Resistance

Eastman: Kodar PCTG 5445

Kodar PCTG copolyester 5445 is not suggested for use in applications requiring outdoor exposure.

Reference: *Kodar PCTG Copolyester 5445,* supplier technical report (MB-94/August 1985) - Eastman Plastics, 1988.

Chapter 26

Polycyclohexylenedimethylene Ethylene Terephthalate

Outdoor Weather Resistance

Eastman: Kodar PETG 6763

Kodar PETG copolyester 6763 does not contain a U.V. inhibitor, and it is not suggested for use in applications involving continuous long term exposure.

Reference: *Kodar PETG Copolyester 6763,* supplier technical report (MB-80F/June 1988) - Eastman Plastics, 1988.

Chapter 27

Liquid Crystal Polyester

Outdoor Weather Resistance

Hoechst AG: Vectra (chemical type: wholly aromatic copolyester)

Vectra, like other plastics, changes in the course of time on exposure to weathering. The main reason for this is UV radiation, which causes a white deposit of degraded material to form on the surface (chalking), with consequent reduction in gloss, color change and deterioration in mechanical properties.

After artificial weathering for 2000 hours, moldings made from Vectra still retained more than 90% of their initial mechanical property values. After one year's outdoor weathering a slight white deposit was detected.

Reference: *Vectra Polymer Materials,* supplier design guide (B 121 BR E 9102/014) - Hoechst AG, 1991.

TABLE 51: Accelerated Weathering with a Xenon Arc Lamp of Hoechst AG Vectra Liquid Crystal Polymer.

Material Family	Liquid Crystal Polyester			
Material Supplier/ Grade	Hoechst AG Vectra A950	Hoechst AG Vectra A130	Hoechst AG Vectra B950	Hoechst AG Vectra A540
Reference Number	70	70	70	70
MATERIAL COMPOSITION				
glass fiber reinforcement		30%		
mineral filler				40%
EXPOSURE CONDITIONS				
Exposure Type	accelerated weathering			
Exposure Test Method	ASTM D2565			
Exposure Apparatus	xenon arc lamp			
Exposure Note	water spray for 18 minutes every 202 minutes			
Exposure Temperature (°C)	125	125	125	125
Exposure Time (days)	83.3	83.3	83.3	83.3
PROPERTIES RETAINED (%)				
Tensile Modulus	90	100	93	95
Flexural Modulus	100	100	95	100
Tensile Strength	97	97	100	100
Flexural Strength	100	100	100	100
Notched Izod Impact Strength	91	100	100	100
Heat Deflection Temperature	92	94	99	93

Chapter 28

Polyarylate

Outdoor Weather Resistance

Amoco Performance Products: Ardel D-100 (features: transparent)

Ardel polyarylate offers outstanding weatherability. It provides the best known UV protection as a coating, laminate or coextruded film. Surpassing PET in UV resistance, it has even been added to that polymer as a UV stabilizer.

Reference: *Ardel Polyarylate - The Tough Weatherable Thermoplastic,* supplier marketing literature (F-47141C) - Union Carbide Corporation.

Accelerated Artificial Weathering Resistance

Amoco Performance Products: Ardel D-100 (features: transparent)

In accelerated weathering tests (XW Weatherometer, Fadeometer and Xenon lamp), polyarylate shows better retention of gloss, much lower haze and better retention of light transmission than a number of polycarbonate samples. Ardel Polyarylate proved to be as tough as polycarbonate after 8,000 hours exposure. While the polycarbonate lost over 90 percent of its original notched Izod toughness after 2,000 hours, polyarylate samples showed only a slight loss after 8,000 hours. There were only minor changes in tensile modulus and yield strength, with minor decreases in yield elongation and significant decreases in break elongation.

Reference: *Ardel Polyarylate - The Tough Weatherable Thermoplastic,* supplier marketing literature (F-47141C) - Union Carbide Corporation.

TABLE 52: Accelerated Weathering in an XW Weatherometer of DuPont Ardel Polyarylate.

Material Family	Polyarylate			
Material Supplier/ Grade	Amoco Performance Products Ardel D-100			
Features	transparent			
Reference Number	145	145	145	145
EXPOSURE CONDITIONS				
Exposure Type	accelerated weathering			
Exposure Apparatus	XW weatherometer			
Exposure Time (days)	41.7	83.3	166.7	333.3
PROPERTIES RETAINED (%)				
Notched Izod Impact Strength	97	92	79	66
Tensile Impact Strength		57	65	60
Heat Deflection Temperature	94.8	96.6	95.4	92

GRAPH 146: Carbon Arc XW Weatherometer Exposure Time vs. Haze of Polyarylate

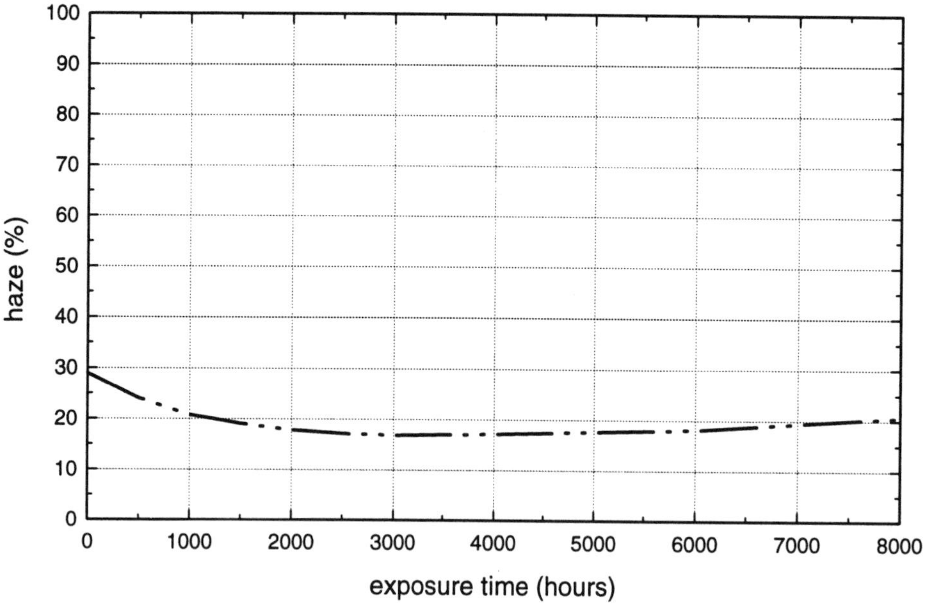

—··—··	Amoco Ardel D-100 Polyarylate (transparent)
Reference No.	145

Chapter 29

Polyimide

Outdoor Weather Resistance

DuPont: Kapton (product form: film)

In the earth's atmosphere, there is a synergistic effect upon Kapton when directly exposed to some combinations of ultraviolet radiation, oxygen, and water. This effect is shown as a loss of elongation when Kapton is exposed in Florida test panels. Kapton also shows a loss of elongation as a function of exposure time in an Atlas weatherometer. Design considerations should recognize this phenomenon.

Reference: *Du Pont Faxed Correspondence,* supplier technical report (P10-2123) - Du Pont Company, 1994.

Ube: Upimol R (features: general purpose grade); **Upimol S** (features: high heat grade)

Upimol is stable when exposed to sunshine.

Reference: *Upimol Polyimide Shape,* supplier technical report - Ube Industries.

Indoor UV Light Resistance

Ube: Upimol R (features: general purpose grade); **Upimol S** (features: high heat grade)

Upimol is stable when exposed to UV light.

Reference: *Upimol Polyimide Shape,* supplier technical report - Ube Industries.

GRAPH 147: Florida Outdoor Weathering Exposure Time vs. Ultimate Elongation of Polyimide

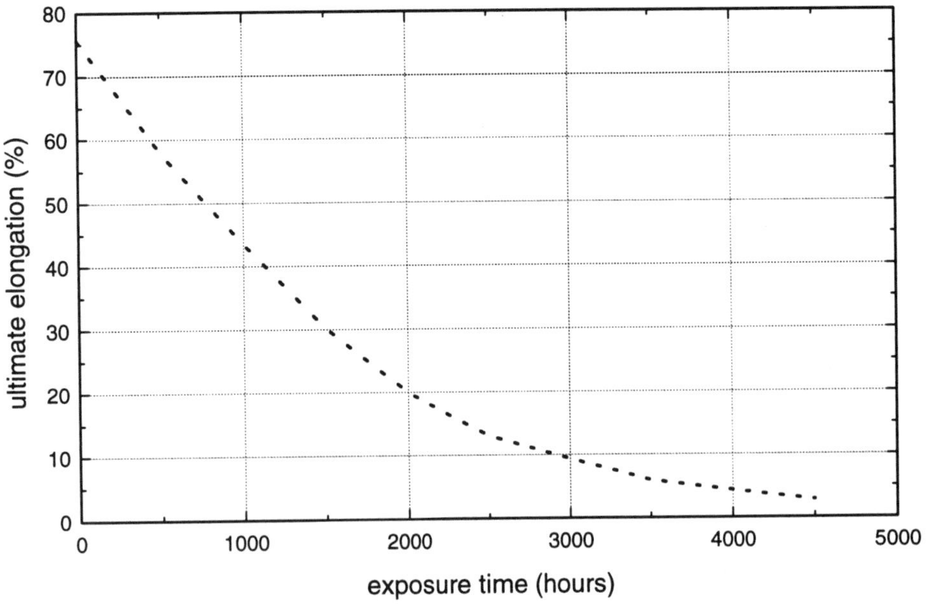

....	DuPont Kapton Polyimide (film)
Reference No.	187

GRAPH 148: Atlas Weatherometer Exposure Time vs. Ultimate Elongation of Polyimide

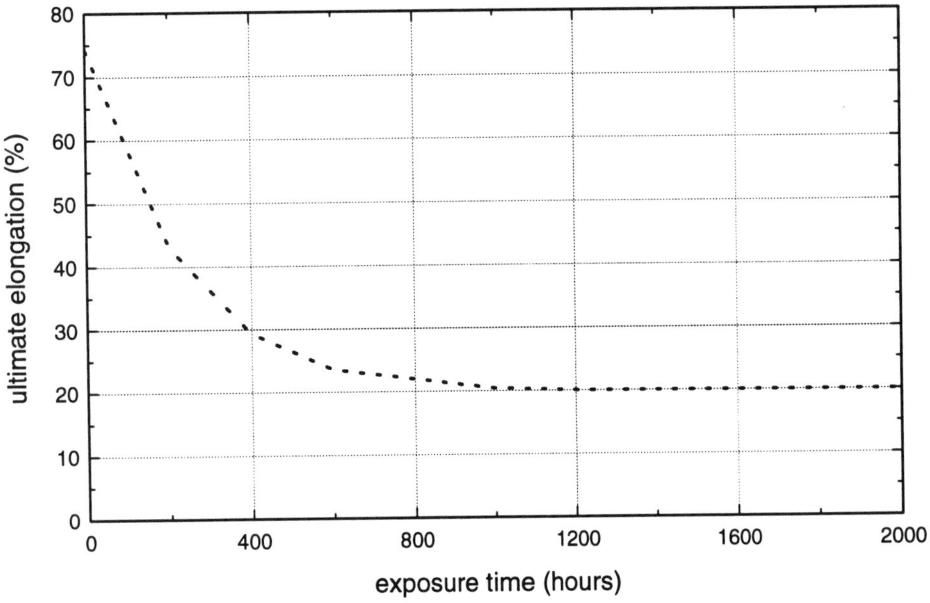

....	DuPont Kapton Polyimide (film)
Reference No.	187

GRAPH 149: Sunshine Weatherometer Exposure Time vs. Elongation Retained of Polyimide

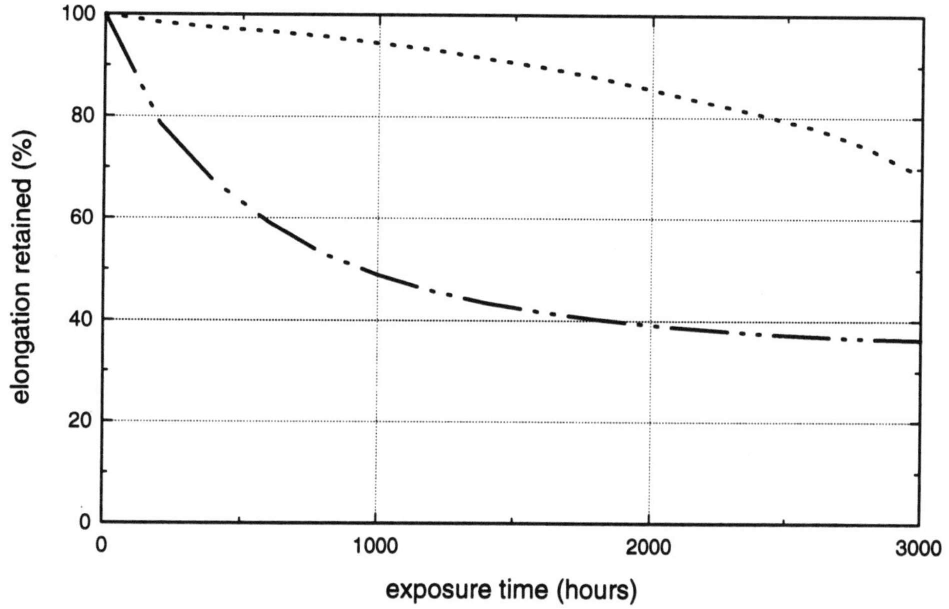

···············	Ube Upilex R Polyimide (gen. purp. grade)
—··—··	Ube Upilex S Polyimide (high heat grade)
Reference No.	97

GRAPH 150: Sunshine Weatherometer Exposure Time vs. Flexural Strength Retained of Polyimide

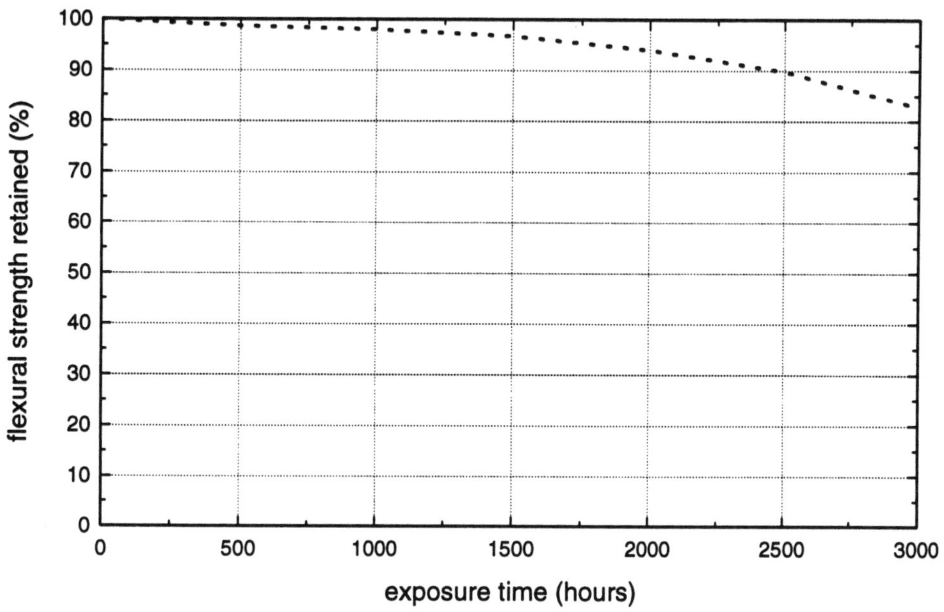

···············	Ube Upimol R Polyimide (gen. purp. grade); temp. 63±3 °C; spray 12min/1 cycle (60 min.)
Reference No.	123

GRAPH 151: Sunshine Weatherometer Exposure Time vs. Tensile Strength Retained of Polyimide

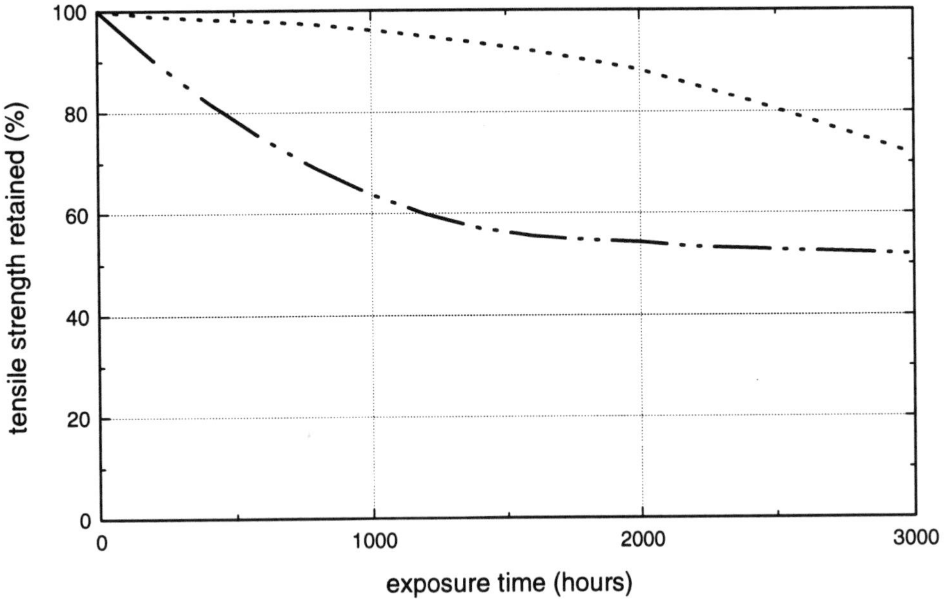

··············	Ube Upilex R Polyimide (gen. purp. grade)
—·—··	Ube Upilex S Polyimide (high heat grade)
Reference No.	97

GRAPH 152: UV-CON Exposure Time vs. Flexural Strength Retained of Polyimide

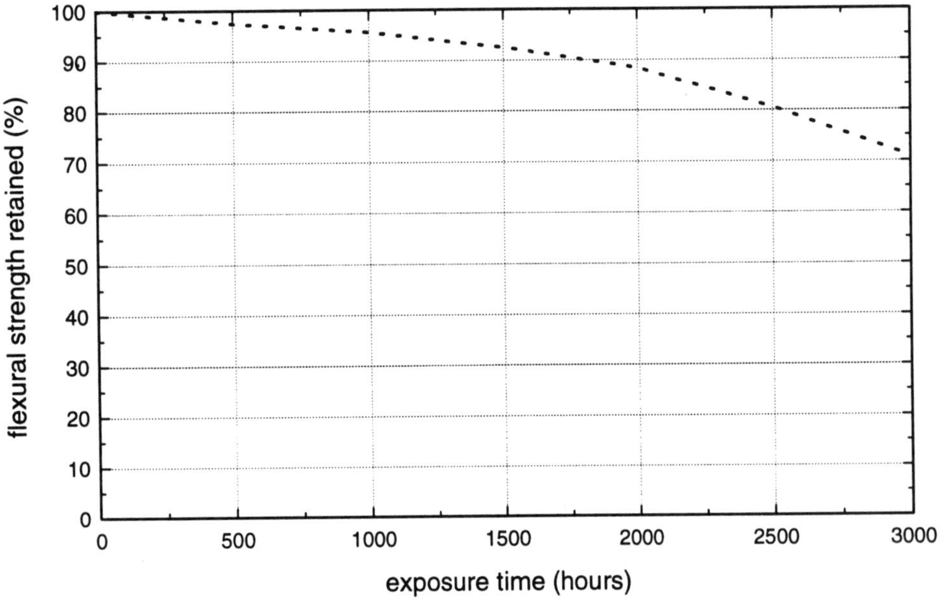

··············	Ube Upimol R Polyimide (gen. purp. grade); temp. 63±3 °C; spray 12min/1 cycle (60 min.)
Reference No.	123

Chapter 30

Polyamideimide

Outdoor Weather Resistance

Amoco Performance Products: Torlon 4203L

Torlon 4301 (features: wear resistant; material compostion: 20% graphite powder, 3% fluorocarbon)

Torlon molding polymers are exceptionally resistant to degradation by ultraviolet light. The bearing grades, such as 4301, contain graphite powder which renders the material black and screens UV radiation. These grades are even more resistant to degradation from outdoor exposure.

Reference: *Torlon Engineering Polymers / Design Manual,* supplier design guide (F-49893) - Amoco Performance Products.

Accelerated Artificial Weathering Resistance

Amoco Performance Products: Torlon 4203L

Torlon 4203L did not degrade after 6,000 hours of weatherometer exposure which is roughly equivalent to five years of outdoor exposure.

Reference: *Torlon Engineering Polymers / Design Manual,* supplier design guide (F-49893) - Amoco Performance Products.

GRAPH 153: **Weatherometer Exposure Time vs. Elongation of Polyamideimide**

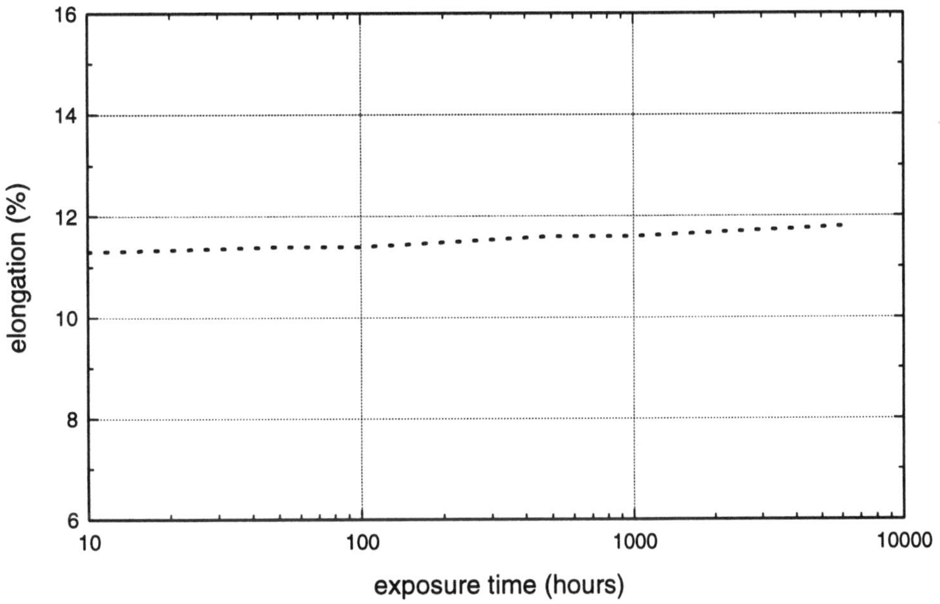

....	Amoco Torlon 4203L PAI
Reference No.	20

GRAPH 154: **Weatherometer Exposure Time vs. Tensile Strength of Polyamideimide**

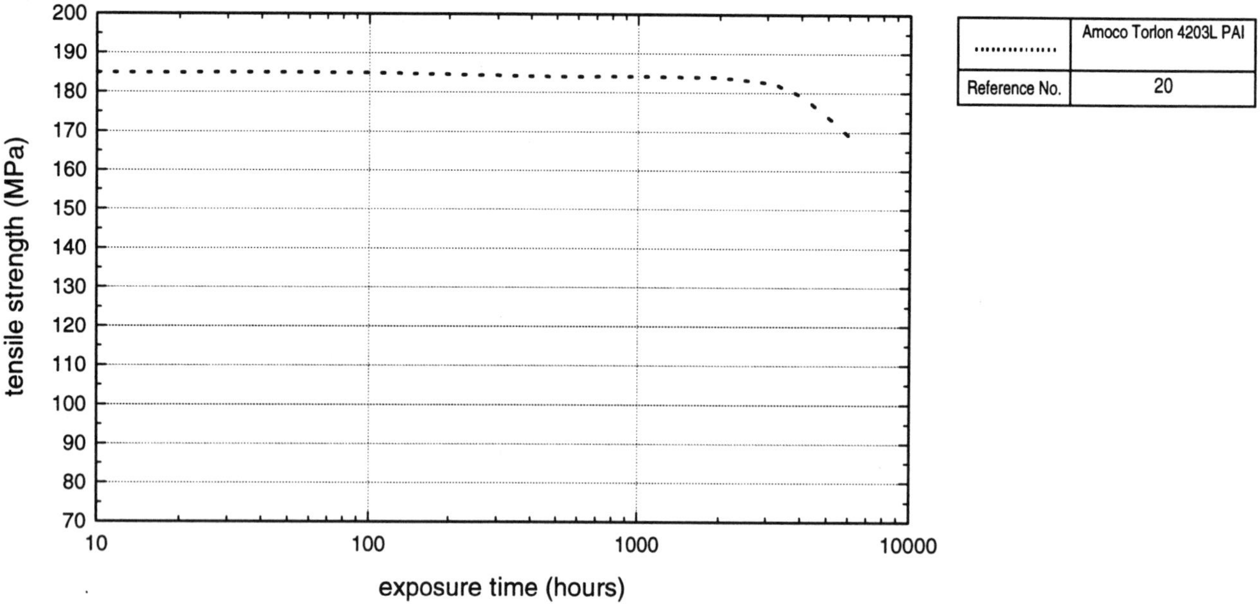

Chapter 31

Polyetherimide

UV Resistance

GE Plastics: Ultem 1000 (features: natural resin, transparent, amber tint)

Ultem resin is inherently resistant to UV radiation without the addition of stabilizers. Exposure to 1,000 hours of xenon arc weatherometer irradiation (0.35 W/m^2 irradiance @ 340 nm, 63 °C) produces a negligible change in the tensile strength of the resin.

Reference: *Ultem Design Guide,* supplier design guide (ULT-201G (6/90) RTB) - General Electric Company, 1990.

GRAPH 155: Xenon Arc Weatherometer Exposure Time vs. Tensile Strength of Polyetherimide

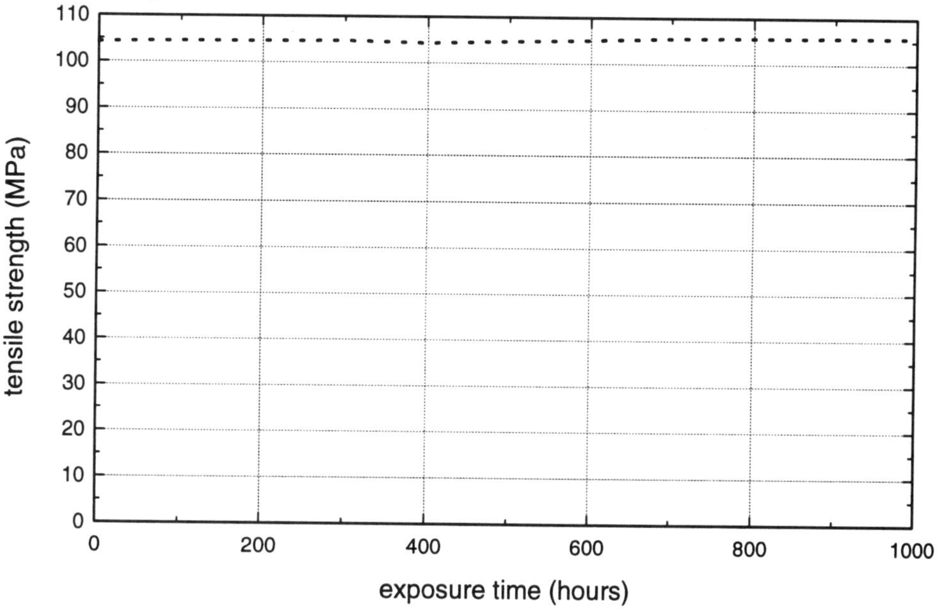

....	GE Ultem 1000 PEI (natural resin, transparent, amber tint); 0.35 W/m² irradience at 340 nm, 63 °C
Reference No.	51

Chapter 32

Polyetheretherketone

Outdoor Weather Resistance

Victrex USA: Victrex PEEK

Like most linear polyaromatics "Victrex" PEEK suffers from the effects of UV degradation during outdoor weathering. However, testing in the UK has shown that this effect is minimal over a twelve month period for both natural and pigmented moldings. In more extreme weathering conditions, painting or pigmenting will protect the polymer from excessive property degradation.

Reference: *Victrex PEEK,* supplier design guide (VK2/0586) - ICI Advanced Materials, 1986.

TABLE 53: Outdoor Weathering in the United Kingdom of ICI Victrex Polyetheretherketone.

Material Family	Polyetheretherketone											
Material Supplier/ Grade	Victrex USA Victrex PEEK 450G				Victrex USA Victrex PEEK 450G				Victrex USA Victrex PEEK 450G			
Features	natural resin				black color				white color			
Reference Number	77	77	77	77	77	77	77	77	77	77	77	77
MATERIAL COMPOSITION												
black pigment					1-2%	1-2%	1-2%	1-2%				
white pigment									1-2%	1-2%	1-2%	1-2%
EXPOSURE CONDITIONS												
Exposure Type	outdoor weathering											
Exposure Country	United Kingdom											
Exposure Time (days)	91	182	273	365	91	182	273	365	91	182	273	365
PROPERTIES RETAINED (%)												
Tensile Strength	100.7	100.7	96.5	96.5	100.7	98.6	100	100	100	100.7	98.6	100.7

TABLE 54: Effect of Pigments on Outdoor Weathering in the United Kingdom of ICI Victrex USA Victrex Polyetheretherketone.

Material Family	Polyetheretherketone											
Material Supplier/ Grade	Victrex USA Victrex PEEK 450G				Victrex USA Victrex PEEK 450G				Victrex USA Victrex PEEK 450G			
Features	yellow color				green color				blue color			
Reference Number	77	77	77	77	77	77	77	77	77	77	77	77
MATERIAL COMPOSITION												
blue pigment									1-2%	1-2%	1-2%	1-2%
green pigment					1-2%	1-2%	1-2%	1-2%				
yellow pigment	1-2%	1-2%	1-2%	1-2%								
EXPOSURE CONDITIONS												
Exposure Type	outdoor weathering											
Exposure Country	United Kingdom											
Exposure Time (days)	91	182	273	365	91	182	273	365	91	182	273	365
PROPERTIES RETAINED (%)												
Tensile Strength	100	100	97.9	96.5	98.6	98.6	95.8	96.5	100.7	100	97.9	98.6

Chapter 33

Polyaryletherketone

Outdoor Weather Resistance

BASF AG: Ultrapek

In common with most polymers composed of aromatic building blocks, uncolored Ultrapek is affected by radiation at the ultraviolet end of the solar spectrum. However, a comparatively long exposure period elapses before the effect, i.e. slight yellowing, becomes visible and no adverse effects are exerted on the mechanical properties of moldings.

Reference: *Ultrapek Product Line, Properties, Processing,* supplier design guide (B 607 e/10.92) - BASF Aktiengesellschaft, 1992.

Chapter 34

Polyethylene

Outdoor Weather Resistance

BASF AG: Lupolen

If polyethylene is exposed for long periods outdoors, it is degraded by the concerted action of atmospheric oxygen and radiation at the UV end of the solar spectrum. The degradation leads to a deterioration in impact resistance and ultimate elongation, and possibly, to discoloration.

The life of moldings exposed outdoors depends not only on the properties of the material, but also to a large extent on the thickness of the moldings and the processing conditions.

The outdoor stability can be increased by a factor of two to ten by finishing the polyethylene with special light stabilizers. Good results are obtained by the HALS type (hindered amine light stabilizers), in some cases in combinations with benzotriazone compounds.

Best outdoor performance is achieved by adding special grades of carbon black. Mass fractions of 2-3% of carbon black uniformly distributed in the matrix improves the outdoor performance by a factor higher than 15.

White and chromatic pigments may improve the light stability of polyethylene but can equally well impair it.

Reference: *Lupolen Polyethylene And Novolen Polypropylene Product Line, Properties, Processing,* supplier design guide (B 579 e / 4.92) - BASF Aktiengesellschaft, 1992.

TABLE 55: Effect of UV Stabilizers and UV Absorbers on Outdoor Weathering of Polyethylene Greenhouse Film.

Material Family	Polyethylene											
Product Form	greenhouse film											
Reference Number	209	209	209	209	209	209	209	209	209	209	209	209
MATERIAL CHARACTERISTICS												
Sample Thickness (mm)	0.102	0.152	0.204	0.102	0.152	0.204	0.102	0.152	0.204	0.102	0.152	0.204
MATERIAL COMPOSITION												
Cyanox 2777 (antioxidant - American Cyanamid)	0.025-0.07%	0.025-0.07%	0.025-0.07%	0.025-0.07%	0.025-0.07%	0.025-0.07%	0.025-0.07%	0.025-0.07%	0.025-0.07%	0.025-0.07%	0.025-0.07%	0.025-0.07%
Cyasorb UV 531 (UV absorber - American Cyanamid)	0.18%	0.15%	0.13%	0.20%	0.18%	0.15%	0.35%	0.30%	0.25%		0.35%	0.30%
Cyasorb UV-3346 (UV stabilizer - American Cyanamid)	0.35%	0.30%	0.25%	0.40%	0.35%	0.30%	0.70%	0.60%	0.50%		0.70%	0.60%
EXPOSURE CONDITIONS												
Exposure Type	outdoor weathering											
Exposure Note	exposure time represents desired service life for greenhouse film produced from the polyethylene / additive combination at the given thickness											
Exposure Time (days)	182	182	182	365	365	365	730	730	730	1095	1095	1095
RESULTS OF EXPOSURE												
Service Life (days)	182	182	182	365	365	365	730	730	730	not recommended	1095	1095

TABLE 56: Effect of UV Stabilizer Amount on Outdoor Weathering of Polyethylene Greenhouse Film.

Material Family	Polyethylene								
Product Form	greenhouse film								
Reference Number	209	209	209	209	209	209	209	209	209
MATERIAL CHARACTERISTICS									
Sample Thickness (mm)	0.127	0.127	0.127	0.127	0.064	0.064	0.064	0.032	0.032
MATERIAL COMPOSITION									
Cyasorb UV-3346 (UV stabilizer - American Cyanamid)	0.13%	0.25%	0.50%	1%	0.25%	0.50%	1%	0.50%	1%
EXPOSURE CONDITIONS									
Exposure Type	outdoor weathering								
Exposure Note	exposure time represents desired service life for greenhouse film produced from the polyethylene / additive combination at the given thickness								
Exposure Time (days)	182	365	730	1095	182	365	730	182	365
RESULTS OF EXPOSURE									
Service Life (days)	182	365	730	1095	182	365	730	182	365

TABLE 57: Outdoor Weathering in California and Pennsylvania of Glass Reinforced Polyethyelene.

Material Family	Polyethylene							
Material Supplier	LNP Engineering Plastics							
Reference Number	215	215	215	215	215	215	215	215
MATERIAL COMPOSITION								
carbon black	1%	1%	1%	1%	1%	1%	1%	1%
glass fiber reinforcement	30%	30%	30%	30%	30%	30%	30%	30%
EXPOSURE CONDITIONS								
Exposure Type	outdoor weathering				outdoor weathering			
Exposure Location	Los Angeles, California	Los Angeles, California	Los Angeles, California	Los Angeles, California	Philadelphia, Pennsylvania	Philadelphia, Pennsylvania	Philadelphia, Pennsylvania	Philadelphia, Pennsylvania
Exposure Country	USA	USA	USA	USA	USA	USA	USA	USA
Exposure Test Method	ASTM D1435	ASTM D1435	ASTM D1435	ASTM D1435	ASTM D1435	ASTM D1435	ASTM D1435	ASTM D1435
Exposure Time (days)	91	182	365	730	91	182	365	730
PROPERTIES RETAINED (%)								
Tensile Strength	98	98.8	98	104	94	95	102	105
Notched Izod Impact Strength	100	92.9			100	92.9		
Unnotched Izod Impact Strength	89	83.3			86	84.8		

Chapter 35

Low Density Polyethylene

GRAPH 156: **Xenon Weatherometer Exposure Time vs. Elongation Retained of Low Density Polyethylene**

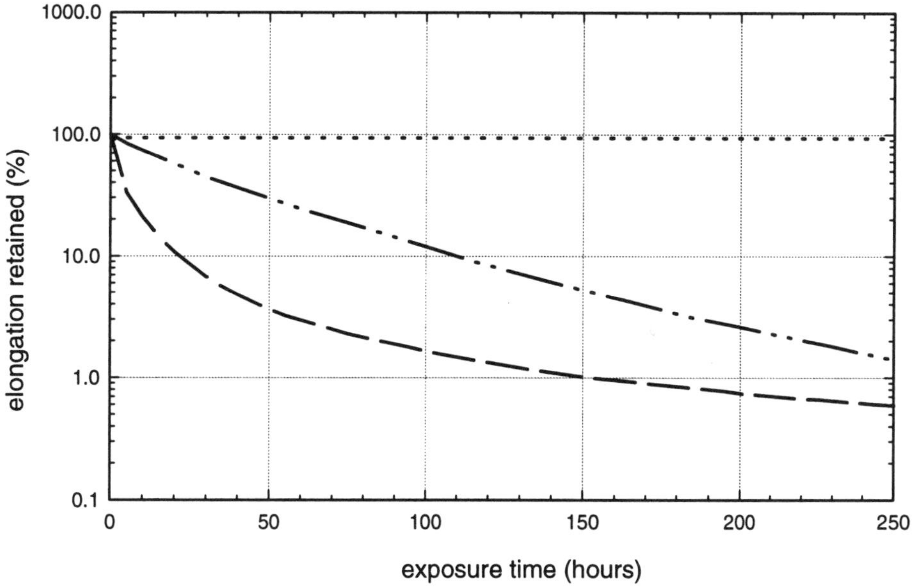

···············	LDPE (control)
—·—··	LDPE (w/ photodegradable additive)
– – – –	Starch LDPE (biodegradable; 10% ECOSTARplus)
Reference No.	168

GRAPH 157: **Composting Exposure Time vs. Elongation Retained of Low Density Polyethylene**

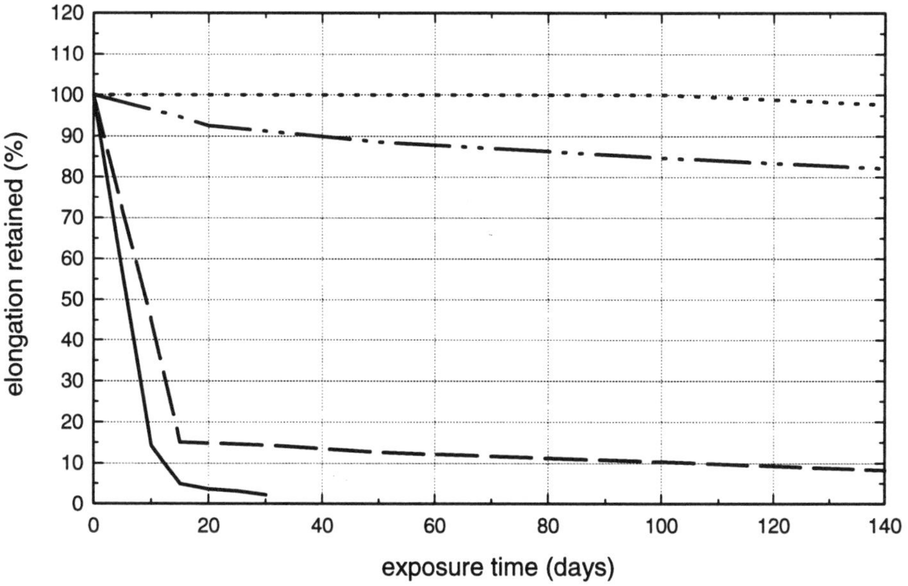

···············	LDPE (control); LDPE with no additives
—·—··	LDPE (w/ photodegradable additive)
– – – –	Starch LDPE (biodegradable; 10% ECOSTARplus); compost site; film brittles as elongation falls below 95%
———	Starch LDPE (biodegradable; 10% ECOSTARplus); laboratory simulation; film brittles as elongation falls below 95%
Reference No.	168

Chapter 36

High Density Polyethylene

Effect of Carbon Black on Weatherability

Phillips: Marlex (density: 0.95 g/cm^3; material compostion: 2.5% furnace black); **Marlex** (density: 0.95 g/cm^3; material compostion: 2.5% channel black)

The protection that even relatively low levels of carbon black imparts to the polymer is so great that no other light stabilizers or UV absorbers are required. Several theories have been advanced to explain this phenomenon. Schonhorn and Luongo stated that the photo-oxidative stabilization of HDPE filled with carbon black is due not only to the light shield capability of carbon black, but also to its moderately low surface energy. Another possibility is, that since antioxidant properties of surface phenolic groups on carbon black have been well characterized, increased stability through interruption of chain propagation may be obtained. Regardless, compounds containing only 0.5% carbon black have been exposed for 10,000 hours in the weatherometer with no loss in tensile strength.

Of all the carbon blacks on the market, channel black is thought to offer the most protection to the polymer. For this reason, two blends were prepared: one contained 2.5% channel black, the other 2.5% furnace black. Tensile specimens containing these formulations were exposed on EMMA in Arizona for eight months. Neither compound exhibited any change in tensile strength during this period. Similarly, samples containing 2.5% furnace black have been exposed in the weatherometer for greater than 25,000 hours without any brittleness or loss in physical properties. Carbon blacks are in a class by themselves with respect to the UV protection they offer.

Reference: *Marlex Polyethylene Weatherability,* supplier technical report (TIB3 (78-89 02)) - Phillips 66 Company, 1989.

Effect of White Pigments on Weatherability

Phillips: Marlex (density: 0.95 g/cm^3)

The weathering resistance of several types of white colorants, Zinc oxide, an exterior grade rutile TiO_2, an indoor rutile TiO_2, as well as an anatase TiO_2 were compared. In all instances, 2% pigment was used in combination with 0.5% of a UV absorber. The anatase TiO_2 and indoor rutile TiO_2 were totally ineffective in protecting HDPE; when compared to an unpigmented version, have poorer performance than natural stabilized resin. However, with exterior grade TiO_2 types, UV protection is somewhat improved.

Zinc oxide on the other hand, provides excellent UV protection to polyethylene. Tensile strength retention of the zinc oxide formulation is significantly better than one containing 2% TiO_2. For best weathering results, zinc oxide can be used, provided its hiding power is sufficient for the intended application. For high opacity film and thin-walled containers, titanium oxide is a better choice as the tint strength of zinc oxide is too low to provide sufficient opaqueness. Weathering performance of an exterior grade TiO_2, without UV absorber, and at three different pigment concentrations show that one formulation containing 2% TiO_2 has only 50% of the weathering resistance of 2% TiO_2 with 0.5% hydroxybenzophenone.

Such systems can be improved by the use of nickel and hindered amine light stabilizers. Systems with nickel complex light and hindered amine stabilizers are not significantly affected after 2,000 hours of weatherometer exposure, as compared to the formulation containing hydroxybenzophenone absorber.

Reference: *Marlex Polyethylene Weatherability,* supplier technical report (TIB3 (78-89 02)) - Phillips 66 Company, 1989.

Effect of Color Pigments on Weatherability

Phillips: Marlex

It has been demonstrated that various pigments have vastly different effects on high density polyethylene. They are important both in an aesthetic sense and as stabilizers. Some pigments such as carbon black, iron oxide, phthalocyanine blue, chrome green, and various red and orange pigments afford good and sometimes excellent protection to the resin, while pigments such as TiO_2, ultramarine blue, and cadmium yellow offer little protection. In some cases these pigments may even be detrimental to the resin. It is interesting to note that pigments of the same general type act quite differently with respect to the protection they afford the resin.

For pigments other than carbon black, data supports the hypothesis that (1) variations in absorption of light in ultraviolet/near ultraviolet wave lengths and (2) interaction between pigments and antioxidants and/or UV stabilizers are the major causes of variations in weatherability. Both good and bad side effects must be considered. For instance, carbon black and iron oxide are difficult to disperse, although they impart excellent UV resistance, while phthalocyanine blue and green pigments cause warpage of some parts molded from polyethylene.

Mercury-cadmium pigments tend to darken upon extended exposure to UV light. Further, many color formulations depend upon a white colorant with good hiding power to obtain the desired color. The only white colorant with strong tint strength is TiO_2, and although materials such as zinc oxide or barium sulfate impart excellent UV resistance, their hiding power is only about 30% of TiO_2. This, of course, restricts the use of these materials in white or pastel formulations.

To succesfully meet application requirements, the proper combination of antioxidants, UV absorbers or stabilizers, and pigments must be determined. Failure to consider all three groups of additives could result in production of compounds with inadequate stability. For this reason, the design of the compound is as important as either the part design or the end use application. All of these factors must be considered in developing a satisfactory part for outdoor applications.

Reference: *Marlex Polyethylene Weatherability,* supplier technical report (TIB3 (78-89 02)) - Phillips 66 Company, 1989.

Effect of Yellow Pigments on Weatherability

Phillips: Marlex (density: 0.95 g/cm^3)

To study the effect of yellow pigments on HDPE weatherability, three were selected and incorporated at a 1% concentration in an ethylene-butene copolymer containing 0.5% of a UV stabilizer. The pigments chosen were cadmium yellow, lithopone yellow and a coated molybdate.

After weatherometer exposure for 8,000 hours, the cadmium yellow had the best performance, followed by coated molybdate, then lithopone yellow. Increasing the concentration of coated molybdate and lithopone yellow improved the weathering performance of the compound, but increased cadmium yellow concentration decreased its overall weathering effectiveness. Since this phenomenon has been demonstrated repeatedly, an assumption can be made that a reaction must occur between pigment and stabilizer at higher pigment concentrations. This apparently does not happen, or at least not as much with coated moybdate or lithopone yellow pigment. The results of interaction of UV absorber and pigment, illustrated by the effect of UV sabilizer in 0.95 density polyethylene resin systems containing 1% and 2% cadmium yellow pigments, show that when no stabilizer is used the 2% system is somewhat better than the 1% cadmium yellow system. However, formulations with stabilizer having the opposite effect occur.

To further study this interaction between stabilizers and cadmium yellow in a polyethylene system, samples were prepared containing two nickel complexes furnished by two suppliers. These nickel complexes are known to perform as light stabilizers, whereas hydroxybenzophenone was brittle at 10,000 hours exposure. However, the 2% cadmium yellow system containing both the nickel Complex A and B exhibited only a modest decrease in tensile strength after this same exposure period. The hindered amine light stabilizer (HALS) would appear to be marginally better than the nickel complex but does not exhibit the green color inherent with nickel stabilizers. This further illustrates the complex interrelationships between stabilizers and pigments.

Both the lithopone and cadmium yellow will fade during extended outdoor exposure. Although this is not a problem with single pigment color formulations, the color change can be significant when cadmium yellow is combined with a more light-stable pigment, such as ultramarine blue, to produce a green color.

Reference: *Marlex Polyethylene Weatherability,* supplier technical report (TIB3 (78-89 02)) - Phillips 66 Company, 1989.

Effect of Red Pigments on Weatherability

Phillips: Marlex (density: 0.95 g/cm^3)

It is apparent after 10,000 hours exposure that the weathering performance of three commonly used red pigments (quinacridone red, mercury-cadmium red and lithopone red) that the 1% quinacridone red formulation is considerably better than either of the other two on 0.95 density stabilized polyethylene resins. There appears to be less difference between the lithopone red and mercury-cadmium red than between these pigments and the quinacridone red.

The same relationship persists at a 2% pigment level. The 2% quinacridone red is still marginally better than 2% mercury-cadmium red, and both are considerably better than 2% lithopone red after 10,000 hours exposure. This indicates that as the pigment concentration increases from 1% to 2%, the mercury-cadmium red shows most improvement. Furthermore, 1% quinacridone red provides better stability than 2% lithopone red.

Previous studies indicated that CP cadmium red would offer virtually the same protection against UV degradation as the mercury-cadmium red. Although the quinacridone and cadmium red pigments extend the outdoor weatherability of HDPE, they have limited use due to lack of color stability. The tint strength of both quinacridone and CP cadmium pigments will weaken when accelerated by a high humidity atmosphere. The most light-stable red is a combination of CP cadmium red and mercury cadmium red pigments.

Earlier studies indicate that the iron oxide is excellent for use in high density polyethylene outdoor applications. An unstabilized system of 0.5% iron oxide was virtually unchanged after 2,000 hours in the weatherometer, while the tensile strength of the 0.5% CP cadmium red started to decay considerably. Since both these formulations were unstabilized, this further demonstrates the significant screening effect of iron oxide in polyethylene. From past experience, iron oxide can be considered to be second only to carbon black in its ability to stabilize HDPE against UV degradation.

Reference: *Marlex Polyethylene Weatherability,* supplier technical report (TIB3 (78-89 02)) - Phillips 66 Company, 1989.

Effect of Orange Pigments on Weatherability

Phillips: Marlex (density: 0.95 g/cm^3)

Four pigments at levels of 1 and 2% (coated molybdate, lithopone, CP cadmium and mercury-cadmium) in 0.95 density stabilized polyethylene were exposed for 10,000 hours in a weatherometer. At a concentration of 1%, the CP cadmium orange appears to be 10-20% better than other pigments. Less difference can be seen among the pigments at the 2% level, although cadmium orange is still approximately 10% better than the others. It does appear that the coated molybdate pigment is somewhat better at 1% than at 2%, but this slight increase falls within experimental error range and can be considered negligible. The only pigment to show a significant difference between the two concentrations is lithopone orange. Since this pigment contains less cadmium than CP cadmium, the overall effect is much the same as a reduced level of cadmium.

The effect of antioxidants on UV stabilization of 0.95 density stabilized polyethylene indicates that the antioxidant system plays a major role in outdoor performance of pigmented HDPE formulations. Two different types of antioxidants were compounded with a UV stabilizer and 1 and 2% cadmium orange. Antioxidant B imparted much more resistance to UV degradation than antioxidant A. This was true for either 1 or 2% pigment levels, and illustrates the need for selecting formulations with the proper antioxidants for outdoor applications.

Reference: *Marlex Polyethylene Weatherability,* supplier technical report (TIB3 (78-89 02)) - Phillips 66 Company, 1989.

Effect of Blue and Green Pigments on Weatherability

Phillips: Marlex (density: 0.95 g/cm^3)

One percent levels of phthalocyanine blue, cobalt blue and ultramarine blue were incorporated in an unstabilized polyethlene system and exposed for 2,000 hours in a weatherometer. This test indicated that the phthalocyanine blue pigment provided 2 to 3 times as much UV protection as ultramarine blue. This level of protection in an unstabilized system is quite good. Cobalt blue, however, appears only moderately effective when compared to phthalocyanine blue.

It is also apparent that ultramarine blue imparts little or no protection to the polymer; since the performance of the compound containing ultramarine blue was little better than natural HDPE. The same general trend is found in stabilized, as well as unstabilized systems.

Many green formulations are prepared by combining ultramarine blue and cadmium yellow. Since the UV protection provided by ultramarine blue in HDPE is poor, and only fair with cadmium yellow, it is not surprising that the combination is rather ineffective. Phthalocyanine green, however, imparts excellent UV resistance to polyethylene as do some of the chrome greens. Compounds containing these pigments last longer than 6,000 hours of weatherometer exposure with no loss of tensile strength.

Reference: *Marlex Polyethylene Weatherability,* supplier technical report (TIB3 (78-89 02)) - Phillips 66 Company, 1989.

Effect of Pigment Dispersion on Weatherability

Phillips: Marlex (density: 0.96 g/cm^3)

Pigment dispersion is important in the compounding of any colored resin, since inadequate dispersion can result in poor appearance, increased cost and poor outdoor weatherability. In order to illustrate the effect of dispersion on weathering, three blends containing 0.5% CP cadmium red were prepared. Each of the blends was compounded in order to achieve a good, fair and poor pigment dispersion.

Tensile specimens from these compounds were aged in the weatherometer for 2,000 hours. Tensile strength measured at 2 in./min. showed very little difference between good/fair pigment dispersion. However, measurements at 20 in./min. revealed that as pigment dispersion improved, there was a marked increase in tensile strength retention after UV exposure. As in the former case, the specimen with poor pigment dispersion was much less resistant to degradation than even the good or fair pigment system. These data clearly indicate that pigment dispersion is important to the UV resistance of a compound.

The degree of carbon black dispersion is also a determining factor in effectiveness of pigment for UV protection. Low levels of carbon black, properly dispersed, offer excellent UV protection. The usual condition is poor dispersion, normally compensated by using up to 2.5% black to give ultimate protection.

Reference: *Marlex Polyethylene Weatherability,* supplier technical report (TIB3 (78-89 02)) - Phillips 66 Company, 1989.

Effect of Part Thickness on Weatherability

Phillips: Marlex (features: 3.05 mm thick; pigment: yellow); **Marlex** (features: 1.52 mm thick; pigment: yellow); **Marlex** (features: 2.29 mm thick; pigment: yellow); **Marlex** (features: 0.76 mm thick; pigment: yellow)

The effect of part thickness plays a significant role in outdoor life. Since degradation of a part occurs from the exterior to the interior, the thicker the part, the more time required to penetrate to a depth that affects its integrity. In one test, a 120-mil thick sample has several times the life expectancy of a 30 mil sample in a yellow high density polyethylene tested outdoors in Arizona.

Reference: *Marlex Polyethylene Weatherability,* supplier technical report (TIB3 (78-89 02)) - Phillips 66 Company, 1989.

Biodegradation

Phillips: Marlex

The fungus resistance of Marlex HDPE was determined using a test method described in Military Specification, MIL-E-5272A. In this procedure, tensile bars were sprayed with a suspension containing a specified mixture of five groups of fungi drawn from 16-day old cultures. The exposed tensile specimens were suspended over water in a vessel which was then sealed. To produce 95 ± 5% relative humidity in the chamber, the entire assembly was heated up to 30°C ± 1°C, and maintained for a period of 28 days. After conditioning, the exposed samples were washed in alcohol, wiped dry and tested for tensile strength according to standard ASTM procedures. As a control, unexposed samples were conditioned and tested in the same manner. Marlex HDPE showed no evidence of deterioration or corrosion due to fungus exposure.

Reference: *Engineering Properties Of Marlex Resins,* supplier design guide (TSM-243) - Phillips 66 Company, 1983.

TABLE 58: Effect of Carbon Black Type on Accelerated Outdoor Weathering by EMMA of Phillips Marlex High Density Polyethylene.

Material Family	High Density Polyethylene							
Material Supplier/ Grade	Phillips Marlex							
Reference Number	112	112	112	112	112	112	112	112
MATERIAL CHARACTERISTICS								
Density	0.95 g/cm³	0.95 g/cm³	0.95 g/cm³	0.95 g/cm³	0.95 g/cm³	0.95 g/cm³	0.95 g/cm³	0.95 g/cm³
MATERIAL COMPOSITION								
channel black	2.5%	2.5%	2.5%	2.5%				
furnace black					2.5%	2.5%	2.5%	2.5%
EXPOSURE CONDITIONS								
Exposure Type	accelerated outdoor weathering	accelerated outdoor weathering	accelerated outdoor weathering	accelerated outdoor weathering	accelerated outdoor weathering	accelerated outdoor weathering	accelerated outdoor weathering	accelerated outdoor weathering
Exposure Location	Phoenix, Arizona	Phoenix, Arizona	Phoenix, Arizona	Phoenix, Arizona	Phoenix, Arizona	Phoenix, Arizona	Phoenix, Arizona	Phoenix, Arizona
Exposure Country	USA	USA	USA	USA	USA	USA	USA	USA
Exposure Apparatus	EMMA	EMMA	EMMA	EMMA	EMMA	EMMA	EMMA	EMMA
Exposure Time (days)	61	122	183	244	61	122	183	244
PROPERTY VALUES AFTER EXPOSURE								
Tensile Strength (MPa)	27.6 {n}	27.6 {n}	27.6 {n}	27.6 {n}	27.6 {n}	27.6 {n}	27.6 {n}	27.6 {n}

TABLE 59: Effect of Color Dispersion on Accelerated Weathering in a Weatherometer of Phillips Marlex High Density Polyethylene with 0.5% CP Cadmium Red Pigment.

Material Family	High Density Polyethylene					
Material Supplier/ Grade	Phillips Marlex					
Reference Number	112	112	112	112	112	112
MATERIAL CHARACTERISTICS						
Density	0.95 g/cm³	0.95 g/cm³	0.95 g/cm³	0.95 g/cm³	0.95 g/cm³	0.95 g/cm³
MATERIAL COMPOSITION						
hydroxybenzophenone (UV absorber)	0.5%	0.50%	0.5%	0.5%	0.5%	0.5%
CP Cadmium Red	0.5%	0.5%	0.5%	0.5%	0.5%	0.5%
note	good pigment dispersion	good pigment dispersion	fair pigment dispersion	fair pigment dispersion	poor pigment dispersion	poor pigment dispersion
EXPOSURE CONDITIONS						
Exposure Type	accelerated weathering	accelerated weathering	accelerated weathering	accelerated weathering	accelerated weathering	accelerated weathering
Exposure Test Method	ASTM D1499	ASTM D1499	ASTM D1499	ASTM D1499	ASTM D1499	ASTM D1499
Exposure Apparatus	Atlas weatherometer	Atlas weatherometer	Atlas weatherometer	Atlas weatherometer	Atlas weatherometer	Atlas weatherometer
Exposure Time (days)	83.3	83.3	83.3	83.3	83.3	83.3
PROPERTY VALUES AFTER EXPOSURE						
Tensile Strength (MPa)	22.4 {l}	14.8 {k}	21.4 {l}	11 {k}	6.9 {l}	2.8 {k}

TABLE 60: Effect of Pigments on Accelerated Weathering in a Weatherometer of Phillips Marlex High Density Polyethylene.

Material Family	High Density Polyethylene								
Material Supplier/ Grade	Phillips Marlex								
Reference Number	112	112	112	112	112	112	112	112	112
MATERIAL CHARACTERISTICS									
Density	0.95 g/cm³	0.95 g/cm³	0.95 g/cm³	0.95 g/cm³	0.95 g/cm³	0.95 g/cm³	0.95 g/cm³	0.95 g/cm³	0.95 g/cm³
MATERIAL COMPOSITION									
hydroxybenzophenone (UV absorber)	0.5%	0.5%	0.5%	0.5%	0.5%	0.5%	0.5%	0.5%	0.5%
coated molybdate orange pigment	1%	2%							
lithopone orange pigment			1%	2%					
mercadmium orange pigment							0.5%	1%	2%
EXPOSURE CONDITIONS									
Exposure Type	accelerated weathering	accelerated weathering	accelerated weathering	accelerated weathering	accelerated weathering	accelerated weathering	accelerated weathering	accelerated weathering	accelerated weathering
Exposure Test Method	ASTM D1499	ASTM D1499	ASTM D1499	ASTM D1499	ASTM D1499	ASTM D1499	ASTM D1499	ASTM D1499	ASTM D1499
Exposure Apparatus	Atlas weatherometer	Atlas weatherometer	Atlas weatherometer	Atlas weatherometer	Atlas weatherometer	Atlas weatherometer	Atlas weatherometer	Atlas weatherometer	Atlas weatherometer
Exposure Time (days)	416.7	416.7	416.7	416.7	416.7	416.7	416.7	416.7	416.7
PROPERTY VALUES AFTER EXPOSURE									
Tensile Strength (MPa)	24.1 {k}	22.1 {k}	17.2 {k}	24.1 {k}	24.1 {k}	17.9 {k}	6.2 {k}	22.1 {k}	24.1 {k}

TABLE 61: Effect of UV Stabilizers on Accelerated Weathering in a Weatherometer of Phillips Marlex High Density Polyethylene with 2% Cadmium Yellow Pigment.

Material Family	High Density Polyethylene			
Material Supplier/ Grade	Phillips Marlex			
Reference Number	112	112	112	112
MATERIAL CHARACTERISTICS				
Density	0.95 g/cm³	0.95 g/cm³	0.95 g/cm³	0.95 g/cm³
MATERIAL COMPOSITION				
hydroxybenzophenone (UV absorber)	0.5%			
UV stabilizer		nickel complex A	nickel complex B	hindered amine light stabilizer (HALS)
cadmium yellow pigment	2%	2%	2%	2%
EXPOSURE CONDITIONS				
Exposure Type	accelerated weathering	accelerated weathering	accelerated weathering	accelerated weathering
Exposure Test Method	ASTM D1499	ASTM D1499	ASTM D1499	ASTM D1499
Exposure Apparatus	Atlas weatherometer	Atlas weatherometer	Atlas weatherometer	Atlas weatherometer
Exposure Time (days)	416.7	416.7	416.7	416.7
PROPERTY VALUES AFTER EXPOSURE				
Tensile Strength (MPa)	2.07 {k}	23.4 {k}	26.2 {k}	29 {k}

TABLE 62: Effect of Yellow Pigments on Accelerated Weathering in a Weatherometer of Phillips Marlex High Density Polyethylene.

Material Family	High Density Polyethylene					
Material Supplier/ Grade	Phillips Marlex					
Reference Number	112	112	112	112	112	112
MATERIAL CHARACTERISTICS						
Density	0.95 g/cm³	0.95 g/cm³	0.95 g/cm³	0.95 g/cm³	0.95 g/cm³	0.95 g/cm³
MATERIAL COMPOSITION						
hydroxybenzophenone (UV absorber)	0.5%	0.5%	0.5%	0.5%	0.5%	0.5%
cadmium yellow pigment			1%			2%
coated molybdate yellow pigment		1%			2%	
lithopone yellow pigment	1%			2%		
EXPOSURE CONDITIONS						
Exposure Type	accelerated weathering	accelerated weathering	accelerated weathering	accelerated weathering	accelerated weathering	accelerated weathering
Exposure Test Method	ASTM D1499	ASTM D1499	ASTM D1499	ASTM D1499	ASTM D1499	ASTM D1499
Exposure Apparatus	Atlas weatherometer	Atlas weatherometer	Atlas weatherometer	Atlas weatherometer	Atlas weatherometer	Atlas weatherometer
Exposure Time (days)	333.3	333.3	333.3	333.3	333.3	333.3
PROPERTY VALUES AFTER EXPOSURE						
Tensile Strength (MPa)	15.1 {k}	22.8 {k}	24.8 {k}	24.1 {k}	26.2 {k}	11 {k}

TABLE 63: Effect of UV Stabilizers on Accelerated Weathering in a Weatherometer of Phillips Marlex High Density Polyethylene with 2% Titanium Dioxide.

Material Family	High Density Polyethylene			
Material Supplier/ Grade	Phillips Marlex			
Reference Number	112	112	112	112
MATERIAL CHARACTERISTICS				
Density	0.95 g/cm³	0.95 g/cm³	0.95 g/cm³	0.95 g/cm³
MATERIAL COMPOSITION				
hydroxybenzophenone (UV absorber)	0.5%			
UV stabilizer		nickel complex A	nickel complex B	hindered amine light stabilizer (HALS)
exterior grade TiO_2	2%	2%	2%	2%
EXPOSURE CONDITIONS				
Exposure Type	accelerated weathering	accelerated weathering	accelerated weathering	accelerated weathering
Exposure Test Method	ASTM D1499	ASTM D1499	ASTM D1499	ASTM D1499
Exposure Apparatus	Atlas weatherometer	Atlas weatherometer	Atlas weatherometer	Atlas weatherometer
Exposure Time (days)	83.3	83.3	83.3	83.3
PROPERTY VALUES AFTER EXPOSURE				
Tensile Strength (MPa)	13.8 {k}	26.9 {k}	27.9 {k}	34.1 {k}

TABLE 64: Effect of Antioxidants and UV Absorber on Accelerated Weathering in a Xenon Weatherometer of Green High Density Polyethylene.

Material Family	High Density Polyethylene					
Product Form	injection molded plaque					
Features	unstabilized, green color	green color	green color	green color	green color	green color
Reference Number	207	207	207	207	207	207
MATERIAL CHARACTERISTICS						
Sample Thickness (mm)	1.52	1.52	1.52	1.52	1.52	1.52
MATERIAL COMPOSITION						
Ultranox 626 (antioxidant - General Electric)			0.125 phr			0.125 phr
Weston 619 (antioxidant - General Electric)		0.125 phr			0.125 phr	
Cyasorb UV 531 (UV absorber - American Cyanamid)				0.125 phr	0.125 phr	0.125 phr
pigment	green	green	green	green	green	green
EXPOSURE CONDITIONS						
Exposure Type	accelerated weathering					
Exposure Apparatus	xenon weatherometer					
Exposure Note	exposure time indicates time required to cause surface crazing and cracking					
Exposure Time (days)	36	54	75	104	92	142
SURFACE AND APPEARANCE						
Visual Appearance	surface crazed and cracked	surface crazed and cracked	surface crazed and cracked	surface crazed and cracked	surface crazed and cracked	surface crazed and cracked
Crazing	occurs at exposure time	occurs at exposure time	occurs at exposure time	occurs at exposure time	occurs at exposure time	occurs at exposure time

TABLE 65: Fungus Resistance of Phillips Marlex High Density Polyethylene.

Material Family	High Density Polyethylene
Material Supplier	Phillips Marlex
Reference Number	101
EXPOSURE CONDITIONS	
Exposure Type	fungus
Exposure Test Method	MIL-E-5272A
Exposure Note	tensile bars were sprayed with a suspension containing a specified mixture of five groups of fungi drawn from 16 day old cultures
Culture	mixture of chaetonium globosum, rhizopus higricans, aspergillus flavus, penicillium funiculosum, fusarium moniliforme
Relative Humidity (%)	95±5
Exposure Temperature (°C)	30±1
Exposure Time (days)	28
PROPERTIES RETAINED (%)	
Tensile Strength	100
Elongation	100

GRAPH 158: **Arizona Outdoor Weathering Exposure Time vs. Tensile Strength of High Density Polyethylene**

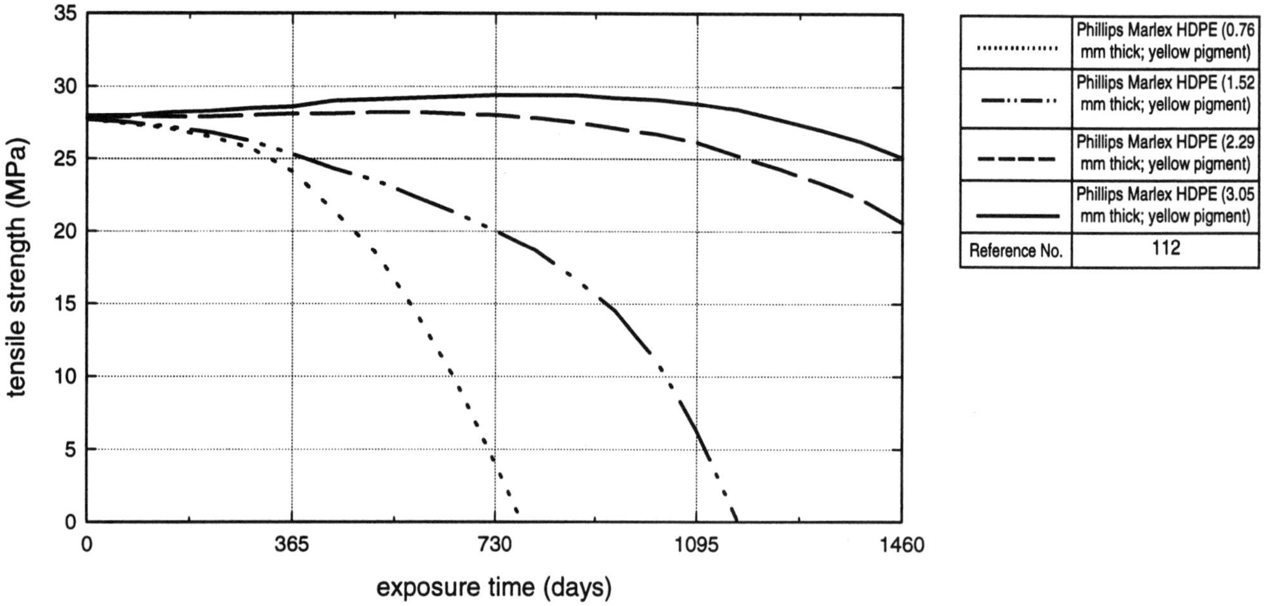

GRAPH 159: **Weatherometer Exposure Time vs. Tensile Strength of High Density Polyethylene**

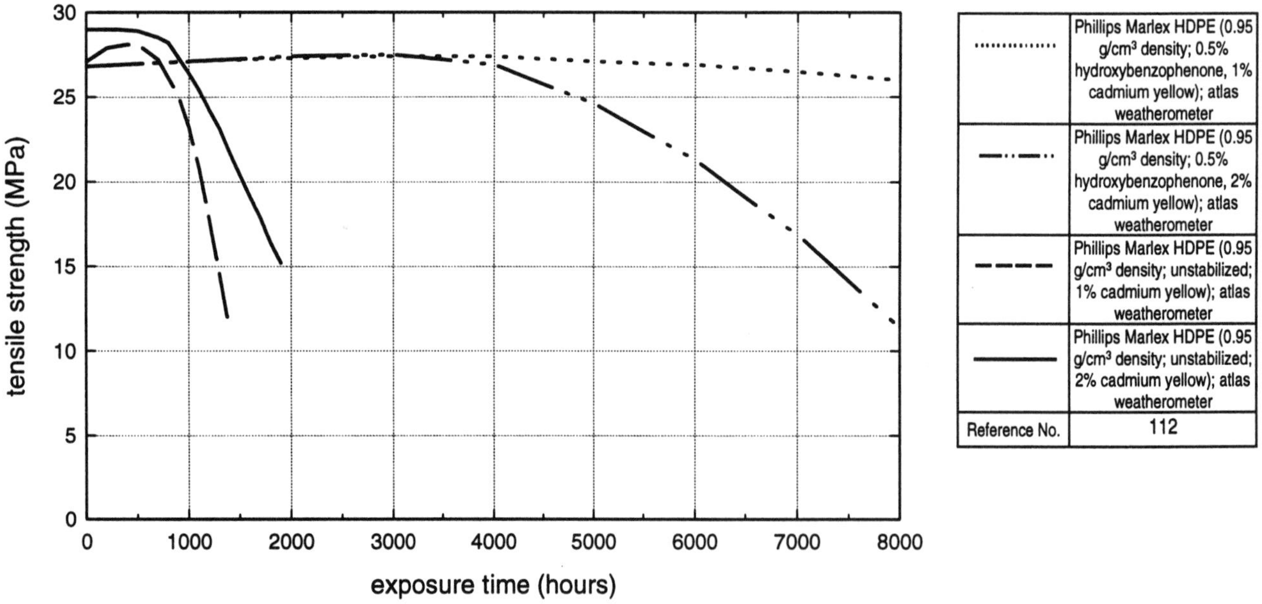

GRAPH 160: Weatherometer Exposure Time vs. Tensile Strength of High Density Polyethylene

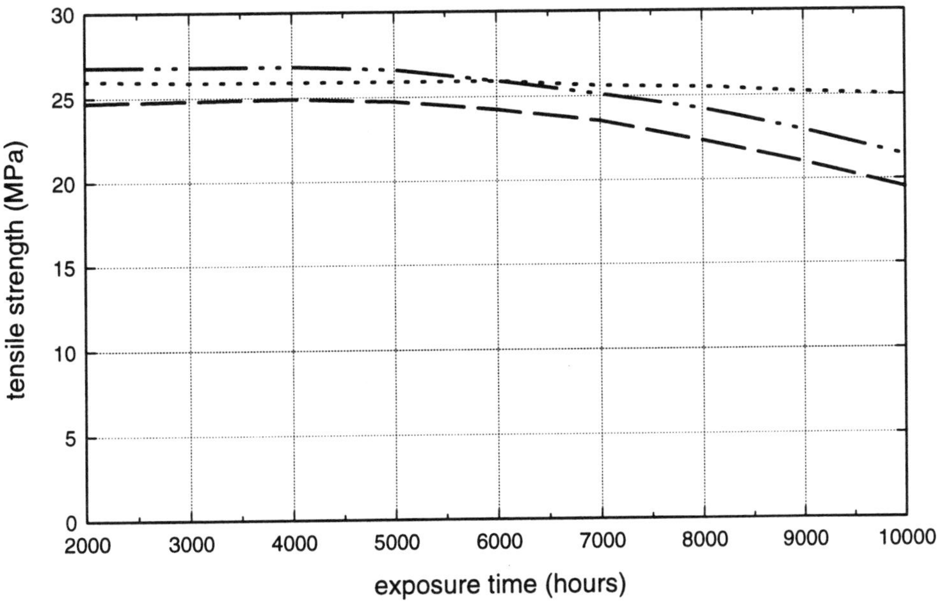

···············	Phillips Marlex HDPE (0.95 g/cm³ density; 0.5% hydroxybenzophenone, 1% quinacridone red pigment)
—··—··	Phillips Marlex HDPE (0.95 g/cm³ density; 0.5% hydroxybenzophenone, 1% mercadmium red)
— — — —	Phillips Marlex HDPE (0.95 g/cm³ density; 0.5% hydroxybenzophenone, 1% lithopone red)
Reference No.	112

GRAPH 161: Weatherometer Exposure Time vs. Tensile Strength of High Density Polyethylene

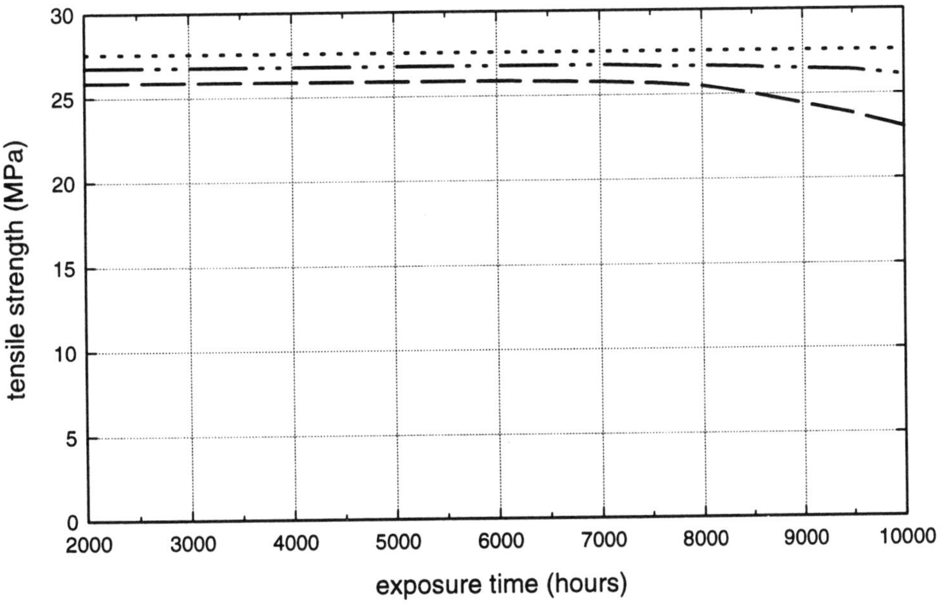

···············	Phillips Marlex HDPE (0.95 g/cm³ density; 0.5% hydroxybenzophenone, 2% quinacridone red pigment)
—··—··	Phillips Marlex HDPE (0.95 g/cm³ density; 0.5% hydroxybenzophenone, 2% mercadmium red)
— — — —	Phillips Marlex HDPE (0.95 g/cm³ density; 0.5% hydroxybenzophenone, 2% lithopone red)
Reference No.	112

GRAPH 162: Weatherometer Exposure Time vs. Tensile Strength of High Density Polyethylene

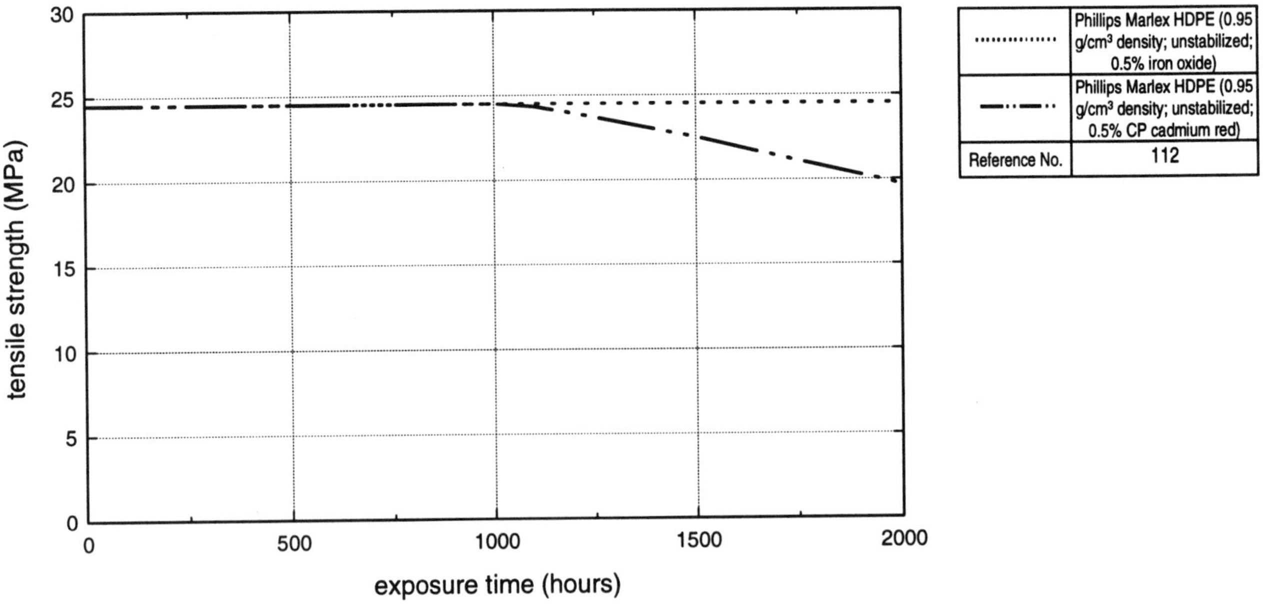

GRAPH 163: Weatherometer Exposure Time vs. Tensile Strength of High Density Polyethylene

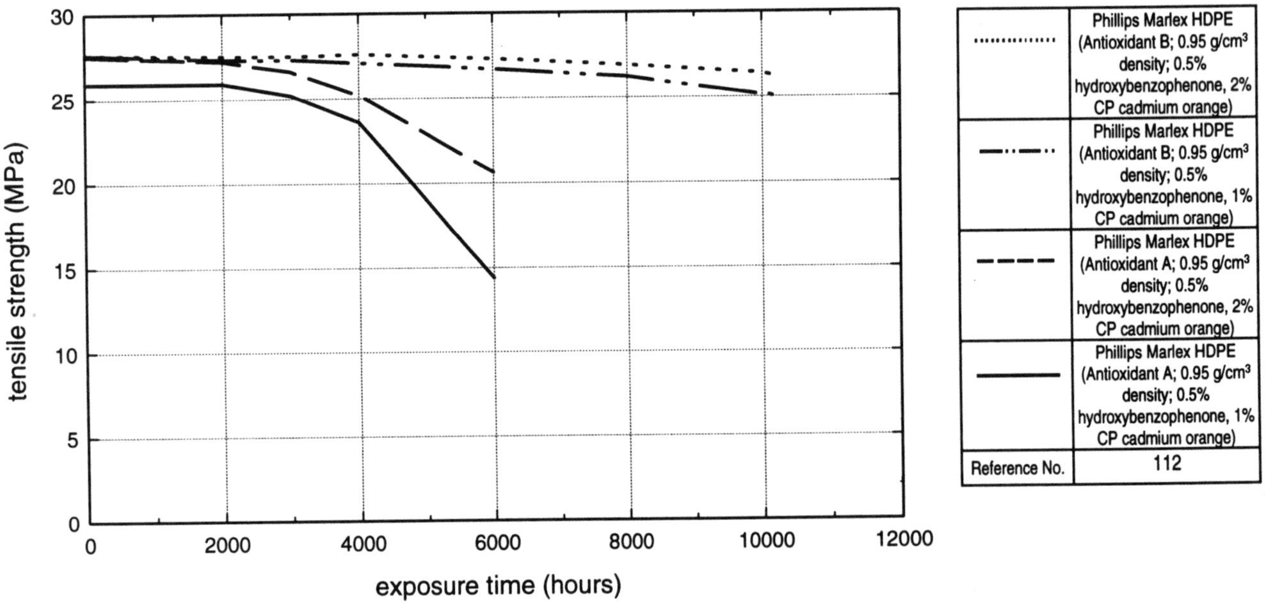

GRAPH 164: **Weatherometer Exposure Time vs. Tensile Strength of High Density Polyethylene**

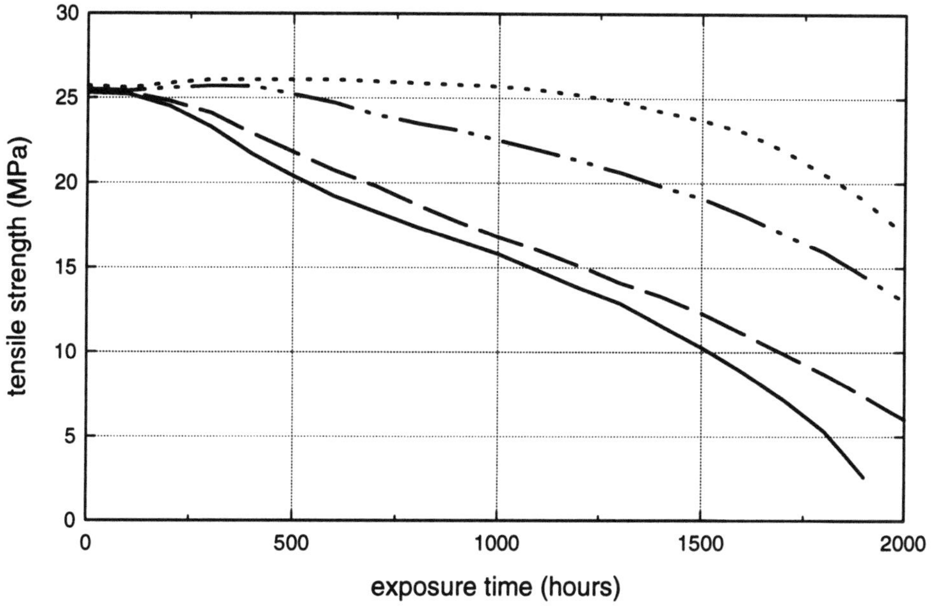

··············	Phillips Marlex HDPE (0.95 g/cm³ density; unstabilized; 1% phthalocyanine blue)
—··—··	Phillips Marlex HDPE (0.95 g/cm³ density; unstabilized; 1% cobalt blue)
— — — —	Phillips Marlex HDPE (0.95 g/cm³ density; unstabilized; 1% ultramarine blue)
————	Phillips Marlex HDPE (0.95 g/cm³ density; natural resin, unstabilized)
Reference No.	112

GRAPH 165: **Weatherometer Exposure Time vs. Tensile Strength of High Density Polyethylene**

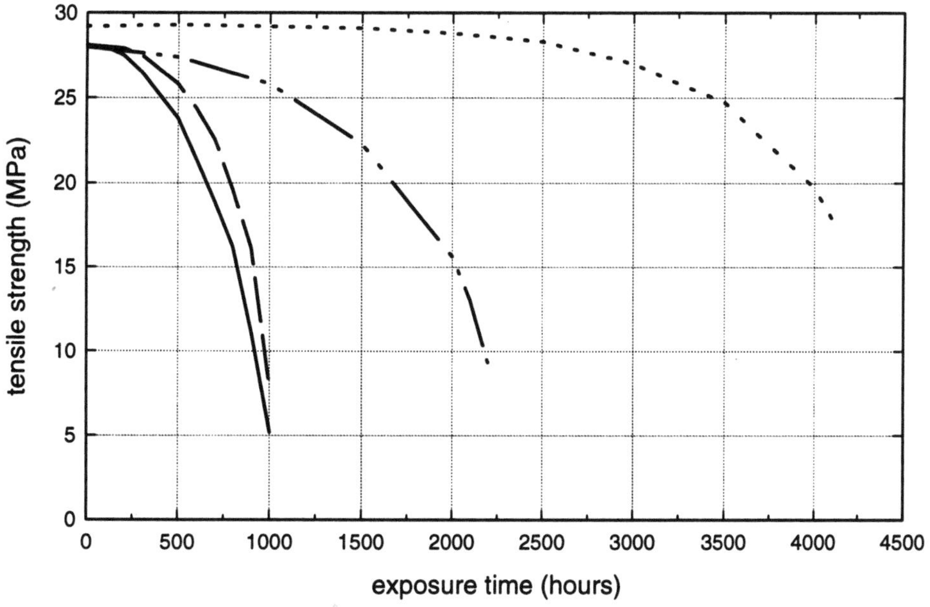

··············	Phillips Marlex HDPE (0.95 g/cm³ density; 0.5% hydroxybenzophenone, 2% zinc oxide)
—··—··	Phillips Marlex HDPE (0.95 g/cm³ density; 0.5% hydroxybenzophenone, 2% exterior TiO_2)
— — — —	Phillips Marlex HDPE (0.95 g/cm³ density; 0.5% hydroxybenzophenone, 2% indoor TiO_2)
————	Phillips Marlex HDPE (0.95 g/cm³density; 0.5% hydroxybenzophenone, 2% anatase TiO_2)
Reference No.	112

GRAPH 166: Weatherometer Exposure Time vs. Tensile Strength of High Density Polyethylene

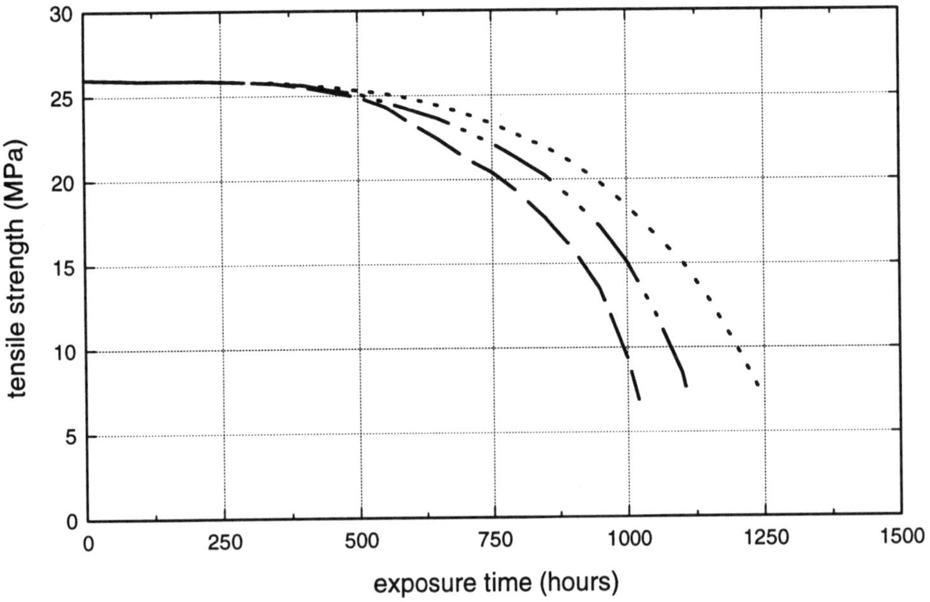

············	Phillips Marlex HDPE (0.95 g/cm³ density; unstabilized; 2% exterior TiO_2)
—··—	Phillips Marlex HDPE (0.95 g/cm³ density; unstabilized; 1% exterior TiO_2)
— — —	Phillips Marlex HDPE (0.95 g/cm³ density; unstabilized; 0.5% exterior TiO_2)
Reference No.	112

GRAPH 167: Weatherometer Exposure Time vs. Tensile Strength of High Density Polyethylene

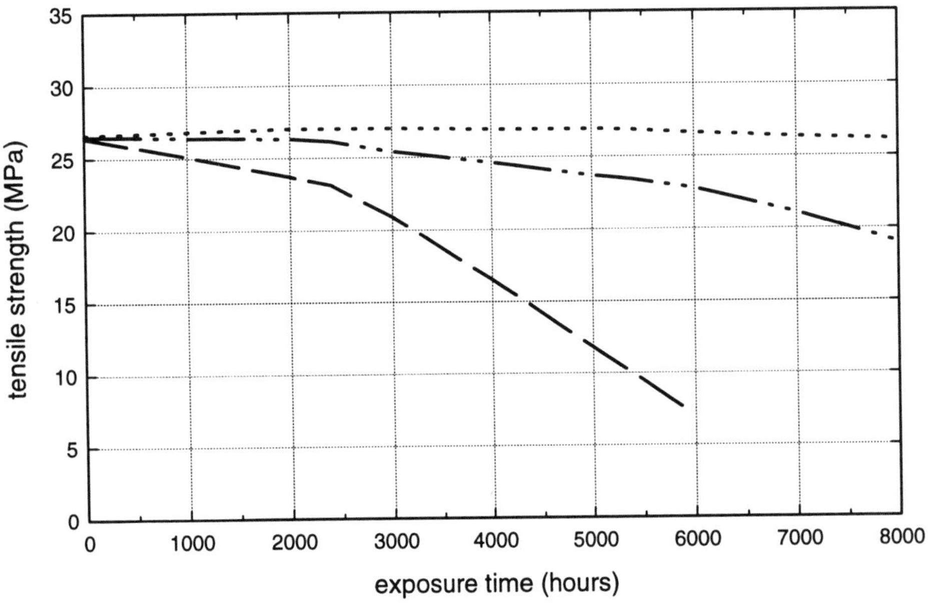

············	Phillips Marlex HDPE (0.95 g/cm³ density; 1% exterior TiO_2, HALS)
—··—	Phillips Marlex HDPE (0.95 g/cm³ density; 1% exterior TiO_2, benzotriazole)
— — —	Phillips Marlex HDPE (0.95 g/cm³ density; 0.5% hydroxybenzophenone, 1% exterior TiO_2)
Reference No.	112

GRAPH 168: Weatherometer Exposure Time vs. Tensile Strength of High Density Polyethylene

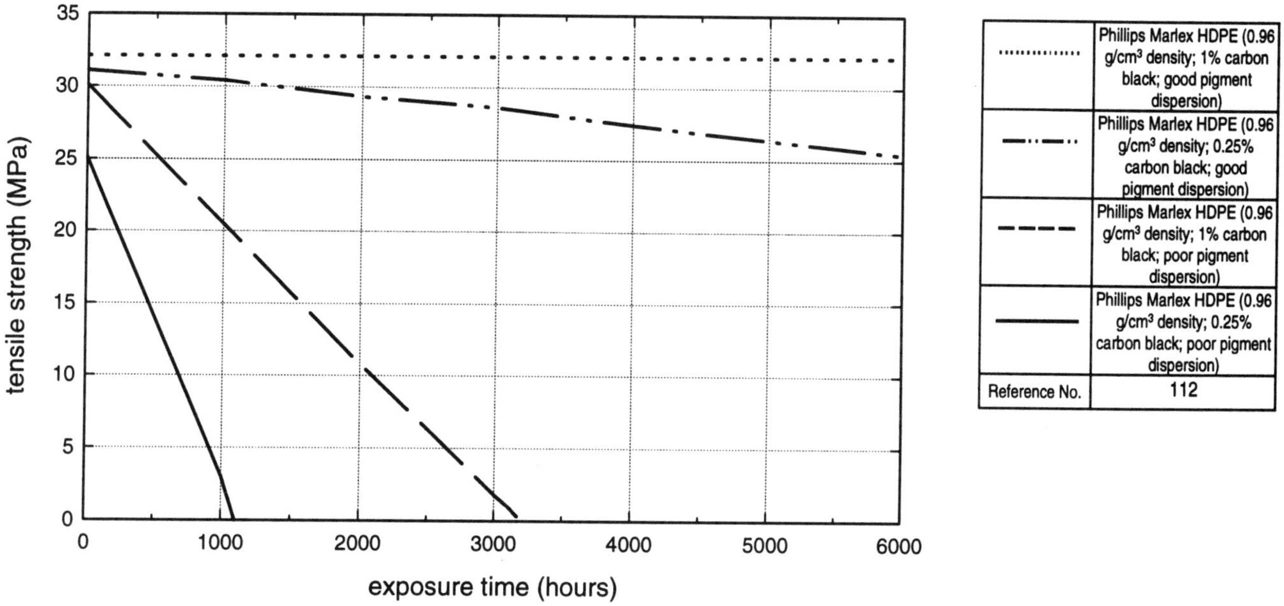

Chapter 37

Ultrahigh Molecular Weight Polyethylene

UV Resistance

Hoechst Celanese: Hostalen GUR 412; Hostalen GUR 412LS (features: UV stabilized)

Molecular degradation may be prevented through the addition of suitable light stabilizers. The light (LS) stabilized GUR 412 samples evidenced no degradation even after a four week exposure period; in other words, their physical characteristics were preserved. The property values after exposure were determined for test specimens 1.3, 10 and 20 mm thick after exposure to a Xenon-lamp device.

Reference: *Hostalen GUR - Effects Of Heat And Light Aging,* supplier technical report (HCC Rev 1/90) - Hoechst Celanese Corporation, 1990.

Hoechst Celanese: Hostalen GUR

The addition of light-absorbing substances provides UV light resistance, with 2.5% carbon black being the most commonly used additive. When the finished product cannot be black, satisfactory UV resistance, which is a minimum of 5 years, can be obtained by 0.5 wt% stabilizer.

Reference: Stein, Harvey L., *Ultrahigh Molecular Weight Polyethylenes (UHMWPE),* Engineered Materials Handbook, Vol. 2, Engineering Plastics, reference book - ASM International, 1988.

Chapter 38

Polyethylene Copolymer

Outdoor Weather Resistance

BASF AG: Lupolen V (Ethylene Vinyl Acetate Polyethylene Copolymer); Lucalen A (Polyethylene Acrylic Acid Copolymer); Lucalen T (Polyethylene Ionomer Copolymer)

These materials are resistant to radiation in the visible spectrum. If polyethylene and its copolymers are exposed for long periods outdoors, they are degraded by radiation at the UV end of the solar spectrum and by atmospheric oxygen. They are also degraded by other light sources with a high proportion of UV radiation. The degradation mechanism is oxidation favored by high temperatures and leads to a deterioration in the mechanical properties and ultimately, to destruction of the material.

If moldings are intended for outdoor use, they must be adequately protected from UV radiation. By far the best UV stability is achieved by adding special grades of carbon black. Proportions of 2-3% improve the UV stability by a factor of 10 to 15. Lupolen and Lucalen grades containing carbon black are included in the base product line and can be recognized from their nomenclature, to which the term "black 413" has been appended. White and chromatic pigments may also improve the UV stability of polyethylene but can also adversely affect it.

If moldings in the natural color or in other hues have to display excellent outdoor performance and fastness to light, the copolymers can be supplied on request with special light stabilizers. Good results are obtained with the HALS type, in some cases in combination with benzotriazole compounds. They can increase the resistance to weathering by a factor of about 2-4, the extent depending on the conditions. The presence of special UV stabilizers is indicated in the nomenclature by a five digit color numeral commencing with the digit 6.

Reference: *Lupolen, Lucalen Product Line, Properties, Processing,* supplier design guide (B 581 e/(8127) 10.91) - BASF Aktiengesellschaft, 1991.

TABLE 66: Effect of UV Stabilizers on Accelerated Weathering in a Xenon Arc Weatherometer of Ethylene Vinyl Acetate Polyethylene Copolymer Greenhouse Film.

Material Family	Ethylene Vinyl Acetate Polyethylene Copolymer					
Product Form	greenhouse film					
Reference Number	146	146	146	146	146	146
MATERIAL COMPOSITION						
Chimassorb 944 (UV stabilizer - American Cyanamid)	0.5%			0.5%		
Cyasorb UV-3346 (UV stabilizer - American Cyanamid)		0.5%			0.5%	
UV 1084 (UV stabilizer - American Cyanamid)			1%			1%
UV 531 (UV stabilizer - American Cyanamid)			0.5%			0.5%
EXPOSURE CONDITIONS						
Exposure Type	accelerated weathering			accelerated weathering		
Exposure Apparatus	xenon arc weatherometer			xenon arc weatherometer		
Exposure Note	Weekly treatment with POUNCE solution [active ingredient: a synthetic PYRETHROID-PERMETHRIN, 3-phenoxy benzyl (±) cis, trans-3-(2,2-dichlorovinyl)-2,2-dimethyl cyclopropane]. Film sprayed on non-exposed side.			untreated		
Exposure Time (days)	121	125	225	>250	>250	>271
PROPERTIES RETAINED (%)						
Elongation	50	50	50	66	84	79

Chapter 39

Polypropylene

Effect of White Pigments on Weatherability

PP (material compostion: 40% Microcal C110S (calcium carbonate - ECC Int.))

Poor quality, impure calcium carbonate filler can adversely affect weatherabiliity. Therefore, generally only fine, relatively pure products are used. Microcal C110S performs favorably on the weathering of polypropylene with a fine competitive limestone product in respect to the change in key properties after 18 weeks exposure in a QUV weatherometer.

Reference: *Microcal Spa C110S For Polypropylene,* supplier technical report (APP033 PI) - ECC International, 1993.

TABLE 67: Effect of Antioxidants on Outdoor Weathering in Florida and Puerto Rico of Polypropylene.

Material Family	Polypropylene										
Product Form	200/16 denier natural multifilament samples										
Reference Number	207	207	207	207	207	207	207	207	207	207	207
MATERIAL COMPOSITION											
calcium stearate	0.05 phr	0.05 phr	0.05 phr	0.05 phr	0.05 phr	0.05 phr	0.05 phr	0.05 phr	0.05 phr	0.05 phr	0.05 phr
Good-rite 3114 (antioxidant - American Cyanamid)	0.1 phr	0.1 phr	0.1 phr								
Irganox 1010 (antioxidant - Ciba Geigy)							0.08 phr	0.08 phr	0.08 phr	0.08 phr	0.08 phr
Irganox 1076 (antioxidant - Ciba Geigy)				0.1 phr	0.1 phr	0.1 phr					
TBPP (antioxidant)										0.1 phr	0.3 phr
Ultranox 626 (antioxidant - General Electric)		0.15 phr			0.15 phr			0.1 phr	0.3 phr		
Weston 619 (antioxidant - General Electric)			0.15 phr			0.15 phr					
EXPOSURE CONDITIONS											
Exposure Type	outdoor weathering						outdoor weathering				
Exposure Location							Florida				
Exposure Country	Puerto Rico						USA				
Exposure Note	total radiation is radiation required to reach 50% retention of initial tensile strength						total radiation is radiation required to reach 50% retention of initial				
Total Radiation (Langleys)	43,000	65,000	72,000	43,000	69,000	54,000	24,000	39,000	47,000	26,000	24,000
PROPERTIES RETAINED (%)											
Tensile Strength	50	50	50	50	50	50	50	50	50	50	50

TABLE 68: Outdoor Weathering in California and Pennsylvania of Glass Reinforced Polypropylene.

Material Family	Polypropylene							
Material Supplier	LNP Engineering Plastics							
Reference Number	215	215	215	215	215	215	215	215

MATERIAL COMPOSITION

carbon black	1%	1%	1%	1%	1%	1%	1%	1%
glass fiber reinforcement	30%	30%	30%	30%	30%	30%	30%	30%

EXPOSURE CONDITIONS

Exposure Type	outdoor weathering				outdoor weathering			
Exposure Location	Los Angeles, California	Los Angeles, California	Los Angeles, California	Los Angeles, California	Philadelphia, Pennsylvania	Philadelphia, Pennsylvania	Philadelphia, Pennsylvania	Philadelphia, Pennsylvania
Exposure Country	USA	USA	USA	USA	USA	USA	USA	USA
Exposure Test Method	ASTM D1435	ASTM D1435	ASTM D1435	ASTM D1435	ASTM D1435	ASTM D1435	ASTM D1435	ASTM D1435
Exposure Time (days)	91	182	365	730	91	182	365	730

PROPERTIES RETAINED (%)

Tensile Strength	99	100	105	105	93	94	107	106
Notched Izod Impact Strength	100	100			91	100		
Unnotched Izod Impact Strength	108	104			121	85.7		

TABLE 69: Effect of Stabilizers and Antioxidants on Outdoor Weathering in Puerto Rico of Polypropylene.

Material Family	Polypropylene										
Product Form	200/16 denier natural multifilament samples										
Reference Number	207	207	207	207	207	207	207	207	207	207	207
MATERIAL COMPOSITION											
calcium stearate	0.05 phr	0.05 phr	0.05 phr	0.05 phr	0.05 phr	0.05 phr	0.05 phr	0.05 phr	0.05 phr	0.05 phr	0.05 phr
P-EPQ (antioxidant)						0.15 phr					
Good-rite 3114 (antioxidant - American Cyanamid)	0.1 phr	0.1 phr	0.1 phr	0.1 phr	0.1 phr	0.1 phr	0.1 phr	0.1 phr	0.1 phr	0.1 phr	0.1 phr
Ultranox 626 (antioxidant - General Electric)			0.15 phr						0.1 phr		0.1 phr
Weston TNPP (antioxidant - General Electric)					0.15 phr						
Weston 619 (antioxidant - General Electric)				0.15 phr							
Cyasorb UV 531 (UV absorber - American Cyanamid)	0.15 phr	0.3 phr	0.15 phr	0.15 phr	0.15 phr	0.15 phr					
Tinuvin 144 (UV stabilizer - Ciba Geigy)										0.4 phr	0.3 phr
Tinuvin 770 (hindered amine light stabilizer - Ciba Geigy)							0.3 phr	0.4 phr	0.3 phr		
EXPOSURE CONDITIONS											
Exposure Type	outdoor weathering						outdoor weathering				
Exposure Country	Puerto Rico						Puerto Rico				
Exposure Note	total radiation is radiation required to reach 50% retention of initial tensile strength						total radiation is radiation required to reach 70% retention of initial tensile strength, experiment terminated due to lack of further test samples				
Total Radiation (Langleys)	48,000	59,000	93,000	113,000	55,000	59,000	176,000	>191,000	>191,000	150,000	150,000
PROPERTIES RETAINED (%)											
Tensile Strength	50	50	50	50	50	50	70	70	70	70	70

TABLE 70: Effect of ECC International Microcal Calcium Carbonate on Accelerated Weathering in QUV of Polypropylene.

Material Family	Polypropylene	
Reference Number	159	159

MATERIAL COMPOSITION

calcium carbonate (limestone)		40%
Microcal C110S (calcium carbonate - ECC International)	40%	

EXPOSURE CONDITIONS

Exposure Type	accelerated weathering	accelerated weathering
Exposure Apparatus	QUV	QUV
Exposure Apparatus Note	B lamps	B lamps
Exposure Time (days)	126	126
Exposure Note	4 hours UV, 4 hours moisture	4 hours UV, 4 hours moisture

SURFACE AND APPEARANCE

Δa Color Change (%)	-0.26	-0.2
Δb Color Change (%)	-0.24	0.15
ΔL Color Change (%)	0.38	0.55
Gloss @ 60° Change (%)	-0.3	-0.8

Chapter 40

Polypropylene Copolymer

GRAPH 169: Outdoor Exposure Time vs. Chip Impact Strength of Polypropylene Copolymer

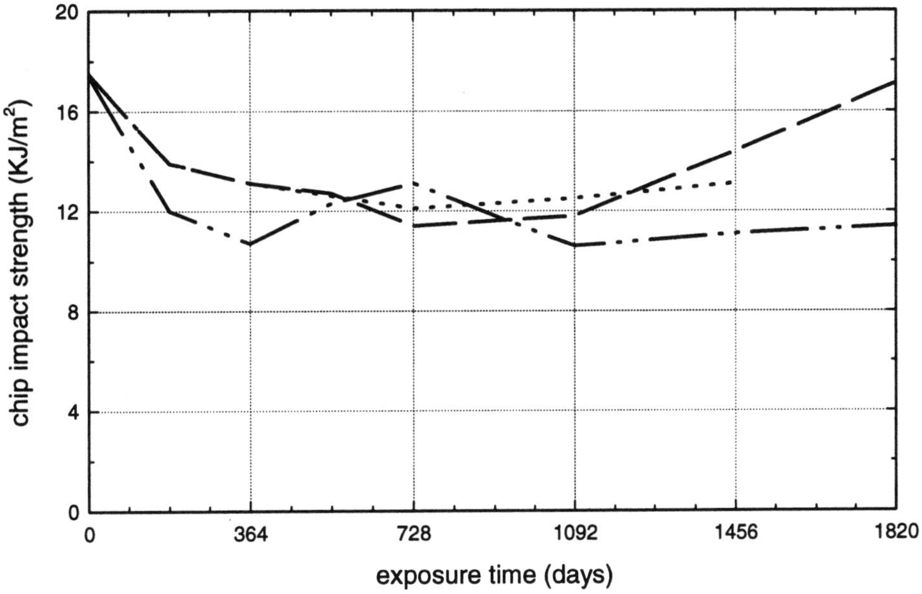

···········	Goodyear LPP30 PP Copol.; Arizona exposure
—··—··—	Goodyear LPP30 PP Copol.; Florida exposure
— — — —	Goodyear LPP30 PP Copol.; Ohio exposure
Reference No.	172

GRAPH 170: Outdoor Exposure Time vs. Delta E Color Change of Polypropylene Copolymer

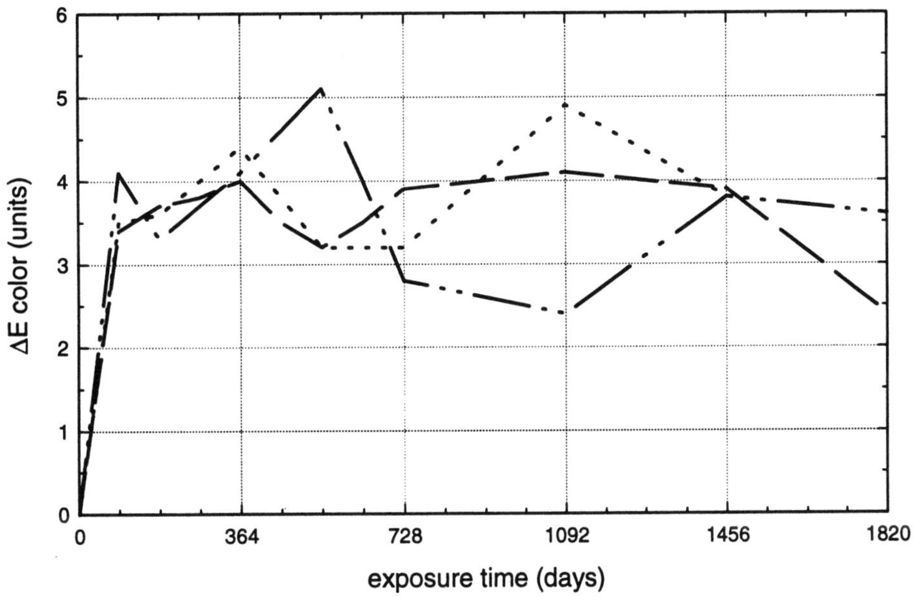

···········	Goodyear LPP30 PP Copol.; Arizona exposure
—··—··—	Goodyear LPP30 PP Copol.; Florida exposure
— — — —	Goodyear LPP30 PP Copol.; Ohio exposure
Reference No.	172

GRAPH 171: Outdoor Exposure Time vs. Flexural Strength of Polypropylene Copolymer

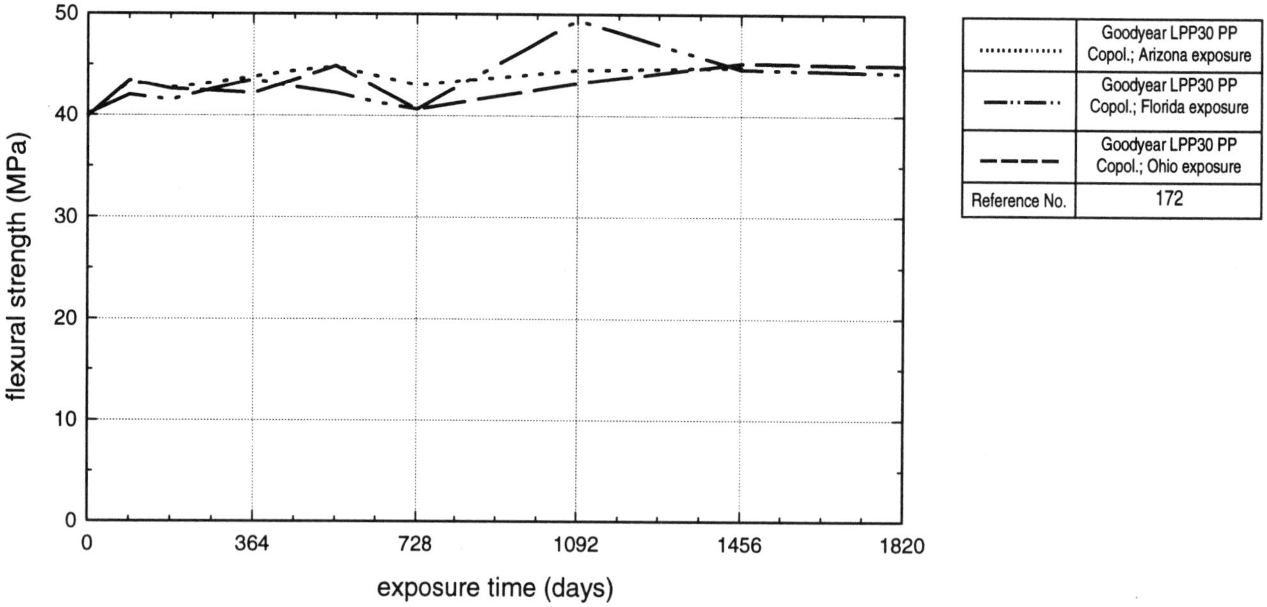

GRAPH 172: Outdoor Exposure Time vs. Tangent Modulus of Polypropylene Copolymer

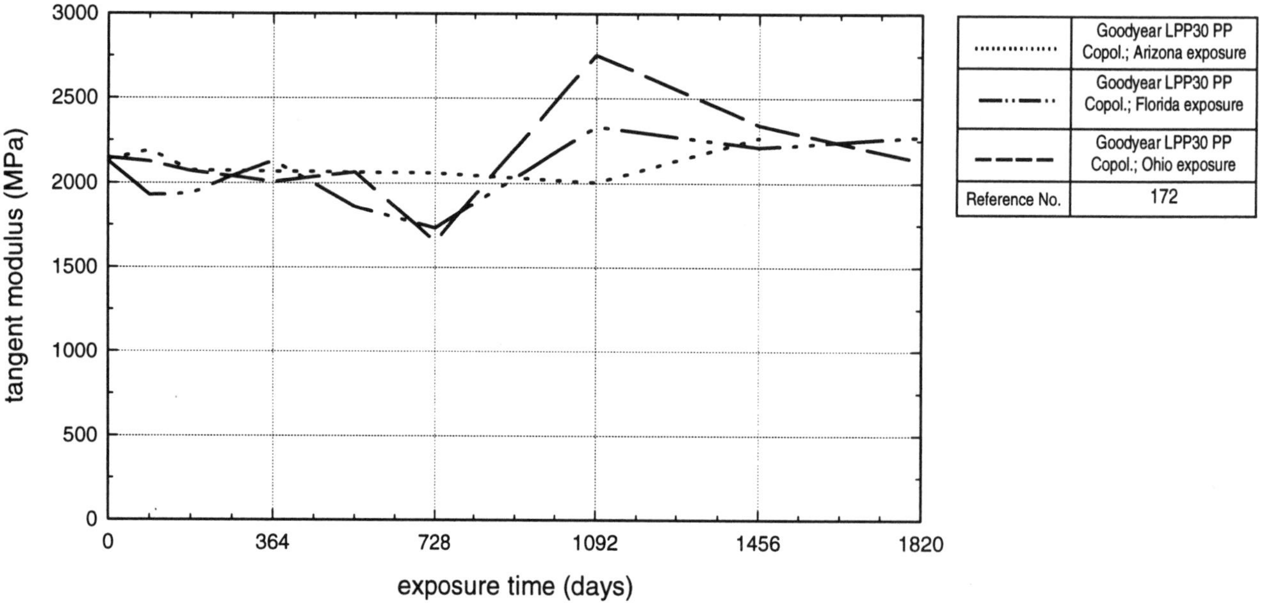

GRAPH 173: **Outdoor Exposure Time vs. Tensile Strength of Polypropylene Copolymer**

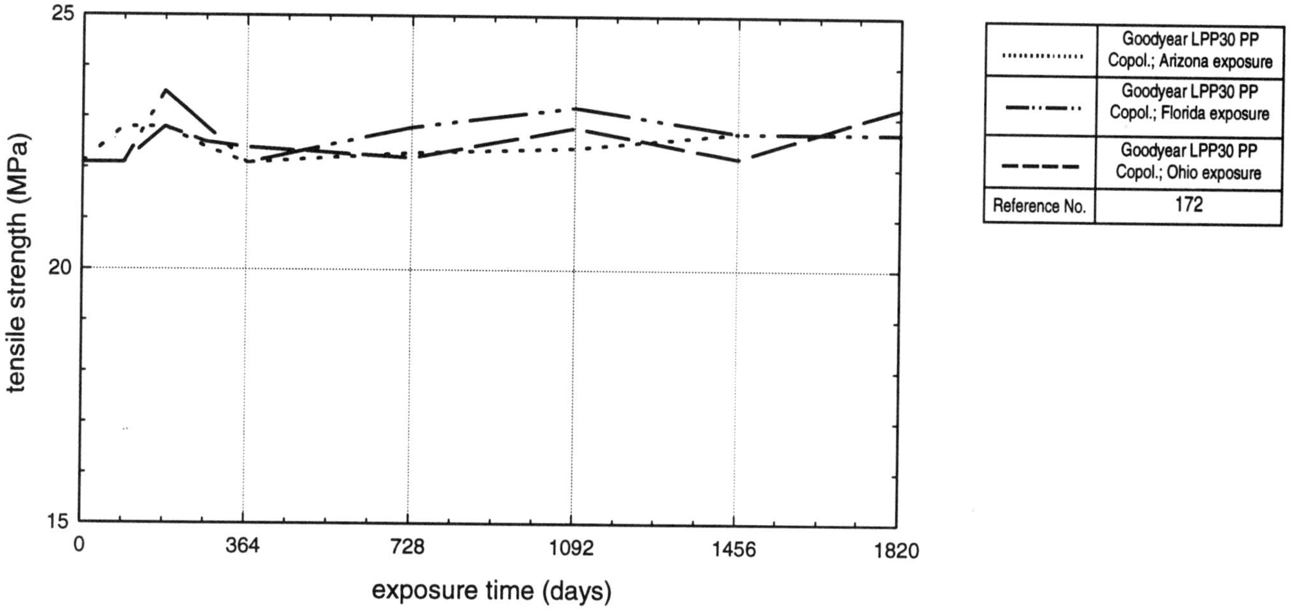

Chapter 41

Polymethylpentene

Weather Resistance

Mitsui: TPX (features: transparent)

The weatherability of TPX is comparable with that of polypropylene. Although TPX is susceptible to UV deterioration, this can be virtually eliminated by adding UV stabilizers (MSW 303).

Reference: *"TPX" Polymethylpentene,* supplier design guide (88.06.3000.Cl.) - Mitsui Petrochemical Industries, Ltd., 1986.

GRAPH 174: **Weatherometer Exposure Time vs. Izod Impact Strength Retained of Polymethylpentene**

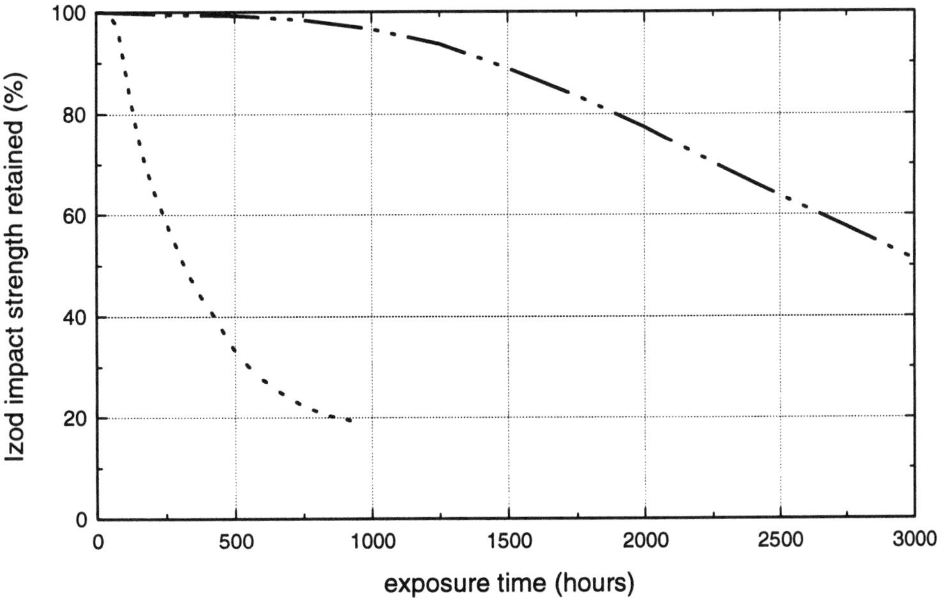

················	Mitsui TPX RT18 PMP (transparent)
—··—··	Mitsui TPX RT18 PMP (transparent; w/ weathering agent)
Reference No.	226

Chapter 42

Polyphenylene Sulfide

Microbiologic Attack

Phillips: Ryton R4 (features: unstabilized; material compostion: 40% glass fiber reinforcement); **Ryton R8** (reinforcement: glass and mineral)

An independent laboratory tested Ryton R-4 and R-8 to determine their resistance to attack by fungus under conditions of open exposure in high humidity, warm atmosphere as well as in the presence of inorganic salts. Samples of Ryton R-4 and R-8 were fungus resistant when tested as specified.

Test standards complied with the procedure outlined in Military Standard 810B Method 508 Procedure 1. Fifteen injection molded plaques were exposed to inoculum on both sides. The plaques were positioned in such a manner as to facilitate equal air circulation. After inoculation, the test specimens were incubated for 28 days with an interim examination after 14 days of incubation. The composite spore suspension was composed of 7-14 day cultures of fungi having confirmed viability.

The viable fugi cultures were as follows:

aspergillus niger
aspergillus flavus
aspergillus versicolor
penicillium funiculosum
chaetonium globosum

Ryton R-4 & R-8	Open Exposure	Mineral Salt Medium
14th Day	No growth	No growth
28th Day	No growth	No growth
Environmental Controls: Cork, Cotton Filter Paper, Leather		
14th Day	Heavy growth	Copious growth
28th Day	Copious growth	Copious growth

Reference: *Ryton Polyphenylene Sulfide Compounds Engineering Properties,* supplier design guide (TSM-266) - Phillips Chemical Company, 1983.

TABLE 71: Accelerated Weathering in an Atlas Weatherometer of Phillips Ryton Polyphenylene Sulfide.

Material Family	Polyphenylene Sulfide						
Material Supplier/ Grade	Phillips Ryton R4				Phillips Ryton R4		
Features	unstabilized						
Reference Number	140	140	140	140	140	140	140
MATERIAL COMPOSITION							
carbon black					2%	2%	2%
glass fiber reinforcement	40%	40%	40%	40%	40%	40%	40%
EXPOSURE CONDITIONS							
Exposure Type	accelerated weathering						
Exposure Apparatus	Atlas weatherometer				Atlas weatherometer		
Exposure Time (days)	83.3	250	333	417	83.3	250	333
PROPERTIES RETAINED (%)							
Tensile Strength	91.3	92.2	86.1	63.5	99.2	99.2	96.7
Elongation	109.1	125.5	111.8	54.5	87.5	77.5	79.2
SURFACE AND APPEARANCE							
Surface Erosion (mm)				0.33			0.51

Chapter 43

General Purpose Polystyrene

Outdoor Weather Resistance

BASF AG: Polystyrol (features: transparent)

Radiation at the ultraviolet end of the solar spectrum is responsible for damage to Polystyrol articles exposed outdoors. The aging thus caused becomes evident by a decrease in mechanical strength and a gradual change in appearance, i.e. in yellowing and loss of gloss. The outdoor performance of Polystyrol colored to dark shades is better than that obtained with pale or transparent hues. For these reasons, Polystyrol cannot be recommended for articles that have to be exposed outdoors for long periods.

Reference: *Polystyrol Product Line, Properties, Processing,* supplier design guide (B 564 e/2.93) - BASF Aktiengesellschaft, 1993.

Indoor UV Light Resistance

BASF AG: Polystyrol 158 K UV (features: uncolored, UV stabilized, transparent, opaque); **Polystyrol 165 H UV** (features: uncolored, UV stabilized, transparent, opaque); **Polystyrol 168 N UV** (features: uncolored, UV stabilized, transparent, opaque)

The resistance to yellowing displayed by UV stabilized Polystyrol is roughly twice as high as that of unstabilized material. Examples of applications for Polystyrol 168 N UV, 165 H UV, and 158 K UV are light diffusers. All these products can be supplied in transparent and opaque shades.

Reference: *Polystyrol Product Line, Properties, Processing,* supplier design guide (B 564 e/2.93) - BASF Aktiengesellschaft, 1993.

BASF AG: Polystyrol (features: transparent)

Under the normal conditions of light and temperature encountered indoors, Polystyrol moldings retain their appearance and perform their functions efficiently for many years.

Reference: *Polystyrol Product Line, Properties, Processing,* supplier design guide (B 564 e/2.93) - BASF Aktiengesellschaft, 1993.

TABLE 72: Outdoor Weathering in California and Pennsylvania of Glass Reinforced General Purpose Polystyrene.

Material Family	General Purpose Polystyrene							
Material Supplier	LNP Engineering Plastics							
Reference Number	215	215	215	215	215	215	215	215
MATERIAL COMPOSITION								
carbon black	1%	1%	1%	1%	1%	1%	1%	1%
glass fiber reinforcement	30%	30%	30%	30%	30%	30%	30%	30%
EXPOSURE CONDITIONS								
Exposure Type	outdoor weathering				outdoor weathering			
Exposure Location	Los Angeles, California	Los Angeles, California	Los Angeles, California	Los Angeles, California	Philadelphia, Pennsylvania	Philadelphia, Pennsylvania	Philadelphia, Pennsylvania	Philadelphia, Pennsylvania
Exposure Country	USA	USA	USA	USA	USA	USA	USA	USA
Exposure Test Method	ASTM D1435	ASTM D1435	ASTM D1435	ASTM D1435	ASTM D1435	ASTM D1435	ASTM D1435	ASTM D1435
Exposure Time (days)	91	182	365	730	91	182	365	730
PROPERTIES RETAINED (%)								
Tensile Strength	91	85.3	89	91	86	88	94	85
Notched Izod Impact Strength	122	100			100	100		
Unnotched Izod Impact Strength	83	83.3			96	87.5		

GRAPH 175: Fadeometer Exposure Time vs. Yellowness Index of General Purpose Polystyrene

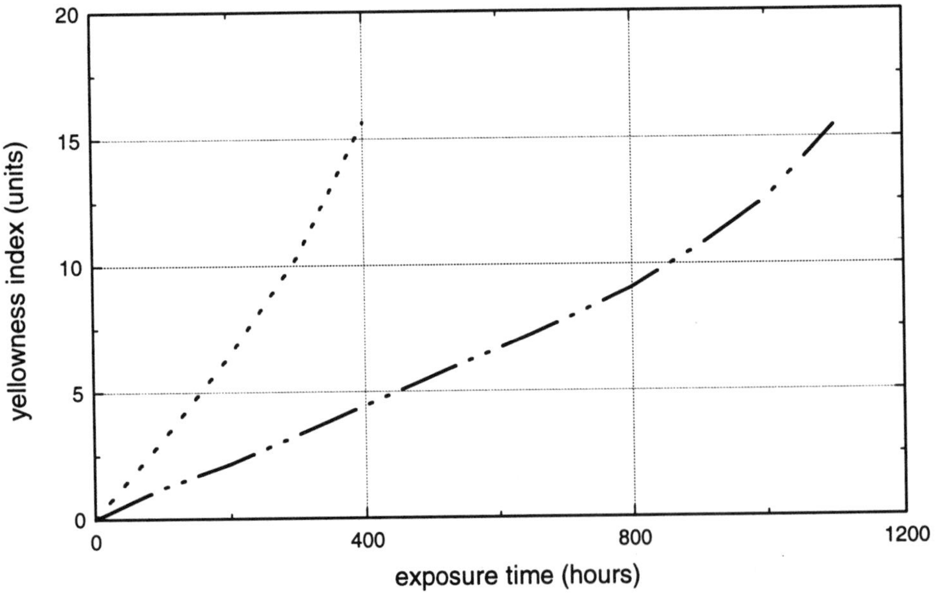

....	GPPS (transparent); Atlas fadeometer
—··—··	GPPS (UV stabilized, transparent); Atlas fadeometer
Reference No.	230

GRAPH 176: Fluorescent Lamp Exposure Time vs. Yellowness Index of General Purpose Polystyrene

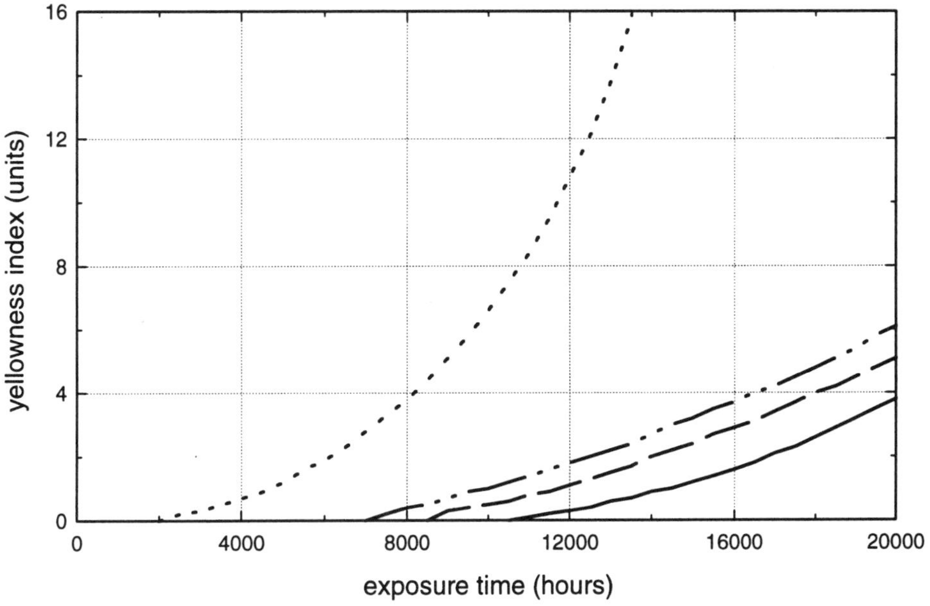

	BASF AG Polystyrol 168 N 013 GPPS (transparent); 2 40 watt fluorescent lamps 1 cm from sample
	BASF AG Polystyrol 168 N UV 72819 GPPS (uncolored, UV stabilized, transparent); 2 40 watt fluorescent lamps 1 cm from sample
	BASF AG Polystyrol 168 N UV 72819 GPPS (uncolored, UV stabilized, transparent); 2 40 watt fluorescent lamps 3 cm from sample
	BASF AG Polystyrol 168 N UV 72819 GPPS (uncolored, UV stabilized, transparent); 2 40 watt fluorescent lamps 6 cm from sample
Reference No.	26

Chapter 44

Impact Polystyrene

Outdoor Weather Resistance

BASF AG: Polystyrol

Radiation at the ultraviolet end of the solar spectrum is responsible for damage to Polystyrol articles exposed outdoors. The aging thus caused becomes evident by a decrease in mechanical strength and a gradual change in appearance, i.e. in yellowing and loss of gloss. The outdoor performance of Polystyrol colored to dark shades is better than that obtained with pale or transparent hues. For these reasons, Polystyrol cannot be recommended for articles that have to be exposed outdoors for long periods.

Reference: *Polystyrol Product Line, Properties, Processing,* supplier design guide (B 564 e/2.93) - BASF Aktiengesellschaft, 1993.

Dow Chemical: Styron 6075 (features: natural resin, flame retardant); Styron (features: unmodified resin)

Styron polystyrene is not considered a weather-resistant plastic. Outdoor weathering is a process involving the interaction of atmosphere, sunlight (UV light), humidity, temperature, wind and precipitation. Continuous long-term outdoor exposure of polystyrene parts results in both polymer discoloration and reduction in material strength and toughness.

Unmodified HIPS resins usually experience greater change from outdoor exposure than do general purpose polystyrene formulations. HIPS resins usually show less change than resins modified with ignition-resistant chemical additives.

Light colors are not recommended for Styron 6000 Series resins unless molded surfaces are painted or otherwise protected from exposure to direct sunlight. Black or brown colors may be used for unpainted or untreated parts and will improve stability.

Reference: *Styron 6000 Ignition Resistant Polystyrene Resins,* supplier marketing literature (301-01673-192R SMG) - Dow Chemical Company, 1992.

IPS

In contrast to ASA, high impact polystyrene articles exposed outdoors undergo changes in mechanical properties and color after a comparatively short period of time. Solar radiation, particularly at the ultraviolet end of the spectrum, acts together with atmospheric oxygen to cause embrittlement and yellowing. These changes occur mainly in the butadiene elastomer.

Reference: *Luran S Acrylonitrile Styrene Acrylate Product Line, Properties, Processing,* supplier design guide (B 566 e / 11.90) - BASF Aktiengesellschaft, 1990.

Indoor UV Light Resistance

BASF AG: Polystyrol

Under the normal conditions of light and temperature encountered indoors, Polystyrol moldings retain their appearance and perform their functions efficiently for many years.

Reference: *Polystyrol Product Line, Properties, Processing,* supplier design guide (B 564 e/2.93) - BASF Aktiengesellschaft, 1993.

GRAPH 177: Fadeometer Exposure Time vs. Yellowness Index of Impact Polystyrene

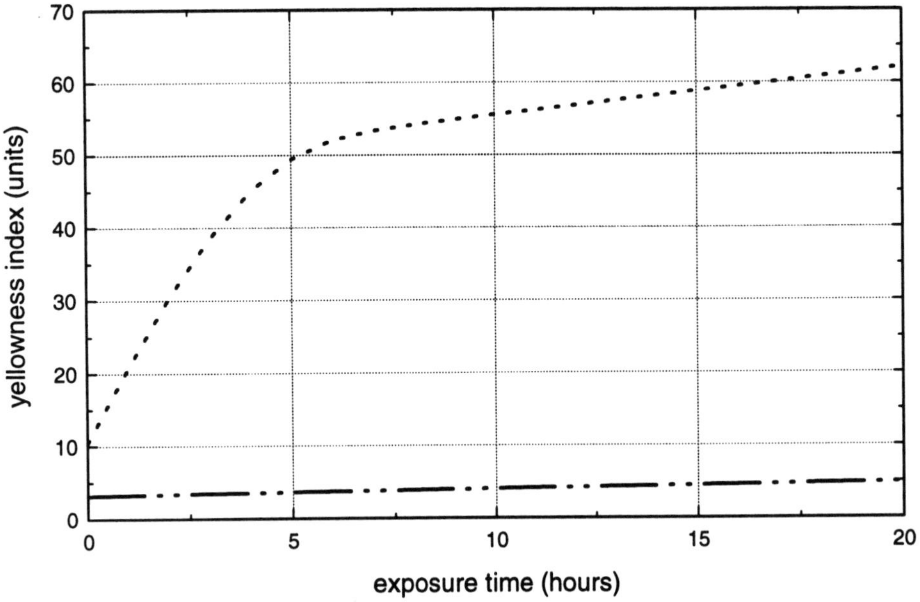

···············	Dow Styron 6075 IPS (natural resin, flame retardant)
—··—··	Dow Styron IPS (unmodified resin)
Reference No.	196

Chapter 45

Polysulfone

Outdoor Weather Resistance

Amoco Performance Products: Udel (features: transparent, amber tint)

Except at very low thicknesses, polysulfone absorbs almost totally in the ultraviolet portion of the spectrum. This accounts for the polymer's rather limited weathering resistance, i.e., its tendency to darken and lose toughness upon outdoor exposure for extended periods of time.

Reference: *The Radiation Response of Udel Polysulfone,* supplier technical report (Number: 101) - Amoco Performance Products, Inc.

TABLE 73: Outdoor Weathering in California and Pennsylvania of Glass Reinforced Polysulfone.

Material Family	Polysulfone							
Material Supplier	LNP Engineering Plastics							
Reference Number	215	215	215	215	215	215	215	215
MATERIAL COMPOSITION								
carbon black	1%	1%	1%	1%	1%	1%	1%	1%
glass fiber reinforcement	30%	30%	30%	30%	30%	30%	30%	30%
EXPOSURE CONDITIONS								
Exposure Type	outdoor weathering				outdoor weathering			
Exposure Location	Los Angeles, California	Los Angeles, California	Los Angeles, California	Los Angeles, California	Philadelphia, Pennsylvania	Philadelphia, Pennsylvania	Philadelphia, Pennsylvania	Philadelphia, Pennsylvania
Exposure Country	USA	USA	USA	USA	USA	USA	USA	USA
Exposure Test Method	ASTM D1435	ASTM D1435	ASTM D1435	ASTM D1435	ASTM D1435	ASTM D1435	ASTM D1435	ASTM D1435
Exposure Time (days)	91	182	365	730	91	182	365	730
PROPERTIES RETAINED (%)								
Tensile Strength	96	94.2	95	95	91	95	92	91
Notched Izod Impact Strength	100	90.9			100	90.9		
Unnotched Izod Impact Strength	100	88.2			98	90.2		

GRAPH 178: Xenon Arc Weatherometer Exposure Time vs. Tensile Strength of Polysulfone

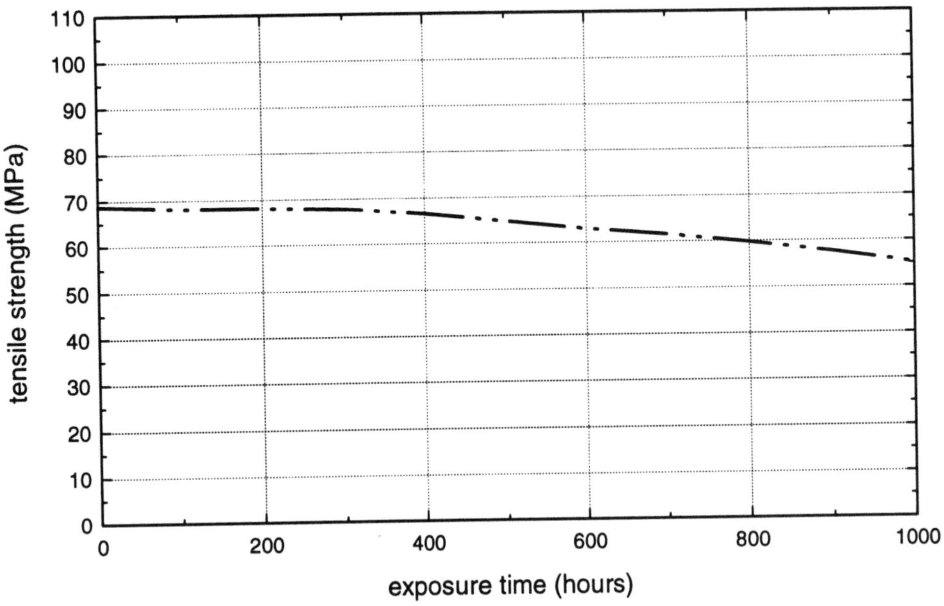

	Polysulfone; 0.35 W/m² irradience at 340 nm, 63°C
Reference No.	51

Chapter 46

Polyethersulfone

Outdoor Weather Resistance

BASF AG: Ultrason E (features: transparent, amber tint); **Ultrason E** (features: transparent, amber tint)

In common with most other aromatic polymers, Ultrason gives rise to moldings that yellow and embrittle quite rapidly if they are exposed outdoors. The adverse effect of ultraviolet radiation can be alleviated by adding activated carbon black to the Ultrason. More effective protection is given by surface coating or metallizing.

Reference: *Ultrason E, Ultrason S Product Line, Properties, Processing,* supplier design guide (B 602 e/10.92) - BASF Aktiengesellschaft, 1992.

GRAPH 179: Xenon Arc Weatherometer Exposure Time vs. Tensile Strength of Polyethersulfone

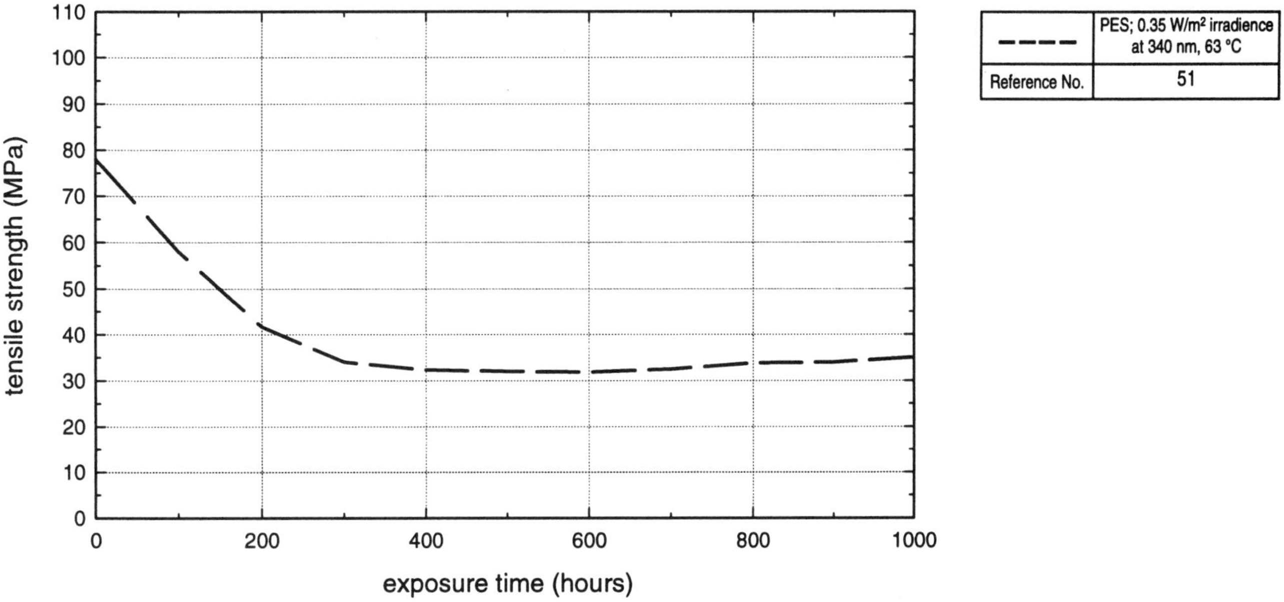

Chapter 47

Styrene Acrylonitrile Copolymer

Outdoor Weather Resistance

BASF AG: Luran (features: transparent)

The mechanical properties of Luran specimens deteriorate after one to two years' outdoor exposure (at an angle of 45° facing south in Ludwigshafen, Germany). The extent to which the mechanical properties are impaired depends on the nature of the specimen and the test procedure. Other consequences of outdoor exposure are yellowing and a rough surface. The Luran resins are also offered in a UV-stabilized form. It can be seen that the rate of decrease in flexural strength is much less for Luran resins containing ultra-violet stabilizers. Another advantage of ultra-violet stabilization is that the color retention is considerably improved.

Reference: *Luran Product Line, Properties, Processing,* supplier design guide (B 565 e/10.83) - BASF Aktiengesellschaft, 1983.

TABLE 74: Outdoor Weathering in Arizona of Dow Chemical Tyril Styrene Acrylonitrile Copolymer.

Material Family	Styrene Acrylonitrile Copolymer
Material Supplier/ Grade	Dow Chemical Tyril 1020
Features	transparent, UV stabilized
Reference Number	195
EXPOSURE CONDITIONS	
Exposure Type	outdoor exposure
Exposure Location	Arizona
Exposure Country	USA
Exposure Time (days)	1095
PROPERTIES RETAINED (%)	
Tensile Strength	101
Elongation	100
SURFACE AND APPEARANCE	
Δ Yellowness Index	5.5
Haze Retained (%)	143.2
Luminous Transmittance Retained (%)	99.8

GRAPH 180: Arizona Outdoor Weathering Exposure Time vs. Yellowness Index of Styrene Acrylonitrile Copolymer

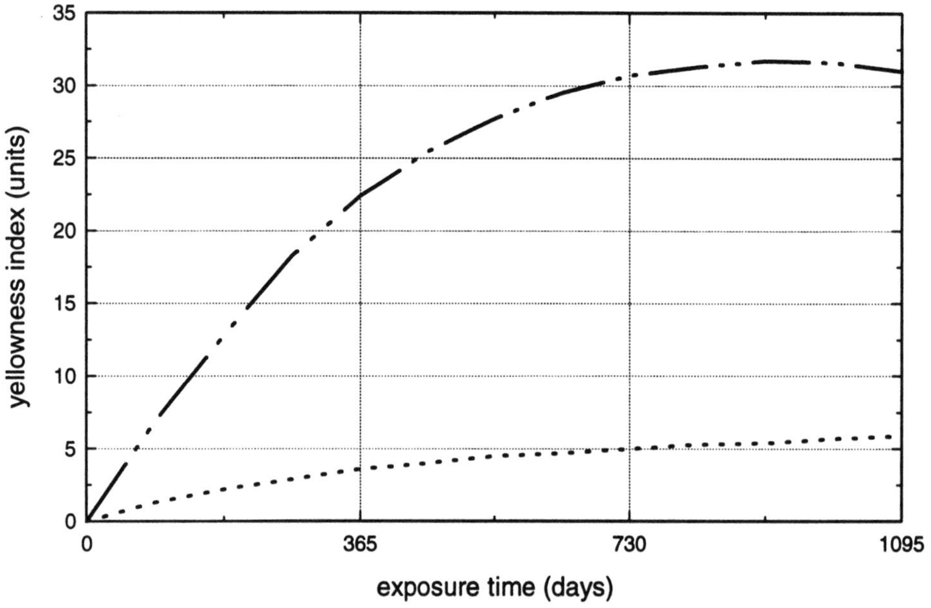

···············	Dow Tyril 1020 SAN (UV stabilized, transparent)
—·· —··	Dow Tyril SAN (transparent, gen. purp. grade)
Reference No.	195

GRAPH 181: UV-CON Accelerated Weathering Exposure Time vs. Yellowness Index of Styrene Acrylonitrile Copolymer

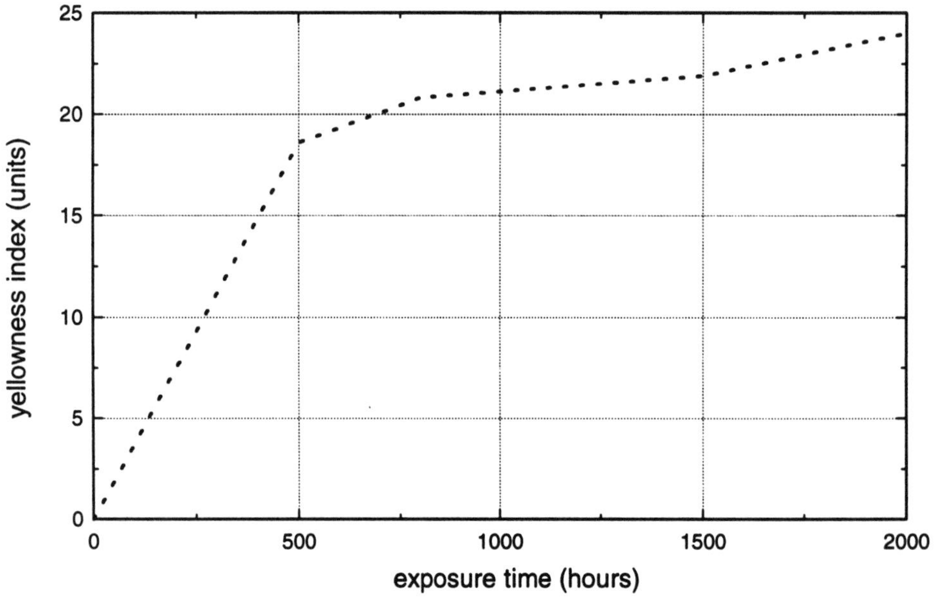

···············	SAN (transparent)
Reference No.	231

Chapter 48

Olefin Modified Styrene Acrylonitrile Copolymer

Outdoor Weather Resistance

Dow Chemical: Rovel 401 (features: natural resin, extrusion grade); **Rovel 701** (features: natural resin, injection molding grade)

The effect of outdoor exposure has been monitored by means of an unnotched "chip impact" test. Chip impact is an Izod-type test in which a 1/2-inch wide sample strip (typically 0.762 to 3.175 mm thick) is supported at one end and struck on the face with a pendulum hammer. This test is very sensitive to degradation (embrittlement, formation of microcracks) of the surface layer of plastic samples such as can occur from UV exposure and other environmental attack. The chip impact test is ideal for measuring weatherability. The results of this combination of severe exposure and critical test procedure indicate that products made of Rovel can have a long service life despite rigorous outdoor environments.

Reference: *Rovel Weatherable Polymers,* supplier technical report (301-621-285) - Dow Chemical Company, 1985.

Dow Chemical: Rovel 501 (features: high impact, white color, weatherable)

After three years of Florida aging white Rovel 501 retained 80% of its drop weight impact properties. Aged white and brown Rovel samples showed a minimum of change in Hunter color meter readings.

Reference: *Rovel Weatherable High Impact Polymers,* supplier marketing literature (301-622-285) - Dow Chemical Company, 1985.

TABLE 75: Outdoor Weathering in Florida of Dow Chemical Rovel Olefin Modified Styrene Acrylonitrile Copolymer.

Material Family	Olefin Modified Styrene Acrylonitrile Copolymer	
Material Supplier/ Grade	Dow Chemical Rovel	Dow Chemical Rovel
Features	high impact, UV stabilized, white color	high impact, UV stabilized, brown color
Reference Number	116	116
EXPOSURE CONDITIONS		
Exposure Type	outdoor weathering	outdoor weathering
Exposure Location	Florida	Florida
Exposure Country	USA	USA
Exposure Note	45° angle south	45° angle south
Exposure Time (days)	365	365
SURFACE AND APPEARANCE		
Δa Color Change	0.48 {c}	-0.66 {c}
Δb Color Change	-0.12 {c}	0.23 {c}
ΔL Color Change	-0.86 {c}	0.65 {c}
ΔE Color Change	0.99 {c}	0.95 {c}

TABLE 76: Outdoor Weathering in Florida of White Dow Chemical Rovel Olefin Modified Styrene Acrylonitrile Copolymer.

Material Family	Olefin Modified Styrene Acrylonitrile Copolymer
Material Supplier/ Grade	Dow Chemical Rovel
Features	white color
Reference Number	194

EXPOSURE CONDITIONS

Exposure Type	outdoor weathering
Exposure Location	Florida
Exposure Country	USA
Exposure Note	45° south
Exposure Time (days)	365

SURFACE AND APPEARANCE

ΔE Color	0.26
Yellowness Index	14.2 {q}
Δ Yellowness Index	-0.4 {q}
Yellowness Index	11.7 {r}
Δ Yellowness Index	-0.2 {r}
Whiteness Index	45.4
Δ Whiteness Index	1.0

GRAPH 182: Florida Outdoor Weathering Exposure Time vs. Drop Weight Impact Retained of Olefin Modified Styrene Acrylonitrile Copolymer

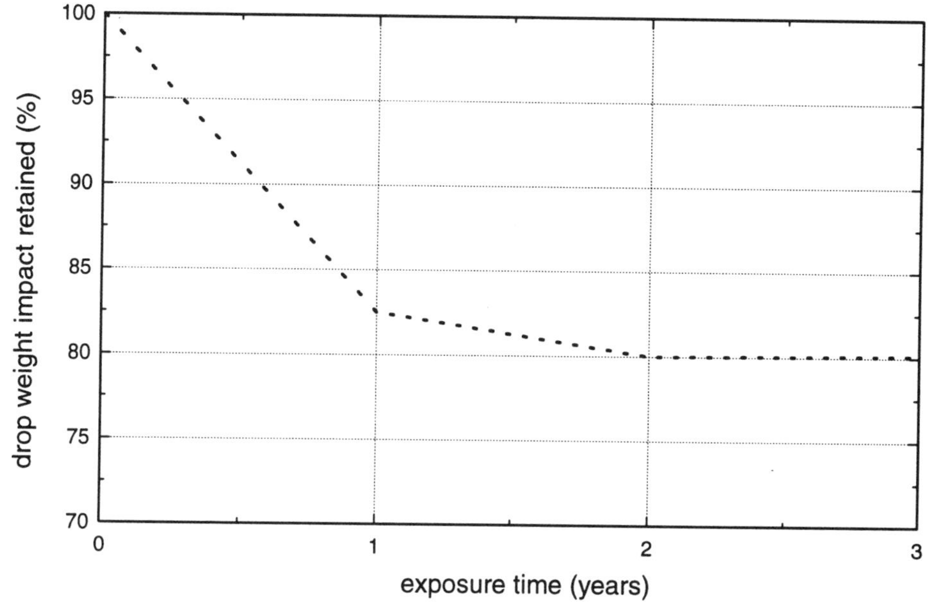

....	Dow Rovel 501 OSA (high impact, white, weatherable); 45° angle facing south
Reference No.	116

GRAPH 183: Florida Weathering Exposure Time vs. Chip Impact Strength of Olefin Modified Styrene Acrylonitrile Copolymer

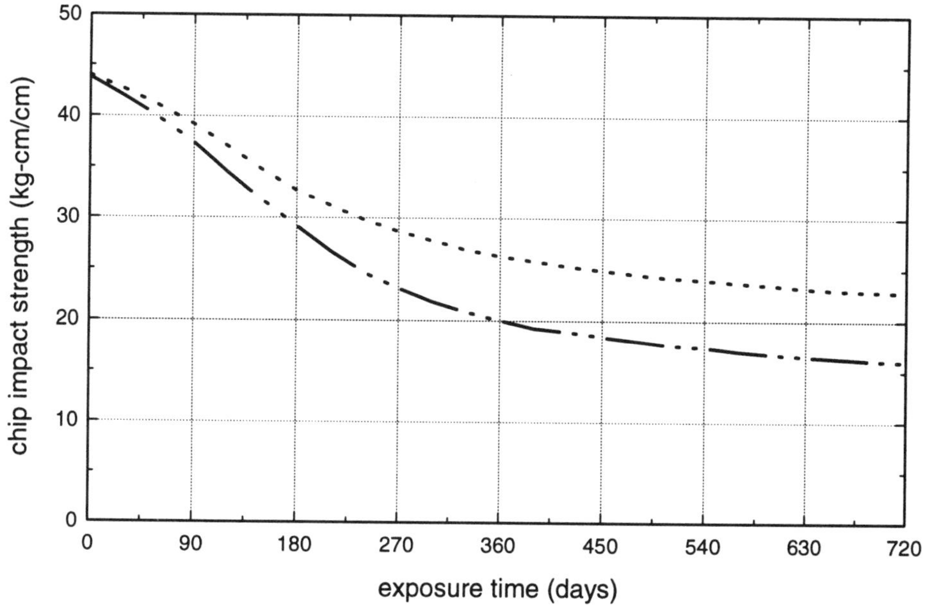

....	Dow Rovel 401 OSA (natural resin, extrusion grade); 45° angle facing south
—·—··	Dow Rovel 701 OSA (natural resin, inj. mold. grade); 45° angle facing south
Reference No.	194

Chapter 49

Styrene Butadiene Copolymer

Outdoor Weather Resistance

Phillips: K-Resin KR01 (features: transparent, unstabilized); **K-Resin KR03** (features: impact modified, transparent; material compostion: 0.5% Tinuvin 770 (hindered amine light stabilizer - Ciba Geigy)); **K-Resin** (features: impact modified, transparent, unstabilized); **K-Resin** (features: impact modified, transparent; material compostion: 0.5% Tinuvin P (UV absrober - Ciba Geigy))

K-Resin copolymers are styrenic copolymers, and with long term exposure to direct sunlight will yellow and ultimately craze and embrittle. Previous weatherometer testing shows K-Resin susceptibility to yellowing and physical property deterioration induced by UV light. With the addition of appropriate UV stabilizers, the estimated life of a K-Resin part may be extended. For general reference, K-Resin will withstand UV exposure for approximately 3 months before excessive brittleness and yellowing are evident. Modifying K-Resin with commercially available UV stabilizers will extend the outdoor life of the material up to 18 months depending on the stabilizer package used.

Reference: *K-Resin SB Copolymers UV Stabilization - Plastics Technical Center Report #412,* supplier technical report (778-93 K 01) - Phillips Petroleium Company, 1994.

Indoor UV Light Resistance

Phillips: K-Resin KR01 (features: transparent, unstabilized); **K-Resin KR03** (features: impact modified, transparent, unstabilized); **K-Resin** (features: impact modified, transparent; material compostion: 0.5% Tinuvin P (UV absrober - Ciba Geigy)); **K-Resin** (features: impact modified, transparent; material compostion: 0.5% Tinuvin 770 (hindered amine light stabilizer - Ciba Geigy))

K-Resin polymers are used in many display applications that are often subjected to reduced levels of UV light, by indirect sunlight or by fluorescent lighting. A study was conducted to determine the extent of K-Resin yellowing under those conditions.

Four samples evaluated in the test included KR01, KR03 and KR03 with two different UV stabilizers, Tinuvin P and Tinuvin 770. Samples were prepared by adding 0.5% of each UV stabilizer to a K-Resin sample and dryblending the sample prior to extrusion and pelletizing. Tinuvin P is a UV absorber made by Ciba Geigy. The Tinuvin 770 is a hindered amine stabilizer also supplied by Ciba Geigy. Clarity of these blends was very good, but there is some slight yellow color developed when the UV stabilizers are initially added to K-Resin polymer.

These samples were compression molded into 15.24 mm plaques. One set was placed in a UV-Con tester, approximately 4" from a bank of 4 fluorescent lights. Another set was placed on a window ledge, exposed to indirect sunlight. A third set was placed in a dark container so the specimens were not exposed to any light source. Hunter "b" color was measured initially and at intervals of 3, 6, 12, 18 and 24 months.

As expected, the non UV stabilized KR01 and KR03 responded similarly in all circumstances. The UV stabilized polymers demonstrated improved resistance to yellowing when exposed to UV light sources. When no light source is present, none of the K-Resin samples appreciably yellowed over the two year period.

In the UV-Con, which was the most severe test, addition of a UV additive substantially improved performance. KR01 and KR03 unmodified samples discolored significantly between 6 and 12 months, and continued to further discolor with extended exposure. The Tinuvin P modified KR03 outperformed Tinuvin 770 modified KR03, after two years having about half as much yellow color development.

In indirect sunlight, the unmodified KR01 and KR03 samples discolored most significantly after 12 months. The KR03 sample containing 0.5% Tinuvin P performed better, with only minimal yellowing over the two-year test period. The Tinuvin 770 modified KR03 did not perform as well as the Tinuvin P modified KR03 discoloring significantly after 12 months.

K-Resin polymers, which will yellow when subjected to long-term direct exposure to sunlight can also yellow in less severe UV exposure. That yellowing can be significantly reduced by the addition of UV stabilizers. Tinuvin P is more effective than Tinuvin 770 for fluorescent and indirect light exposure. K-Resin parts stored in dark storage showed no significant yellowing over at least a two-year period.

Reference: *K-Resin SB Copolymers UV Stabilization - Plastics Technical Center Report #412,* supplier technical report (778-93 K 01) - Phillips Petroleium Company, 1994.

Chapter 50

Styrene Butadiene Block Copolymer

Accelerated Artificial Weathering Resistance

BASF AG: Styrolux 656 C (features: unstabilized); **Styrolux** (features: UV stabilized); **Styrolux** (features: transparent)

High-energy solar radiation damages Styrolux articles exposed outdoors. They turn yellow, and their mechanical properties subsequently deteriorate.

The time before yellowing commences can be substantially lengthened by means of ultraviolet stabilizers. Specimens were exposed in the Xenotest 1200 under the conditions specified in DIN 53 387 - a filter system corresponding to solar radiation, a black panel temperature of 45°C, a relative humidity of 65%, and a dry period of 102 minutes in the cycle. Styrolux 656 C unstabilized showed visible color change after 14 days exposure in the Xenotest. After 120 days, the sample was a bright yellow. UV stabilized Styrolux began to show visible color change after 60 days exposure in the Xenotest. After 120 days, the specimen was a very dull yellow.

Reference: *Styrolux Product Line, Properties, Processing,* supplier design guide (B 583 e/(950) 12.91) - BASF Aktiengesellschaft, 1992.

Chapter 51

Polyvinyl Chloride

Outdoor Weather Resistance

Geon Company: Geon 85856 (features: weatherable, exterior grade); **Geon 85890** (features: weatherable, exterior grade); **Geon 85891** (features: weatherable, exterior grade); **Geon 87371** (features: indoor UV stable); **Geon**

Many common plastic articles are used outdoors. These include vinyl house siding and vinyl gutters that have performed successfully for over twenty-five years. Vinyl, when formulated and processed properly, has excellent weatherability.

BF Goodrich has incorporated this weathering technology into high flow and/or high impact, injection molding compounds. These compounds are comparable with polycarbonate and exceed modified PPO and ABS for physical property retention and colors change.

Outdoor exposure subjects all plastics to three harmful effects: sunlight, moisture, and heating. Sunlight is composed of a spectrum of wavelengths and the higher energy ultraviolet wavelengths of 290-400 nanometers cause some disruption of the chemical structure. Vinyl is most sensitive to a wavelength of 320 nm, which can lead to loss of hydrogen chloride from the polymer backbone, leaving double-bonds that contribute to some yellowing. Disruption of the chemical structure also leads to oxidation and formation of peroxides, aldehydes, ketones, and alcohols.

Moisture accelerates the rate of decomposition by eroding the by-products of weathering. While vinyl does not support the growth of fungus and mildew, moisture and dirt may permit some growth on the surface of the plastic.

Thermal distortion often occurs because of differential expansion, due to temperature gradients which cause hot portions to expand more than cooler portions of the product. With the proper formulation and part design, problems of thermal distortion can be minimized. Heat build-up in applications exposed to the sun, softening temperature, and coefficient of thermal expansion are all influenced by the formulation.

Plastics that have weathered usually have some change in color as well as some change in physical properties.

BF Goodrich performs weathering tests according to ASTM D 1435 in Arizona (hot and dry), in South Florida (hot and humid), and in Ohio (temperate, with industrial pollution present). Samples are inclined at 45°, facing south, without insulative backing.

Slight changes occur in the physical properties of weatherable Geon Vinyl during weathering exposure. The tensile strength of Geon Vinyl at yield normally increases after weathering and there is actually an increase in tensile strength at yield during weathering in Arizona. The weatherability of Rigid Geon Vinyl is indicated by the retention of elongation and higher ductility after weathering. Weatherable Geon Compounds retain excellent dropped dart toughness after exposure in Arizona, Florida, and Ohio. They also have very good color fastness; however, white vinyl normally increases in yellowness for about one year then bleaches to a less yellow color. It should be noted that Geon 87371 is not designed for outdoor use and loses toughness steadily.

Reference: *Engineering Design Guide To Rigid Geon Custom Injection Molding Vinyl Compounds,* supplier design guide (CIM-020) - BFGoodrich Geon Vinyl Division, 1989.

Huls AG: Vestolit BAU (chemical type: graft polymer; product form: profile extrusion)

The degree of light fastness and weather resistance is a decisive criterion for PVC windows, as has been shown in numerous tests and documented in wide-ranging technical publications. The weather resistance of finished PVC windows depends, however, on a number of factors. The use of light-fast modifiers, e.g. acrylate-based, is the first essential. Graft polymers perform better in this respect because they are more homogeneous than mechanically mixed, so-called polyblends, albeit with a chemically identical or similar base. The use of approved stabilizing systems is necessary, but a graft polymer produced by the suspension method is always more easy to stabilize than, for example, an unmodified PVC produced by the emulsion process. Despite this, emulsion PVC is often used commercially to replace some of the suspension PVC.

Profile extrusion also plays a critical role. The extrudability of Vestolit Bau has a direct and positive effect on the light fastness and weather resistance of the PVC windows made from it.

Reference: *Vestolit BAU For World-Wide Windows,* supplier technical report (1083e/May 1987/bu) - Huls AG, 1987.

PVC (features: white color, exterior grade)

After three years of Florida aging, white exterior grade PVC retained 56% of its drop weight impact properties. Aged white and brown UV stabilized PVC showed a meaningful change in Hunter color meter readings.

Reference: *Rovel Weatherable High Impact Polymers,* supplier marketing literature (301-622-285) - Dow Chemical Company, 1985.

Indoor UV Light Resistance

Geon Company: Geon 87371 (features: indoor UV stable)

Geon 87371 is a light stable compound used in many indoor applications. Results illustrate the excellent color retention of this compound in HPUV exposure for 300 hours at 42°C (108°F). Plastics which are used indoors for business machines, appliances, or furniture are exposed to fluorescent light that is less severe than direct sunlight. The HPUV test can be used to simulate the effect of fluorescent lighting and filtered sunlight. The HPUV test uses two lamps simultaneously, a cool white fluorescent lamp and a filtered sunlamp test.

Reference: *Engineering Design Guide To Rigid Geon Custom Injection Molding Vinyl Compounds,* supplier design guide (CIM-020) - BFGoodrich Geon Vinyl Division, 1989.

Effect of Bacterial Attack and Fungus

Geon Company: Geon Duracap (product form: capstock)

Evaluation of weathered DuraCap samples do not indicate any fungus growth or show any evidence of mildew. The fungus and mildew resistance of capstock is equivalent to rigid PVC.

Reference: *Duracap Vinyl Capstock Compounds,* supplier marketing literature (DC-001) - BFGoodrich Geon Vinyl Division, 1988.

TABLE 77: Outdoor Weathering in Arizona, Florida and Ohio of Gold Geon Company Geon Polyvinyl Chloride.

Material Family	Polyvinyl Chloride								
Material Supplier/ Grade	Geon Company Geon 86101 gold 070								
Features	capstock compound, gold color								
Reference Number	192	192	192	192	192	192	192	192	192
EXPOSURE CONDITIONS									
Exposure Type	outdoor weathering			outdoor weathering			outdoor weathering		
Exposure Location	Avon Lake, Ohio			Arizona			Florida		
Exposure Country	USA			USA			USA		
Exposure Time (days)	183	365	730	183	365	730	183	365	730
SURFACE AND APPEARANCE									
Visual Appearance	slight change (pass)	slight change (pass)	no to slight change (pass)	no to slight change (pass)	no to slight change (pass)	slight change (pass)	slight change (pass)	no to slight change (pass)	no to slight change (pass)
ΔE Color Change	0.5	2.4	4.8	0.4	1.2	1.5	1.6	2.7	1.5

TABLE 78: Outdoor Weathering in Arizona, Florida and Ohio of Yellow Geon Company Geon Polyvinyl Chloride.

Material Family	Polyvinyl Chloride								
Material Supplier/ Grade	Geon Company Geon 86101 yellow 650								
Features	capstock compound, yellow color								
Reference Number	192	192	192	192	192	192	192	192	192
EXPOSURE CONDITIONS									
Exposure Type	outdoor weathering			outdoor weathering			outdoor weathering		
Exposure Location	Avon Lake, Ohio			Arizona			Florida		
Exposure Country	USA			USA			USA		
Exposure Time (days)	183	365	730	183	365	730	183	365	730
SURFACE AND APPEARANCE									
Visual Appearance	no to slight change (pass)	slight change (pass)	no to slight change (pass)	no change from original	no to slight change (pass)	slight change (pass)	no to slight change (pass)	no to slight change (pass)	no change from original
ΔE Color Change	1.1	3.8	5.1	1.2	1.8	1.9	2.3	3.1	1.2

TABLE 79: Outdoor Weathering in Arizona, Florida and Ohio of White Geon Company Geon Polyvinyl Chloride.

Material Family	Polyvinyl Chloride								
Material Supplier/ Grade	Geon Company Geon 86101 white 145								
Features	capstock compound, white color								
Reference Number	192	192	192	192	192	192	192	192	192
EXPOSURE CONDITIONS									
Exposure Type	outdoor weathering			outdoor weathering			outdoor weathering		
Exposure Location	Avon Lake, Ohio			Arizona			Florida		
Exposure Country	USA			USA			USA		
Exposure Time (days)	183	365	730	183	365	730	183	365	730
SURFACE AND APPEARANCE									
Visual Appearance	slight change (pass)	slight change (pass)	slight change (pass)	no to slight change (pass)	no change from original	no to slight change (pass)	no to slight change (pass)	no to slight change (pass)	no to slight change (pass)
ΔE Color Change	2.2	4.2		0.2	1.3		1.8	2.7	

TABLE 80: Outdoor Weathering in Arizona, Florida and Ohio of Olive Geon Company Geon Polyvinyl Chloride.

Material Family	Polyvinyl Chloride								
Material Supplier/ Grade	Geon Company Geon 86101 olive 080								
Features	capstock compound, olive color								
Reference Number	192	192	192	192	192	192	192	192	192
EXPOSURE CONDITIONS									
Exposure Type	outdoor weathering			outdoor weathering			outdoor weathering		
Exposure Location	Avon Lake, Ohio			Arizona			Florida		
Exposure Country	USA			USA			USA		
Exposure Time (days)	183	365	730	183	365	730	183	365	730
SURFACE AND APPEARANCE									
Visual Appearance	no to slight change (pass)	slight change (pass)	slight change (pass)	no to slight change (pass)	slight change (pass)	slight change (pass)	slight change (pass)	slight change (pass)	no to slight change (pass)
ΔE Color Change	1.3	2.6	3.6	0.9	0.9	1.2	2.2	3.1	1.2

TABLE 81: Outdoor Weathering in Arizona, Florida and Ohio of White Geon Company Geon Polyvinyl Chloride.

Material Family	Polyvinyl Chloride								
Material Supplier/ Grade	Geon Company Geon 86101 white 138								
Features	capstock compound, white color								
Reference Number	192	192	192	192	192	192	192	192	192
EXPOSURE CONDITIONS									
Exposure Type	outdoor weathering			outdoor weathering			outdoor weathering		
Exposure Location	Avon Lake, Ohio			Arizona			Florida		
Exposure Country	USA			USA			USA		
Exposure Time (days)	183	365	730	183	365	730	183	365	730
SURFACE AND APPEARANCE									
Visual Appearance	slight change (pass)	slight change (pass)	slight change (pass)	no change from original	no change from original	slight change (pass)	no to slight change (pass)	no to slight change (pass)	no change from original
ΔE Color Change	3.1	5.7		1.3	2	2.5	1	2.7	1.5

TABLE 82: Outdoor Weathering in Arizona, Florida and Ohio of Tan Geon Company Geon Polyvinyl Chloride.

Material Family	Polyvinyl Chloride								
Material Supplier/ Grade	Geon Company Geon 86101 tan 360								
Features	capstock compound, tan color								
Reference Number	192	192	192	192	192	192	192	192	192
EXPOSURE CONDITIONS									
Exposure Type	outdoor weathering			outdoor weathering			outdoor weathering		
Exposure Location	Avon Lake, Ohio			Arizona			Florida		
Exposure Country	USA			USA			USA		
Exposure Time (days)	183	365	730	183	365	730	183	365	730
SURFACE AND APPEARANCE									
Visual Appearance	no to slight change (pass)	slight change (pass)	no to slight change (pass)	no to slight change (pass)	no change from original	slight change (pass)	slight change (pass)	no to slight change (pass)	slight change to change (pass)
ΔE Color Change	1.9	3.1	3.8	0.7	1.7	0.5	4.4	2.9	0.2

TABLE 83: Outdoor Weathering in Arizona, Florida and Ohio of Red Geon Company Geon Polyvinyl Chloride.

Material Family	Polyvinyl Chloride								
Material Supplier/ Grade	Geon Company Geon 86101 red 730								
Features	capstock compound, red color								
Reference Number	192	192	192	192	192	192	192	192	192
EXPOSURE CONDITIONS									
Exposure Type	outdoor weathering			outdoor weathering			outdoor weathering		
Exposure Location	Avon Lake, Ohio			Arizona			Florida		
Exposure Country	USA			USA			USA		
Exposure Time (days)	183	365	730	183	365	730	183	365	730
SURFACE AND APPEARANCE									
Visual Appearance	no to slight change (pass)	slight change (pass)	change (pass)	slight change (pass)	slight change (pass)	slight change to change (pass)	slight change to change (pass)	slight change (pass)	slight change to change (pass)
ΔE Color Change	3.9	1.5	5.7	7.7	9.8	13.7	8.1	6.6	8.1

TABLE 84: Outdoor Weathering in Arizona, Florida and Ohio of Green Geon Company Geon Polyvinyl Chloride.

Material Family	Polyvinyl Chloride								
Material Supplier/ Grade	Geon Company Geon 86101 green 530								
Features	capstock compound, green color								
Reference Number	192	192	192	192	192	192	192	192	192
EXPOSURE CONDITIONS									
Exposure Type	outdoor weathering			outdoor weathering			outdoor weathering		
Exposure Location	Avon Lake, Ohio			Arizona			Florida		
Exposure Country	USA			USA			USA		
Exposure Time (days)	183	365	730	183	365	730	183	365	730
SURFACE AND APPEARANCE									
Visual Appearance	no to slight change (pass)	slight change (pass)	slight change (pass)	no change from original	slight change (pass)	slight change (pass)	slight change (pass)	slight change (pass)	slight change (pass)
ΔE Color Change	3.6	5	6.4	2.4	3.4	5.6	4.9	4.7	5.8

TABLE 85: **Outdoor Weathering in Arizona, Florida and Ohio of Grey Geon Company Geon Polyvinyl Chloride.**

Material Family	Polyvinyl Chloride								
Material Supplier/ Grade	Geon Company Geon 86101 gray 240								
Features	capstock compound, grey color								
Reference Number	192	192	192	192	192	192	192	192	192
EXPOSURE CONDITIONS									
Exposure Type	outdoor weathering			outdoor weathering			outdoor weathering		
Exposure Location	Avon Lake, Ohio			Arizona			Florida		
Exposure Country	USA			USA			USA		
Exposure Time (days)	183	365	730	183	365	730	183	365	730
SURFACE AND APPEARANCE									
Visual Appearance	slight change (pass)	slight change (pass)	slight change (pass)	no to slight change (pass)	slight change (pass)	slight change (pass)	slight change (pass)	slight change (pass)	slight change (pass)
ΔE Color Change	2.5	4.5	4.8	1.7	2.4	4.2	2.3	3.9	3.8

TABLE 86: **Outdoor Weathering in Arizona, Florida and Ohio of Brown Geon Company Geon Polyvinyl Chloride.**

Material Family	Polyvinyl Chloride							
Material Supplier/ Grade	Geon Company Geon 86101 brown 382							
Features	brown color, capstock compound							
Reference Number	192	192	192	192	192	192	192	192
EXPOSURE CONDITIONS								
Exposure Type	outdoor weathering			outdoor weathering		outdoor weathering		
Exposure Location	Avon Lake, Ohio			Arizona		Florida		
Exposure Country	USA			USA		USA		
Exposure Time (days)	183	365	730	365	730	183	365	730
SURFACE AND APPEARANCE								
Visual Appearance	no to slight change (pass)	slight change (pass)	slight change (pass)	slight change (pass)	slight change to change (pass)	no to slight change (pass)	slight change (pass)	slight change to change (pass)
ΔE Color Change	1.7	1.8		2.5	3.9	1.1	3.2	3.2

TABLE 87: Outdoor Weathering in Arizona, Florida and Ohio of Ivory Geon Company Geon Polyvinyl Chloride.

Material Family	Polyvinyl Chloride								
Material Supplier/ Grade	Geon Company Geon 86101 ivory 035								
Features	capstock compound, ivory color								
Reference Number	192	192	192	192	192	192	192	192	192

EXPOSURE CONDITIONS

Exposure Type	outdoor weathering			outdoor weathering			outdoor weathering		
Exposure Location	Avon Lake, Ohio			Arizona			Florida		
Exposure Country	USA			USA			USA		
Exposure Time (days)	183	365	730	183	365	730	183	365	730

SURFACE AND APPEARANCE

Visual Appearance	no to slight change (pass)	slight change (pass)	no to slight change (pass)	no to slight change (pass)	no to slight change (pass)	no to slight change (pass)	no to slight change (pass)	no to slight change (pass)	no change from original
ΔE Color Change	1.5	4.1	4.9	0.1	1.5	2.1	1.6	2.2	1.8

TABLE 88: Outdoor Weathering in Arizona, Florida and Ohio of Brown Geon Company Geon Polyvinyl Chloride.

Material Family	Polyvinyl Chloride							
Material Supplier/ Grade	Geon Company Geon 86101 brown 372							
Features	brown color, capstock compound							
Reference Number	192	192	192	192	192	192	192	192

EXPOSURE CONDITIONS

Exposure Type	outdoor weathering			outdoor weathering		outdoor weathering		
Exposure Location	Avon Lake, Ohio			Arizona		Florida		
Exposure Country	USA			USA		USA		
Exposure Time (days)	183	365	730	365	730	183	365	730

SURFACE AND APPEARANCE

Visual Appearance	no to slight change (pass)	slight change (pass)	slight change (pass)	no to slight change (pass)	slight change to change (pass)	no to slight change (pass)	slight change (pass)	slight change (pass)
ΔE Color Change	0.8	0.5		2	4	1.3	3.5	3.6

TABLE 89: Outdoor Weathering in Arizona, Florida and Ohio of Blue Geon Company Geon Polyvinyl Chloride.

Material Family	Polyvinyl Chloride							
Material Supplier/ Grade	Geon Company Geon 86101 blue 460							
Features	blue color, capstock compound							
Reference Number	192	192	192	192	192	192	192	192

EXPOSURE CONDITIONS

Exposure Type	outdoor weathering			outdoor weathering		outdoor weathering		
Exposure Location	Avon Lake, Ohio			Arizona		Florida		
Exposure Country	USA			USA		USA		
Exposure Time (days)	183	365	730	365	730	183	365	730

SURFACE AND APPEARANCE

Visual Appearance	slight change (pass)	slight change (pass)	slight change (pass)	slight change (pass)	slight change (pass)	no to slight change (pass)	no to slight change (pass)	slight change (pass)
ΔE Color Change	0.6	3.3		2.2	3.5	1.5	2	2.3

TABLE 90: Outdoor Weathering in Arizona, Florida and Ohio of Black Geon Company Geon Polyvinyl Chloride.

Material Family	Polyvinyl Chloride								
Material Supplier/ Grade	Geon Company Geon 86101 black 288								
Features	black color, capstock compound								
Reference Number	192	192	192	192	192	192	192	192	192

EXPOSURE CONDITIONS

Exposure Type	outdoor weathering			outdoor weathering			outdoor weathering		
Exposure Location	Avon Lake, Ohio			Arizona			Florida		
Exposure Country	USA			USA			USA		
Exposure Time (days)	183	365	730	183	365	730	183	365	730

SURFACE AND APPEARANCE

Visual Appearance	slight change (pass)	slight change (pass)	slight change (pass)	slight change (pass)	slight change to change (pass)	slight change (pass)	slight change (pass)	slight change (pass)	slight change (pass)
ΔE Color Change	1.8	2.1	1.6	0.7	3.3	5.6	6.4	4.7	4.7

TABLE 91: Outdoor Weathering in Florida of Polyvinyl Chloride.

Material Family	Polyvinyl Chloride	
Features	UV stabilized, white color	UV stabilized, brown color
Reference Number	116	116
EXPOSURE CONDITIONS		
Exposure Type	outdoor weathering	outdoor weathering
Exposure Location	Florida	Florida
Exposure Country	USA	USA
Exposure Note	45° angle south	45° angle south
Exposure Time (days)	365	365
SURFACE AND APPEARANCE		
Δa Color	0.3 {c}	-0.4 {c}
Δb Color	-0.8 {c}	0.1 {c}
ΔL Color	1.6 {c}	1.8 {c}
ΔE Color	1.8 {c}	1.8 {c}

TABLE 92: Outdoor Weathering in California and Pennsylvania of Glass Reinforced Polyvinyl Chloride.

Material Family	Polyvinyl Chloride							
Material Supplier	LNP Engineering Plastics							
Reference Number	215	215	215	215	215	215	215	215
MATERIAL COMPOSITION								
carbon black	1%	1%	1%	1%	1%	1%	1%	1%
glass fiber reinforcement	15%	15%	15%	15%	15%	15%	15%	15%
EXPOSURE CONDITIONS								
Exposure Type	outdoor weathering				outdoor weathering			
Exposure Location	Los Angeles, California	Los Angeles, California	Los Angeles, California	Los Angeles, California	Philadelphia, Pennsylvania	Philadelphia, Pennsylvania	Philadelphia, Pennsylvania	Philadelphia, Pennsylvania
Exposure Country	USA	USA	USA	USA	USA	USA	USA	USA
Exposure Test Method	ASTM D1435	ASTM D1435	ASTM D1435	ASTM D1435	ASTM D1435	ASTM D1435	ASTM D1435	ASTM D1435
Exposure Time (days)	91	182	365	730	91	182	365	730
PROPERTIES RETAINED (%)								
Tensile Strength	100	99.2	103	102	97	97.3	104	107
Notched Izod Impact Strength	112	100			100	100		
Unnotched Izod Impact Strength	83	91.3			100	101		

TABLE 93: Accelerated Indoor Exposure by HPUV of Geon Company Geon Polyvinyl Chloride.

Material Family	Polyvinyl Chloride
Material Supplier/ Grade	Geon Company Geon 87371
Features	indoor UV stable
Reference Number	189

EXPOSURE CONDITIONS

Exposure Type	accelerated indoor exposure
Exposure Apparatus	HPUV
Exposure Temperature (°C)	42
Exposure Time (days)	12.5

SURFACE AND APPEARANCE

Δa Color	-0.03
Δb Color	-0.05
ΔL Color	0.01
ΔE Color	0.06

TABLE 94: Accelerated Indoor Exposure by HPUV of Indoor UV Stable, White Geon Company Geon Polyvinyl Chloride.

Material Family	Polyvinyl Chloride	
Material Supplier/ Grade	Geon Company Geon HTX 6210 - white 160	Geon Company Geon HTX 6220 - white 160
Features	indoor UV stable, white color	indoor UV stable, white color
Reference Number	191	191

EXPOSURE CONDITIONS

Exposure Type	accelerated indoor exposure	accelerated indoor exposure
Exposure Test Method	ASTM D4674	ASTM D4674
Exposure Apparatus	HPUV	HPUV
Exposure Time (days)	12.5	12.5

SURFACE AND APPEARANCE

ΔE Color	<0.5	<0.5

GRAPH 184: **Arizona Outdoor Weathering Exposure Time vs. Drop Dart Impact Strength of Polyvinyl Chloride**

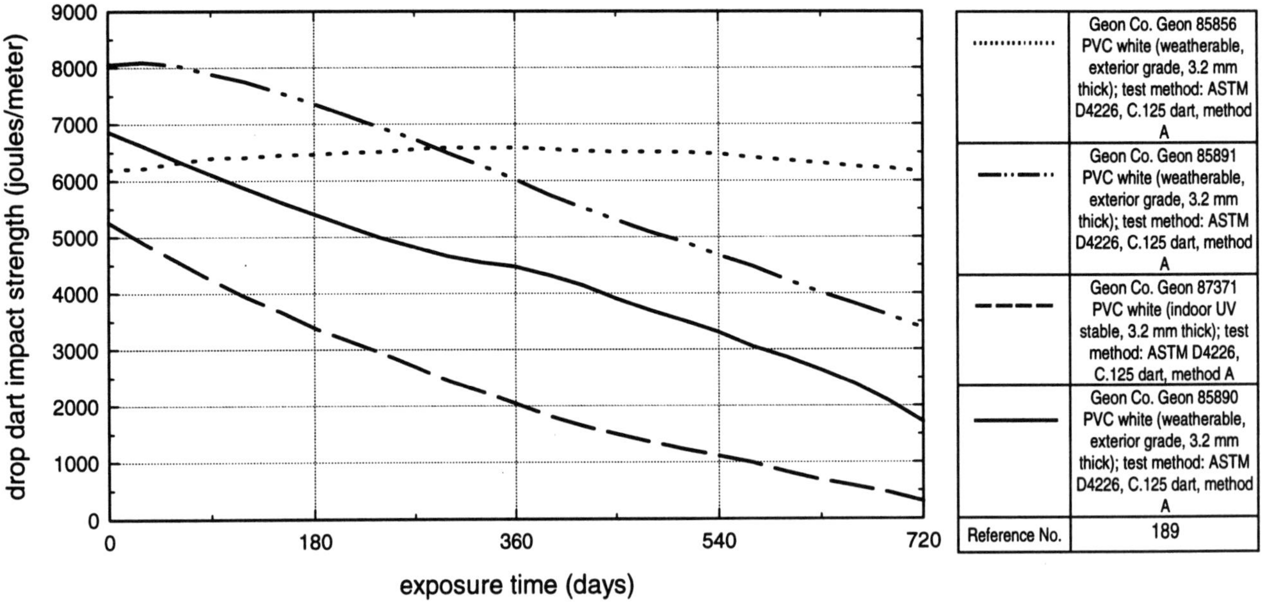

GRAPH 185: **Arizona Outdoor Weathering Exposure Time vs. Elongation of Polyvinyl Chloride**

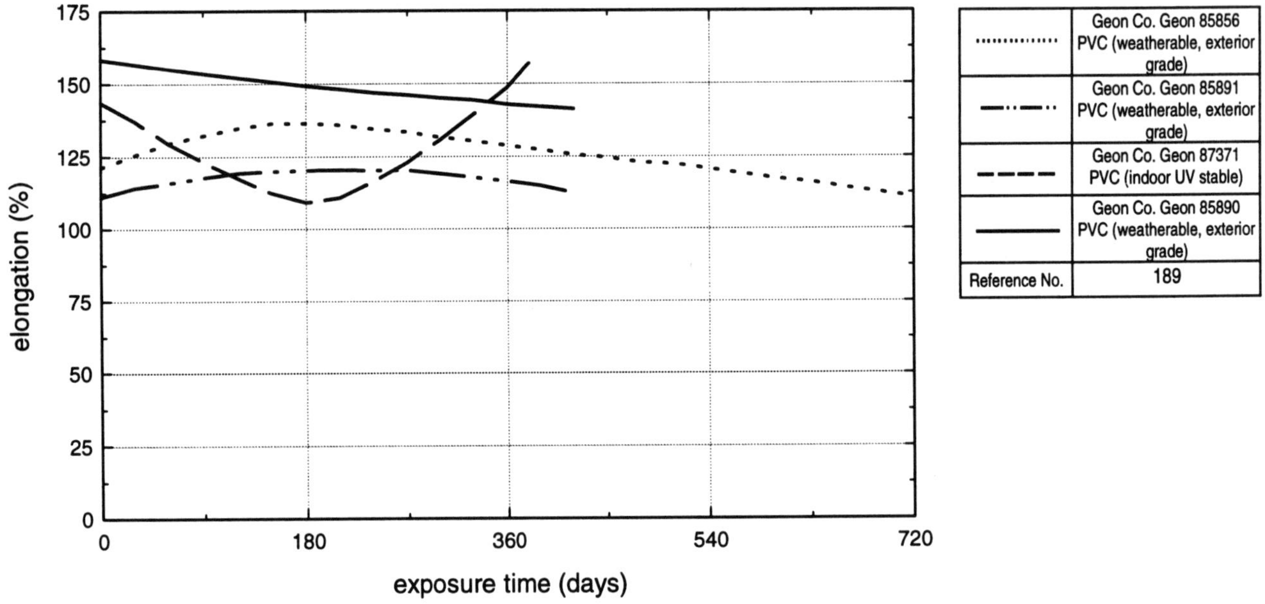

GRAPH 186: Arizona Outdoor Weathering Exposure Time vs. Elongation of Polyvinyl Chloride

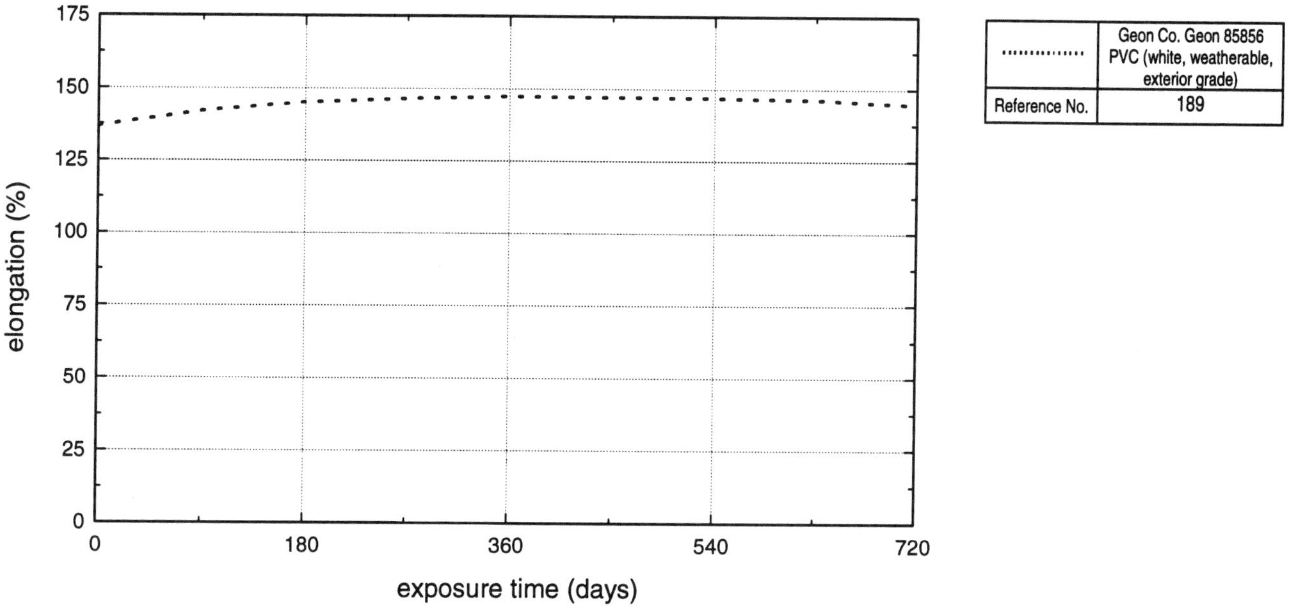

GRAPH 187: Arizona Outdoor Weathering Exposure Time vs. Tensile Strength at Yield of Polyvinyl Chloride

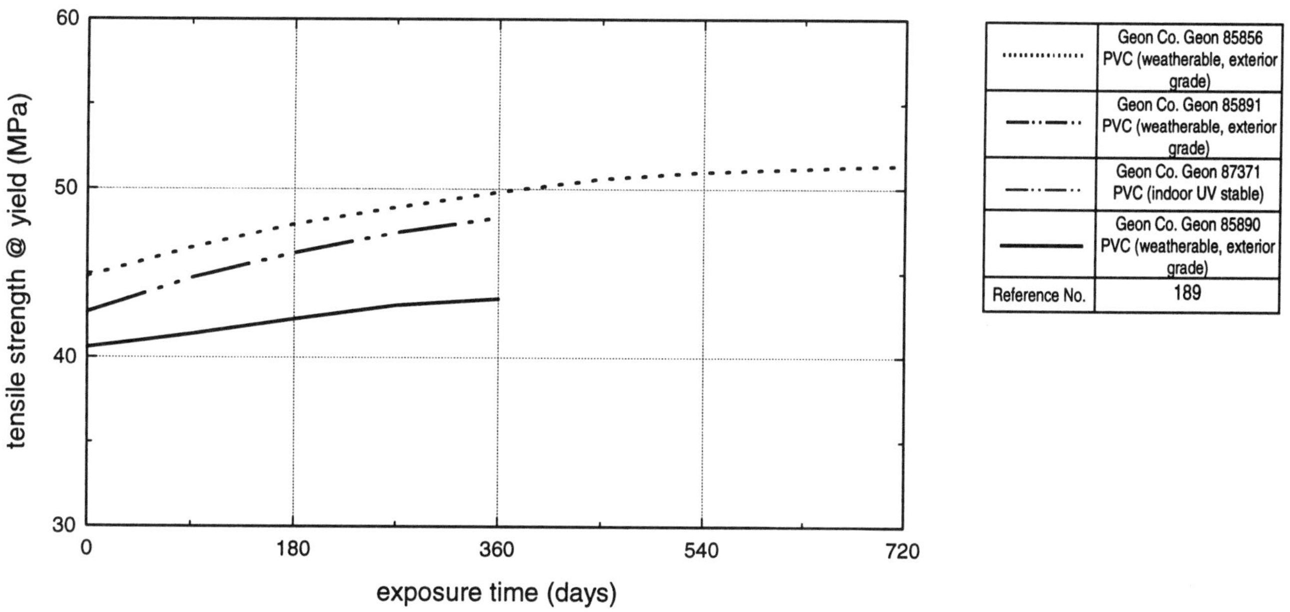

GRAPH 188: Arizona Outdoor Weathering Exposure Time vs. Delta E Color Change of Polyvinyl Chloride

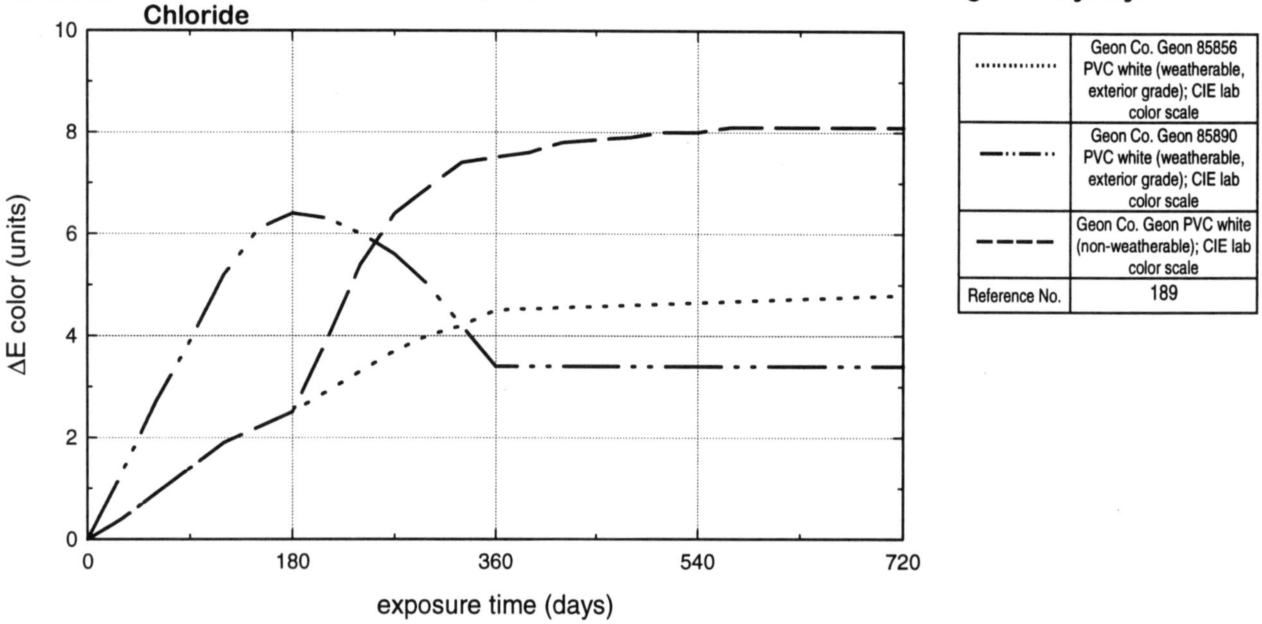

GRAPH 189: Arizona, Florida and Ohio Outdoor Weathering Exposure Time vs. Drop Dart Impact Strength of Polyvinyl Chloride

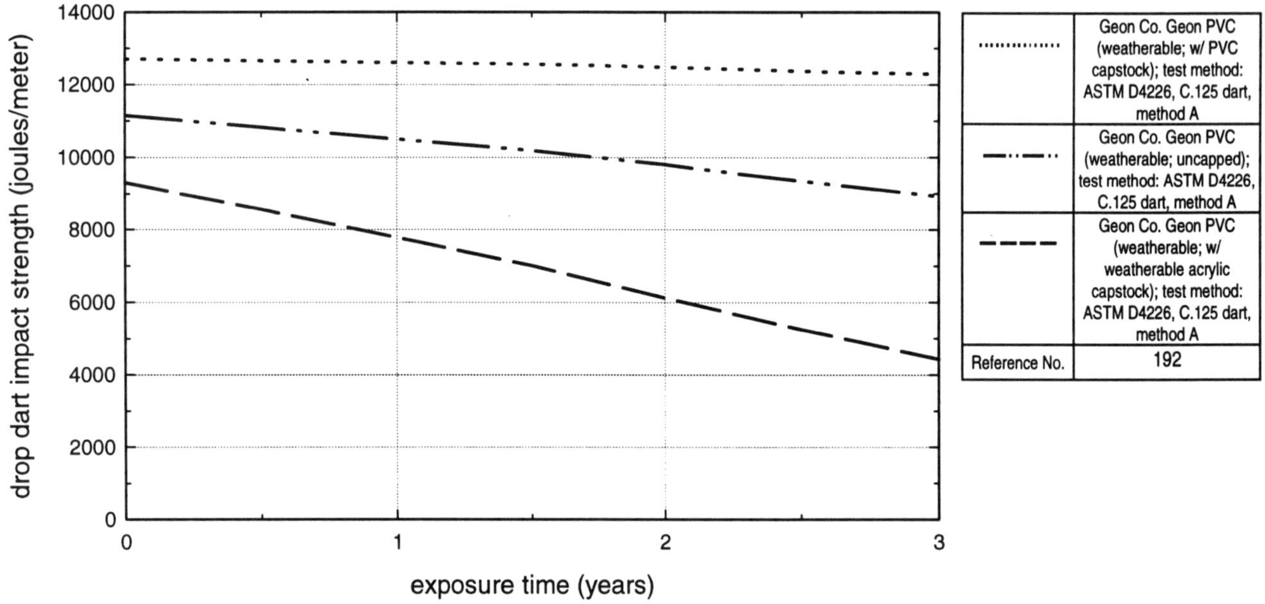

GRAPH 190: Florida Outdoor Weathering Exposure Time vs. Drop Dart Impact Strength of Polyvinyl Chloride

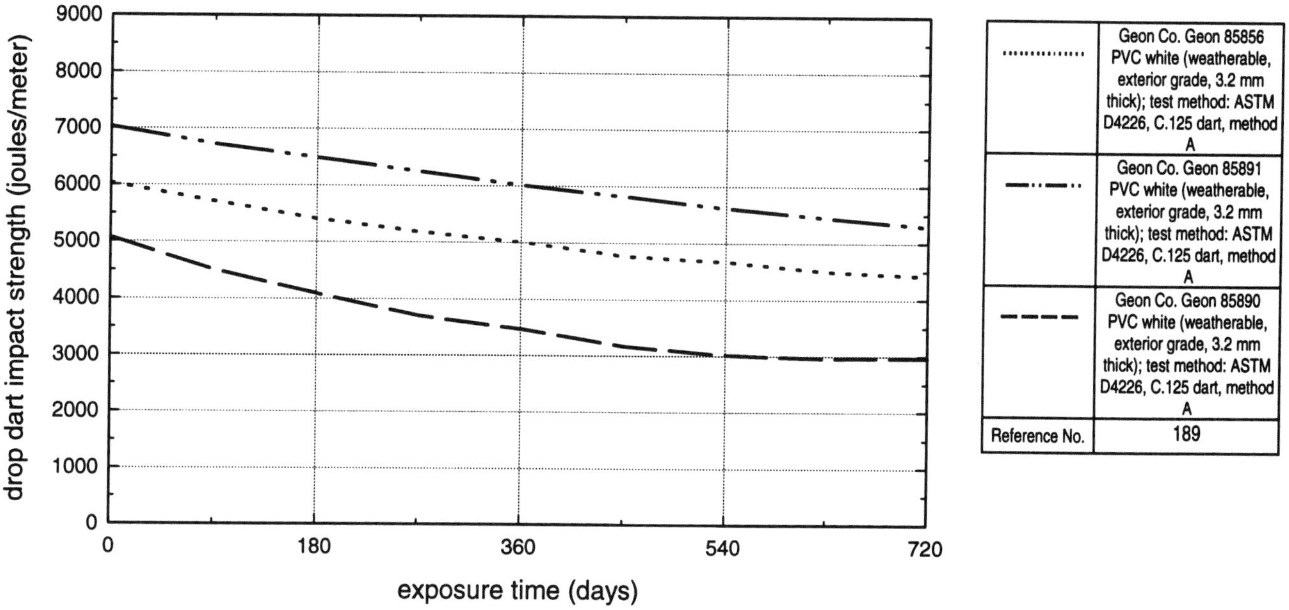

GRAPH 191: Florida Outdoor Weathering Exposure Time vs. Drop Weight Impact Retained of Polyvinyl Chloride

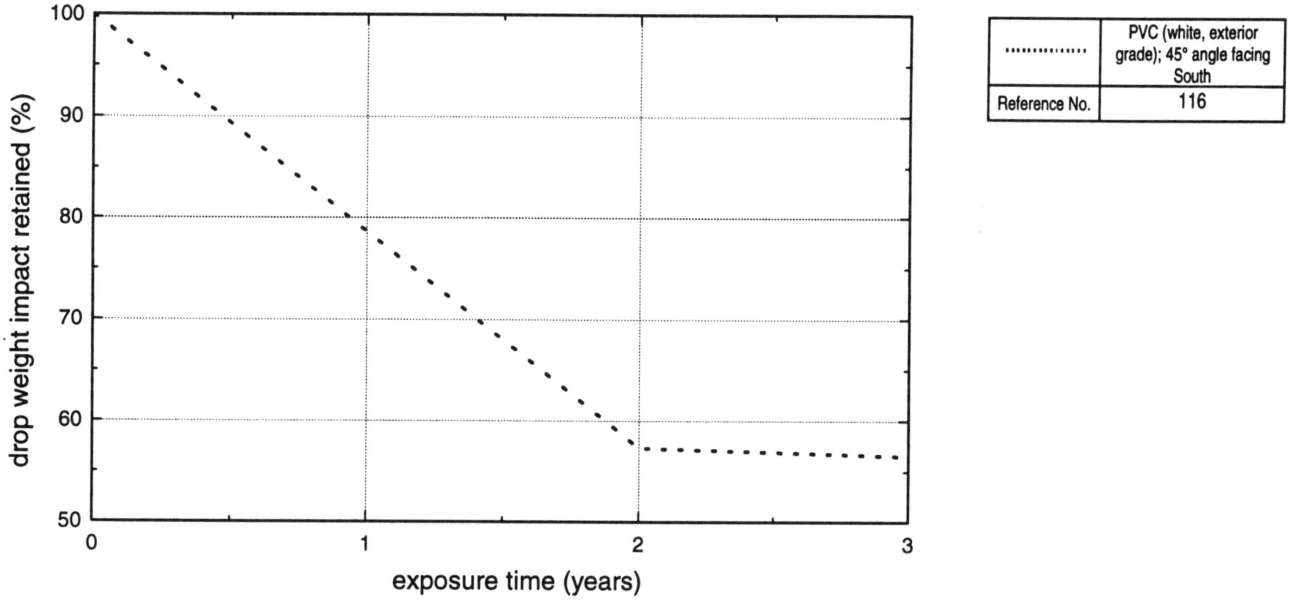

GRAPH 192: Florida Outdoor Weathering Exposure Time vs. Delta E Color Change of Polyvinyl Chloride

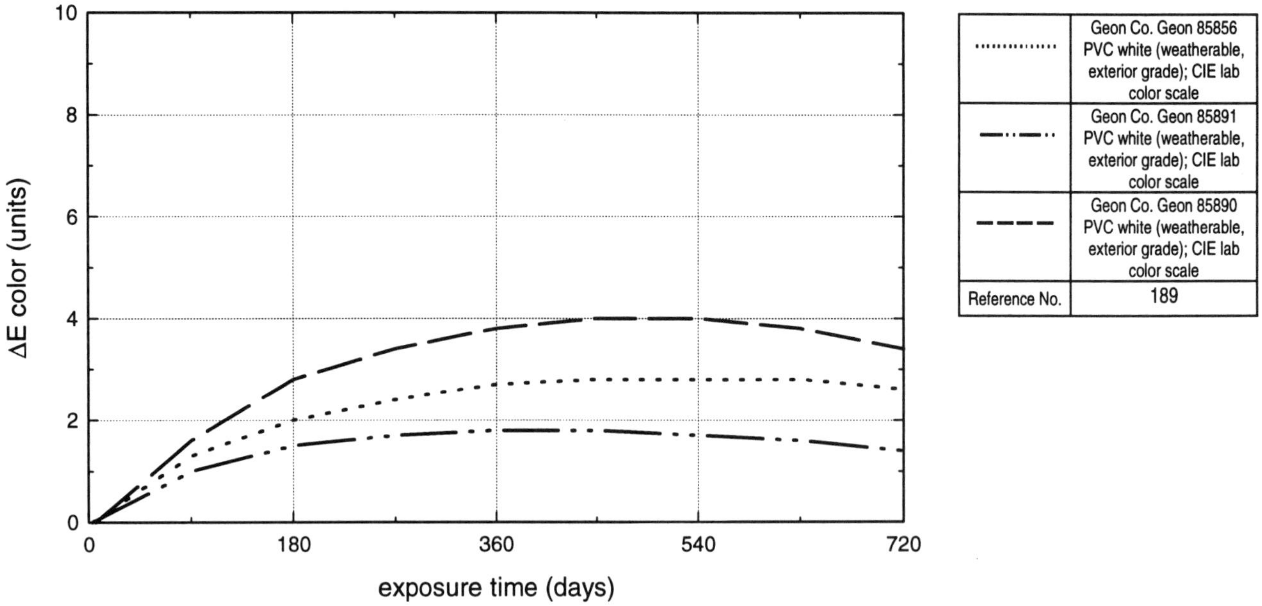

GRAPH 193: Ohio Outdoor Weathering Exposure vs. Delta E Color Change of Polyvinyl Chloride

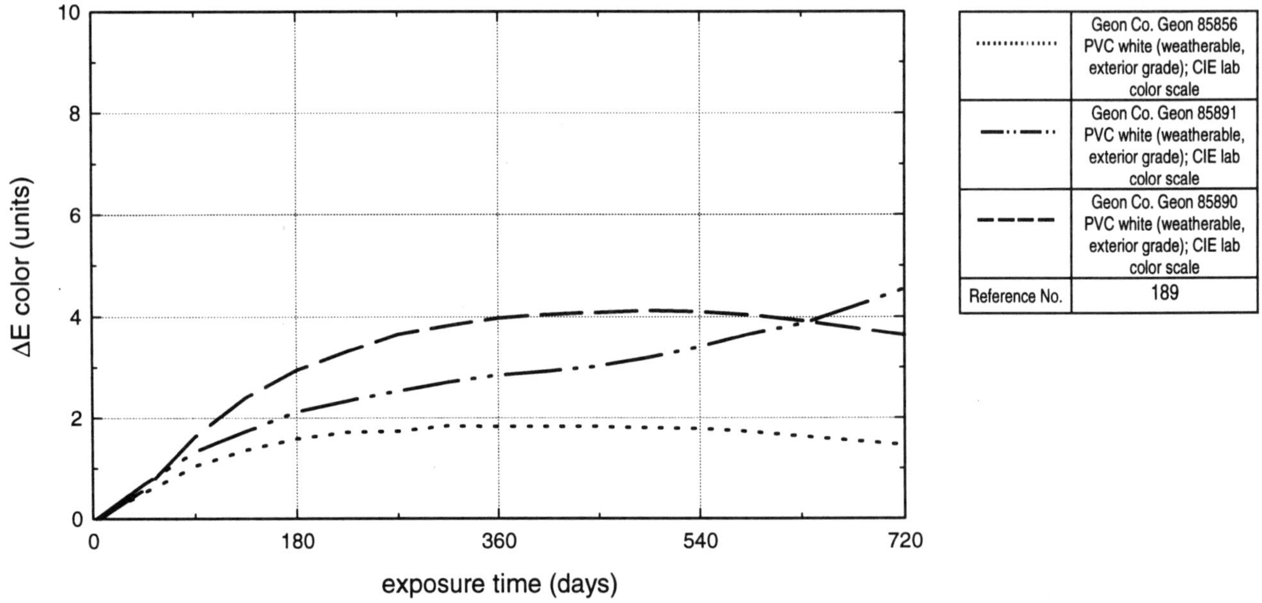

GRAPH 194: Ohio Outdoor Weathering Exposure vs. Drop Dart Impact Strength of Polyvinyl Chloride

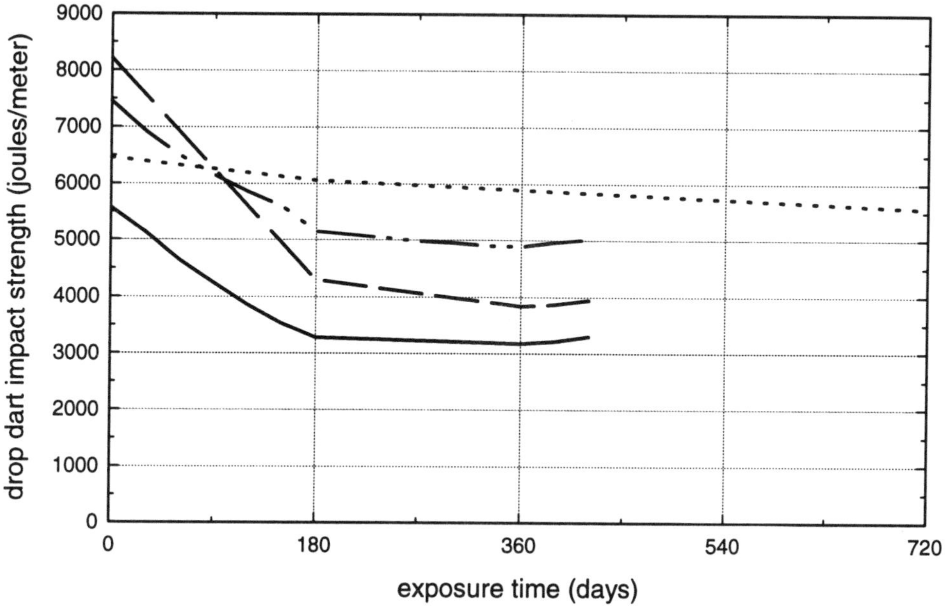

....	Geon Co. Geon 85856 PVC white (weatherable, exterior grade, 3.2 mm thick); test method: ASTM D4226, C.125 dart, method A
—··—··	Geon Co. Geon 85891 PVC white (weatherable, exterior grade, 3.2 mm thick); test method: ASTM D4226, C.125 dart, method A
– – – –	Geon Co. Geon 87371 PVC white (indoor UV stable, 3.2 mm thick); test method: ASTM D4226, C.125 dart, method A
———	Geon Co. Geon 85890 PVC white (weatherable, exterior grade, 3.2 mm thick); test method: ASTM D4226, C.125 dart, method A
Reference No.	189

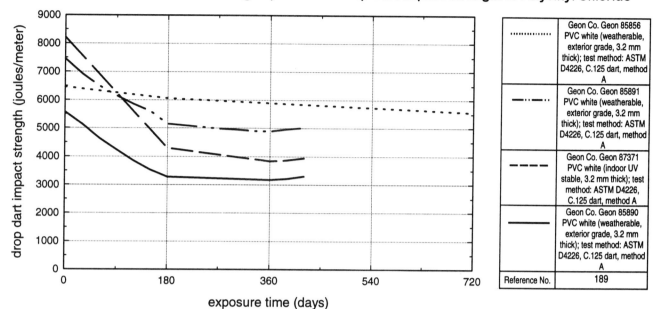

GRAPH 195: Weatherometer Exposure Time vs. Tensile Strength at Yield of Polyvinyl Chloride

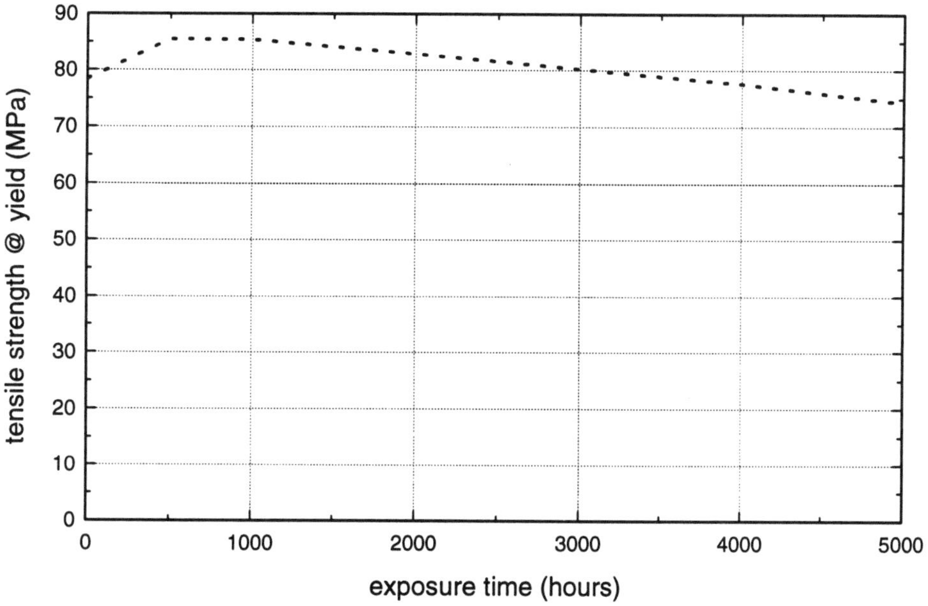

....	Geon Co. Fiberloc 801GR20 PVC (20% glass fiber)
Reference No.	190

Chapter 52

CHLORINATED POLYVINYL CHLORIDE

GRAPH 196: **Florida Outdoor Weathering Exposure Time vs. Drop Weight Impact Retained of Chlorinated Polyvinyl Chloride**

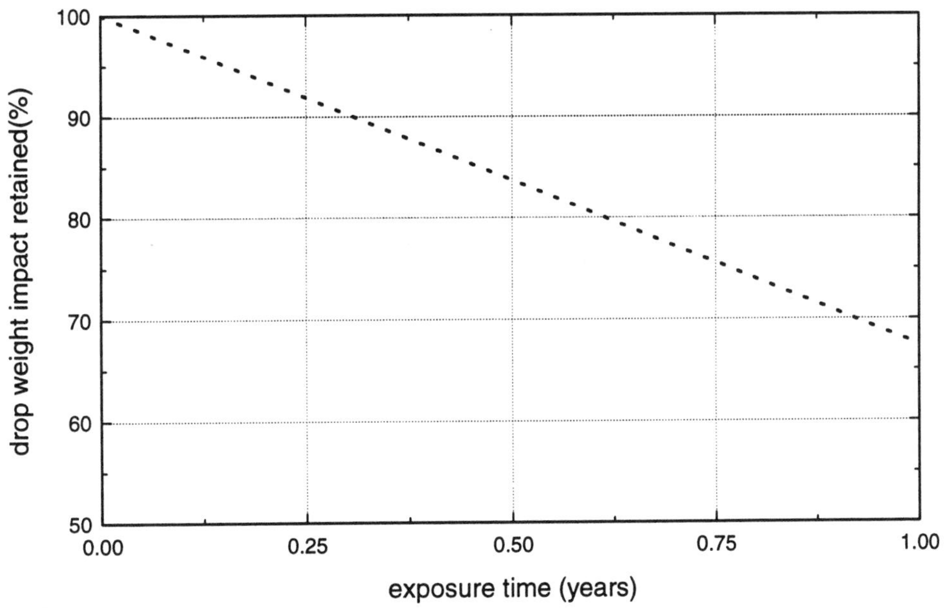

···············	CPVC (PVC capped); 45° angle facing South
Reference No.	116

Chapter 53

ABS Polyvinyl Chloride Alloy

Indoor UV Light Resistance

Monsanto: Triax CBE/1 (features: flame retardant)

Triax CBE/1 offers UV stability in a flame retardant alloy, without the problems of plate-out and bloom. Business machine enclosures in Triax CBE/1 will show only minimal change in color or fade over time, unlike modified PPO, FR ABS and other flame retardant materials. Its natural UV stability, without the use of additives has been demonstrated in extensive laboratory testing.

Reference: *Introducing Superior UV Stability With Good Looks That Last In Business Machine Housings.*, supplier marketing literature (7110) - Monsanto Chemical Company, 1990.

TABLE 95: **Accelerated Weathering in a QUV of Novatec Novaloy 9000 ABS Polyvinyl Chloride Alloy.**

Material Family	ABS Polyvinyl Chloride Alloy			
Material Supplier/ Grade	Novatec Novaloy 9000			
Reference Number	128	128	128	128
EXPOSURE CONDITIONS				
Exposure Type	accelerated weathering			
Exposure Apparatus	QUV			
Exposure Temperature (°C)	50	50	50	50
Exposure Time (days)	20.83	41.7	62.5	83.3
Exposure Note	both sides of specimens exposed; 4 hours of light and 4 hours of humidity	both sides of specimens exposed; 4 hours of light and 4 hours of humidity	both sides of specimens exposed; 4 hours of light and 4 hours of humidity	both sides of specimens exposed; 4 hours of light and 4 hours of humidity
PROPERTIES RETAINED (%)				
Notched Izod Impact Strength	79.3	70.2	55.3	36.2
SURFACE AND APPEARANCE				
Δa Color Change				(36% greener) (f)
Δb Color Change				(53% yellower) (f)
ΔL Color Change				(18% lighter) (f)

GRAPH 197: **Accelerated Indoor UV Exposure Time vs. Delta E Color Change of ABS Polyvinyl Chloride Alloy**

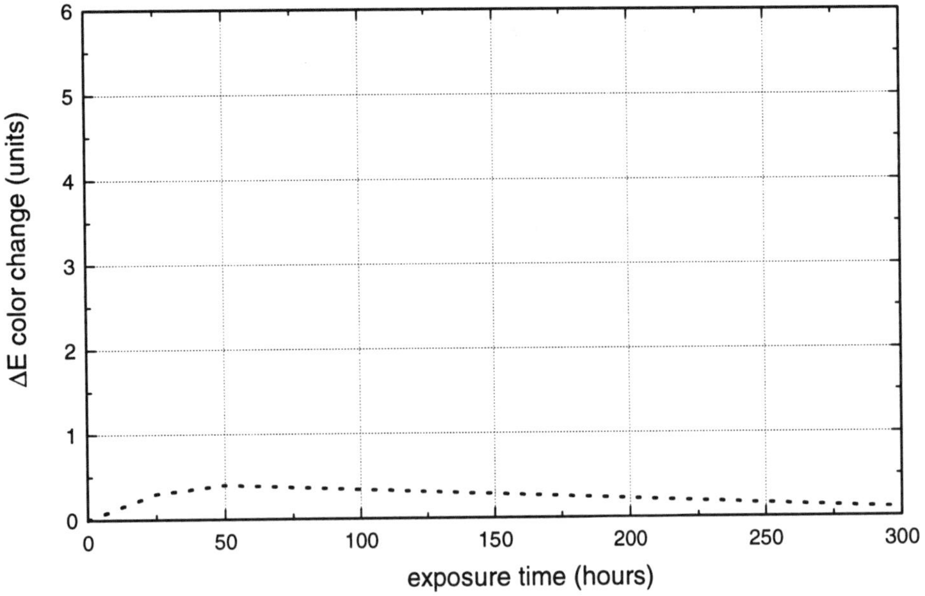

	Monsanto Triax CBE/1 ABS PVC Alloy (flame retardant)
Reference No.	232

Chapter 54

Acrylic PVC Alloy

Outdoor Weather Resistance

Kleerdex: Kydex 100 (product form: sheet)

Based on limited exposure testing, Kydex demonstrates adequate weathering resistance for many applications. However, it is currently not recommended for outdoor use in applications where retention of original appearance is critical.

Reference: *Physical Properties Kydex 100 Acrylic PVC Alloy Sheet,* supplier technical report (KC-89-03) - Kleerdex Company, 1989.

Kleerdex: Kydex 150 (product form: sheet); **Kydex 160** (product form: sheet); **Kydex 510** (product form: sheet); **Kydex 550** (features: weatherability; product form: sheet)

Kydex 150 and 160 are not recommended for outdoor useage where the product will be exposed to direct sunlight, although it has been used outdoors with varying results. If Kydex 150 or 160 must be used outdoors, better results can be realized if a lighter color is used, and the Kydex is located on the northern face of the structure.

It is important to note that the major effect of UV exposure on Kydex 150 and 160 is a color change after prolonged exposure to sunlight. All other physical properties of Kydex sheet remain unaffected.

Kydex 550 and Kydex 510 are grades which are recommended for UV exposure applications. Kydex 550 is available in limited colors. Kydex 510 is available in Kleerdex's 33 standard colors, as well as custom colors. Both of these formulations offer superior resistance to UV rays, and will show no significant color degradation when exposed to UV rays.

Reference: *Supplier Written Correspondence,* Kleerdex Company, 1994.

Chapter 55

Polycarbonate ABS Alloy

Indoor UV Light Resistance

Miles: Bayblend FR 1439 (features: flame retardant); **Bayblend FR 1440** (features: flame retardant); **Bayblend FR 1441** (features: flame retardant)

Bayblend FR 1400 resins virtually resist discoloration and fading caused by ultraviolet (UV) radiation from fluorescent lighting or filtered sunlight in an office environment. Parts molded of Bayblend FR resins benefit from the advantages of an integral color and finish.

Due to the UV resistance of Bayblend FR resins, parts can be molded which retain good mechanical properties without light-protective coatings. Inherent color stability is a significant feature for business machine and electronics housings which are exposed to fluorescent and window light.

Reference: *Bayblend FR Resins For Business Machines And Electronics,* supplier marketing literature (55-D808(5)J 313-10/88) - Mobay Corporation, 1988.

TABLE 96: Accelerated Indoor Exposure by HPUV and Xeneon Arc Weatherometer of Dow Chemical Pulse Polycarbonate ABS Alloy.

Material Family	Polycarbonate ABS Alloy	
Material Supplier/ Grade	Dow Chemical Pulse 1745	
Features	flame retardant, UV stabilized	
Reference Number	199	199
EXPOSURE CONDITIONS		
Exposure Type	accelerated indoor exposure	accelerated indoor exposure
Exposure Apparatus	HPUV	xenon arc weatherometer
Exposure Test Method	ASTM 4674	ASTM D5549
Exposure Test Method Note	Hewlett Packard test	IBM test
Exposure Time (days)	12.5	12.5
SURFACE AND APPEARANCE		
ΔE Color	<1.5	<1.0

TABLE 97: Accelerated Indoor Exposure by HPUV of General Electric Cycoloy Polycarbonate ABS Alloy.

Material Family	Polycarbonate ABS Alloy		
Material Supplier/ Grade	GE Plastics Cycoloy C2800	GE Plastics Cycoloy C2950	GE Plastics Cycoloy C2950HF
Features	flame retardant; Hewlett Packard parchment white color	flame retardant; Hewlett Packard parchment white color	flame retardant; Hewlett Packard parchment white color
Reference Number	176	176	176

MATERIAL COMPOSITION

Material Composition Note	no brominated or chlorinated additives	no brominated or chlorinated additives	

EXPOSURE CONDITIONS

Exposure Type	Accelerated Indoor UV Light		
Exposure Apparatus	HPUV	HPUV	HPUV
Exposure Note	3 year office exposure simulation	3 year office exposure simulation	3 year office exposure simulation

SURFACE AND APPEARANCE

ΔE Color	0.49	0.49	0.49

Chapter 56

Starch Modified Low Density Polyethylene

GRAPH 197a: Xenon Weatherometer Exposure Time vs. Elongation Retained of Low Density Polyethylene

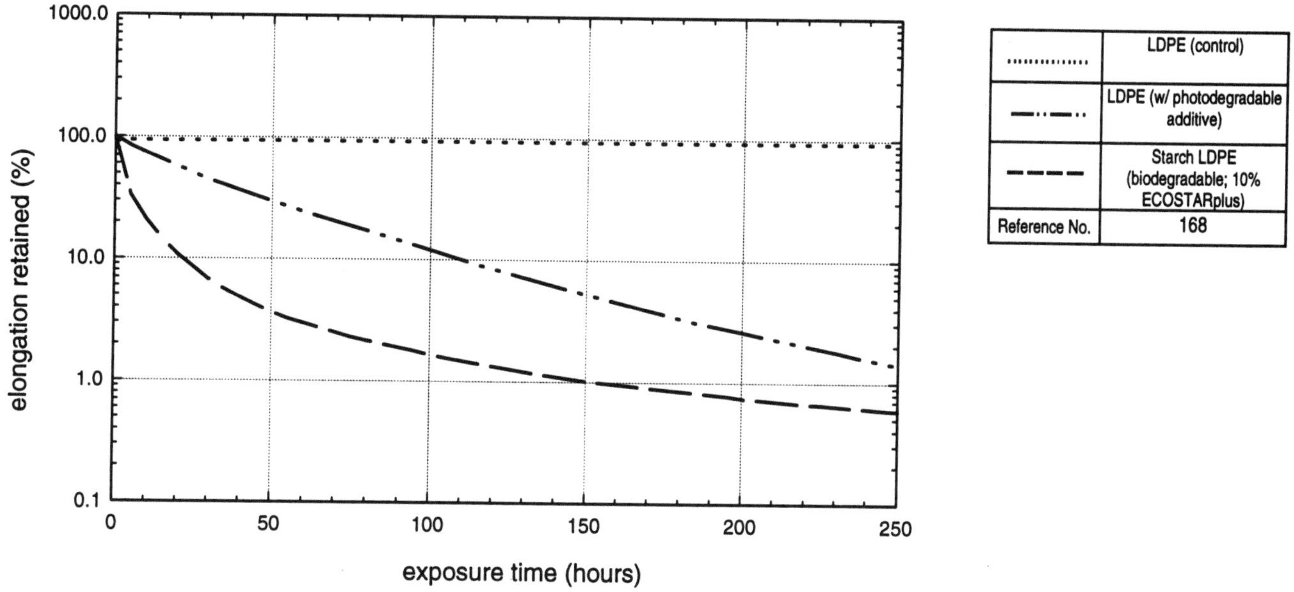

GRAPH 198: Composting Exposure Time vs. Elongation Retained of Low Density Polyethylene

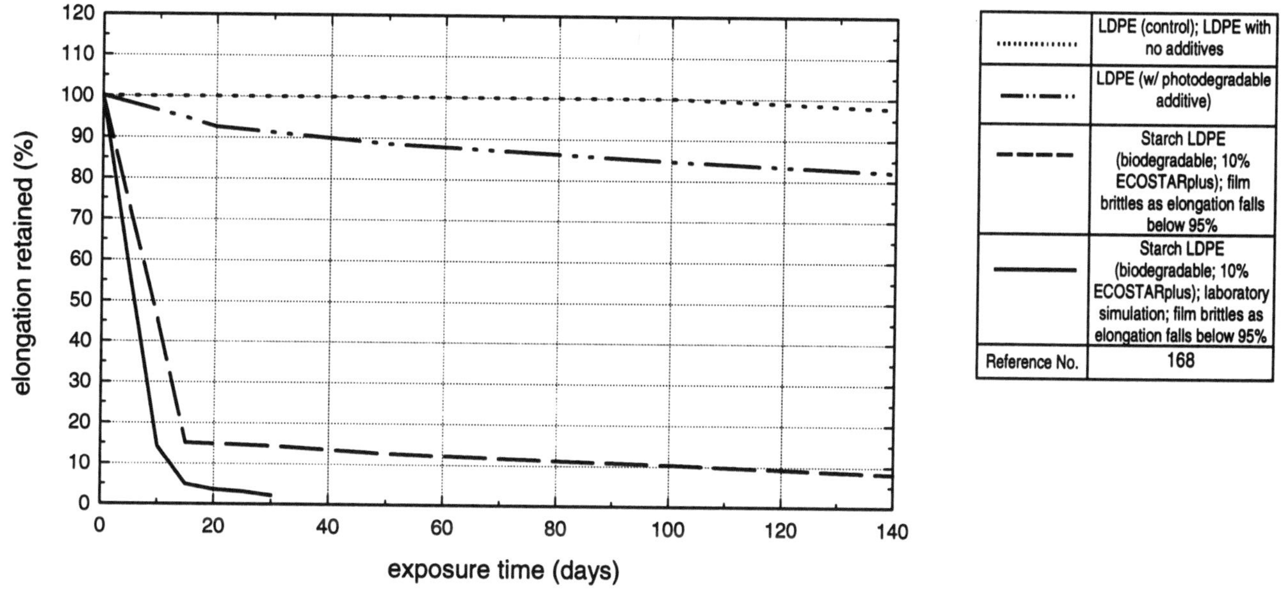

Chapter 57

Starch Modified Polyethylene Alloy

GRAPH 199: Burial Time vs. Starch Granules Digested of Starch Modified Polyethylene Alloy

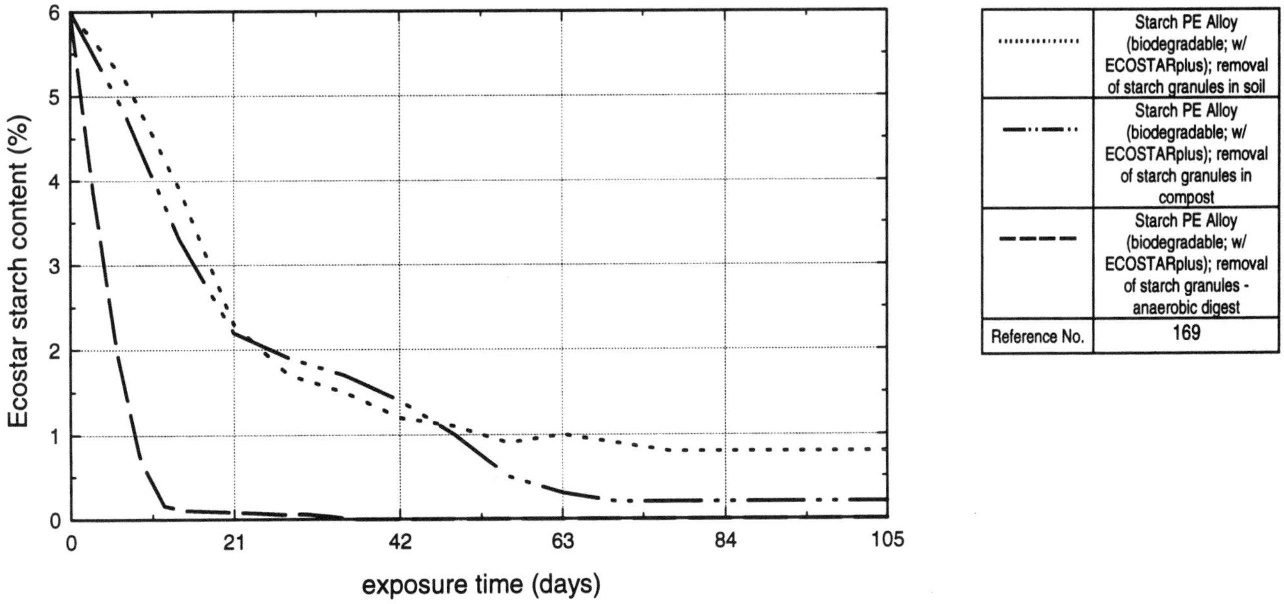

Chapter 58

Starch Synthetic Resin Alloy

Biodegradation

Novamont: Mater-Bi (features: biodegradable)

Mater-Bi can be assimilated to a biomass. It can mineralize aerobically, generating CO_2 and H_2O, and can also mineralize aerobically, releasing methane, or be part of compost made of urban solid garbage and therefore become a fertilizer to produce new agricultural products.

The biodegradation of Mater-Bi products is facilitated by their capacity to swell when in contact with water. Speed of biodegradation, as in all insoluble products, is affected by the physical state of samples (crystallinity, orientation, porosity, etc.).

Reference: *Mater-Bi - The Latest Plastic Material Introduces The True Value Of Biodegradability. Today, 1991.*, supplier marketing literature - Novamont, 1991.

Starch Synthetic Resin Alloy (additives: starch (ECOSTARplus); features: biodegradable)

The use of starch as a cost effective "accelerant" for degrading plastics was developed in the 1970's. Scientists found that when starch is included in the structure of the polymer, the plastic degrades from microbiological attack. However, standard starch was unsuitable due to its high moisture content which caused foaming during processing and its hydrophilic surface which made it incompatible with hydrophobic polymers. This led to the discovery of a technique to modify the hydrophilic starch surface and reduce the moisture content so that it could be combined with plastic resin in a standard polymer process typically run at temperatures above 160°C, resulting in products with good physical strength properties.

Plastics made with the Ecostarplus system incorporate a modified starch dried to less than 1% moisture content (compared to the 10-12% in normal starch). The surface of the starch is modified to make it hydrophobic, facilitating its mixing with polymers. This provides the basic system for accelerating the biodegradation of the plastic. A chemical "accelerant" is then added which not only further enhances biodegredation but also adds another significant feature to the plastic - photodegradability. In summary, the Ecostarplus system in plastics accelerates the degradation process not only in the biological environment due to starch and other additives, but upon exposure to sunlight.

Degradation Mechanism Plastics utilizing the Ecostarplus system begin to degrade when the starch granules in the polymer are attacked by microorganisms, such as fungi and bacteria. The removal of the starch granules weakens the polymer matrix. Degradation is accelerated further by the formation of peroxides from the autoxidant in the system as a result of interaction with catalysts also present in the Ecostarplus system. All of this reduces the molecular weight of the polymer causing brittleness in the plastic which in turn then breaks up into small particles. These remaining particles are then digested by microorganisms. The degree to which the Ecostarplus system accelerates the degradation of the plastic depends on its exposure to a variety of natural conditions. An independent study by the "TNO of the Netherlands" states that "the degradation speed of the Ecostarplus foil in the bio-climate will be of a magnitude 100 or more higher than in the case that the polymer material does not contain the special additive."

Ecostarplus maximizes its effectiveness by combining both biodegradable and photodegradable technologies. While products which are only photodegradable can be useful in addressing the problem of plastic litter, they are ineffective in soil, have no benefit in composting and are only partially effective in agricultural applications such as mulch film. Ecostarplus is effective in all of these situations because it accelerates the breakdown of plastics products when exposed to soil and/or sunlight.

Disposal The benefits of accelerated degradation are available when the waste stream created by plastics and other forms of waste are disposed of at biologically active sites such as composting, agricultural applications or where plastic litter is a concern. As is true for many waste products, Ecostar additives do not provide a benefit if the intended disposal site for a customer's product is a biologically inactive landfill or airtight container, since the waste stream is not exposed to the needed bioclimate and microorganisms. It has been proven through several studies, however, that this is equally true for substances such as paper and even products in our food chains.

Applications Many polymers such as Polyethlenes, Polypropylenes & Polystyrenes are compounded with Ecostarplus to enhance the properties of the polymer to degrade.

Studies of Biodegradation and Photodegradation Composting trials by the Fraunhoffer Institute in Germany showed oxidative breakdown in film containing Ecostarplus within 6 weeks. This oxidation indicates the commencement of embrittlement. Brittleness is the point at which the film has lost its integrity and fragments.

Iowa State University has conducted composting trials that showed 70% reduction in molecular weight within three months for films containing Ecostarplus. Reduction in molecular weight also correlates with the embrittlement of film.

Soil burial tests have been carried out at the Chemicals Inspection and Testing Institute in Japan, showing 95% loss of elongation and resulting brittleness in polyethylene samples containing Ecostarplus after 8 months. They also found a reduction in molecular weight distribution and oxidation of the polymer during these tests.

By burying low density polyethylene film samples containing 20% Ecostarplus in soil, DiPeco, a Swiss research firm, observed a 95% reduction in elongation within 18 months.

A comprehensive survey by the TNO of the Netherlands shows that films containing Ecostarplus have, prior to entering the bio-climate, good processing and positive strength properties. Further, they state that "the degradation speed of the Ecostarplus foil in the bioclimate will be of a magnitude 100 or more higher than in the case of a polymer material that does not contain the special additive." On the ultimate fate of the Ecostarplus system product they state that "the amount of carbon dioxide generated was more than could be attributed to the starch and other natural materials, showing that at least some of the polymer in the short time of the test was metabolized to carbon dioxide."

In October 1991, an independent study on the degradation of bags containing Ecostarplus was undertaken at the GV.A Kopie Co. in Germany that showed "at the end of May 1992 the bags were badly decomposed." and "remaining parts disintegrated by light cultivation into hardly recognizable parts."

Reference: *ECOSTARplus Leads The Way,* supplier written correspondence - Ecostar International L.P., 1993.

TABLE 98: Soil Burial and Fungus Resistance of Biodegradable Novamont Mater-Bi Starch Synthetic Resin Alloy.

Material Family	Starch Synthetic Resin Alloy			
Material Supplier/ Grade	Novamont Mater-Bi AF05H			
Features	biodegradable			
Reference Number	239	239	239	239
EXPOSURE CONDITIONS				
Exposure Type	soil burial			fungus
Exposure Test Method				ASTM G21-70
Exposure Time (days)	7	21	49	
CHANGE IN PHYSICAL CHARACTERISTICS				
Weight Change (%)	>-39	>-50	>-55	
RESULTS OF EXPOSURE				
Growth Rate				4

Chapter 59

Diallyl Phthalate Resin

TABLE 99: Outdoor Weathering in New Jersey of Filled and Reinforced Diallyl Phthalate Resin.

Material Family	Diallyl Phthalate Resin									
Reference Number	126	126	126	126	126	126	126	126	126	126
MATERIAL CHARACTERISTICS										
U.S. Government Designation	MDG	MDG	MDG	SDG	SDG	SDI-5	SDI-5	SDI-30	SDI-30	SDI-30
MATERIAL COMPOSITION										
Reinforcement				short glass fiber	short glass fiber	acrylic fiber	acrylic fiber	polyethylene terephthalate fiber	polyethylene terephthalate fiber	polyethylene terephthalate fiber
Filler	mineral	mineral	mineral							
EXPOSURE CONDITIONS										
Exposure Type	outdoor weathering									
Exposure Location	New Jersey									
Exposure Country	USA									
Exposure Time (days)	182	365	548	365	548	182	365	182	365	548
PROPERTIES RETAINED (%)										
Tensile Modulus	81.4	93.2	105.1	85.7	99.3	103	96	107.3	90.3	100.3
Tensile Strength	85	101.5	106.1	91.5	106.9	104	112	98.3	101.4	95.4
Elongation	95	88.4	97.4	96.5	103.5	105	115	92.6	113.2	94.1
Work to Produce Failure	101.4	108.7	107.2	94.7	115.4	111	135	105.6	121.6	88

Chapter 60

Thermoset Polyester

Weather Resistance

Kalwall Sunlite (features: transparent, 0.99 mm thick; manufacturing method: continuous layup; product form: acrylic coated sheet)

In March, 1977, it was decided to evaluate plastics currently used in glazing, signs, solar collectors, skylights and various other applications, and their relative physical properties before and after weathering exposure tests. Samples of the various materials to be tested were purchased from Industrial and Sign Distribution Companies located in Cincinnati. The sample of Kalwall Sunlite was colorless (clear) and had a nominal thickness of 0.99 mm.

The material showed a change that was visually objectionable and had a high yellowness index, after exposure to three separate weathering tests: twin carbon arc weatherometer, outside EMMAQUA test in Arizona, and Florence, KY outside exposure at a 45° angle facing south. The change in yellowness index, although significant, is not particularly objectionable. The samples exposed in the twin carbon arc weatherometer for 3,000 hours were tested for tensile strength, light transmission, and both Charpy and Izod impact (notched). Kalwall Sunlite showed an increase in tensile strength from 75.79 MPa to 90.26 MPa and a significant loss (11%) in light transmittance. Kalwall could not be tested by either Charpy or Izod methods.

Reference: *Comparison Of Plastics Used In Glazing, Signs, Skylights And Solar Collector Applications - Technical Bulletin 143,* competitor's technical report (ADARIS 50-1037-01) - Aristech Chemical Corporation, 1989.

TABLE 100: Outdoor Weathering in Kentucky, Accelerated Outdoor Weathering by EMMAQUA and Accelerated Weathering in a Carbon Arc Weatherometer of Kalwall Sunlite Acrylic Coated Polyester.

Material Family	Polyester		
Material Supplier/ Grade	Kalwall Sunlite		
Features	transparent, acrylic coated, solar panel material		
Manufacturing Method	continuous layup		
Product Form	polyester/ glass laminate		
Reference Number	152	152	152

MATERIAL CHARACTERISTICS

Sample Thickness (mm)	0.99	0.99	0.99

EXPOSURE CONDITIONS

Exposure Type	accelerated outdoor weatheing	accelerated weathering	outdoor weathering
Exposure Location	Arizona		Florence, Kentucky
Exposure Country	USA		USA
Exposure Apparatus	EMMAQUA	carbon arc weatherometer	
Exposure Note			45° angle south
Exposure Time (days)	365	125	730

PROPERTIES RETAINED (%)

Tensile Strength		119.1	

SURFACE AND APPEARANCE

Δ Yellowness Index	38.67	24.16	7.87
Luminous Transmittance Retained (%)		87.5	

GRAPH 200: **Florence, Kentucky Outside Weathering Exposure Time vs. Yellowness Index of Polyester**

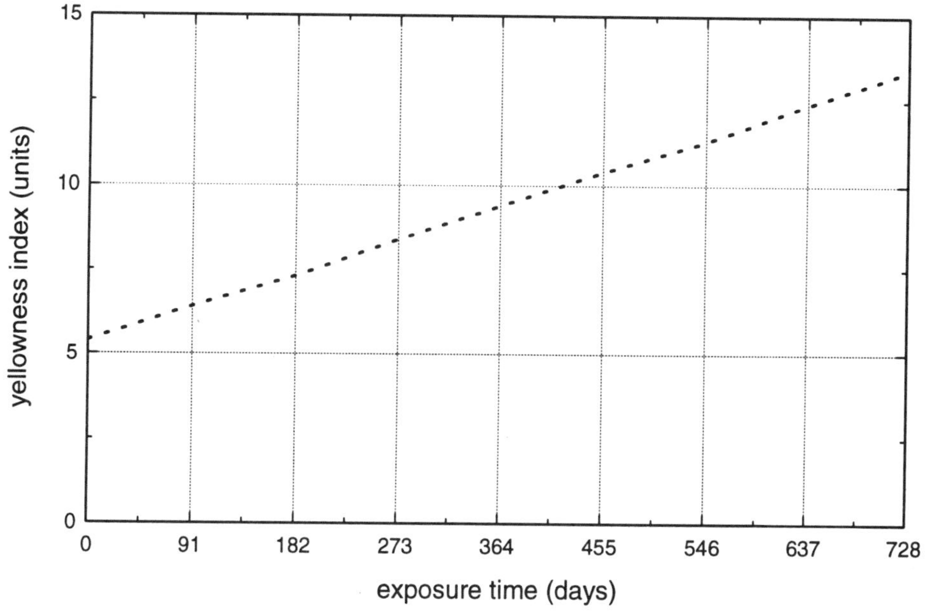

....	Kalwall Sunlite Polyester (transparent, 0.99 mm thick; continuous layup; acrylic coated sheet); test lab: Aristech Chemical Corp.
Reference No.	152

GRAPH 201: **EMMAQUA Arizona Accelerated Weathering Exposure Time vs. Yellowness Index of Polyester**

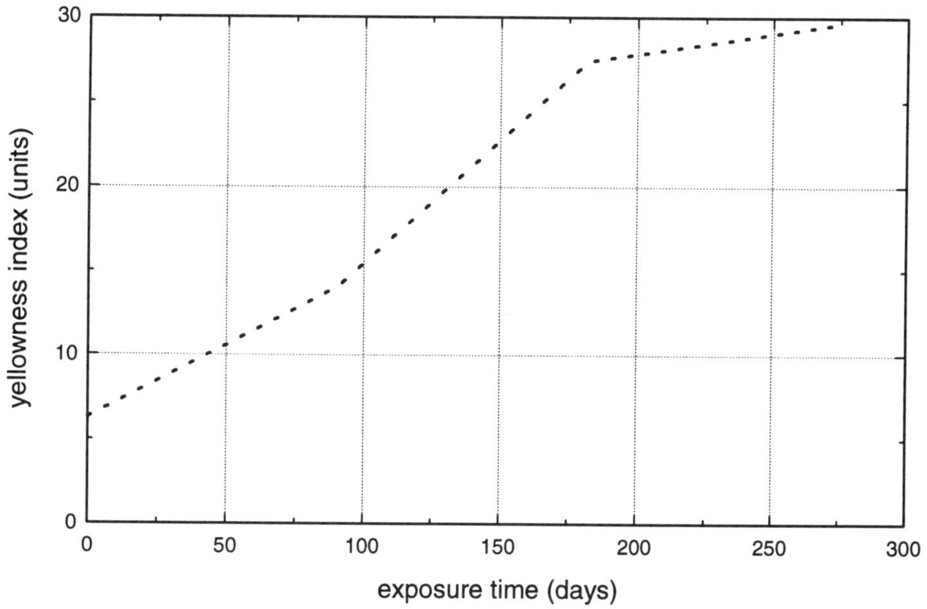

....	Kalwall Sunlite Polyester (transparent, 0.99 mm thick; continuous layup; acrylic coated sheet); test lab: DSET
Reference No.	152

GRAPH 202: Twin Carbon Arc Weatherometer Exposure Time vs. Yellowness Index of Polyester

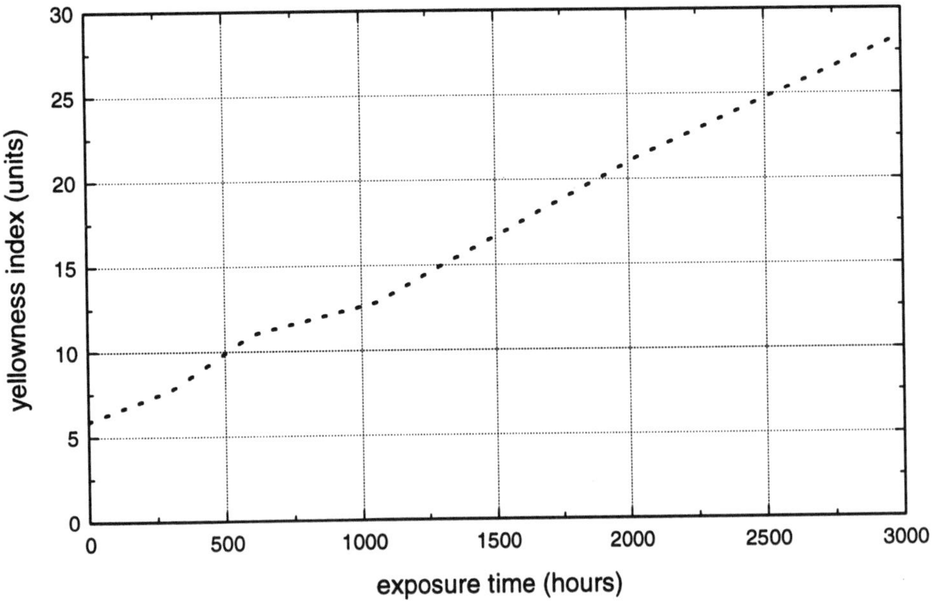

··············	Kalwall Sunlite Polyester (transparent, 0.99 mm thick; continuous layup; acrylic coated sheet); test lab: Aristech Chemical
Reference No.	152

Chapter 61

Polyurethane Reaction Injection Molding System

Weather Resistance

Recticel: Colo-Fast LM 161 (chemical type: aliphatic polyurethane; features: low modulus elastomer)

LM-161 has been subjected to a number of artificial weathering tests, as well as outdoor exposure in Florida. Typically, there is a slight increase in gloss of a medium gloss sample in most tests without polishing the sample. This is due to smoothing of the tiny irregularities in the surface (faithfully reproducing the mold texture) which gave the sample its original medium gloss level. On extremely long exposure, longer than required by automotive specifications, there is some dulling of the surface, but this can be restored by polishing with common automotive cleaners. No waxing is needed.

After artificial exposure there is little change in gloss even without washing or polishing. Gloss increases slightly in samples exposed for 9 months in Florida. These samples must be washed to remove the dirt that accumulates under outdoor conditions. No surface cracking or discoloration was observed on artificial weathering or Florida exposure. Outdoor exposure in Florida is continuing.

Samples without pigment have also been exposed to artificial weathering conditions. Even without pigment, the samples show only slight discoloration and gloss change, demonstrating the inherent weatherability of this gasket material. When light colored pigments are used in this system, the urethane will not cause the gasket to discolor or lose gloss. Systems with light colored pigments are now under test.

Reference: Carver, T. Granville, Kubizne, Peter J., Huys, Dirk, *Reaction Injection Molded Modular Window Gaskets Using Light Stable Aliphatic Polyurethane,* Polyurethanes: Exploring New Horizons - Proceedings Of The SPI 30th Annual Technical / Marketing Conference, conference proceedings - The Society Of the Plastics Industry, Inc.

Outdoor Weather Resistance

Miles: Baydur; Bayflex

Good resistance to aging and weathering is characteristic of RIM parts, though extended exposure to the sun's UV rays may result in a color shift at the surface. (The end use of the part should determine whether or not the surface needs a protective coating, e.g., UV-stable paints.)

Reference: *Bayflex And Baydur RIM Polyurethane Systems For Unparalleled Design Freedom With RIM,* supplier marketing literature (53-D601(10)L) - Mobay Corporation, 1986.

TABLE 101: Accelerated Weathering in a Xenon Arc Weatherometer of Recticel Colo-Fast Polyurethane Reaction Injection Molding System.

Material Family	Polyurethane Reaction Injection Molding System								
Material Supplier/ Grade	Recticel Colo-Fast SPR 5 BL			Recticel Colo-Fast SPR 30 BL			Recticel Colo-Fast SPR 50 BL		
Features	in-mold skin			in-mold skin			in-mold skin		
Chemical Type	aliphatic polyurethane			aliphatic polyurethane			aliphatic polyurethane		
Reference Number	221	221	221	221	221	221	221	221	221
EXPOSURE CONDITIONS									
Exposure Type	accelerated weathering			accelerated weathering			accelerated weathering		
Exposure Apparatus	xenon arc weatherometer			xenon arc weatherometer			xenon arc weatherometer		
Exposure Note	ES 25 apparatus; GM specifications			ES 25 apparatus; GM specifications			ES 25 apparatus; GM specifications		
Exposure Cycle Note	back plate temperature: 88°C; cycle: 3.8 hours light, 1 hour dark, no water spray			back plate temperature: 88°C; cycle: 3.8 hours light, 1 hour dark, no water spray			back plate temperature: 88°C; cycle: 3.8 hours light, 1 hour dark, no water spray		
Exposure Time (days)	20.8	41.7	66.7	20.8	41.7	66.7	20.8	41.8	66.7
SURFACE AND APPEARANCE									
ΔE Color Change	0.25-0.6 {s}	0.3-1.5 {s}	0.5-1.8 {s}	0.2-0.5 {s}	0.35-0.95 {s}	0.5-1.6 {s}	0.2-0.5 {s}	0.35-0.95 {s}	0.5-1.6 {s}
Gloss @ 60° Retained (%)	90-125 {t}	90-120 {t}	85-120 {t}	80-135 {t}	75-130 {t}	70-135 {t}	75-140 {t}	60-145 {t}	50-150 {t}
Surface Degradation	microscopic evaluation OK, visual evaluation OK	microscopic evaluation OK to limited cracking, visual evaluation OK	microscopic evaluation shows limited cracking, visual evaluation is borderline OK	microscopic evaluation OK, visual evaluation OK	microscopic evaluation shows limited cracking, visual evaluation OK	microscopic evaluation shows limited cracking, visual evaluation is borderline OK	microscopic evaluation OK, visual evaluation OK	microscopic evaluation shows limited cracking, visual evaluation OK	microscopic evaluation shows limited cracking, visual evaluation is borderline OK

TABLE 102: Accelerated Weathering in a Xenon Arc Weatherometer of Recticel Colo-Fast Polyurethane Reaction Injection Molding System.

Material Family	Polyurethane Reaction Injection Molding System								
Material Supplier/ Grade	Recticel Colo-Fast SPR 5 BL			Recticel Colo-Fast SPR 30 BL			Recticel Colo-Fast SPR 50 BL		
Features	in-mold skin			in-mold skin			in-mold skin		
Chemical Type	aliphatic polyurethane			aliphatic polyurethane			aliphatic polyurethane		
Reference Number	221	221	221	221	221	221	221	221	221

EXPOSURE CONDITIONS

Exposure Type	accelerated weathering			accelerated weathering			accelerated weathering		
Exposure Apparatus	xenon arc weatherometer			xenon arc weatherometer			xenon arc weatherometer		
Exposure Note	ES 25 apparatus; Japanese specifications			ES 25 apparatus; Japanese specifications			ES 25 apparatus; Japanese specifications		
Exposure Cycle Note	back plate temperature: 68°C; cycle: 12 minutes light, 48 minutes dark, continuous exposure			back plate temperature: 68°C; cycle: 12 minutes light, 48 minutes dark, continuous exposure			back plate temperature: 68°C; cycle: 12 minutes light, 48 minutes dark, continuous exposure		
Exposure Time (days)	20.8	41.7	66.7	20.8	41.7	66.7	20.8	41.8	66.7

SURFACE AND APPEARANCE

ΔE Color Change	0.2-0.6 {s}	0.45-0.9 {s}	0.95-1.6 {s}	0.2-0.3 {s}	0.5-0.9 {s}	0.75-0.95 {s}	0.2-0.3 {s}	0.5-0.9 {s}	0.75-0.95 {s}
Gloss @ 60° Retained (%)	90-125 {t}	90-140 {t}	90-220 {t}	120-130 {t}	160-190 {t}	250-320 {t}	120-130 {t}	160-190 {t}	250-320 {t}
Surface Degradation	microscopic evaluation OK, visual evaluation OK	microscopic evaluation shows limited cracking, visual evaluation OK	microscopic evaluation shows limited cracking, visual evaluation is borderline OK	microscopic evaluation OK, visual evaluation OK	microscopic evaluation shows limited cracking, visual evaluation OK	microscopic evaluation shows limited cracking, visual evaluation is borderline OK	microscopic evaluation OK, visual evaluation OK	microscopic evaluation shows limited cracking, visual evaluation OK	microscopic evaluation shows limited cracking, visual evaluation is borderline OK

TABLE 103: Accelerated Weathering in a Fadeometer of Recticel Colo-Fast Polyurethane Reaction Injection Molding System.

Material Family	Polyurethane Reaction Injection Molding System		
Material Supplier/ Grade	Recticel Colo-Fast SPR 5 BL	Recticel Colo-Fast SPR 30 BL	Recticel Colo-Fast SPR 50 BL
Features	in-mold skin	in-mold skin	in-mold skin
Chemical Type	aliphatic polyurethane	aliphatic polyurethane	aliphatic polyurethane
Reference Number	221	221	221
EXPOSURE CONDITIONS			
Exposure Type	accelerated weathering		
Exposure Apparatus	fadeometer		
Exposure Cycle Note	back plate temperature: 86-88°C; no water spray, continuous exposure		
Relative Humidity (%)	50	50	50
Exposure Time (days)	20.8	20.8	20.8
SURFACE AND APPEARANCE			
ΔE Color Change	0.2 {s}	0.25 {s}	0.25 {s}
Gloss @ 60° Retained (%)	95 {t}	100 {t}	100 {t}
Surface Degradation	microscopic evaluation OK, visual evaluation OK	microscopic evaluation OK, visual evaluation OK	microscopic evaluation OK, visual evaluation OK

GRAPH 203: Florida Outdoor Weathering Exposure Time vs. Delta b Color Scale of Polyurethane

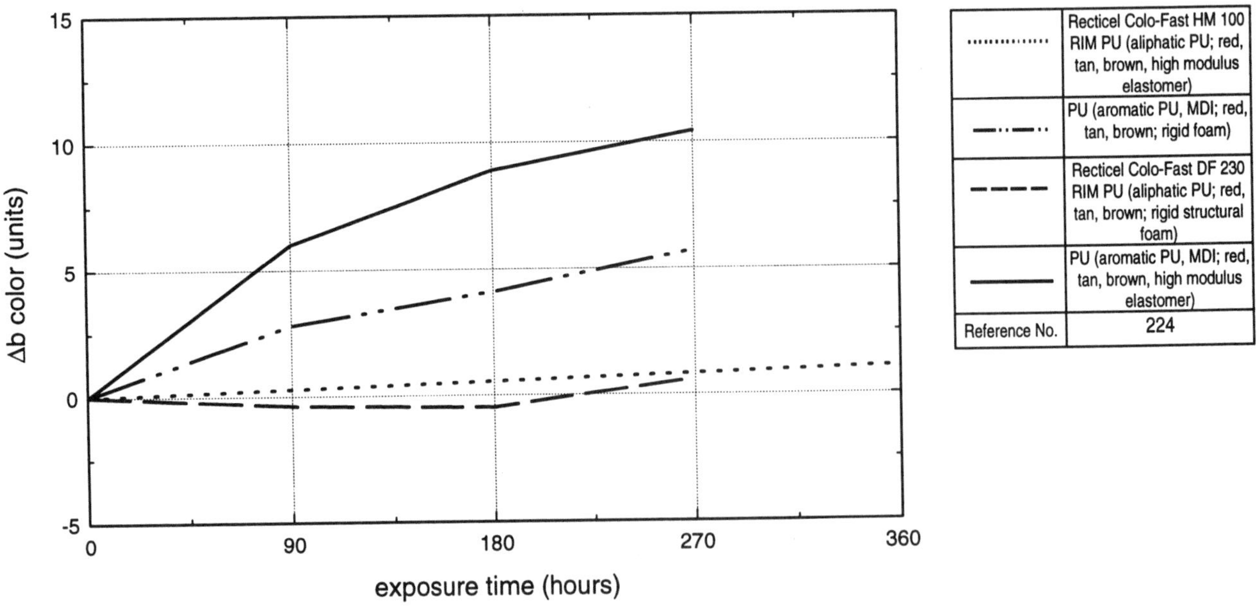

GRAPH 204: QUV Exposure Time vs. Gloss Retained of Polyurethane Reaction Injection Molding System

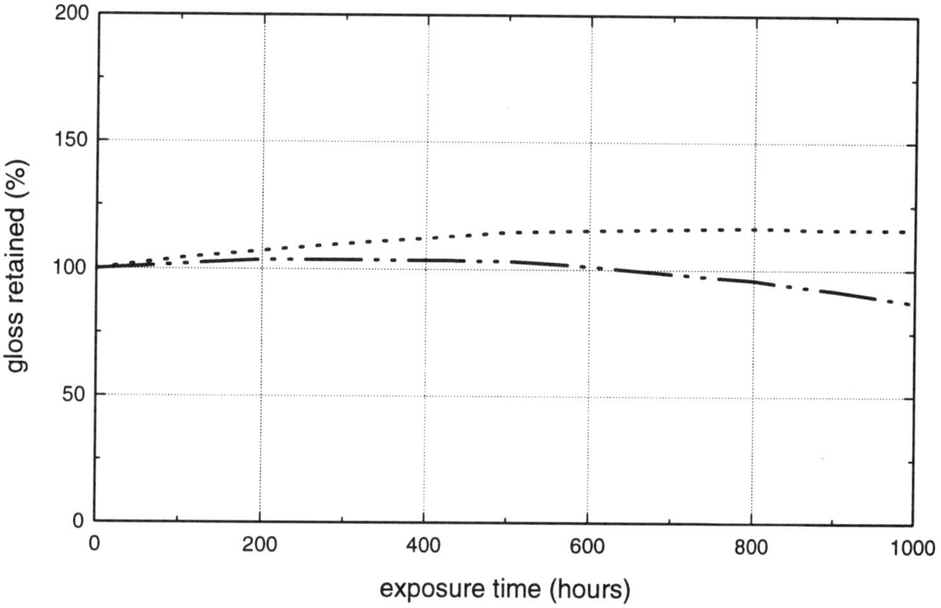

....	Recticel Colo-Fast LM 161 RIM PU (low modulus elastomer, black); washed and polished
—··—··	Recticel Colo-Fast LM 161 RIM PU (low modulus elastomer, black); as weathered
Reference No.	224

GRAPH 205: Sunshine Carbon Arc Weatherometer Exposure Time vs. Gloss Retained of Polyurethane Reaction Injection Molding System

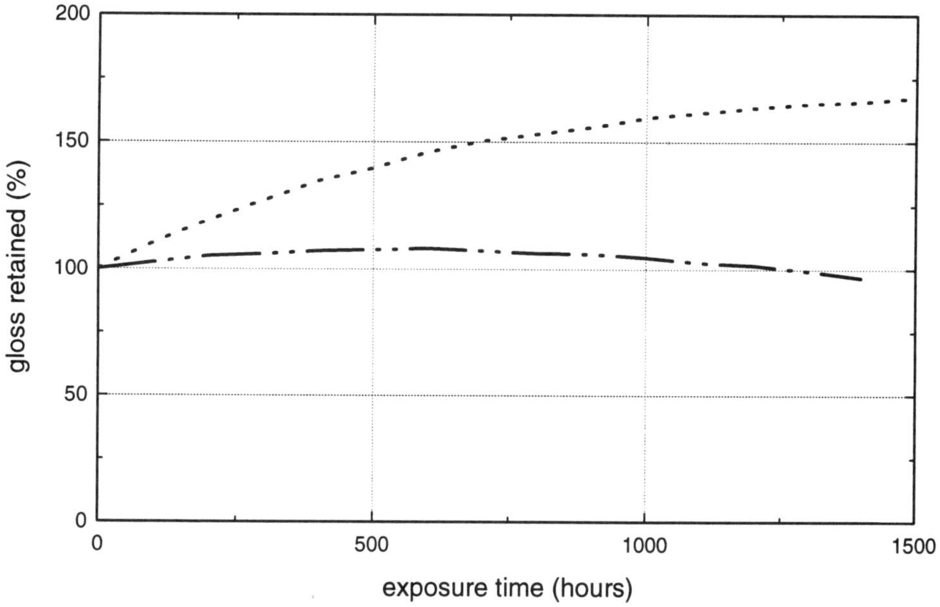

....	Recticel Colo-Fast LM 161 RIM PU (low modulus elastomer, black); washed and polished
—··—··	Recticel Colo-Fast LM 161 RIM PU (low modulus elastomer, black); as weathered
Reference No.	224

Chapter 62

Chlorinated Polyethylene Elastomer

Weather Resistance

Dow Chemical: Tyrin

Sunlight, ozone and oxygen can commonly cause environmental degradation of many elastomers by attacking both saturated and (especially) unsaturated sites along the polymer chain. Because Tyrin products have a saturated molecular backbone, they are not as susceptible to such attacks as are many other elastomers.

Based on both in-service use and accelerated laboratory aging tests, Tyrin elastomers can provide excellent long-term outdoor weathering performance.

Reference: *Tyrin CPE Elastomers,* supplier design guide (306-00094-1190 RJD) - Dow Chemical Company, 1990.

Ozone Resistance

Dow Chemical: Tyrin

Long-term exposures of properly formulated rubbers made with Tyrin elastomers have shown no type of cracking when tested in an ozone atmosphere or when tested in an ozone atmosphere or when exposed to outdoor weathering.

Reference: *Tyrin CPE Elastomers,* supplier design guide (306-00094-1190 RJD) - Dow Chemical Company, 1990.

Chapter 63

Olefinic Thermoplastic Elastomer

Weather Resistance

Advanced Elastomer Systems: Santoprene

For UV exposure, black Santoprene grades perform better than colorable grades and softer grades perform better than harder grades. UV stable black Santoprene grades or pigmented UV stable colorable Santoprene grades are good choices to replace EPDM, Neoprene, Hypalon, silicone, or PVC in outdoor applications where appearance is not a major consideration. Colorable UV stable grades will provide tensile property retention equivalent to, or better than, black EPDM and Neoprene, but color shifts may be more noticeable.

Addition of pigments can change performance greatly, usually for the better. Pigments vary widely as to their ability to enhance UV stability. Some pigments, such as organic dyes, are themselves not very stable to UV. UV stable colorable Santoprene grades are better than GP colorable, but not as good as UV stable black grades for UV exposure. For continuous outdoor exposure, always use a black grade instead of a colorable grade, if possible. Never use a GP colorable grade unless the part is truly massive, heavily pigmented, and surface appearance after a few months is not important.

Reference: *Weatherability Of Santoprene Rubber Compared To Other Materials,* supplier technical report (TCD01588) - Advanced Elastomer Systems, 1988.

Evode Plastics: Forprene

Both the elastomer and the thermoplastic are olefinic. The crosslinking system used for the elastomer is highly efficient and produces a tightly bound network. This and the olefinic nature of the raw materials ensure the Forprene compound has excellent weather and ozone resistance.

Reference: *Forprene By S.O.F.TER.,* supplier marketing literature (RDS 049/9240) - Evode Plastics.

UV Resistance

Dow Chemical: Engage (features: 23 Mooney (ML 1+4 @ 125°C))

ENGAGE POE's have improved UV stability resistance and high temperature stability.

UV stability was determined by measuring the formation of carbonyl groups (due to chain scission) and the resulting decrease in molecular weight. When compared with EPDM and EPM, the ENGAGE POE demonstrated superior natural UV stability, forming very few carbonyl groups and showing very little decrease in molecular weight.

Reference: *Engage Polyolefin Elastomers,* supplier marketing literature (305-01995-1293 SMG) - Dow Chemical Company, 1993.

Outdoor Weather Resistance

Advanced Elastomer Systems: Santoprene 101-64 (features: black color, general purpose grade, 64 Shore A hardness); **Santoprene 103-40** (features: black color, general purpose grade, 40 Shore D hardness); **Santoprene 121-67** (features: UV stabilized, black color, 67 Shore A hardness); **Santoprene 121-73** (features: UV stabilized, black color, 73 Shore A hardness); **Santoprene 121-80** (features: UV stabilized, black color, 80 Shore A hardness); **Santoprene 123-40** (features: UV stabilized, black color, 40 Shore D hardness)

Santoprene rubber was exposed to conventional Arizona aging by DSET Laboratories, Inc. (Phoenix, AZ) at a 5° tilt from the horizontal. This tilt is preferable to 0°, as it allows for some drainage and dirt wash off during rains. Direct exposures are intended for materials which will be used outdoors and subjected to all elements of weather. The exposure period consisted of 6, 12, 24 and 48 months. Santoprene rubber black grades show a ΔE color change of less than 3 for the 48 month exposure. The harder grades of black UV stabilized and general purpose Santoprene rubber show a slight deterioration in hardness over the 48 month period. The softer grades retain their hardness. All grades tested maintained reasonably good tensile strength over the 48 months.

Santoprene rubber was also exposed to conventional Arizona aging with spray. This exposure method is the same as conventional aging, except a water spray is used to induce moisture weathering conditions. The introduction of moisture plays an important role in improving both the relevancy and reproducibility of the weathering test results. Spray nozzles are mounted above the face of the rack at points distributed to insure uniform wetting of the entire exposure area. Distilled water is sprayed for 4 hours preceding sunrise to soak the samples, and then twenty times during the day in 15 second bursts. The purpose of the wetting is twofold. First, the introduction of water in the otherwise arid climate induces and accelerates some degradation modes which do not occur as rapidly, if at all, without moisture. Second, a thermal shock causes a reduction in specimen surface temperatures as much as 14 °C (25 °F). This results in physical stresses which accelerate the degradation process. Of the specimens tested for color change, over a 48 month period, black UV grades of Santoprene rubber show a considerable improvement over general purpose Santoprene. The ΔE color change for most of the black UV grades remains at less than 3. The softer grades of Santoprene rubber show minor hardness changes. The harder grades do continue to increase in hardness with exposure time. Most of the materials maintain a significant tensile strength during the exposure period. Both the black UV stable and general purpose grades of Santoprene rubber maintain most of their elongation during exposure.

Santoprene rubbers were also exposed to conventional Florida aging. This test method is a real time exposure by DSET Laboratories, Inc. (Homestead, FL) at a 5° tilt from the horizontal. Since this location has a much higher average humidity, this exposure is harsher on some materials. Humidity has a great effect on the color of the harder grades of general purpose Santoprene rubber. However, it has a minor effect on black UV stable grades, especially over longer exposures. While, most of the hardness changes are minor, the harder grades of Santoprene rubber, both general purpose and black UV stable, show the most increase. Most of the materials sustain their tensile strength with exposure. Softer grades of general purpose Santoprene rubber do not retain their elongation properties as well as black UV stable grades.

Conventional Florida aging with spray was used to test Santoprene rubbers. This exposure method is the same as the conventional aging, except water spray is used to induce moisture weathering conditions. Moisture is introduced the same in Florida as it is in Arizona. The color change for black UV stable grades of Santoprene rubber is very small. General purpose harder grades of Santoprene rubber have a large ΔE. Significant hardness changes occur in the harder grades of general purpose and black UV stable grades. Tensile strength retention is good for most of the materials tested. Santoprene rubber general purpose grades retain their tensile strength similar to black UV stable grades. General purpose grades show a continuing decrease in elongation with exposure time.

Reference: *Weathering Of Santoprene Thermoplastic Rubber Black Ultraviolet Grades,* supplier technical report (TCD00592) - Advanced Elastomer Systems, 1992.

Advanced Elastomer Systems: Santoprene 101-64 (features: black color, general purpose grade, 64 Shore A hardness); **Santoprene 103-40** (features: black color, general purpose grade, 40 Shore D hardness); **Santoprene 121-67** (features: UV stabilized, black color, 67 Shore A hardness); **Santoprene 121-73** (features: UV stabilized, black color, 73 Shore A hardness); **Santoprene 121-80** (features: UV stabilized, black color, 80 Shore A hardness); **Santoprene 123-40** (features: UV stabilized, black color, 40 Shore D hardness)

Santoprene rubber black UV grades were developed to meet weather resistance requirements. These grades offer excellent long term performance in outdoor exposure applications, as well as rubber properties similar to or better than chloroprene (neoprene), EPDM and chlorosulfonated polyethylene thermoset rubbers. These grades, rather than the general purpose grades of Santoprene rubber, should be used for applications involving direct exposure to sunlight or any other strong UV source where a black material is required.

Black UV grades should be selected for those applications requiring several years of service life outdoors. Such applications are exterior automotive, window glazings, electrical jackets, etc. UV grades of Santoprene rubber should always be considered for outdoor applications.

Reference: *Weathering Of Santoprene Thermoplastic Rubber Black Ultraviolet Grades,* supplier technical report (TCD00592) - Advanced Elastomer Systems, 1992.

Advanced Elastomer Systems: Santoprene 101-73 (features: black color, 73 Shore A hardness, 3.2 mm thick; manufacturing method: compression molding)

After two years' exposure in Florida, both direct and under glass, severe chalking of Santoprene was a surprise. The material is based on polymers - EPDM and PP - which are considered good for weatherability.

Reference: *Florida Weathering Of Chemigum TPE,* supplier technical report (TPE 06-0292/498900-2/92) - Goodyear Chemicals, 1992.

DuPont: Alcryn

Many applications for Alcryn melt-processible rubber involve outdoor exposure, where it is subjected to the deteriorating effects of sunlight, water (often acidic), and ozone. Like most organic polymers, the polymer base for Alcryn is slowly degraded by direct exposure to high energy UV light. Available weathering data indicate that Alcryn will provide outstanding outdoor durability if it is protected against UV degradation by pigments such as carbon black, titanium dioxide or stable organic colors, plus organic UV absorbers or stabilizers. Data clearly demonstrate that properly pigmented Alcryn, in black or colored compounds, will give long, reliable service in outdoor applications.

Reference: *Alcryn Weather Resistance Guide,* supplier technical report (196240A) - Du Pont Company, 1990.

Accelerated Outdoor Weathering Resistance

Advanced Elastomer Systems: Santoprene 101-64 (features: black color, general purpose grade, 64 Shore A hardness); **Santoprene 103-40** (features: black color, general purpose grade, 40 Shore D hardness); **Santoprene 121-67** (features: UV stabilized, black color, 67 Shore A hardness); **Santoprene 121-73** (features: UV stabilized, black color, 73 Shore A hardness); **Santoprene 121-80** (features: UV stabilized, black color, 80 Shore A hardness); **Santoprene 123-40** (features: UV stabilized, black color, 40 Shore D hardness)

Outdoor accelerated exposure testing was done using equatorial mount with mirrors for acceleration (EMMA) on Santoprene rubber. This test method uses natural sunlight and special reflecting mirrors to concentrate the sunlight to the intensity of about eight suns. A blower which directs air over and under the samples is used to cool the specimens. This limits the increase in surface temperatures of most materials to 10 °C (18 °F) above the maximum surface temperature that is reached by identically mounted samples exposed to direct sunlight at the same time and locations without concentration. DSET Laboratories designed and maintains this equipment. The exposure period was a total of 6 and 12 months, which has been correlated to about 2-1/2 and 5 years of actual aging in a Florida environment. The general purpose grades of Santoprene rubber show a greater color change than that of black UV stable grades. The material with the best color stability is Santoprene rubber 121-73. Elongation of harder grades of general purpose Santoprene rubber exhibit significant deterioration during the exposure period; the soft general purpose and UV stable grades continue to retain elongational properties.

Specimens were also exposed to an equatorial mount with mirrors for acceleration plus water (EMMAQUA). This is an accelerated weathering test method which uses the same apparatus as the EMMA, except water spray is used to induce moisture weathering conditions. Softer, black UV stabilized grades of Santoprene showed only minor changes in color. The rest of the grades had major changes. The hardness change was minor. All grades suffered a significant loss in tensile strength and general purpose grades do not retain their tensile strength as well as black UV stable grades. Elongation retention was very difficult for harder grades and the softer grades performed better in this area.

Reference: *Weathering Of Santoprene Thermoplastic Rubber Black Ultraviolet Grades,* supplier technical report (TCD00592) - Advanced Elastomer Systems, 1992.

Accelerated Artificial Weathering Resistance

Advanced Elastomer Systems: Santoprene 101-64 (features: black color, general purpose grade, 64 Shore A hardness); **Santoprene 103-40** (features: black color, general purpose grade, 40 Shore D hardness); **Santoprene 121-67** (features: UV stabilized, black color, 67 Shore A hardness); **Santoprene 121-73** (features: UV stabilized, black color, 73 Shore A hardness); **Santoprene 121-80** (features: UV stabilized, black color, 80 Shore A hardness); **Santoprene 123-40** (features: UV stabilized, black color, 40 Shore D hardness); **Santoprene 123-50**

Artificial accelerated weathering devices, such as Atlas UV-Con or QUV testing, often produce results that do not correlate to actual outdoor exposure of a material. Some materials, like Santoprene black UV stable grades, deteriorate noticeably in extended UV-Con tests but show only modest changes in actual outdoor use.

Under xenon arc weatherometer testing (to General Motors standard TM30-2 and SAE J 1885), most UV grades of Santoprene rubber retain 90% or more of their tensile properties after 3,000 hours and 80% or more after 5,000 hours. The color change for general purpose grades was significant, while the rest of the grades had a ΔE of less than 3. All Santoprene grades showed only minor changes in hardness and good elongation retention. Retention of tensile strength was generally good, with black UV stable grades having increased tensile strength retention over general purpose grades. Xenon arc testing according to SAE J 1960 showed that for the softer, black UV stable grades, the color retention is excellent, up to 2,500 KJ/m^2 measured at 340 nm. The hardest grade of this material, 123-50, does show significant color change with exposure. The pass/fail point is usually $\Delta E < 3$ at 2,500 KJ/m^2. Gray scale rating of the same samples, using the same exposure conditions and time, again shows significant color change for the hardest material (123-50).

Reference: *Weathering Of Santoprene Thermoplastic Rubber Black Ultraviolet Grades*, supplier technical report (TCD00592) - Advanced Elastomer Systems, 1992.

Effect of Carbon Black on Weatherability

DuPont: Alcryn 1060BK (additives: carbon black; features: 62 Shore A hardness); **Alcryn 1070BK** (additives: carbon black; features: 68 Shore A hardness); **Alcryn 1080BK** (additives: carbon black; features: 76 Shore A hardness); **Alcryn 1160BK** (additives: carbon black; features: 62 Shore A hardness); **Alcryn 1170BK** (additives: carbon black; features: 68 Shore A hardness); **Alcryn 1180BK** (additives: carbon black; features: 76 Shore A hardness); **Alcryn**

Black grades of Alcryn contain more than enough carbon black to provide outstanding weather resistance. Representative samples of the extrusion types showed little change in properties or surface integrity after accelerated weathering tests or after two years outdoors in Florida.

The weatherability of Alcryn compares favorably to that of the major oil-resistant thermoset elastomers. In accelerated tests, Alcryn 1170BK (1070BK) resists weather and ozone about as well as Hypalon synthetic rubber, and has equivalent weather resistance and much better ozone resistance than Neoprene. Both Hypalon and Neoprene are widely used for their durability in outdoor applications. Alcryn also shows considerably better weather and ozone resistance than nitrile rubber. This is an important point, because while nitrile rubber of medium acrylontrile content has the same excellent oil resistance as Alcryn, it is generally considered unsuitable for applications involving outdoor exposure.

Reference: *Alcryn Weather Resistance Guide*, supplier technical report (196240A) - Du Pont Company, 1990.

Effect of Color Pigments on Weatherability

DuPont: Alcryn 3055NC (features: natural resin, 57 Shore A hardness); **Alcryn 3065NC** (features: natural resin, 64 Shore A hardness); **Alcryn 3155NC** (features: natural resin, 57 Shore A hardness); **Alcryn 3165NC** (features: natural resin, 64 Shore A hardness); **Alcryn**

The neutral types of Alcryn, as manufactured, are not compounded for weather resistance. They are colorable bases which must be pigmented for extended outdoor durability. When adequately pigmented in white or a variety of bright colors, the neutral grades appear to weather as well as the black grades. Alcryn compares very favorably to colored EPDM and cross-linked PVC/NBR used in the construction industry. White Alcryn can also be achieved with excellent color and property retention. The color pigments used in studies were added via commercial color concentrates of the type used with polyvinyl chloride (PVC). In pigmenting Alcryn, color concentrates based on PVC are strongly recommended to insure good compatibility and the least impact on desirable properties. Color concentrates based on polyethylene (PE) or ethylene/vinyl acetate (EVA) resins will cause poor knit strength in molded parts of Alcryn. As a guideline, pigment concentrations normally used to prevent UV degradation of PVC are a good starting point for use in Alcryn.

Reference: *Alcryn Weather Resistance Guide,* supplier technical report (196240A) - Du Pont Company, 1990.

Ozone Resistance

Advanced Elastomer Systems: Santoprene

The results of both ASTM D1171 and ASTM D518 ozone testing indicate that Santoprene rubber exhibits outstanding resistance to ozone. Its ozone resistance at least equals that of EPDM and several competitive TPEs now available in the marketplace.

Reference: *Ozone Resistance Of Santoprene Rubber,* supplier technical report (TCD01787) - Advanced Elastomer Systems, 1986.

BP Chemicals: TPR 1600

TPR 1700

The results of both ASTM D1171 and ASTM D518 ozone testing indicate that TPR rubber exhibits outstanding resistance to ozone.

Reference: *Ozone Resistance Of Santoprene Rubber,* supplier technical report (TCD01787) - Advanced Elastomer Systems, 1986.

Dow Chemical: Engage (cure: 10 min. @ 175°C; material compostion: 100 phr Engage, 35 phr paraffinic plasticizer, 85 phr N-550 carbon black, 5 phr zinc oxide, 10 phr magnesium oxide, 0.5 phr zinc stearate, 4 phr antioxidant, 1 phr co-agent, 10 phr dicumyl peroxide (40%))

Formulated ENGAGE POE's are extremely resistant to ozone attack. This is largely due to the saturated backbones of ENGAGE POE's, which are less susceptible to environmental degradation than, for example, materials having double bonds in their backbone, such as IIR, SBR and NR.

No cracking occurred after a compound made with ENGAGE POE was tested at 100 pphm ozone.

Reference: *Engage Polyolefin Elastomers,* supplier marketing literature (305-01995-1293 SMG) - Dow Chemical Company, 1993.

DuPont: Alcryn

Alcryn is virtually immune to ozone attack.

Reference: *Alcryn Weather Resistance Guide,* supplier technical report (196240A) - Du Pont Company, 1990.

Evode Plastics: Forprene

Both the elastomer and the thermoplastic are olefinic. The crosslinking system used for the elastomer is highly efficient and produces a tightly bound network. This and the olefinic nature of the raw materials ensure the Forprene compound has excellent weather and ozone resistance.

Reference: *Forprene By S.O.F.TER.,* supplier marketing literature (RDS 049/9240) - Evode Plastics.

Technor Apex: Telcar 332; Telcar 942

The results of both ASTM D1171 and ASTM D518 ozone testing indicate that Telcar rubber exhibits outstanding resistance to ozone.

Reference: *Ozone Resistance Of Santoprene Rubber,* supplier technical report (TCD01787) - Advanced Elastomer Systems, 1986.

Effect of Bacterial Attack and Fungus

DuPont: Alcryn

Fungus growth, observed in some outdoor exposures in warm, moist climates, can be prevented by the addition of a suitable fungicide. Vinyzene SB-1-ELV (supplied by Morton International Corp.), a commercial antimicrobial additive recommended for use in PVC, prevented mildew and other fungus growth on Alcryn. Other commercial fungicides may be equally effective.

Manufacturers who have used Alcryn in outdoor applications report no fungus growth and no change in appearance or performance of Alcryn parts after outdoor exposures of up to 20 months.

Reference: *Alcryn Weather Resistance Guide,* supplier technical report (196240A) - Du Pont Company, 1990.

TABLE 104: Outdoor Weathering, Accelerated Outdoor Weathering by EMMAQUA and Accelerated Weathering with a Xenon Arc Weatherometer of Black Advanced Elastomer Systems Santoprene Olefinic Thermoplastic Elastomer.

Material Family	Olefinic Thermoplastic Elastomer					
Material Supplier/ Grade	Advanced Elastomer Systems Santoprene 101-73			Advanced Elastomer Systems Santoprene 103-40		
Features	black color			black color		
Reference Number	157	157	157	157	157	157
MATERIAL CHARACTERISTICS						
Shore A hardness	73	73	73			
Shore D hardness				40	40	40
EXPOSURE CONDITIONS						
Exposure Type	accelerated outdoor weathering	accelerated weathering	outdoor weathering	accelerated outdoor weathering	accelerated weathering	outdoor weathering
Exposure Location	Phoenix, Arizona		Phoenix, Arizona	Phoenix, Arizona		Phoenix, Arizona
Exposure Country	USA		USA	USA		USA
Exposure Apparatus	EMMAQUA	xenon arc weatherometer		EMMAQUA	xenon arc weatherometer	
Exposure Time (days)	365	216	365	365	216	365
Exposure Note			water spray added			water spray added
PROPERTIES RETAINED (%)						
Tensile Strength	89.7	83.4	98.8	69.3	48.8	93.0
Elongation	61.9	78.8	75.6	58.4	12.3	83.4
SURFACE AND APPEARANCE						
ΔE Color Change	5.5	4.1	4.1	14.9	18.1	14.5

TABLE 105: Outdoor Weathering, Accelerated Outdoor Weathering by EMMAQUA and Accelerated Weathering in a Xenon Arc Weatherometer of Black, UV Stabilized Advanced Elastomer Systems Santoprene Olefinic Thermoplastic Elastomer.

Material Family	Olefinic Thermoplastic Elastomer					
Material Supplier/ Grade	Advanced Elastomer Systems Santoprene 121-73			Advanced Elastomer Systems Santoprene 123-40		
Features	black color, UV stabilized			black color, UV stabilized		
Reference Number	157	157	157	157	157	157
MATERIAL CHARACTERISTICS						
Shore A hardness	73	73	73			
Shore D hardness				40	40	40
EXPOSURE CONDITIONS						
Exposure Type	accelerated outdoor weathering	accelerated weathering	outdoor weathering	accelerated outdoor weathering	accelerated weathering	outdoor weathering
Exposure Location	Phoenix, Arizona		Phoenix, Arizona	Phoenix, Arizona		Phoenix, Arizona
Exposure Country	USA		USA	USA		USA
Exposure Apparatus	EMMAQUA	xenon arc weatherometer		EMMAQUA	xenon arc weatherometer	
Exposure Time (days)	365	216	365	365	216	365
Exposure Note			water spray added			water spray added
PROPERTIES RETAINED (%)						
Tensile Strength	94.9	85.4	117.8	95.5	79.8	96.9
Elongation	68.1	75.4	67.8	66.8	75.0	77.3
SURFACE AND APPEARANCE						
ΔE Color Change	3.0	2.0	1.9	8.4	9.2	7.2

TABLE 106: Outdoor Weathering in Arizona With and Without Water Spray of Black, General Purpose Advanced Elastomer Systems Santoprene Olefinic Thermoplastic Elastomer.

Material Family	Olefinic Thermoplastic Elastomer				Olefinic Thermoplastic Elastomer			
Material Supplier/ Grade	Advanced Elastomer Systems Santoprene 103-40				Advanced Elastomer Systems Santoprene 103-40			
Features	black color, general purpose				black color, general purpose			
Sample Form	injection molded slab				injection molded slab			
Reference Number	120				120			
MATERIAL CHARACTERISTICS								
Shore D hardness	40	40	40	40	40	40	40	40
Sample Thickness (mm)	3.1	3.1	3.1	3.1	3.1	3.1	3.1	3.1
Sample Width (mm)	76.2	76.2	76.2	76.2	76.2	76.2	76.2	76.2
Sample Length (mm)	127	127	127	127	127	127	127	127
EXPOSURE CONDITIONS								
Exposure Type	outdoor weathering				outdoor weathering			
Exposure Location	Phoenix, Arizona				Phoenix, Arizona			
Exposure Country	USA				USA			
Test Lab	DSET Laboratories				DSET Laboratories			
Exposure Note	5° angle				5° angle; with water spray (distilled water sprayed for 4 hours preceeding sunrise and then 20 times during the day in 15 second bursts)			
Exposure Time (days)	182	365	730	1461	182	365	730	1461
PROPERTIES RETAINED (%)								
Tensile Strength	95 {ac}	96 {ac}	92 {ac}	96 {ac}	92 {ac}	93 {ac}	97 {ac}	99 {ac}
Elongation	95 {ac}	95 {ac}	91 {ac}	94 {ac}	93 {ac}	91 {ac}	97 {ac}	97 {ac}
CHANGE IN PHYSICAL CHARACTERISTICS								
Hardness Change (units)	4 {ae}	3 {ae}	3 {ae}	8 {ae}	3 {ae}	1 {ae}	2 {ae}	7 {ae}
SURFACE AND APPEARANCE								
ΔE Color Change	3 {ab}	7.3 {ab}	8.3 {ab}	11.9 {ab}	7.6 {ab}	4.8 {ab}	6.1 {ab}	8.3 {ab}

TABLE 107: Outdoor Weathering in Arizona and Accelerated Outdoor Weathering by EMMAQUA of Colorable Advanced Elastomer Systems Olefinic Thermoplastic Elastomer.

Material Family	Olefinic Thermoplastic Elastomer											
Material Supplier/ Grade	Advanced Elastomer Systems Santoprene 201-73	Advanced Elastomer Systems Santoprene 203-40	Advanced Elastomer Systems Santoprene 221-73-W157	Advanced Elastomer Systems Santoprene 223-50-W157	Advanced Elastomer Systems Santoprene 221-73-W157	Advanced Elastomer Systems Santoprene 223-50-W157	Advanced Elastomer Systems Santoprene 201-73	Advanced Elastomer Systems Santoprene 203-40	Advanced Elastomer Systems Santoprene 221-73-W157	Advanced Elastomer Systems Santoprene 223-50-W157	Advanced Elastomer Systems Santoprene 221-73-W157	Advanced Elastomer Systems Santoprene 223-50-W157
Features	neutral color	neutral color	neutral color, UV stabilized	neutral color, UV stabilized	grey color, UV stabilized	brown color, UV stabilized	neutral color	neutral color	neutral color, UV stabilized	neutral color, UV stabilized	grey color, UV stabilized	brown color, UV stabilized
Reference Number	157	157	157	157	157	157	157	157	157	157	157	157
MATERIAL CHARACTERISTICS												
Shore A hardness	73		73		73		73		73		73	
Shore D hardness		40		50		50		40		50		50
EXPOSURE CONDITIONS												
Exposure Type	outdoor weathering						accelerated outdoor weathering					
Exposure Location	Arizona						Arizona					
Exposure Country	USA						USA					
Exposure Apparatus							EMMAQUA					
Exposure Time (days)	365						365					
Exposure Note	water spray added											
PROPERTIES RETAINED (%)												
Tensile Strength	19.3	11.8	82.7	75.4	75.9	85.1	0.0	0.0	81.0	46.0	68.0	68.4
Elongation	48.7	1.5	50.5	63.6	70.4	68.6	0.0	0.0	49.5	16.7	68.5	68.4
SURFACE AND APPEARANCE												
ΔE Color Change	17.8	35.2	17.9	12.4	12.1	8.4	>40.0	>40.0	17.2	16.2	13.0	17.6

TABLE 108: **Outdoor Weathering in Arizona and Florida of Black Advanced Elastomer Systems Santoprene Olefinic Thermoplastic Elastomer.**

Material Family	Olefinic Thermoplastic Elastomer									
Material Supplier/ Grade	Advanced Elastomer Systems Santoprene 591-65W175				Advanced Elastomer Systems Santoprene 121-73				Advanced Elastomer Systems Santoprene 121-67	Advanced Elastomer Systems Santoprene 121-73
Features	black color				black color, UV stabilized				black color, UV stabilized	black color, UV stabilized
Reference Number	155	155	155	155	156	156	156	156	155	155

MATERIAL CHARACTERISTICS

Shore A hardness	65	65	65	65	73	73	73	73	67	73

EXPOSURE CONDITIONS

Exposure Type	outdoor weathering									
Exposure Location	Arizona				Arizona				Florida	Florida
Exposure Country	USA				USA				USA	USA
Exposure Time (days)	183	365	730	1461	183	365	730	1461		

PROPERTIES RETAINED (%)

Tensile Strength									83 {ac}	81 {ac}
Elongation									80 {ac}	86 {ac}

CHANGE IN PHYSICAL CHARACTERISTICS

Hardness Change (units)									0 {ae}	-3 {ae}

SURFACE AND APPEARANCE

ΔE Color Change	3.6 (samples washed)	2.94 (samples washed)	1.4 (samples washed)	1.17 (samples washed)	4.62 (samples washed)	3.29 (samples washed)	2 (samples washed)	1.58 (samples washed)		

TABLE 109: Outdoor Weathering in Arizona and Florida of Colored Advanced Elastomer Systems Santoprene Olefinic Thermoplastic Elastomer.

Material Family	Olefinic Thermoplastic Elastomer							
Material Supplier/ Grade	Advanced Elastomer Systems Santoprene 221-73				Advanced Elastomer Systems Santoprene 221-73			
Features	grey color, UV stabilized				brown color, UV stabilized			
Reference Number	155	155	155	155	155	155	155	155
MATERIAL CHARACTERISTICS								
Shore A hardness	73	73	73	73	73	73	73	73
EXPOSURE CONDITIONS								
Exposure Type	outdoor weathering				outdoor weathering			
Exposure Location	Arizona	Arizona	Arizona	Arizona	Arizona	Arizona	Arizona	Arizona
Exposure Country	USA	USA	USA	USA	USA	USA	USA	USA
Exposure Time (days)	183	365	730	1461	183	365	730	1461
SURFACE AND APPEARANCE								
ΔE Color Change	3.99 (samples washed)	1.29 (samples washed)	6.39 (samples washed)	8.06 (samples washed)	10.53 (samples washed)	4.92 (samples washed)	12.25 (samples washed)	14.24 (samples washed)

TABLE 110: Outdoor Weathering in Arizona of Black Advanced Elastomer Systems Santoprene Olefinic Thermoplastic Elastomer.

Material Family	Olefinic Thermoplastic Elastomer									
Material Supplier/ Grade	Advanced Elastomer Systems Santoprene 101-73		Advanced Elastomer Systems Santoprene 103-40		Advanced Elastomer Systems Santoprene 121-73		Advanced Elastomer Systems Santoprene 123-40		Advanced Elastomer Systems Santoprene 521-70	
Features	black color, general purpose grade		black color, general purpose grade		black color, UV stabilized		black color, UV stabilized		black color, UV stabilized	
Reference Number	119	119	119	119	119	119	119	119	119	119
MATERIAL CHARACTERISTICS										
Sample Thickness (mm)	0.76	0.76	0.76	0.76	0.76	0.76	0.76	0.76	0.76	0.76
Shore A hardness	73	73			73	73			70	70
Shore D hardness			40	40			40	40		
EXPOSURE CONDITIONS										
Exposure Type	outdoor weathering									
Exposure Location	Phoenix, Arizona		Phoenix, Arizona		Phoenix, Arizona		Phoenix, Arizona		Phoenix, Arizona	
Exposure Country	USA		USA		USA		USA		USA	
Exposure Time (days)	182	365	182	365	182	365	182	365	182	365
Test Lab	DSET Laboratories	DSET Laboratories	DSET Laboratories	DSET Laboratories	DSET Laboratories	DSET Laboratories	DSET Laboratories	DSET Laboratories	DSET Laboratories	DSET Laboratories
PROPERTIES RETAINED (%)										
100% Modulus	135	135	128	132	142	152	150	165	103	125
Tensile Strength	87	90	91	92	98	93	97	105	62	71
Elongation @ Break	68	73	78	77	74	59	74	73	64	57
SURFACE AND APPEARANCE										
ΔE Color Change	3.5 {b}	4 {b}	10.8 {b}	17.5 {b}	0.3 {b}	3 {b}	7.2 {b}	9.2 {b}	1.3 {b}	1.5 {b}

TABLE 111: Outdoor Weathering in Arizona With and Without Water Spray of Black, General Purpose Advanced Elastomer Systems Santoprene Olefinic Thermoplastic Elastomer.

Material Family	Olefinic Thermoplastic Elastomer	Olefinic Thermoplastic Elastomer
Material Supplier/ Grade	Advanced Elastomer Systems Santoprene 101-64	Advanced Elastomer Systems Santoprene 101-64
Features	black color, general purpose	black color, general purpose
Sample Form	injection molded slab	injection molded slab
Reference Number	120	120

MATERIAL CHARACTERISTICS

Shore A hardness	64	64	64	64	64	64	64	64
Sample Thickness (mm)	3.1	3.1	3.1	3.1	3.1	3.1	3.1	3.1
Sample Width (mm)	76.2	76.2	76.2	76.2	76.2	76.2	76.2	76.2
Sample Length (mm)	127	127	127	127	127	127	127	127

EXPOSURE CONDITIONS

Exposure Type	outdoor weathering				outdoor weathering			
Exposure Location	Phoenix, Arizona				Phoenix, Arizona			
Exposure Country	USA				USA			
Test Lab	DSET Laboratories				DSET Laboratories			
Exposure Note	5° angle				5° angle; with water spray (distilled water sprayed for 4 hours preceeding sunrise and then 20 times during the day in 15 second bursts)			
Exposure Time (days)	182	365	730	1461	182	365	730	1461

PROPERTIES RETAINED (%)

Tensile Strength	75 {ac}	74 {ac}	75 {ac}	75 {ac}	74 {ac}	77 {ac}	70 {ac}	75 {ac}
Elongation	69 {ac}	72 {ac}	74 {ac}	73 {ac}	72 {ac}	74 {ac}	73 {ac}	76 {ac}

CHANGE IN PHYSICAL CHARACTERISTICS

Hardness Change (units)	2 {ae}	2 {ae}	3 {ae}	3 {ae}	1 {ae}	1 {ae}	2 {ae}	2 {ae}

SURFACE AND APPEARANCE

ΔE Color Change	5.2 {ab}	3.1 {ab}	1.7 {ab}	1.5 {ab}	4.5 {ab}	4.1 {ab}	1.4 {ab}	1.7 {ab}

TABLE 112: Outdoor Weathering in Arizona With and Without Water Spray of Black, UV Stabilized Advanced Elastomer Systems Santoprene Olefinic Thermoplastic Elastomer.

Material Family	Olefinic Thermoplastic Elastomer				Olefinic Thermoplastic Elastomer			
Material Supplier/ Grade	Advanced Elastomer Systems Santoprene 121-73				Advanced Elastomer Systems Santoprene 121-73			
Features	black color, UV stabilized				black color, UV stabilized			
Sample Form	injection molded slab				injection molded slab			
Reference Number	120				120			
MATERIAL CHARACTERISTICS								
Shore A hardness	73	73	73	73	73	73	73	73
Sample Thickness (mm)	3.1	3.1	3.1	3.1	3.1	3.1	3.1	3.1
Sample Width (mm)	76.2	76.2	76.2	76.2	76.2	76.2	76.2	76.2
Sample Length (mm)	127	127	127	127	127	127	127	127
EXPOSURE CONDITIONS								
Exposure Type	outdoor weathering				outdoor weathering			
Exposure Location	Phoenix, Arizona				Phoenix, Arizona			
Exposure Country	USA				USA			
Test Lab	DSET Laboratories				DSET Laboratories			
Exposure Note	5° angle				5° angle; with water spray (distilled water sprayed for 4 hours preceeding sunrise and then 20 times during the day in 15 second bursts)			
Exposure Time (days)	182	365	730	1461	182	365	730	1461
PROPERTIES RETAINED (%)								
Tensile Strength	85 {ac}	78 {ac}	79 {ac}	77 {ac}	58 {ac}	77 {ac}	82 {ac}	84 {ac}
Elongation	84 {ac}	81 {ac}	80 {ac}	83 {ac}	75 {ac}	78 {ac}	82 {ac}	87 {ac}
CHANGE IN PHYSICAL CHARACTERISTICS								
Hardness Change (units)	2 {ae}	3 {ae}	-1 {ae}	4 {ae}	4 {ae}	-4 {ae}	2 {ae}	4 {ae}
SURFACE AND APPEARANCE								
ΔE Color Change	4.6 {ab}	3.3 {ab}	2 {ab}	1.6 {ab}	3.2 {ab}	3.7 {ab}	3.2 {ab}	1.6 {ab}

TABLE 113: Outdoor Weathering in Arizona With and Without Water Spray of Black, UV Stabilized Advanced Elastomer Systems Santoprene Olefinic Thermoplastic Elastomer.

Material Family	Olefinic Thermoplastic Elastomer				Olefinic Thermoplastic Elastomer			
Material Supplier/ Grade	Advanced Elastomer Systems Santoprene 121-80				Advanced Elastomer Systems Santoprene 121-80			
Features	black color, UV stabilized				black color, UV stabilized			
Sample Form	injection molded slab				injection molded slab			
Reference Number	120				120			
MATERIAL CHARACTERISTICS								
Shore A hardness	80	80	80	80	80	80	80	80
Sample Thickness (mm)	3.1	3.1	3.1	3.1	3.1	3.1	3.1	3.1
Sample Width (mm)	76.2	76.2	76.2	76.2	76.2	76.2	76.2	76.2
Sample Length (mm)	127	127	127	127	127	127	127	127
EXPOSURE CONDITIONS								
Exposure Type	outdoor weathering				outdoor weathering			
Exposure Location	Phoenix, Arizona				Phoenix, Arizona			
Exposure Country	USA				USA			
Test Lab	DSET Laboratories				DSET Laboratories			
Exposure Note	5° angle				5° angle; with water spray (distilled water sprayed for 4 hours preceeding sunrise and then 20 times during the day in 15 second bursts)			
Exposure Time (days)	182	365	730	1461	182	365	730	1461
PROPERTIES RETAINED (%)								
Tensile Strength	80 {ac}	77 {ac}	75 {ac}	63 {ac}	76 {ac}	76 {ac}	79 {ac}	80 {ac}
Elongation	80 {ac}	79 {ac}	75 {ac}	77 {ac}	79 {ac}	82 {ac}	83 {ac}	87 {ac}
CHANGE IN PHYSICAL CHARACTERISTICS								
Hardness Change (units)	3 {ae}	5 {ae}	4 {ae}	5 {ae}	3 {ae}	4 {ae}	5 {ae}	5 {ae}
SURFACE AND APPEARANCE								
ΔE Color Change	1.8 {ab}	1.1 {ab}	1.4 {ab}	1.7 {ab}	2.3 {ab}	1.1 {ab}	0.3 {ab}	2.4 {ab}

TABLE 114: Outdoor Weathering in Arizona With and Without Water Spray of Black, UV Stabilized Advanced Elastomer Systems Santoprene Olefinic Thermoplastic Elastomer.

Material Family	Olefinic Thermoplastic Elastomer				Olefinic Thermoplastic Elastomer			
Material Supplier/ Grade	Advanced Elastomer Systems Santoprene 123-40				Advanced Elastomer Systems Santoprene 123-40			
Features	black color, UV stabilized				black color, UV stabilized			
Sample Form	injection molded slab				injection molded slab			
Reference Number	120				120			
MATERIAL CHARACTERISTICS								
Shore D hardness	40	40	40	40	40	40	40	40
Sample Thickness (mm)	3.1	3.1	3.1	3.1	3.1	3.1	3.1	3.1
Sample Width (mm)	76.2	76.2	76.2	76.2	76.2	76.2	76.2	76.2
Sample Length (mm)	127	127	127	127	127	127	127	127
EXPOSURE CONDITIONS								
Exposure Type	outdoor weathering				outdoor weathering			
Exposure Location	Phoenix, Arizona				Phoenix, Arizona			
Exposure Country	USA				USA			
Test Lab	DSET Laboratories				DSET Laboratories			
Exposure Note	5° angle				5° angle; with water spray (distilled water sprayed for 4 hours preceeding sunrise and then 20 times during the day in 15 second bursts)			
Exposure Time (days)	182	365	730	1461	182	365	730	1461
PROPERTIES RETAINED (%)								
Tensile Strength	97 {ac}	94 {ac}	99 {ac}	97 {ac}	96 {ac}	97 {ac}	101 {ac}	100 {ac}
Elongation	94 {ac}	92 {ac}	98 {ac}	94 {ac}	97 {ac}	95 {ac}	96 {ac}	98 {ac}
CHANGE IN PHYSICAL CHARACTERISTICS								
Hardness Change (units)	5 {ae}	6 {ae}	5 {ae}	10 {ae}	5 {ae}	5 {ae}	6 {ae}	9 {ae}
SURFACE AND APPEARANCE								
ΔE Color Change	0.3 {ab}	1.4 {ab}	2.7 {ab}	2.8 {ab}	1.4 {ab}	1.5 {ab}	0.7 {ab}	2 {ab}

TABLE 115: Outdoor Weathering in Arizona with Water Spray Added of Black Advanced Elastomer Systems Olefinic Thermoplastic Elastomer.

Material Family	Olefinic Thermoplastic Elastomer									
Material Supplier/ Grade	Advanced Elastomer Systems Santoprene 101-73		Advanced Elastomer Systems Santoprene 103-40		Advanced Elastomer Systems Santoprene 121-73		Advanced Elastomer Systems Santoprene 123-40		Advanced Elastomer Systems Santoprene 521-70	
Features	black color, general purpose grade		black color, general purpose grade		black color, UV stabilized		black color, UV stabilized		black color, UV stabilized	
Reference Number	119	119	119	119	119	119	119	119	119	119
MATERIAL CHARACTERISTICS										
Sample Thickness (mm)	0.76	0.76	0.76	0.76	0.76	0.76	0.76	0.76	0.76	0.76
Shore A hardness	73	73			73	73			70	70
Shore D hardness			40	40			40	40		
EXPOSURE CONDITIONS										
Exposure Type	outdoor weathering									
Exposure Location	Phoenix, Arizona		Phoenix, Arizona		Phoenix, Arizona		Phoenix, Arizona		Phoenix, Arizona	
Exposure Country	USA		USA		USA		USA		USA	
Exposure Time (days)	182	365	182	365	182	365	182	365	182	365
Exposure Note	water spray added	water spray added	water spray added	water spray added	water spray added	water spray added	water spray added	water spray added	water spray added	water spray added
Test Lab	DSET Laboratories	DSET Laboratories	DSET Laboratories	DSET Laboratories	DSET Laboratories	DSET Laboratories	DSET Laboratories	DSET Laboratories	DSET Laboratories	DSET Laboratories
PROPERTIES RETAINED (%)										
100% Modulus	139	139	129	122	159	179	198	141	108	70
Tensile Strength	92	98	92	93	106	117	113	97	66	59
Elongation @ Break	68	75	77	83	68	68	60	77	61	86
SURFACE AND APPEARANCE										
ΔE Color Change	3.2 {b}	4.1 {b}	16.7 {b}	14.5 {b}	2.7 {b}	2 {b}	8 {b}	7.2 {b}	1.8 {b}	3.4 {b}

TABLE 116: Outdoor Weathering in Arizona With and Without Water Spray of Advanced Elastomer Systems Santoprene Olefinic Thermoplastic Elastomer.

Material Family	Olefinic Thermoplastic Elastomer				Olefinic Thermoplastic Elastomer			
Material Supplier/ Grade	Advanced Elastomer Systems Santoprene 121-67				Advanced Elastomer Systems Santoprene 121-67			
Features	black color, UV stabilized				black color, UV stabilized			
Sample Form	injection molded slab				injection molded slab			
Reference Number	120				120			
MATERIAL CHARACTERISTICS								
Shore A hardness	67	67	67	67	67	67	67	67
Sample Thickness (mm)	3.1	3.1	3.1	3.1	3.1	3.1	3.1	3.1
Sample Width (mm)	76.2	76.2	76.2	76.2	76.2	76.2	76.2	76.2
Sample Length (mm)	127	127	127	127	127	127	127	127
EXPOSURE CONDITIONS								
Exposure Type	outdoor weathering				outdoor weathering			
Exposure Location	Phoenix, Arizona				Phoenix, Arizona			
Exposure Country	USA				USA			
Test Lab	DSET Laboratories				DSET Laboratories			
Exposure Note	5° angle				5° angle; with water spray (distilled water sprayed for 4 hours preceeding sunrise and then 20 times during the day in 15 second bursts)			
Exposure Time (days)	182	365	730	1461	182	365	730	1461
PROPERTIES RETAINED (%)								
Tensile Strength	88 {ac}	83 {ac}	87 {ac}	75 {ac}	83 {ac}	89 {ac}	90 {ac}	
Elongation	83 {ac}	84 {ac}	86 {ac}	93 {ac}	77 {ac}	88 {ac}	89 {ac}	
CHANGE IN PHYSICAL CHARACTERISTICS								
Hardness Change (units)	-1 {ae}	-1 {ae}	-1 {ae}	1 {ae}	-1 {ae}	-1 {ae}	-1 {ae}	
SURFACE AND APPEARANCE								
ΔE Color Change	3.6 {ab}	2.9 {ab}	1.4 {ab}	1.2 {ab}	4.5 {ab}	0.8 {ab}	2.7 {ab}	

TABLE 117: Outdoor Weathering in Florida of Black Advanced Elastomer Systems Olefinic Thermoplastic Elastomer.

Material Family	Olefinic Thermoplastic Elastomer	
Material Supplier/ Grade	Advanced Elastomer Systems Santoprene 101-73	
Features	black color, dull surface	
Manufacturing Method	compression molding	
Reference Number	175	175

MATERIAL CHARACTERISTICS

Shore A hardness	73	73
Sample Thickness (mm)	3.2	3.2

EXPOSURE CONDITIONS

Exposure Type	outdoor weathering	
Exposure Location	Florida	Florida
Exposure Country	USA	USA
Exposure Note	direct exposure; 5° south	under glass exposure; 5° south
Exposure Time (days)	730	730

PROPERTIES RETAINED (%)

100% Modulus	78	101
Tensile Strength	90	83
Elongation @ Break	107	77

CHANGE IN PHYSICAL CHARACTERISTICS

Shore A Hardness Change (units)	3	4

SURFACE AND APPEARANCE

Gloss @ 60° Retained (%)	35.5 {g}	29 {g}
ΔE Color	1.03 {h}	0.66 {h}
Visual Observation (as weathered)	severe chalking	chalking
Visual Observation (washed)	slight chalking, no cracks	clean, no cracks

TABLE 118: Outdoor Weathering in Florida With and Without Water Spray of Black, General Purpose Advanced Elastomer Systems Santoprene Olefinic Thermoplastic Elastomer.

Material Family	Olefinic Thermoplastic Elastomer				Olefinic Thermoplastic Elastomer			
Material Supplier/ Grade	Advanced Elastomer Systems Santoprene 101-64				Advanced Elastomer Systems Santoprene 101-64			
Features	black color, general purpose				black color, general purpose			
Sample Form	injection molded slab				injection molded slab			
Reference Number	120				120			
MATERIAL CHARACTERISTICS								
Shore A hardness	64	64	64	64	64	64	64	64
Sample Thickness (mm)	3.1	3.1	3.1	3.1	3.1	3.1	3.1	3.1
Sample Width (mm)	76.2	76.2	76.2	76.2	76.2	76.2	76.2	76.2
Sample Length (mm)	127	127	127	127	127	127	127	127
EXPOSURE CONDITIONS								
Exposure Type	outdoor weathering				outdoor weathering			
Exposure Location	Homestead, Florida				Homestead, Florida			
Exposure Country	USA				USA			
Test Lab	DSET Laboratories				DSET Laboratories			
Exposure Note	5° angle				5° angle; with water spray (distilled water sprayed for 4 hours preceeding sunrise and then 20 times during the day in 15 second bursts)			
Exposure Time (days)	182	365	730	1461	182	365	730	1461
PROPERTIES RETAINED (%)								
Tensile Strength	78 {ac}	80 {ac}	69 {ac}	73 {ac}	79 {ac}		71 {ac}	71 {ac}
Elongation	78 {ac}	80 {ac}	75 {ac}	78 {ac}	83 {ac}		69 {ac}	75 {ac}
CHANGE IN PHYSICAL CHARACTERISTICS								
Hardness Change (units)	0 {ae}	1 {ae}	2 {ae}	1 {ae}	0 {ae}		1 {ae}	3 {ae}
SURFACE AND APPEARANCE								
ΔE Color Change	6.3 {ab}	4.4 {ab}	1.2 {ab}	1.7 {ab}	6.5 {ab}		2.6 {ab}	1.43 {ab}

TABLE 119: Outdoor Weathering in Florida With and Without Water Spray of Black, General Purpose Advanced Elastomer Systems Santoprene Olefinic Thermoplastic Elastomer.

Material Family	Olefinic Thermoplastic Elastomer	Olefinic Thermoplastic Elastomer
Material Supplier/ Grade	Advanced Elastomer Systems Santoprene 103-40	Advanced Elastomer Systems Santoprene 103-40
Features	black color, general purpose	black color, general purpose
Sample Form	injection molded slab	injection molded slab
Reference Number	120	120

MATERIAL CHARACTERISTICS

Shore D hardness	40	40	40	40	40	40	40	40
Sample Thickness (mm)	3.1	3.1	3.1	3.1	3.1	3.1	3.1	3.1
Sample Width (mm)	76.2	76.2	76.2	76.2	76.2	76.2	76.2	76.2
Sample Length (mm)	127	127	127	127	127	127	127	127

EXPOSURE CONDITIONS

Exposure Type	outdoor weathering				outdoor weathering			
Exposure Location	Homestead, Florida				Homestead, Florida			
Exposure Country	USA				USA			
Test Lab	DSET Laboratories				DSET Laboratories			
Exposure Note	5° angle				5° angle; with water spray (distilled water sprayed for 4 hours preceeding sunrise and then 20 times during the day in 15 second bursts)			
Exposure Time (days)	182	365	730	1461	182	365	730	1461

PROPERTIES RETAINED (%)

Tensile Strength	98 {ac}	97 {ac}	95 {ac}	98 {ac}	96 {ac}		97 {ac}	65 {ac}
Elongation	100 {ac}	96 {ac}	97 {ac}	99 {ac}	98 {ac}		97 {ac}	48 {ac}

CHANGE IN PHYSICAL CHARACTERISTICS

Hardness Change (units)	2 {ae}	3 {ae}	8 {ae}	5 {ae}	0 {ae}		7 {ae}	7 {ae}

SURFACE AND APPEARANCE

ΔE Color Change	2.7 {ab}	9.1 {ab}	11.2 {ab}	11.7 {ab}	4.8 {ab}		8.5 {ab}	16.2 {ab}

TABLE 120: Outdoor Weathering in Florida With and Without Water Spray of Black, UV Stabilized Advanced Elastomer Systems Santoprene Olefinic Thermoplastic Elastomer.

Material Family	Olefinic Thermoplastic Elastomer				Olefinic Thermoplastic Elastomer			
Material Supplier/ Grade	Advanced Elastomer Systems Santoprene 121-80				Advanced Elastomer Systems Santoprene 121-80			
Features	black color, UV stabilized				black color, UV stabilized			
Sample Form	injection molded slab				injection molded slab			
Reference Number	120				120			
MATERIAL CHARACTERISTICS								
Shore A hardness	80	80	80	80	80	80	80	80
Sample Thickness (mm)	3.1	3.1	3.1	3.1	3.1	3.1	3.1	3.1
Sample Width (mm)	76.2	76.2	76.2	76.2	76.2	76.2	76.2	76.2
Sample Length (mm)	127	127	127	127	127	127	127	127
EXPOSURE CONDITIONS								
Exposure Type	outdoor weathering				outdoor weathering			
Exposure Location	Homestead, Florida				Homestead, Florida			
Exposure Country	USA				USA			
Test Lab	DSET Laboratories				DSET Laboratories			
Exposure Note	5° angle				5° angle; with water spray (distilled water sprayed for 4 hours preceeding sunrise and then 20 times during the day in 15 second bursts)			
Exposure Time (days)	182	365	730	1461	182	365	730	1461
PROPERTIES RETAINED (%)								
Tensile Strength	84 {ac}	81 {ac}	78 {ac}	87 {ac}	84 {ac}		80 {ac}	87 {ac}
Elongation	88 {ac}	83 {ac}	84 {ac}	92 {ac}	89 {ac}		82 {ac}	90 {ac}
CHANGE IN PHYSICAL CHARACTERISTICS								
Hardness Change (units)	4 {ae}	2 {ae}	5 {ae}	3 {ae}	4 {ae}		5 {ae}	3 {ae}
SURFACE AND APPEARANCE								
ΔE Color Change	2.3 {ab}	2.1 {ab}	0.4 {ab}	2.4 {ab}	0.4 {ab}		0.5 {ab}	2.1 {ab}

TABLE 121: Outdoor Weathering in Florida With and Without Water Spray of Black, UV Stabilized Advanced Elastomer Systems Santoprene Olefinic Thermoplastic Elastomer.

Material Family	Olefinic Thermoplastic Elastomer				Olefinic Thermoplastic Elastomer			
Material Supplier/ Grade	Advanced Elastomer Systems Santoprene 121-73				Advanced Elastomer Systems Santoprene 121-73			
Features	black color, UV stabilized				black color, UV stabilized			
Sample Form	injection molded slab				injection molded slab			
Reference Number	120				120			
MATERIAL CHARACTERISTICS								
Shore A hardness	73	73	73	73	73	73	73	73
Sample Thickness (mm)	3.1	3.1	3.1	3.1	3.1	3.1	3.1	3.1
Sample Width (mm)	76.2	76.2	76.2	76.2	76.2	76.2	76.2	76.2
Sample Length (mm)	127	127	127	127	127	127	127	127
EXPOSURE CONDITIONS								
Exposure Type	outdoor weathering				outdoor weathering			
Exposure Location	Homestead, Florida				Homestead, Florida			
Exposure Country	USA				USA			
Test Lab	DSET Laboratories				DSET Laboratories			
Exposure Note	5° angle				5° angle; with water spray (distilled water sprayed for 4 hours preceeding sunrise and then 20 times during the day in 15 second bursts)			
Exposure Time (days)	182	365	730	1461	182	365	730	1461
PROPERTIES RETAINED (%)								
Tensile Strength	82 {ac}	85 {ac}	81 {ac}	88 {ac}	77 {ac}		81 {ac}	84 {ac}
Elongation	86 {ac}	87 {ac}	86 {ac}	93 {ac}	80 {ac}		82 {ac}	86 {ac}
CHANGE IN PHYSICAL CHARACTERISTICS								
Hardness Change (units)	1 {ae}	-1 {ae}	-3 {ae}	3 {ae}	3 {ae}		2 {ae}	4 {ae}
SURFACE AND APPEARANCE								
ΔE Color Change	5.6 {ab}	5 {ab}	1.5 {ab}	1.3 {ab}	3.9 {ab}		3.4 {ab}	1.6 {ab}

TABLE 122: Outdoor Weathering in Florida With and Without Water Spray of Black, UV Stabilized Advanced Elastomer Systems Santoprene Olefinic Thermoplastic Elastomer.

Material Family	Olefinic Thermoplastic Elastomer				Olefinic Thermoplastic Elastomer			
Material Supplier/ Grade	Advanced Elastomer Systems Santoprene 123-40				Advanced Elastomer Systems Santoprene 123-40			
Features	black color, UV stabilized				black color, UV stabilized			
Sample Form	injection molded slab				injection molded slab			
Reference Number	120				120			
MATERIAL CHARACTERISTICS								
Shore D hardness	40	40	40	40	40	40	40	40
Sample Thickness (mm)	3.1	3.1	3.1	3.1	3.1	3.1	3.1	3.1
Sample Width (mm)	76.2	76.2	76.2	76.2	76.2	76.2	76.2	76.2
Sample Length (mm)	127	127	127	127	127	127	127	127
EXPOSURE CONDITIONS								
Exposure Type	outdoor weathering				outdoor weathering			
Exposure Location	Homestead, Florida				Homestead, Florida			
Exposure Country	USA				USA			
Test Lab	DSET Laboratories				DSET Laboratories			
Exposure Note	5° angle				5° angle; with water spray (distilled water sprayed for 4 hours preceeding sunrise and then 20 times during the day in 15 second bursts)			
Exposure Time (days)	182	365	730	1461	182	365	730	1461
PROPERTIES RETAINED (%)								
Tensile Strength	97 {ac}	97 {ac}	98 {ac}	96 {ac}	99 {ac}		101 {ac}	103 {ac}
Elongation	98 {ac}	99 {ac}	96 {ac}	94 {ac}	99 {ac}		97 {ac}	100 {ac}
CHANGE IN PHYSICAL CHARACTERISTICS								
Hardness Change (units)	4 {ae}	6 {ae}	9 {ae}	7 {ae}	4 {ae}		9 {ae}	8 {ae}
SURFACE AND APPEARANCE								
ΔE Color Change	1.2 {ab}	0.9 {ab}	0.9 {ab}	3.2 {ab}	0.5 {ab}		1.3 {ab}	2.6 {ab}

TABLE 123: Outdoor Weathering in Florida With and Without Water Spray of Advanced Elastomer Systems Santoprene Olefinic Thermoplastic Elastomer.

Material Family	Olefinic Thermoplastic Elastomer				Olefinic Thermoplastic Elastomer			
Material Supplier/ Grade	Advanced Elastomer Systems Santoprene 121-67				Advanced Elastomer Systems Santoprene 121-67			
Features	black color, UV stabilized				black color, UV stabilized			
Sample Form	injection molded slab				injection molded slab			
Reference Number	120				120			
MATERIAL CHARACTERISTICS								
Shore A hardness	67	67	67	67	67	67	67	67
Sample Thickness (mm)	3.1	3.1	3.1	3.1	3.1	3.1	3.1	3.1
Sample Width (mm)	76.2	76.2	76.2	76.2	76.2	76.2	76.2	76.2
Sample Length (mm)	127	127	127	127	127	127	127	127
EXPOSURE CONDITIONS								
Exposure Type	outdoor weathering				outdoor weathering			
Exposure Location	Homestead, Florida				Homestead, Florida			
Exposure Country	USA				USA			
Test Lab	DSET Laboratories				DSET Laboratories			
Exposure Note	5° angle				5° angle; with water spray (distilled water sprayed for 4 hours preceeding sunrise and then 20 times during the day in 15 second bursts)			
Exposure Time (days)	182	365	730	1461	182	365	730	1461
PROPERTIES RETAINED (%)								
Tensile Strength	89 {ac}	93 {ac}	62 {ac}	93 {ac}	89 {ac}		85 {ac}	93 {ac}
Elongation	87 {ac}	92 {ac}	80 {ac}	96 {ac}	90 {ac}		86 {ac}	93 {ac}
CHANGE IN PHYSICAL CHARACTERISTICS								
Hardness Change (units)	-2 {ae}	-2 {ae}	0 {ae}	1 {ae}	0 {ae}		1 {ae}	1 {ae}
SURFACE AND APPEARANCE								
ΔE Color Change	5.4 {ab}	4.8 {ab}	1.1 {ab}	1 {ab}	3.2 {ab}		3 {ab}	1.8 {ab}

TABLE 124: Accelerated Outdoor Weathering by EMMA and EMMAQUA and Accelerated Weathering in a Xenon Arc Weatherometer of Black, UV Stabilized Advanced Elastomer Systems Santoprene Olefinic Thermoplastic Elastomer.

Material Family	Olefinic Thermoplastic Elastomer					Olefinic Thermoplastic Elastomer		
Material Supplier/ Grade	Advanced Elastomer Systems Santoprene 121-73					Advanced Elastomer Systems Santoprene 121-73		
Features	black color, UV stabilized					black color, UV stabilized		
Sample Form	injection molded slab					injection molded slab		
Reference Number	120					120		
MATERIAL CHARACTERISTICS								
Shore A hardness	73	73	73	73	73	73	73	73
Sample Thickness (mm)	3.1	3.1	3.1	3.1	3.1	3.1	3.1	3.1
Sample Width (mm)	76.2	76.2	76.2	76.2	76.2	76.2	76.2	76.2
Sample Length (mm)	127	127	127	127	127	127	127	127
EXPOSURE CONDITIONS								
Exposure Type	accelerated weathering		accelerated outdoor weathering			accelerated outdoor weathering		
Exposure Apparatus	xenon arc weatherometer		EMMA			EMMAQUA		
Exposure Note	temperature: 89°C; 3.8 hrs in light at 50% RH, 1 hr in dark at 100% RH; test method: SAE J1885; GM TM30-2							
Exposure Time (days)	125	208	182	365	730	182	365	730
Energy @ 340nm (kJ/m^2)	4703	7838						
Energy @ <380nm (MJ/m^2)	403	672						
PROPERTIES RETAINED (%)								
100% Modulus	105 {ac}	102 {ac}						
Tensile Strength	78 {ac}	73 {ac}	82 {ac}	85 {ac}	95 {ac}	88 {ac}	87 {ac}	55 {ac}
Elongation	75 {ac}	72 {ac}	72 {ac}	81 {ac}	52 {ac}	85 {ac}	66 {ac}	68 {ac}
CHANGE IN PHYSICAL CHARACTERISTICS								
Hardness Change (units)	2 {ae}	2 {ae}	2 {ae}	2 {ae}	-3 {ae}	3 {ae}	4 {ae}	-5{ae}
SURFACE AND APPEARANCE								
ΔE Color Change	2 {ab}	2.1 {ab}	3.4 {ab}	2.3 {ab}	2.5 {ab}	3.6 {ab}	3 {ab}	2.1 {ab}

TABLE 125: Accelerated Outdoor Weathering by EMMA and EMMAQUABlack, General Purpose Advanced Elastomer Systems Santoprene Olefinic Thermoplastic Elastomer.

Material Family	Olefinic Thermoplastic Elastomer			Olefinic Thermoplastic Elastomer		
Material Supplier/ Grade	Advanced Elastomer Systems Santoprene 103-40			Advanced Elastomer Systems Santoprene 103-40		
Features	black color, general purpose			black color, general purpose		
Sample Form	injection molded slab			injection molded slab		
Reference Number	120			120		
MATERIAL CHARACTERISTICS						
Shore D hardness	40	40	40	40	40	40
Sample Thickness (mm)	3.1	3.1	3.1	3.1	3.1	3.1
Sample Width (mm)	76.2	76.2	76.2	76.2	76.2	76.2
Sample Length (mm)	127	127	127	127	127	127
EXPOSURE CONDITIONS						
Exposure Type	accelerated outdoor weathering			accelerated outdoor weathering		
Exposure Apparatus	EMMA			EMMAQUA		
Exposure Time (days)	182	365	730	182	365	730
PROPERTIES RETAINED (%)						
Tensile Strength	86 {ac}	82 {ac}	58 {ac}	88 {ac}	75 {ac}	11 {ac}
Elongation	69 {ac}	65 {ac}	6 {ac}	75 {ac}	61 {ac}	1 {ac}
CHANGE IN PHYSICAL CHARACTERISTICS						
Hardness Change (units)	3 {ae}	3 {ae}	2 {ae}	2 {ae}	2 {ae}	-3{ae}
SURFACE AND APPEARANCE						
ΔE Color Change	15 {ab}	14.6 {ab}	12.2 {ab}	15 {ab}	14.9 {ab}	9.5 {ab}

TABLE 126: Accelerated Outdoor Weathering by EMMAQUA of Black Advanced Elastomer Systems Santoprene Olefinic Thermoplastic Elastomer.

Material Family	Olefinic Thermoplastic Elastomer									
Material Supplier/ Grade	Advanced Elastomer Systems Santoprene 101-73		Advanced Elastomer Systems Santoprene 103-40		Advanced Elastomer Systems Santoprene 121-73		Advanced Elastomer Systems Santoprene 123-40		Advanced Elastomer Systems Santoprene 521-70	
Features	black color, general purpose grade		black color, general purpose grade		black color, UV stabilized		black color, UV stabilized		black color, UV stabilized	
Reference Number	119	119	119	119	119	119	119	119	119	119
MATERIAL CHARACTERISTICS										
Sample Thickness (mm)	0.76	0.76	0.76	0.76	0.76	0.76	0.76	0.76	0.76	0.76
Shore A hardness	73	73			73	73			70	70
Shore D hardness			40	40			40	40		
EXPOSURE CONDITIONS										
Exposure Type	accelerated outdoor weathering									
Exposure Location	Phoenix, Arizona		Phoenix, Arizona		Phoenix, Arizona		Phoenix, Arizona		Phoenix, Arizona	
Exposure Country	USA		USA		USA		USA		USA	
Exposure Apparatus	EMMAQUA	EMMAQUA	EMMAQUA	EMMAQUA	EMMAQUA	EMMAQUA	EMMAQUA	EMMAQUA	EMMAQUA	EMMAQUA
Exposure Time (days)	182	365	182	365	182	365	182	365	182	365
Exposure Note	correlates to approximately 2.5 years actual Florida aging	correlates to approximately 5 years actual Florida aging	correlates to approximately 2.5 years actual Florida aging	correlates to approximately 5 years actual Florida aging	correlates to approximately 2.5 years actual Florida aging	correlates to approximately 5 years actual Florida aging	correlates to approximately 2.5 years actual Florida aging	correlates to approximately 5 years actual Florida aging	correlates to approximately 2.5 years actual Florida aging	correlates to approximately 5 years actual Florida aging
PROPERTIES RETAINED (%)										
100% Modulus	155	122	120	122	147	115	164	122	101	80
Tensile Strength	87	90	82	69	96	94	83	95	54	53
Elongation @ Break	77	62	72	59	88	68	75	67	71	60
SURFACE AND APPEARANCE										
ΔE Color Change	6.8 {b}	5.6 {b}	15.1 {b}	14.9 {b}	3.6 {b}	3.1 {b}	9.5 {b}	8.4 {b}	1.8 {b}	2.2 {b}

TABLE 127: Accelerated Outdoor Weathering in Arizona by EMMA and EMMAQUA of Black Advanced Elastomer Systems Santoprene Olefinic Thermoplastic Elastomer.

Material Family	Olefinic Thermoplastic Elastomer									
Material Supplier/ Grade	Advanced Elastomer Systems Santoprene 101-73		Advanced Elastomer Systems Santoprene 103-40		Advanced Elastomer Systems Santoprene 121-73		Advanced Elastomer Systems Santoprene 123-40		Advanced Elastomer Systems Santoprene 521-70	
Features	black color, general purpose grade		black color, general purpose grade		black color, UV stabilized		black color, UV stabilized		black color, UV stabilized	
Reference Number	119	119	119	119	119	119	119	119	119	119
MATERIAL CHARACTERISTICS										
Sample Thickness (mm)	0.76	0.76	0.76	0.76	0.76	0.76	0.76	0.76	0.76	0.76
Shore A hardness	73	73			73	73			70	70
Shore D hardness			40	40			40	40		
EXPOSURE CONDITIONS										
Exposure Type	accelerated outdoor weathering									
Exposure Location	Phoenix, Arizona		Phoenix, Arizona		Phoenix, Arizona		Phoenix, Arizona		Phoenix, Arizona	
Exposure Country	USA		USA		USA		USA		USA	
Exposure Apparatus	EMMA	EMMA	EMMA	EMMA	EMMA	EMMA	EMMA	EMMA	EMMA	EMMA
Exposure Time (days)	182	365	182	365	182	365	182	365	182	365
Exposure Note	correlates to approximately 2.5 years actual Florida aging	correlates to approximately 5 years actual Florida aging	correlates to approximately 2.5 years actual Florida aging	correlates to approximately 5 years actual Florida aging	correlates to approximately 2.5 years actual Florida aging	correlates to approximately 5 years actual Florida aging	correlates to approximately 2.5 years actual Florida aging	correlates to approximately 5 years actual Florida aging	correlates to approximately 2.5 years actual Florida aging	correlates to approximately 5 years actual Florida aging
PROPERTIES RETAINED (%)										
100% Modulus	129	130	126	123	122	119	121	124	81	80
Tensile Strength	83	84	80	77	90	93	79	85	50	49
Elongation @ Break	68	68	68	63	75	84	74	73	64	64
SURFACE AND APPEARANCE										
ΔE Color Change	5.9 {b}	5.9 {b}	15 {b}	15 {b}	3.3 {b}	3.3 {b}	8.2 {b}	8.2 {b}	1 {b}	1 {b}

TABLE 128: Accelerated Outdoor Weathering by EMMA and EMMAQUA and Accelerated Weathering in a Xenon Arc Weatherometer of Black, UV Stabilized Advanced Elastomer Systems Santoprene Olefinic Thermoplastic Elastomer.

Material Family	Olefinic Thermoplastic Elastomer	Olefinic Thermoplastic Elastomer
Material Supplier/ Grade	Advanced Elastomer Systems Santoprene 123-40	Advanced Elastomer Systems Santoprene 123-40
Features	black color, UV stabilized	black color, UV stabilized
Sample Form	injection molded slab	injection molded slab
Reference Number	120	120

MATERIAL CHARACTERISTICS

Shore D hardness	40	40	40	40	40	40	40	40
Sample Thickness (mm)	3.1	3.1	3.1	3.1	3.1	3.1	3.1	3.1
Sample Width (mm)	76.2	76.2	76.2	76.2	76.2	76.2	76.2	76.2
Sample Length (mm)	127	127	127	127	127	127	127	127

EXPOSURE CONDITIONS

Exposure Type	accelerated weathering		accelerated outdoor weathering			accelerated outdoor weathering		
Exposure Apparatus	xenon arc weatherometer		EMMA			EMMAQUA		
Exposure Note	temperature: 89°C; 3.8 hrs in light at 50% RH, 1 hr in dark at 100% RH; test method: SAE J1885; GM TM30-2							
Exposure Time (days)	125	208	182	365	730	182	365	730
Energy @ 340nm (kJ/m^2)	4703	7838						
Energy @ <380nm (MJ/m^2)	403	672						

PROPERTIES RETAINED (%)

100% Modulus	99 {ac}	109 {ac}						
Tensile Strength	94 {ac}	96 {ac}	94 {ac}	100 {ac}	101 {ac}	98 {ac}	114 {ac}	31 {ac}
Elongation	93 {ac}	90 {ac}	80 {ac}	79 {ac}	76 {ac}	82 {ac}	73 {ac}	2 {ac}

CHANGE IN PHYSICAL CHARACTERISTICS

Hardness Change (units)	5 {ae}	7 {ae}	7 {ae}	5 {ae}	3 {ae}	5 {ae}	7 {ae}	0{ae}

SURFACE AND APPEARANCE

ΔE Color Change	2.7 {ab}	2.1 {ab}	8.3 {ab}	8.9 {ab}	8.7 {ab}	9.5 {ab}	8.4 {ab}	7.3 {ab}

TABLE 129: Accelerated Weathering in a Xenon Arc Weatherometer of Black Advanced Elastomer Systems Santoprene Olefinic Thermoplastic Elastomer.

Material Family	Olefinic Thermoplastic Elastomer							
Material Supplier/ Grade	Advanced Elastomer Systems Santoprene 101-73		Advanced Elastomer Systems Santoprene 103-40		Advanced Elastomer Systems Santoprene 121-73		Advanced Elastomer Systems Santoprene 123-40	
Features	black color general, purpose grade		black color general, purpose grade		black color, UV stabilized		black color, UV stabilized	
Reference Number	119	119	119	119	119	119	119	119
MATERIAL CHARACTERISTICS								
Sample Thickness (mm)	0.76	0.76	0.76	0.76	0.76	0.76	0.76	0.76
Shore A hardness	73	73			73	73		
Shore D hardness			40	40			40	40
EXPOSURE CONDITIONS								
Exposure Type	accelerated weathering							
Exposure Test Method	GM, TM30-2	GM, TM30-2	GM, TM30-2	GM, TM30-2	GM, TM30-2	GM, TM30-2	GM, TM30-2	GM, TM30-2
Exposure Apparatus	xenon arc weatherometer	xenon arc weatherometer	xenon arc weatherometer	xenon arc weatherometer	xenon arc weatherometer	xenon arc weatherometer	xenon arc weatherometer	xenon arc weatherometer
Exposure Time (days)	125	208	125	208	125	208	125	208
Exposure Temperature (°C)	89	89	89	89	89	89	89	89
Exposure Note	3.8 hrs in light at 50% RH, 1 hr in dark at 100% RH	3.8 hrs in light at 50% RH, 1 hr in dark at 100% RH	3.8 hrs in light at 50% RH, 1 hr in dark at 100% RH	3.8 hrs in light at 50% RH, 1 hr in dark at 100% RH	3.8 hrs in light at 50% RH, 1 hr in dark at 100% RH	3.8 hrs in light at 50% RH, 1 hr in dark at 100% RH	3.8 hrs in light at 50% RH, 1 hr in dark at 100% RH	3.8 hrs in light at 50% RH, 1 hr in dark at 100% RH
PROPERTIES RETAINED (%)								
100% Modulus	114	115	106		119	118	32	122
Tensile Strength	81	83	56	49	91	86	74	81
Elongation @ Break	74	79	45	12	79	75	75	69
SURFACE AND APPEARANCE								
ΔE Color	2.8 {b}	4 {b}	7.9 {b}	18.1 {b}	1 {b}	2 {b}	5.5 {b}	9.3 {b}

TABLE 130: Accelerated Weathering in a Xenon Arc Weatherometer of Advanced Elastomer Systems Santoprene Olefinic Thermoplastic Elastomer.

Material Family	Olefinic Thermoplastic Elastomer					
Material Supplier/ Grade	Advanced Elastomer Systems Santoprene 201-73	Advanced Elastomer Systems Santoprene 203-40	Advanced Elastomer Systems Santoprene 221-73-W157	Advanced Elastomer Systems Santoprene 223-50-W157	Advanced Elastomer Systems Santoprene 221-73-W157	Advanced Elastomer Systems Santoprene 223-50-W157
Features	neutral color	neutral color	neutral color, UV stabilized	neutral color, UV stabilized	grey color, UV stabilized	brown color, UV stabilized
Reference Number	157	157	157	157	157	157
MATERIAL CHARACTERISTICS						
Shore A hardness	73		73		73	
Shore D hardness		40		50		50
EXPOSURE CONDITIONS						
Exposure Type	accelerated weathering					
Exposure Apparatus	xenon arc weatherometer					
Exposure Time (days)	214					
PROPERTIES RETAINED (%)						
Tensile Strength	7.9	0.0	88.0	63.6	68.8	62.6
Elongation	0.0	0.0	31.5	46.6	66.9	55.0
SURFACE AND APPEARANCE						
ΔE Color Change	>40.0	>40.0	15.7	14.2	9.8	16.0

TABLE 131: Accelerated Weathering in a Xenon Arc Weatherometer of Black, General Purpose Advanced Elastomer Systems Santoprene Olefinic Thermoplastic Elastomer.

Material Family	Olefinic Thermoplastic Elastomer	Olefinic Thermoplastic Elastomer
Material Supplier/ Grade	Advanced Elastomer Systems Santoprene 101-64	Advanced Elastomer Systems Santoprene 101-64
Features	black color, general purpose	black color, general purpose
Sample Form	injection molded slab	injection molded slab
Reference Number	120	120

MATERIAL CHARACTERISTICS

Shore A hardness	64	64
Sample Thickness (mm)	3.1	3.1
Sample Width (mm)	76.2	76.2
Sample Length (mm)	127	127

EXPOSURE CONDITIONS

Exposure Type	accelerated weathering	accelerated weathering
Exposure Apparatus	xenon arc weatherometer	xenon arc weatherometer
Exposure Note	temperature: 89°C; 3.8 hrs in light at 50% RH, 1 hr in dark at 100% RH; test method: SAE J1885; GM TM30-2	temperature: 89°C; 3.8 hrs in light at 50% RH, 1 hr in dark at 100% RH; test method: SAE J1885; GM TM30-2
Exposure Time (days)	125	208
Energy @ 340nm (kJ/m^2)	4703	7838
Energy @ <380nm (MJ/m^2)	403	672

PROPERTIES RETAINED (%)

Tensile Strength	61 {ac}	64 {ac}
Elongation	57 {ac}	67 {ac}

CHANGE IN PHYSICAL CHARACTERISTICS

Hardness Change (units)	1 {ae}	1 {ae}

SURFACE AND APPEARANCE

ΔE Color Change	7.8 {ab}	2.5 {ab}

TABLE 132: Accelerated Weathering in a Xenon Arc Weatherometer of Black, UV Stabilized Advanced Elastomer Systems Santoprene Olefinic Thermoplastic Elastomer.

Material Family	Olefinic Thermoplastic Elastomer	Olefinic Thermoplastic Elastomer
Material Supplier/ Grade	Advanced Elastomer Systems Santoprene 121-67	Advanced Elastomer Systems Santoprene 121-67
Features	black color, UV stabilized	black color, UV stabilized
Sample Form	injection molded slab	injection molded slab
Reference Number	120	120

MATERIAL CHARACTERISTICS

Shore A hardness	67	67
Sample Thickness (mm)	3.1	3.1
Sample Width (mm)	76.2	76.2
Sample Length (mm)	127	127

EXPOSURE CONDITIONS

Exposure Type	accelerated weathering	accelerated weathering
Exposure Apparatus	xenon arc weatherometer	xenon arc weatherometer
Exposure Note	temperature: 89°C; 3.8 hrs in light at 50% RH, 1 hr in dark at 100% RH; test method: SAE J1885; GM TM30-2	temperature: 89°C; 3.8 hrs in light at 50% RH, 1 hr in dark at 100% RH; test method: SAE J1885; GM TM30-2
Exposure Time (days)	125	208
Energy @ 340nm (kJ/m^2)	4703	7838
Energy @ <380nm (MJ/m^2)	403	672

PROPERTIES RETAINED (%)

100% Modulus	107 {ac}	110 {ac}
Tensile Strength	91 {ac}	83 {ac}
Elongation	98 {ac}	93 {ac}

CHANGE IN PHYSICAL CHARACTERISTICS

Hardness Change (units)	3 {ae}	2 {ae}

SURFACE AND APPEARANCE

ΔE Color Change	1.5 {ab}	2 {ab}

TABLE 133: Accelerated Weathering in a Xenon Arc Weatherometer of Black, UV Stabilized Advanced Elastomer Systems Santoprene Olefinic Thermoplastic Elastomer.

Material Family	Olefinic Thermoplastic Elastomer	Olefinic Thermoplastic Elastomer
Material Supplier/ Grade	Advanced Elastomer Systems Santoprene 123-50	Advanced Elastomer Systems Santoprene 123-50
Features	black color, UV stabilized	black color, UV stabilized
Sample Form	injection molded slab	injection molded slab
Reference Number	120	120

MATERIAL CHARACTERISTICS

Shore D hardness	50	50
Sample Thickness (mm)	3.1	3.1
Sample Width (mm)	76.2	76.2
Sample Length (mm)	127	127

EXPOSURE CONDITIONS

Exposure Type	accelerated weathering	accelerated weathering
Exposure Apparatus	xenon arc weatherometer	xenon arc weatherometer
Exposure Note	temperature: 89°C; 3.8 hrs in light at 50% RH, 1 hr in dark at 100% RH; test method: SAE J1885; GM TM30-2	temperature: 89°C; 3.8 hrs in light at 50% RH, 1 hr in dark at 100% RH; test method: SAE J1885; GM TM30-2
Exposure Time (days)	125	208
Energy @ 340nm (kJ/m^2)	4703	7838
Energy @ <380nm (MJ/m^2)	403	672

PROPERTIES RETAINED (%)

100% Modulus	105 {ac}	107 {ac}
Tensile Strength	92 {ac}	95 {ac}
Elongation	93 {ac}	91 {ac}

TABLE 134: Accelerated Weathering in a Xenon Arc Weatherometer of Black, UV Stabilized Advanced Elastomer Systems Santoprene Olefinic Thermoplastic Elastomer.

Material Family	Olefinic Thermoplastic Elastomer	Olefinic Thermoplastic Elastomer
Material Supplier/ Grade	Advanced Elastomer Systems Santoprene 121-87	Advanced Elastomer Systems Santoprene 121-87
Features	black color, UV stabilized	black color, UV stabilized
Sample Form	injection molded slab	injection molded slab
Reference Number	120	120

MATERIAL CHARACTERISTICS

Shore A hardness	87	87
Sample Thickness (mm)	3.1	3.1
Sample Width (mm)	76.2	76.2
Sample Length (mm)	127	127

EXPOSURE CONDITIONS

Exposure Type	accelerated weathering	accelerated weathering
Exposure Apparatus	xenon arc weatherometer	xenon arc weatherometer
Exposure Note	temperature: 89°C; 3.8 hrs in light at 50% RH, 1 hr in dark at 100% RH; test method: SAE J1885; GM TM30-2	temperature: 89°C; 3.8 hrs in light at 50% RH, 1 hr in dark at 100% RH; test method: SAE J1885; GM TM30-2
Exposure Time (days)	125	208
Energy @ 340nm (kJ/m²)	4703	7838
Energy @ <380nm (MJ/m²)	403	672

PROPERTIES RETAINED (%)

100% Modulus	99 {ac}	104 {ac}
Tensile Strength	88 {ac}	87 {ac}
Elongation	89 {ac}	84 {ac}

TABLE 135: Accelerated Weathering in a Xenon Arc Weatherometer of Black, UV Stabilized Advanced Elastomer Systems Santoprene Olefinic Thermoplastic Elastomer.

Material Family	Olefinic Thermoplastic Elastomer						
Material Supplier/ Grade	Advanced Elastomer Systems Santoprene 123-40	Advanced Elastomer Systems Santoprene 123-50	Advanced Elastomer Systems Santoprene 121-58	Advanced Elastomer Systems Santoprene 121-67	Advanced Elastomer Systems Santoprene 121-73	Advanced Elastomer Systems Santoprene 121-80	Advanced Elastomer Systems Santoprene 121-87
Features	black color, UV stabilized	black color, UV stabilized	black color, UV stabilized	black color, UV stabilized	black color, UV stabilized	black color, UV stabilized	black color, UV stabilized
Sample Form	injection molded slab	injection molded slab	injection molded slab	injection molded slab	injection molded slab	injection molded slab	injection molded slab
Reference Number	120	120	120	120	120	120	120
MATERIAL CHARACTERISTICS							
Shore A hardness			58	67	73	80	87
Shore D hardness	40	50					
Sample Thickness (mm)	3.1	3.1	3.1	3.1	3.1	3.1	3.1
Sample Width (mm)	76.2	76.2	76.2	76.2	76.2	76.2	76.2
Sample Length (mm)	127	127	127	127	127	127	127
EXPOSURE CONDITIONS							
Exposure Type	accelerated weathering						
Exposure Apparatus	xenon arc weatherometer						
Exposure Note	40 minutes of light followed by 20 minutes light and front specimen spray, followed by 60 minutes light (irradiance at 340 nm-0.55 W/m^2, black panel temperature-70°C, wet bulb depression-12°C, conditioning water-45°C) , followed by 60 minutes dark with back rack spray and repeating (W/m^2, black panel temperature-38°C, wet bulb depression-0.0°C, conditioning water-40°C).						
Exposure Time (hours)	3	3	3	3	3	3	3
Energy @ 340nm (kJ/m^2)	2500	2500	2500	2500	2500	2500	2500
SURFACE AND APPEARANCE							
Gray Scale	4 {i}	3 {i}	4 {i}	4.5 {i}	4.5 {i}	4.5 {i}	4 {i}
ΔE Color Change	2.7 {ab}	4.7 {ab}	1.2 {ab}	1.9 {ab}		1.9 {ab}	

TABLE 136: Accelerated Weathering in a Xenon Arc Weatherometer of Black, UV Stabilized Advanced Elastomer Systems Santoprene Olefinic Thermoplastic Elastomer.

Material Family	Olefinic Thermoplastic Elastomer	Olefinic Thermoplastic Elastomer
Material Supplier/ Grade	Advanced Elastomer Systems Santoprene 121-80	Advanced Elastomer Systems Santoprene 121-80
Features	black color, UV stabilized	black color, UV stabilized
Sample Form	injection molded slab	injection molded slab
Reference Number	120	120

MATERIAL CHARACTERISTICS

Shore A hardness	80	80
Sample Thickness (mm)	3.1	3.1
Sample Width (mm)	76.2	76.2
Sample Length (mm)	127	127

EXPOSURE CONDITIONS

Exposure Type	accelerated weathering	accelerated weathering
Exposure Apparatus	xenon arc weatherometer	xenon arc weatherometer
Exposure Note	temperature: 89°C; 3.8 hrs in light at 50% RH, 1 hr in dark at 100% RH; test method: SAE J1885; GM TM30-2	temperature: 89°C; 3.8 hrs in light at 50% RH, 1 hr in dark at 100% RH; test method: SAE J1885; GM TM30-2
Exposure Time (days)	125	208
Energy @ 340nm (kJ/m²)	4703	7838
Energy @ <380nm (MJ/m²)	403	672

PROPERTIES RETAINED (%)

100% Modulus	102 {ac}	97 {ac}
Tensile Strength	81 {ac}	74 {ac}
Elongation	78 {ac}	74 {ac}

CHANGE IN PHYSICAL CHARACTERISTICS

Hardness Change (units)	4 {ae}	5 {ae}

SURFACE AND APPEARANCE

ΔE Color Change	2.3 {ab}	0.5 {ab}

TABLE 137: **Accelerated Indoor Exposure in UVCON of Evode Plastics Forprene Olefinic Thermoplastic Elastomer.**

Material Family	Olefinic Thermoplastic Elastomer													
Material Supplier/ Trade Name	Evode Plastics Forprene													
Material Grade	630	631	632	633	634	635	636	637	638	639	640	641	642	643
Reference Number	161	161	161	161	161	161	161	161	161	161	161	161	161	161
MATERIAL CHARACTERISTICS														
Shore A hardness	55	60	65	70	75	80	85	90						
Shore D hardness									40	45	50	55	60	65
EXPOSURE CONDITIONS														
Exposure Type	Accelerated Indoor UV Exposure													
Exposure Apparatus	UV -CON													
Exposure Test Method	ASTM D4329													
Exposure Note	UV cycle: 4 hours on, 4 hours off; condensation cycle: 4 hours on, 4 hours off													
Exposure Temperature (°C)	60	60	60	60	60	60	60	60	60	60	60	60	60	60
Exposure Time (days)	7	7	7	7	7	7	7	7	7	7	7	7	7	7
PROPERTIES RETAINED (%)														
Tensile Strength	95	95	93	94	94	95	95	97	96	96	96	98	98	99
Elongation	94	93	93	93	92	92	93	93	95	96	96	97	97	98
CHANGE IN PHYSICAL CHARACTERISTICS														
Shore A Hardness Change	3	3	2	2	2	2	2	2						
Shore D Hardness Change									1	1	1	0	0	0

TABLE 138: Accelerated Indoor Exposure in Xenon Lamp of Evode Plastics Forprene Olefinic Thermoplastic Elastomer.

Material Family	Olefinic Thermoplastic Elastomer													
Material Supplier/ Trade Name	Evode Plastics Forprene													
Material Grade	630	631	632	633	634	635	636	637	638	639	640	641	642	643
Reference Number	161	161	161	161	161	161	161	161	161	161	161	161	161	161
MATERIAL CHARACTERISTICS														
Shore A hardness	55	60	65	70	75	80	85	90						
Shore D hardness									40	45	50	55	60	65
EXPOSURE CONDITIONS														
Exposure Type	Accelerated UV Light													
Exposure Apparatus	Xenon lamp													
Exposure Temperature (°C)	40	40	40	40	40	40	40	40	40	40	40	40	40	40
Exposure Time (days)	41.7	41.7	41.7	41.7	41.7	41.7	41.7	41.7	41.7	41.7	41.7	41.7	41.7	41.7
RESULTS OF EXPOSURE														
UV Resistance	excellent	excellent	excellent	excellent	excellent	excellent	excellent	excellent	excellent	excellent	excellent	excellent	excellent	excellent

TABLE 139: Ozone Resistance of Evode Plastics Forprene Olefinic Thermoplastic Elastomer.

Material Family	Olefinic Thermoplastic Elastomer													
Material Supplier/ Trade Name	Evode Plastics Forprene													
Material Grade	630	631	632	633	634	635	636	637	638	639	640	641	642	643
Reference Number	161	161	161	161	161	161	161	161	161	161	161	161	161	161
MATERIAL CHARACTERISTICS														
Shore A hardness	55	60	65	70	75	80	85	90						
Shore D hardness									40	45	50	55	60	65
EXPOSURE CONDITIONS														
Exposure Type	ozone													
Exposure Test Method	ASTM D1149													
Exposure Note	100 pphm ozone; sample strained 20%													
Exposure Temperature (°C)	35	35	35	35	35	35	35	35	35	35	35	35	35	35
Exposure Time (hours)	200	200	200	200	200	200	200	200	200	200	200	200	200	200
RESULTS OF EXPOSURE														
Ozone Resistance	no cracks {aj}	no cracks {aj}	no cracks {aj}	no cracks {aj}	no cracks {aj}	no cracks {aj}	no cracks {aj}	no cracks {aj}	no cracks {aj}	no cracks {aj}	no cracks {aj}	no cracks {aj}	no cracks {aj}	no cracks {aj}

TABLE 140: Ozone Resistance of Advanced Elastomer Systems Santoprene Olefinic Thermoplastic Elastomer.

Material Family	Olefinic Thermoplastic Elastomer										
Material Supplier/ Grade	Advanced Elastomer Systems Santoprene 201-55	Advanced Elastomer Systems Santoprene 101-55	Advanced Elastomer Systems Santoprene 271-64	Advanced Elastomer Systems Santoprene 201-73	Advanced Elastomer Systems Santoprene 101-73	Advanced Elastomer Systems Santoprene 201-80	Advanced Elastomer Systems Santoprene 101-80	Advanced Elastomer Systems Santoprene 201-87	Advanced Elastomer Systems Santoprene 101-87	Advanced Elastomer Systems Santoprene 251-80	Advanced Elastomer Systems Santoprene 251-85
Features	neutral color	black color	FDA grade, neutral color	neutral color	black color	neutral color	black color	neutral color	black color	flame retardant, neutral color	flame retardant, neutral color
Reference Number	158	158	158	158	158	158	158	158	158	158	158
MATERIAL CHARACTERISTICS											
Shore A hardness	55	55	64	73	73	80	80	87	87	80	85
EXPOSURE CONDITIONS											
Exposure Type	ozone										
Exposure Test Method	ASTM D1171										
Exposure Apparatus	Monsanto's ozone chamber										
Exposure Specimen Note	extruded triangle profile mounted over a mandrel										
Exposure Time (hours)	336										
Exposure Note	168 hours @ 400 pphm, 168 hours @ 500 pphm										
SURFACE AND APPEARANCE											
Time to Initial Crack (hours)	no cracking	no cracking	no cracking	no cracking	no cracking	no cracking	no cracking	no cracking	no cracking	no cracking	no cracking
Shell Rating (0-10 scale, 0 represents total cracking, 10 represents no cracking)	10	10	10	10	10	10	10	10	10	10	10

TABLE 141: Ozone Resistance of Dow Chemical Engage Olefinic Thermoplastic Elastomer.

Material Family	Olefinic Thermoplastic Elastomer		
Cure	10 min. @ 175°C		
Material Supplier/ Grade	Dow Chemical Engage POE		
Reference Number	204	204	204
MATERIAL COMPOSITION			
Engage	100 phr	100 phr	100 phr
dicumyl peroxide (40%)	10 phr	10 phr	10 phr
magnesium oxide	10 phr	10 phr	10 phr
zinc stearate	0.5 phr	0.5 phr	0.5 phr
paraffinic plasticizer	35 phr	35 phr	35 phr
antioxidant	4 phr	4 phr	4 phr
zinc oxide	5 phr	5 phr	5 phr
N-550 carbon black	85 phr	85 phr	85 phr
co-agent	1 phr	1 phr	1 phr
EXPOSURE CONDITIONS			
Exposure Type	ozone		
Exposure Test Method	ASTM D1149		
Exposure Note	100 pphm ozone; samples at 20% strain		
Exposure Temperature (°C)	40	40	40
Exposure Time (hours)	24	48	72
SURFACE AND APPEARANCE			
Number of Cracks	0	0	0

TABLE 142: Ozone Resistance of Black Advanced Elastomer Systems Santoprene Olefinic Thermoplastic Elastomer.

Material Family	Olefinic Thermoplastic Elastomer			
Material Supplier/ Grade	Advanced Elastomer Systems Santoprene 121-67	Advanced Elastomer Systems Santoprene 121-73	Advanced Elastomer Systems Santoprene 591-65	Advanced Elastomer Systems Santoprene 591-73
Features	black color, UV stabilized	black color, UV stabilized		
Reference Number	156	156	155	155

MATERIAL CHARACTERISTICS

Shore A hardness	67	73	65	73

EXPOSURE CONDITIONS

Exposure Type	ozone			
Exposure Test Method	ASTM D1149	ASTM D1149	ASTM D1149	ASTM D1149
Exposure Temperature (°C)	40	40	40	40
Exposure Note	300 pphm ozone	300 pphm ozone	300 pphm ozone	300 pphm ozone
Exposure Time (hours)	100	100	100	100

RESULTS OF EXPOSURE

Ozone Resistance	excellent	excellent	excellent	excellent

TABLE 143: Ozone Resistance of Advanced Elastomer Systems Santoprene, Technor Apex Telcar and BP Chemicals TPR Olefinic Thermoplastic Elastomer.

Material Family	Olefinic Thermoplastic Elastomer								
Material Supplier/ Grade	Advanced Elastomer Systems Santoprene 201-55	Advanced Elastomer Systems Santoprene 201-73	Advanced Elastomer Systems Santoprene 201-80	Advanced Elastomer Systems Santoprene 201-87	Advanced Elastomer Systems Santoprene 101-64	Technor Apex Telcar 332	Technor Apex Telcar 942	BP Chemicals TPR 1600	BP Chemicals TPR 1700
Features	neutral color	neutral color	neutral color	neutral color	black color				
Reference Number	158	158	158	158	158	158	158	158	158
MATERIAL CHARACTERISTICS									
Shore A hardness	55	73	80	87	64				
EXPOSURE CONDITIONS									
Exposure Type	ozone								
Exposure Test Method	ASTM D518								
Exposure Apparatus	Monsanto's ozone chamber								
Exposure Specimen Note	bent loop specimen								
Exposure Time (hours)	336								
Exposure Note	168 hours @ 400 pphm, 168 hours @ 500 pphm								
SURFACE AND APPEARANCE									
Time to Initial Crack (hours)	no cracking	no cracking	no cracking	no cracking	no cracking	no cracking	no cracking	no cracking	no cracking
Shell Rating (0-10 scale, 0 represents total cracking, 10 represents no cracking)	10	10	10	10	10	10	10	10	10

GRAPH 205a: Xenon Arc Weatherometer Exposure Time vs. Carbonyl Formation - Area Normalized of Olefinic Thermoplastic Elastomer

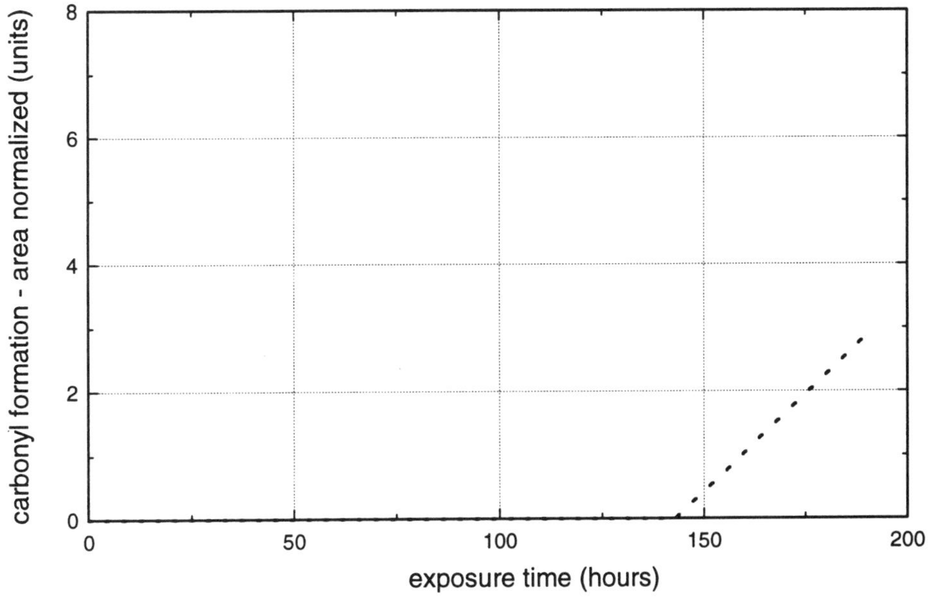

··············	Dow Engage TPO (23 Mooney (ML 1+4 @ 125°C))
Reference No.	204

GRAPH 206: Xenon Arc Weatherometer Exposure Time vs. Decrease in Molecular Weight of Olefinic Thermoplastic Elastomer

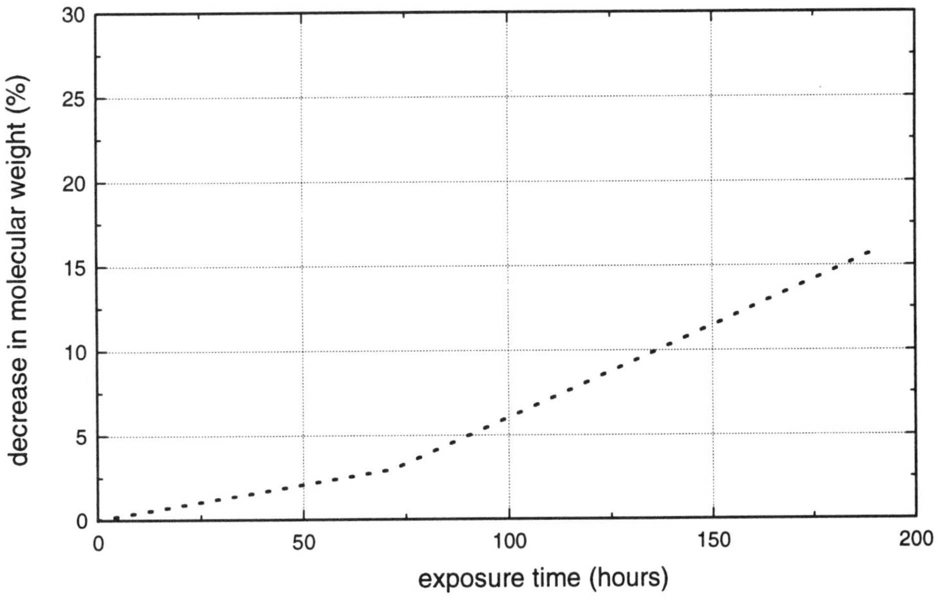

··············	Dow Engage TPO (23 Mooney (ML 1+4 @ 125°C))
Reference No.	204

Chapter 64

Polyester Thermoplastic Elastomer

Outdoor Weather Resistance

DuPont: Hytrel

When Hytrel polyester elastomer is subjected to outdoor exposure, the most serious effect is degradation by ultraviolet (UV) light. Addition of pigments and/or UV stabilizers will provide adequate protection from such exposure.

Reference: *Pigmentation And Weathering Protection Of 'Hytrel',* supplier technical report (HYT-303(R1) / E-73191) - Du Pont Company, 1985.

Effect of Carbon Black on Weatherability

DuPont: Hytrel 4056 (features: 1.9 mm thick; material compostion: 1% SAF black (masterbatch form); product form: injection molded slab); **Hytrel** (features: 1.9 mm thick; material compostion: 0.5% SAF black (masterbatch form); product form: injection molded slab); **Hytrel 6345** (features: 0.19 mm thick, 63 Shore D hardness; material compostion: 1.9% Hytrel 10MS (polycarbodiimide moisture stabilizer, masterbatch); product form: extrusion film); **Hytrel** (features: 0.25 mm thick; material compostion: 1.9% carbon black (masterbatch form), 1.9% Hytrel 10MS (polycarbodiimide moisture stabilizer, masterbatch); product form: extrusion film); **Hytrel** (features: 0.15 mm thick, 63 Shore D hardness; material compostion: 0.8% carbon black (masterbatch form); product form: extrusion film); **Hytrel** (features: 0.13 mm thick, 63 Shore D hardness; material compostion: 1.9% carbon black (masterbatch form), 1.9% Hytrel 10MS (polycarbodiimide moisture stabilizer, masterbatch); product form: extrusion film); **Hytrel HT-X-3803** (features: 1.9 mm thick, 63 Shore D hardness; material compostion: 0.5% SAF black (masterbatch form); product form: injection molded slab); **Hytrel** (features: 1.9 mm thick, 63 Shore D hardness; material compostion: 0% carbon black; product form: injection molded slab)

Like most thermoplastics, Hytrel requires protection against UV degradation on outdoor exposure. The most efficient method is the incorporation of low levels of carbon black. Where non-black products are desired, weather protection is most efficiently obtained by incorporation of UV stabilizers alone (natural) or in combination with low levels of colored pigments.

Specimens, 1.9 mm (0.075") thick, behaved very well under weatherometer and Florida aging with as little as 0.5% black; little additional improvement is achieved at 1% black. As sample thickness is decreased however, higher levels of the protective additive are required. Tests indicate that 1.9% carbon black is required for excellent protection in 0.25 mm (0.010") thick films, with somewhat compromised property retention at lower thicknesses. Some tested compounds contained moisture stabilizing additives such as 1.9% polycarbodiimide in addition to the carbon black, which undoubtedly provided some protection in its own right. However, the contribution is believed to be minimal, with only a slight compromise in property retention predicted for similar black compounds containing no polycarbodiimide.

Reference: *Hytrel Technical Notes - " Weather Protection of Hytrel with Carbon Black",* supplier technical report (I-48) - DuPont Company.

DuPont: Hytrel

The most efficient method of protecting Hytrel from the elements is through the use of low levels of carbon black. Excellent resistance to degradation is observed in weatherometer and Florida aging studies with specimens 1.9 mm (0.075 in) thick and containing 0.5 and 1% carbon black. These levels of carbon black provide good protection for thick sections;

for thinner parts additional protection is needed. For films 0.25 mm [0.010 in] thick up to 3% carbon black is needed for protection from UV light.

A convenient way to incorporate carbon black into Hytrel is to use Hytrel G-40CB, a concentrate containing 25% carbon black. This concentrate is available in pellet form and may be used to color and/or increase the UV light resistance of all types of Hytrel.

Reference: *Pigmentation And Weathering Protection Of 'Hytrel',* supplier technical report (HYT-303(R1) / E-73191) - Du Pont Company, 1985.

Effect of Color Pigments on Weatherability

DuPont: Hytrel

Hytrel can be pigmented or colored using pigment concentrates or master batches as pellet/pellet blends with the virgin polymer. Other techniques such as liquid colorants and dusting-on of powdered pigments have also been used. Hytrel has also been colored by dipping in liquid dye baths. In every case the utility and compatibility of any system used with Hytrel must be determined by trial if no previous experience with it exists.

Following is a brief summary of pigments which have been found to be compatible with Hytrel. It is a guide and is not intended to be all-inclusive.

Black - Most carbon blacks will color Hytrel satisfactorily. MT, SAF and SRF blacks provide good weatherability at the 1% level. The use of Hytrel G-40CB is recommended as an easy means of adding carbon black.

Brown - Iron oxide types, e.g. Mapico brown 1020A (Cities Service Company, Columbian Chemical Co.)

Blue - Phthalocyanines, Monarch Blue (Ciba Geigy Corp.) X-3228 and Monarch Blue X-2925

Green - Chromium oxide hydrates. Also phthalocyanines Monolite Green (ICI Americas) 751.

Red - Quinacridone pigment, Monastral (Ciba Geigy Corp.) RT-791-D and lead chromate, Krolor (Heubach Inc.) pigment color KR-980-D. Color fastness of RT-791-D is better in laboratory tests.

Orange - Lead chromates, Krolor (Ciba Geigy Corp.) KO-786-D and KO-789-D.

Yellow - Lead chromates, Krolor (Ciba Geigy Corp.) KY-781-D and Krolor KY-787-D.

White - Titanium dioxide. TI-Pure R-960 (E.I. du Pont de Nemours & Co., Inc.) titanium dioxide pigment has been found to be better than TI-Pure R-101 in weatherability. Zinc oxide, in the form of AZO-77S (ASARCO, Inc.), has been found to be as good as TiO_2.

NOTE: Cadmium-based pigments should not be used because they are detrimental to Hytrel.

CAUTION: Acidic clays (pH 4.5 to 5.5) have been known to promote decomposition of Hytrel. Compounding of Hytrel with acidic pigments, lubricants or additives should be avoided.

COLOR CONCENTRATES Hytrel is the preferred carrier for use in concentrates to be let down in Hytrel because all types are compatible at all levels. In general, types of Hytrel having lower melting points and/or lower melt viscosity are more easily dispersed in other types of Hytrel when loaded with reinforcing (viscosity increasing) pigments. The protection obtained from added concentrates is highly dependent on the efficiency of mixing. In all cases it is recommended that the particular system be evaluated in the end-use application.

Other polymers are usually not recommended as carriers because of limited compatibility. Ethylene-vinyl acetate and polyvinyl chloride based concentrates are never recommended at any level because they degrade at the temperatures required to process Hytrel.

Reference: *Pigmentation And Weathering Protection Of 'Hytrel',* supplier technical report (HYT-303(R1) / E-73191) - Du Pont Company, 1985.

Effect of UV Stabilizers on Weatherability

DuPont: Hytrel

In applications where non-black products are desired, Hytrel G-20UV should be incorporated to increase the UV light resistance of natural or pigmented parts. Hytrel G-20UV is a concentrate of ultraviolet light stabilizers. It is available in pellet form and may be let down in all grades of Hytrel.

Reference: *Pigmentation And Weathering Protection Of 'Hytrel'*, supplier technical report (HYT-303(R1) / E-73191) - Du Pont Company, 1985.

TABLE 144: Outdoor Weathering in Florida of DuPont Hytrel Polyester Thermoplastic Elastomer.

Material Family	Polyester Thermoplastic Elastomer									
Material Supplier/ Grade	DuPont Hytrel HT-X-3803		DuPont Hytrel HT-X-3803		DuPont Hytrel 4056			DuPont Hytrel 4056		
Product Form	injection molded slab		injection molded slab		injection molded slab			injection molded slab		
Reference Number	111	111	111	111	111	111	111	111	111	111
MATERIAL CHARACTERISTICS										
Shore D hardness	63	63	63	63						
SampleThickness	1.9 mm	1.9 mm	1.9 mm	1.9 mm	1.9 mm	1.9 mm	1.9 mm	1.9 mm	1.9 mm	1.9 mm
MATERIAL COMPOSITION										
carbon black	0%	0%								
SAF black (masterbatch form)			0.5%	0.5%	0.5%	0.5%	0.5%	1%	1%	1%
EXPOSURE CONDITIONS										
Exposure Type	outdoor weathering									
Exposure Location	Florida		Florida		Florida			Florida		
Exposure Country	USA		USA		USA			USA		
Exposure Note	45° south		45° south		45° south			45° south		
Exposure Time (days)	182	365	182	365	182	365	730	182	365	730
PROPERTIES RETAINED (%)										
Tensile Strength	70	58	94	97	96	100	80	97	106	99
Elongation @ Break	13	0.5	93	75	104	112	102	104	110	104
SURFACE AND APPEARANCE										
Surface Appearance Note			good for all black specimens with loss of gloss at extended exposures, but no surface crazing		good for all black specimens with loss of gloss at extended exposures, but no surface crazing			good for all black specimens with loss of gloss at extended exposures, but no surface crazing		

TABLE 145: Effect of Carbon Black on Outdoor Exposure in Florida of DuPont Hytrel Polyester Thermoplastic Elastomer.

Material Family	Polyester Thermoplastic Elastomer							
Material Supplier/ Grade	DuPont Hytrel 40D				DuPont Hytrel 40D			
Product Form	injection molded slab				injection molded slab			
Reference Number	203	203	203	203	203	203	203	203
MATERIAL CHARACTERISTICS								
Shore D hardness	40	40	40	40	40	40	40	40
SampleThickness	1.9 mm	1.9 mm	1.9 mm	1.9 mm	1.9 mm	1.9 mm	1.9 mm	1.9 mm
MATERIAL COMPOSITION								
SAF black (masterbatch form)	0.5%	0.5%	0.5%	0.5%	1%	1%	1%	1%
EXPOSURE CONDITIONS								
Exposure Type	outdoor weathering	outdoor weathering	outdoor weathering	outdoor weathering	outdoor weathering	outdoor weathering	outdoor weathering	outdoor weathering
Exposure Location	Florida	Florida	Florida	Florida	Florida	Florida	Florida	Florida
Exposure Country	USA	USA	USA	USA	USA	USA	USA	USA
Exposure Note	45° angle south	45° angle south	45° angle south	45° angle south	45° angle south	45° angle south	45° angle south	45° angle south
Exposure Time (days)	182	365	730	1826	182	365	730	1826
PROPERTIES RETAINED (%)								
Tensile Strength	95.5 {m}	100 {m}	79.9 {m}	60.4 {m}	97.7 {m}	107 {m}	99 {m}	68.2 {m}
Elongation @ Break	104 {m}	113 {m}	102 {m}	136 {m}	104 {m}	111 {m}	104 {m}	119 {m}
SURFACE AND APPEARANCE								
Surface Appearance Note				very good with loss of gloss but no surface crazing				very good with loss of gloss but no surface crazing

TABLE 146: Effect of Carbon Black Level on Outdoor Weathering in Florida of DuPont Hytrel Polyester Thermoplastic Elastomer.

Material Family	Polyester Thermoplastic Elastomer									
Material Supplier/ Grade	DuPont Hytrel 5556			DuPont Hytrel 5556		DuPont Hytrel 5556		DuPont Hytrel 5556		
Manufacturing Method	compression molding			compression molding		compression molding		compression molding		
Product Form	film			film		film		film		
Reference Number	203	203	203	203	203	203	203	203	203	203
MATERIAL CHARACTERISTICS										
SampleThickness	0.25 mm	0.25 mm	0.25 mm	0.25 mm	0.25 mm	0.25 mm	0.25 mm	0.25 mm	0.25 mm	0.25 mm
MATERIAL COMPOSITION										
SAF (N110) black (masterbatch form, 0.018-0.020 microns)	0.5%	0.5%	0.5%	0.94%	0.94%	1.5%	1.5%	3%	3%	3%
EXPOSURE CONDITIONS										
Exposure Type	outdoor weathering									
Exposure Location	Delaware	Delaware	Delaware	Delaware	Delaware	Delaware	Delaware	Delaware	Delaware	Delaware
Exposure Country	USA	USA	USA	USA	USA	USA	USA	USA	USA	USA
Exposure Time (days)	182	730	1278	182	730	182	730	182	730	1278
PROPERTIES RETAINED (%)										
Tensile Strength	42.3	too brittle to test	too brittle to test	67.4	18.4	73	33	83.3	45.6	40.6
Elongation @ Break	52.1	too brittle to test	too brittle to test	80.6	3.5	86.5	49	87.7	61.6	54.8

TABLE 147: Effect of Carbon Black and Film Thickness on Outdoor Weathering in Florida of DuPont Hytrel Polyester Thermoplastic Elastomer.

Material Family	Polyester Thermoplastic Elastomer												
Material Supplier/ Grade	DuPont Hytrel 6345	DuPont Hytrel 6345				DuPont Hytrel 6345				DuPont Hytrel 6345			
Product Form	extrusion film	extrusion film, black color				extrusion film, black color				extrusion film, black color			
Reference Number	111	111	111	111	111	111	111	111	111	111	111	111	111
MATERIAL CHARACTERISTICS													
Shore D hardness	63	63	63	63	63	63	63	63	63	63	63	63	63
SampleThickness	0.15 mm	0.13 mm	0.13 mm	0.13 mm	0.13 mm	0.19 mm	0.19 mm	0.19 mm	0.19 mm	0.25 mm	0.25 mm	0.25 mm	0.25 mm
MATERIAL COMPOSITION													
carbon black (masterbatch form)	0.80%	1.9%	1.9%	1.9%	1.9%	1.9%	1.9%	1.9%	1.9%	1.9%	1.9%	1.9%	1.9%
Hytrel 10MS (polycarbodiimide moisture stabilizer)		1.9%	1.9%	1.9%	1.9%	1.9%	1.9%	1.9%	1.9%	1.9%	1.9%	1.9%	1.9%
EXPOSURE CONDITIONS													
Exposure Type	outdoor weathering												
Exposure Location	Florida	Florida				Florida				Florida			
Exposure Country	USA	USA				USA				USA			
Exposure Time (days)	182	182	365	730	1096	182	365	730	1096	182	365	730	1096
PROPERTIES RETAINED (%)													
Tensile Strength	21	53	71	39	34	56	64	51	36	72	73	67	62
Elongation @ Break	6.2	108	104	40	26	96	100	56	36	91	100	91	87
SURFACE AND APPEARANCE													
Surface Appearance Note	good for all black specimens with loss of gloss at extended exposures, but no surface crazing												

TABLE 148: Effect of Carbon Black on Accelerated Weathering in a Weatherometer of DuPont Hytrel Polyester Thermoplastic Elastomer.

Material Family	Polyester Thermoplastic Elastomer			
Material Supplier/ Grade	DuPont Hytrel 40D		DuPont Hytrel 40D	
Product Form	injection molded slab		injection molded slab	
Reference Number	203	203	203	203
MATERIAL CHARACTERISTICS				
Shore D hardness	40	40	40	40
Sample Thickness	1.9 mm	1.9 mm	1.9 mm	1.9 mm
MATERIAL COMPOSITION				
SAF black (masterbatch form)	0.5%	0.5%	1%	1%
EXPOSURE CONDITIONS				
Exposure Type	accelerated weathering			
Exposure Apparatus	carbon arc weatherometer	carbon arc weatherometer	carbon arc weatherometer	carbon arc weatherometer
Exposure Time (days)	20.8	41.7	20.8	41.7
PROPERTIES RETAINED (%)				
Tensile Strength	97.8 {m}	84.3 {m}	104 {m}	104 {m}
Elongation @ Break	113 {m}	109 {m}	111 {m}	106 {m}

TABLE 149: Effect of Carbon Black Level on Accelerated Weathering in a Weatherometer of DuPont Hytrel Polyester Thermoplastic Elastomer.

Material Family	Polyester Thermoplastic Elastomer									
Material Supplier/ Grade	DuPont Hytrel 5556		DuPont Hytrel 5556		DuPont Hytrel 5556			DuPont Hytrel 5556		
Manufacturing Method	compression molding		compression molding		compression molding			compression molding		
Product Form	film		film		film			film		
Reference Number	203	203	203	203	203	203	203	203	203	203
MATERIAL CHARACTERISTICS										
SampleThickness	0.25 mm	0.25 mm	0.25 mm	0.25 mm	0.25 mm	0.25 mm	0.25 mm	0.25 mm	0.25 mm	0.25 mm
MATERIAL COMPOSITION										
SAF (N110) black (masterbatch form, 0.018-0.020 microns)	0.5%	0.5%	0.94%	0.94%	1.5%	1.5%	1.5%	3%	3%	3%
EXPOSURE CONDITIONS										
Exposure Type	accelerated weathering									
Exposure Apparatus	carbon arc weatherometer		carbon arc weatherometer		carbon arc weatherometer			carbon arc weatherometer		
Exposure Time (days)	4.2	10.9	4.2	10.9	4.2	12.5	25	4.2	12.5	25
PROPERTIES RETAINED (%)										
Tensile Strength	35	26	69.5	28.5	66.4	38	27	82.8	65.7	46
Elongation @ Break	42.5	14.5	90.3	18.1	97	54	5.4	94.5	82.2	63

TABLE 150: Effect of Carbon Black on Accelerated Weathering in a Weatherometer of DuPont Hytrel Polyester Thermoplastic Elastomer.

Material Family	Polyester Thermoplastic Elastomer									
Material Supplier/ Grade	DuPont Hytrel HT-X-3803		DuPont Hytrel HT-X-3803				DuPont Hytrel 4056		DuPont Hytrel 4056	
Product Form	injection molded slab		injection molded slab				injection molded slab		injection molded slab	
Reference Number	111	111	111	111	111	111	111	111	111	111
MATERIAL CHARACTERISTICS										
Shore D hardness	63	63	63	63	63	63				
SampleThickness	1.9 mm	1.9 mm	1.9 mm	1.9 mm	1.9 mm	1.9 mm	1.9 mm	1.9 mm	1.9 mm	1.9 mm
MATERIAL COMPOSITION										
carbon black	0%	0%								
SAF black (masterbatch form)			0.5%	0.5%	0.5%	0.5%	0.5%	0.5%	1 %	1%
EXPOSURE CONDITIONS										
Exposure Type	accelerated weathering									
Exposure Apparatus	carbon arc weatherometer		carbon arc weatherometer				carbon arc weatherometer		carbon arc weatherometer	
Exposure Time (days)	20.8	41.7	20.8	41.7	83.3	166.7	20.8	41.7	20.8	41.7
PROPERTIES RETAINED (%)										
Tensile Strength	73	30	96	99	100	86	98	84	104	103
Elongation @ Break	2.6	2.6	88	81	93	96	112	108	110	106
SURFACE AND APPEARANCE										
Surface Appearance Note			good for all black specimens with loss of gloss at extended exposures, but no surface crazing				good for all black specimens with loss of gloss at extended exposures, but no surface crazing			

TABLE 151: Soil Burial and Fungus Resistance of DuPont Hytrel Polyester Thermoplastic Elastomer.

Material Family	Polyester Thermoplastic Elastomer						
Material Supplier/ Grade	DuPont Hytrel						
Reference Number	234	234	234	234	234	234	234
MATERIAL CHARACTERISTICS							
Shore D hardness	40	40	40	40	40	40	40
EXPOSURE CONDITIONS							
Exposure Type	soil burial	fungus					
Exposure Country	Panama						
Exposure Test Method		ASTM D1924-63					
Culture		aspergillus niger	aspergillus flavus	aspergillus versicolor	penicillin funiculosum	pullularia pullulans	trichloderma sp.
Exposure Time (days)	365						
PROPERTIES RETAINED (%)							
100% Modulus	99						
300% Modulus	98						
Tensile Strength	82						
Elongation @ Break	82						
Shore D Hardness	98						
RESULTS OF EXPOSURE							
Observed Growth		none	none	very slight, sparse	none	none	none

Chapter 65

Styrenic Thermoplastic Elastomer

Weather Resistance

Shell Chemical: Kraton D (chemical type: styrene butadiene styrene block copolymer (SBS)); **Kraton D** (chemical type: styrene isoprene styrene block copolymer (SIS)))); **Kraton G** (chemical type: styrene ethylene butylene styrene block copolymer (SEBS))

Kraton D series rubbers are unsaturated and are therefore susceptible to attack by oxygen, ozone, and ultraviolet radiation, especially when stressed. As manufactured, they contain sufficient antioxidant to protect them against oxidation during manufacture, shipment, and storage. That level of antioxidant, however, is not sufficient to protect end-use products from degradation, especially in thin films or coatings, nor is it sufficient in view of other unsaturated resins, plasticizers, etc., which may be included in a formulation. Therefore, stabilizers should be added in the formulation to protect the material both during processing and during its service life.

Kraton G series rubbers also contain a low level of phenolic antioxidant as manufactured. Although the stability of these polymers is much better than that of the Kraton D series rubbers, it is good practice to include additional stabilizer in formulated products containing even these more stable polymers.

In choosing a stabilizer package, it should be noted that the rubber network is more susceptible to attack than the polystyrene domains. Thus it is best to use stabilizers which associate primarily with the rubber network.

The two types of rubber midblock - polybutadiene and polyisoprene - in Kraton D series rubbers behave differently when attacked by oxygen, ozone, or UV radiation. Polybutadiene (Kraton SBS rubbers) tends to crosslink with films becoming hard and brittle. Polyisoprene (Kraton SIS rubbers) tends to undergo chain scission, whereby films become soft and tacky. Blends of the two types show less change on aging than does either type alone.

Reference: *Kraton Thermoplastic Rubber,* supplier design guide (SC:198-89) - Shell Chemical Company, 1989.

Shell Chemical: Kraton D (chemical type: styrene isoprene styrene block copolymer (SIS)); **Kraton D** (chemical type: styrene butadiene styrene block copolymer (SBS))

Kraton D rubbers contain an unsaturated rubber midblock which is similar to natural rubber or SBR in resistance to degradation. Thus they can be degraded by oxygen, ozone and ultraviolet light - particularly when under stress. Various antioxidants, antiozonants and ultraviolet (UV) light stabilizers can be used to provide adequate resistance for many applications.

Reference: *Kraton Thermoplastic Rubber,* supplier design guide (SC:198-89) - Shell Chemical Company, 1989.

Shell Chemical: Kraton G (chemical type: styrene ethylene butylene styrene block copolymer (SEBS))

Kraton G rubbers have a saturated, olefin rubber type centerblock, and thus have good resistance to degradation. When properly formulated, they have withstood 4,000 hours of weatherometer exposure with minimal change in properties. They also will pass ozone chamber testing without cracking or loss of properties. Where processing and/or use conditions are expected to be severe, additional stabilizers can be added. Even under very mild conditions, small amounts of stabilizer additives still should be used with Kraton G rubbers.

Reference: *Kraton Thermoplastic Rubber,* supplier design guide (SC:198-89) - Shell Chemical Company, 1989.

UV Resistance

Evode Plastics: Evoprene G (chemical type: styrene ethylene butylene styrene block copolymer (SEBS))

The benefit of SEBS lies in its chemical structure. The block copolymer has a fully saturated midblock (i.e. no double bonds). This is responsible for the generally excellent ozone and UV resistance. All Evoprene G compounds have excellent UV resistance when pigmented black, but some grades require extra stabilization when being used in natural or light colors.

Reference: *Technical Information Evoprene G,* supplier technical report (RDS 028/9240) - Evode Plastics.

Evode Plastics: Evoprene Super G Black 4286 (chemical type: styrene ethylene butylene styrene block copolymer (SEBS); features: black color); **Evoprene Super G Natural 4305** (chemical type: styrene ethylene butylene styrene block copolymer (SEBS); features: natural resin)

The black 4286 is UV stable. However, while the Natural 4305 contains UV stabilizer, these compounds are still likely to show some discoloration in outdoor situations or on prolonged exposure to high temperatures in white or light colors. Caution should therefore be exercised when using Evoprene Super G compounds in other than black applications.

Reference: *Evoprene Super G Thermoplastic Elastomer Compounds,* supplier marketing literature (RDS 050/9240) - Evode Plastics.

Effect of UV Stabilizers on Weatherability

Shell Chemical: Kraton D (chemical type: styrene isoprene styrene block copolymer (SIS)); **Kraton D** (chemical type: styrene butadiene styrene block copolymer (SBS)); **Kraton G** (chemical type: styrene ethylene butylene styrene block copolymer (SEBS))

Degradation by exposure to UV light is denoted by discoloration and surface embrittlement of Kraton rubber compounds. In most indoor applications, this type of degradation is not a problem. However, if direct exposure to sunlight is expected, Kraton D series rubbers must be protected, particularly if the formulations are clear and/or nonpigmented.

One or more stabilizers should be added during compounding, at the level of about 0.5 phr. Tinuvin P has been found effective at that level. A combination of 0.3 phr each of Uvinul 400 and Tinuvin 326 is particularly effective for most products, although in white stocks it can cause slight discoloration.

With opaque products, even without UV stabilizers, the addition of up to five parts of a reflective filler such as titanium dioxide (such as Titanox RA-50, Titanium Pigments Co.) or a light-absorbing filler such as carbon black (such as those marketed by Cabot Corp.) gives excellent protection.

Reference: *Kraton Thermoplastic Rubber,* supplier design guide (SC:198-89) - Shell Chemical Company, 1989.

Ozone Resistance

Evode Plastics: Evoprene G (chemical type: styrene ethylene butylene styrene block copolymer (SEBS))

The benefit of SEBS lies in its chemical structure. The block copolymer has a fully saturated midblock (i.e. no double bonds). This is responsible for the generally excellent ozone resistance.

Reference: *Technical Information Evoprene G,* supplier technical report (RDS 028/9240) - Evode Plastics.

Evode Plastics: Evoprene Super G Black 4286 (chemical type: styrene ethylene butylene styrene block copolymer (SEBS); features: black color); **Evoprene Super G Natural 4305** (chemical type: styrene ethylene butylene styrene block copolymer (SEBS); features: natural resin)

Evoprene Super G grades show no cracks when exposed to BS Spec 903-A43 (100 pphm / 200 hours / 20% strain).

Reference: *Evoprene Super G Thermoplastic Elastomer Compounds,* supplier marketing literature (RDS 050/9240) - Evode Plastics.

Shell Chemical: Kraton 7720G

The results of both ASTM D1171 and ASTM D518 ozone testing indicate that Kraton rubber exhibits outstanding resistance to ozone.

Reference: *Ozone Resistance Of Santoprene Rubber,* supplier technical report (TCD01787) - Advanced Elastomer Systems, 1986.

Shell Chemical: Kraton D (chemical type: styrene isoprene styrene block copolymer (SIS)); **Kraton D** (chemical type: styrene butadiene styrene block copolymer (SBS)); **Kraton G** (chemical type: styrene ethylene butylene styrene block copolymer (SEBS))

Kraton D series rubbers are susceptible to degradation by ozone, particularly when under stress. Degradation is evidenced by surface crazing and hardening, or major cracking.

Antiozonants such as Nickel Dibutyl Dithiocarbamate (NBC) and Pennzone B improve the resistance of Kraton rubbers to ozone, however, they give a green color to the formulation. Ozone Protector 80, a nonstaining antiozonant is also effective but considerably higher loadings are required than with NBC or Pennzone B. Pennzone B is not suitable for hot melt formulations, as it accelerates the crosslinking of the rubber segments.

Reference: *Kraton Thermoplastic Rubber,* supplier design guide (SC:198-89) - Shell Chemical Company, 1989.

TABLE 152: Ozone Resistance of Shell Chemcial Kraton Styrenic Thermoplastic Elastomer.

mat-ds-id	645
Material Family	Styrenic Thermoplastic Elastomer
Material Supplier/ Grade	Shell Chemical Kraton 7720G
Reference Number	158
EXPOSURE CONDITIONS	
Exposure Type	ozone
Exposure Test Method	ASTM D518
Exposure Apparatus	Monsanto's ozone chamber
Exposure Specimen Note	bent loop specimen
Exposure Time (hours)	336
Exposure Note	168 hours @ 400 pphm, 168 hours @ 500 pphm
SURFACE AND APPEARANCE	
Time to Initial Crack (hours)	no cracking
Shell Rating (0-10 scale, 0 represents total cracking, 10 represents no cracking)	10

Chapter 66

Urethane Thermoplastic Elastomer

UV Resistance

BASF: Elastollan

Plastics are chemically degraded when exposed to UV-radiation. The degree of degradation depends on duration and intensity. In the case of polyurethanes, the effect is seen initially as embrittlement of the exposed material surface. This is accompanied by a yellowing in color and a reduction in mechanical properties. Polyester types are more resistant to UV than polyether types. Embrittlement of the material due to irradiation occurs on the exposed surface and the damage has only a certain depth of penetration. For this reason, comparisons can only be made from tests using specimens of the same cross-section. In the case of thicker test pieces, the proportion of the specimen's cross-section damaged by irradiation is smaller. Conversely, the proportion of thinner samples which is damaged is greater. Protection against aging can be achieved by using pigments and/or UV stabilizers.

Reference: *Elastollan Design And Processing Guide,* supplier design guide - BASF Corporation, 1993.

Dow Chemical: Pellethane 2103-80 AEF (chemical type: aromatic isocyanate; features: unstabilized); **Pellethane** (chemical type: aromatic isocyanate, polyester urethane elastomer); **Pellethane** (chemical type: aromatic isocyanate, polyether urethane elastomer)

Thermoplastic polyurethane elastomers (TPU's) are known to have poor color stability when exposed to UV light. Pellethane TPU resins are based on aromatic isocyanates, which absorb ultraviolet radiation. In turn, this can cause the material to become yellow. Depending on thickness, the material may embrittle in very thin films. Degradation upon UV exposure tends to limit the use of TPU's in outdoor applications without the use of stabilizers.

Studies have supported two mechanisms for photodegradation in MDI-based polyurethanes involving photo-oxidation of the aromatic isocyanate and direct photolytic cleavage of the urethane group. Photo-oxidation of the aromatic isocyanate may produce a diquinone imide. Quinone imides are primarily responsible for the rapid yellowing of MDI polyurethane. Subjecting oxygen to aromatic radicals causes formation of semi-quinone and quinone groups, which have strong UV absorption and act as a photostabilizing surface layer that protects the bulk polymer.

Degradation also can be caused by direct photolytic cleavage of the urethane group, which can occur in the absence of oxygen. It has been shown that this cleavage of the urethane group can result in a photo-Fries rearrangement. Further photochemistry of the aromatic amine photo-Fries product can lead to the formation of colored azo products, which account for the photo-induced yellowing in the absence of oxygen. Similar to the semi-quinone and quinone groups, the photo-Fries product is also an efficient UV absorber and is capable of acting as an internal filter to inhibit subsequent photo-oxidation. When irradiated in air, the distribution between the photo-Fries product and the oxidation products was found to be wavelength dependent, with the photo-Fries product being dominant at wavelengths below 330-340 nm.

Marked property loss is not always observed when aromatic TPU's are exposed to UV light. This is due to the equal probability of chain scission and radical recombination processes upon oxidation of the aromatic urethanes.

In one study, infrared (IR) analysis on UV exposed ester-based urethanes showed substantial loss of aromatic structures, urethane structures and methylene group content. The samples had yellowed and lost tensile properties upon UV exposure. The polyester component was found to be more photostable than the MDI component.

On the other hand, the polyether soft segment is not as stable to UV exposure as polyester-based segments. Studies showed that TPU's made using polyester macro glycols maintain properties better after UV exposure than those made with polyether glycols. This study also showed that thicker samples are less affected by UV radiation, due to limited surface penetration.

An aliphatic polyurethane containing a polyether soft segment also underwent rapid photodegradation with catastrophic loss of molecular weight after a short xenon arc exposure of only 231 hours. FTIR analyses confirmed that it was the polyether component that degraded. Numerous other studies support the degradation effects on polyether urethanes due to oxidation, and the improved resistance to oxidation of polyester urethanes.

Reference: *UV Stabilization Of Aromatic Pellethane Thermoplastic Polyurethane Elastomers,* supplier technical report (306-00439-1293 SMG) - Dow Chemical Company, 1993.

Effect of UV Stabilizers on Weatherability

Dow Chemical: Pellethane 2103-80 AEF (material compostion: 0.5% hydroxybenzotriazole (UV absorber), 0.5% hindered amine light stabilizer (HALS)); **Pellethane** (chemical type: aromatic isocyanate; features: unstabilized); **Pellethane** (material compostion: 0.25% hydroxybenzotriazole (UV absorber), 0.25% hindered amine light stabilizer (HALS)); **Pellethane** (chemical type: aromatic isocyanate, polyester urethane elastomer); **Pellethane** (chemical type: aromatic isocyanate, polyether urethane elastomer)

Two methods of UV stabilization are commonly used. One is a UV absorber, a benzotriazole; the other is a UV stabilizer, a hindered amine. Hydroxybenzotriazoles preferentially absorb light in the 300-400 nm wavelength range. They dissipate the light energy by a tautomeric process, which protects the polymer by preventing it from absorbing harmful radiation. Hindered amines, on the other hand, are not UV absorbers. Although the function of this additive is not fully understood, it is widely felt that they act as radical scavengers. Hindered amines, through their formation of nitroxyl radicals, terminate and deactivate alkyl radicals and peroxy radicals, which are known to participate in the photo-oxidation process. In functioning as radical scavengers, the stabilizing species (the nitroxyl radical) is regenerated, and continues to scavenge.

Antioxidants are also commonly used to aid in UV stabilization. Although antioxidants are neither a light stabilizer or UV absorber, they often improve the overall weatherability of the TPU when used in combination with a UV absorber or light stabilizer. They do this by interrupting the free-radical process during photo-oxidation.

It is also very common to color the resin black as well as add components for UV stabilization.

Yellowness and change in physical properties for Pellethane 2103-80AEF resins were measured after exposure to QUV light. UV stabilizers (both a benzotriazole and a hindered amine) were added using concentrates. Addition of 1% and 2% of each concentrate yielded UV stabilizer levels of 0.25% and 0.5% of each concentrate. The addition of UV stabilizers does reduce the rate and extent of yellowing in Pellethane 2103-80AEF resins. Despite the color development, Pellethane 2103-80AEF resins maintain physical strength to a considerable extent after QUV exposure. With no stabilizer, there is about a 30% loss of tensile strength and elongation after 2,000 hours of exposure. With stabilizers, the loss is less than 10%. In another QUV study, it was found that neither a UV absorber nor light stabilizer alone was sufficient to retard discoloration and retain good physical properties. However, synergism did occur when the two were blended together and used as a package.

Reference: *UV Stabilization Of Aromatic Pellethane Thermoplastic Polyurethane Elastomers,* supplier technical report (306-00439-1293 SMG) - Dow Chemical Company, 1993.

Ozone Resistance

BASF: Elastollan

Ozone is highly reactive, especially with organic substances. Rubber-based elastomers are destroyed through cracking under the influence of ozone.

Elastollan, on the other hand, is resistant to ozone and does not deteriorate or lose its elastomeric properties when exposed to ozone.

Reference: *Elastollan Design And Processing Guide,* supplier design guide - BASF Corporation, 1993.

Microbiologic Attack

BASF: Elastollan

Polyester-based thermoplastic polyurethanes are subject to degradation through microbiological attack, if exposed for long periods to high levels of moisture and heat. In such climates, micro-organisms cause break down of the ester linkages and destruction of the elastomer. This attack is initially localized, in contrast to hydrolytic degradation, which affects the whole surface.

Reference: *Elastollan Design And Processing Guide,* supplier design guide - BASF Corporation, 1993.

TABLE 153: Accelerated Weathering in a Fadeometer of BF Goodrich Estane Urethane Thermoplastic Elastomer.

Material Family	Urethane Thermoplastic Elastomer							
Material Supplier/ Grade	BF Goodrich Estane 58202 nat 023				BF Goodrich Estane 58300 nat 021			
Features	natural resin, flame retardant				natural resin, extrusion grade			
Reference Number	198	198	198	198	198	198	198	198
MATERIAL CHARACTERISTICS								
Shore A hardness	82	82	82	82	80	80	80	80
EXPOSURE CONDITIONS								
Exposure Type	accelerated weathering				accelerated weathering			
Exposure Apparatus	fadeometer				fadeometer			
Exposure Test Method	ASTM D1499				ASTM D1499			
Exposure Time (days)	0.83	2.5	4.2	8.3	0.83	2.5	4.2	8.3
PROPERTIES RETAINED (%)								
100% Modulus	100	100	88.9	88.9	100	107.1	100	114.3
300% Modulus	109.1	104.5	100	100	97.6	107.3	102.4	97.6
Tensile Strength	92.8	89.2	84.3	77.1	101.1	101.1	84.3	77.5
Ultimate Elongation	101.6	101.6	104.8	100	97.2	95.8	95.8	100

TABLE 154: Accelerated Weathering in a Fadeometer and a QUV of BF Goodrich Estane Urethane Thermoplastic Elastomer.

Material Family	Urethane Thermoplastic Elastomer					
Material Supplier/ Grade	BF Goodrich Estane 58315 nat 025					
Features	natural resin					
Reference Number	198	198	198	198	198	198
MATERIAL CHARACTERISTICS						
Shore A hardness	85	85	85	85	85	85
EXPOSURE CONDITIONS						
Exposure Type	accelerated weathering				accelerated weathering	
Exposure Apparatus	fadeometer				QUV	
Exposure Test Method	ASTM D1499					
Exposure Time (days)	2.5	4.2	8.3	12.5	8.3	20.8
PROPERTIES RETAINED (%)						
100% Modulus	90.9	106.1	105.1	104.5	104	105
300% Modulus	95.4	105.1	99.7	100	98.3	94.9
Tensile Strength	100.9	104.9	99.9	91.7	81.6	71
Ultimate Elongation	101.8	101.8	104.4	100	102.6	101.8

TABLE 155: Accelerated Weathering in a Fadeometer of BF Goodrich Estane Urethane Thermoplastic Elastomer.

Material Family	Urethane Thermoplastic Elastomer						
Material Supplier/ Grade	BF Goodrich Estane 58315			BF Goodrich Estane 58863 nat 025			
Features				natural resin			
Reference Number	198	198	198	198	198	198	198
MATERIAL CHARACTERISTICS							
Sample Thickness (mm)	0.38	0.38	0.38				
Sample Length (mm)	152	152	152				
Sample Width (mm)	12.7	12.7	12.7				
Shore A hardness (units)	85	85	85	85	85	85	85
EXPOSURE CONDITIONS							
Exposure Type	accelerated weathering			accelerated weathering			
Exposure Apparatus	fadeometer			fadeometer			
Exposure Test Method	ASTM D1499			ASTM D1499			
Exposure Time (days)	4.2	8.3	12.5	0.83	2.5	4.2	8.3
PROPERTIES RETAINED (%)							
100% Modulus	107.2 {w}	115.7 {w}	125.3 {w}	105	110	100	105
300% Modulus	102.8 {w}	105.6 {w}	102.1 {w}	103.2	106.5	106.5	106.5
Tensile Strength	92.3 {w}	96.4 {w}	26.2 {w}	109.1	107.3	100	74.5
Ultimate Elongation	103.6 {w}	101.8 {w}	55.4 {w}	203.2	206.4	213	190.4
Photovolt Reflectance	100	100	91.3				

GRAPH 207: QUV Exposure Time vs. Elongation of Urethane Thermoplastic Elastomer

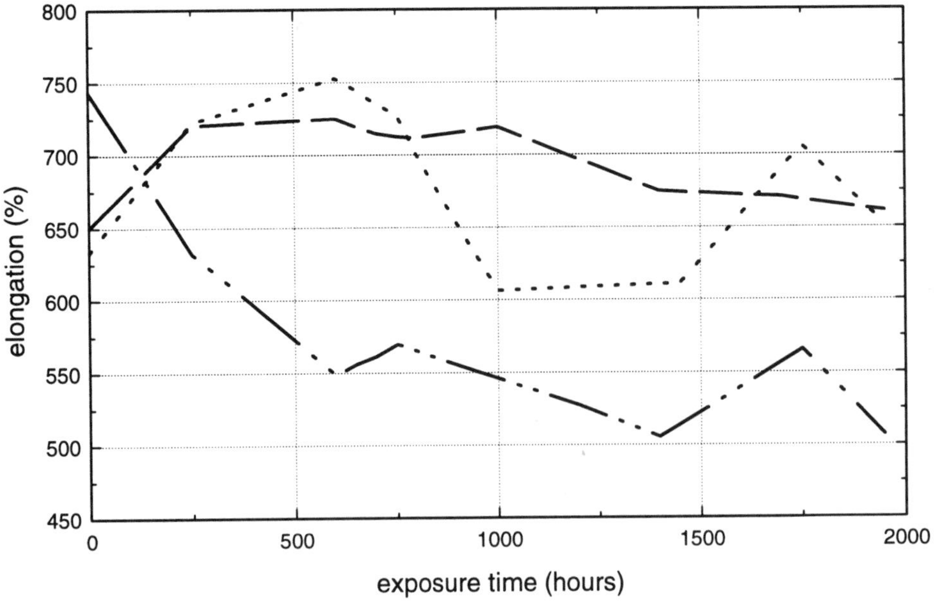

··············	Dow Pellethane 2103-80 AEF TPU (aromatic isocyanate; unstabilized); 16 hours UV @ 70°C, 8 hours condensation @ 70°C
—··—··	Dow Pellethane 2103-80 AEF TPU (0.25% hydroxybenzotriazole, 0.25% HALS); 16 hours UV @ 70°C, 8 hours condensation @ 70°C
— — — —	Dow Pellethane 2103-80 AEF TPU (0.5% hydroxybenzotriazole, 0.5% HALS); 16 hours UV @ 70°C, 8 hours condensation @ 70°C
Reference No.	188

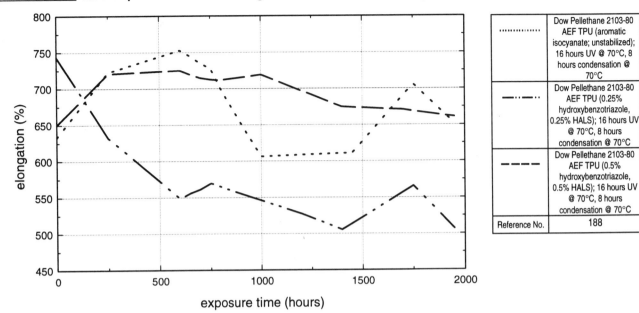

GRAPH 208: QUV Exposure Time vs. Tensile Strength of Urethane Thermoplastic Elastomer

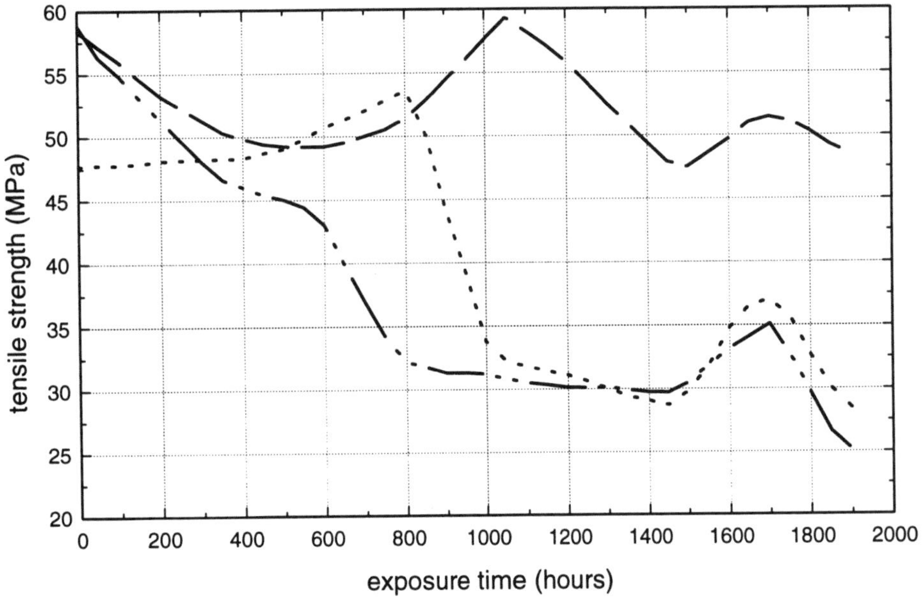

··············	Dow Pellethane 2103-80 AEF TPU (aromatic isocyanate; unstabilized); 16 hours UV @ 70°C, 8 hours condensation @ 70°C
—··—··	Dow Pellethane 2103-80 AEF TPU (0.25% hydroxybenzotriazole, 0.25% HALS); 16 hours UV @ 70°C, 8 hours condensation @ 70°C
— — — —	Dow Pellethane 2103-80 AEF TPU (0.5% hydroxybenzotriazole, 0.5% HALS); 16 hours UV @ 70°C, 8 hours condensation @ 70°C
Reference No.	188

GRAPH 209: QUV Exposure Time vs. Yellowness Index of Urethane Thermoplastic Elastomer

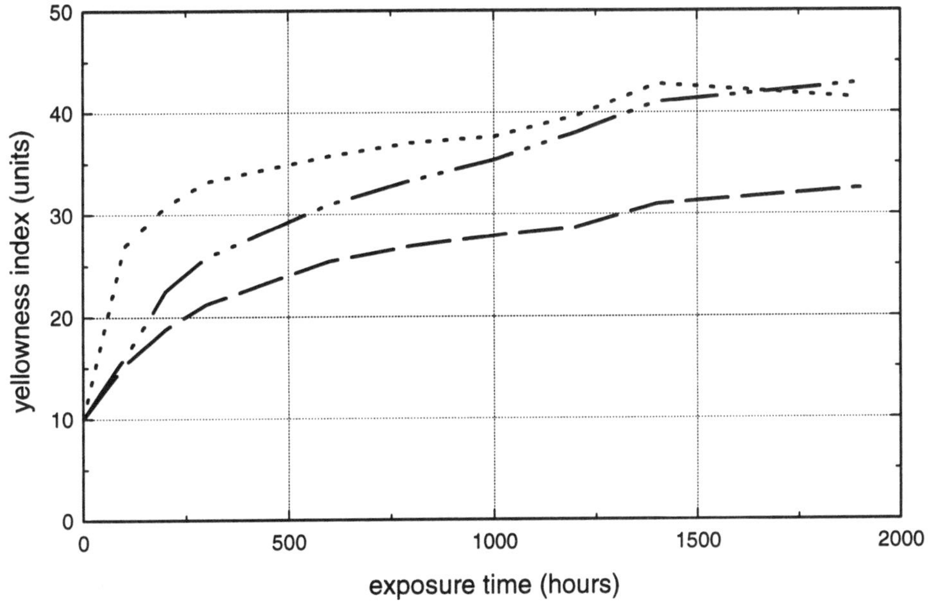

··············	Dow Pellethane 2103-80 AEF TPU (aromatic isocyanate; unstabilized); 16 hours UV @ 70°C, 8 hours condensation @ 70°C
—··—··	Dow Pellethane 2103-80 AEF TPU (0.25% hydroxybenzotriazole, 0.25% HALS); 16 hours UV @ 70°C, 8 hours condensation @ 70°C
— — —	Dow Pellethane 2103-80 AEF TPU (0.5% hydroxybenzotriazole, 0.5% HALS); 16 hours UV @ 70°C, 8 hours condensation @ 70°C
Reference No.	188

GRAPH 210: QUV Exposure Time vs. Yellowness Index of Thermoplastic Polyether Urethane Elastomer

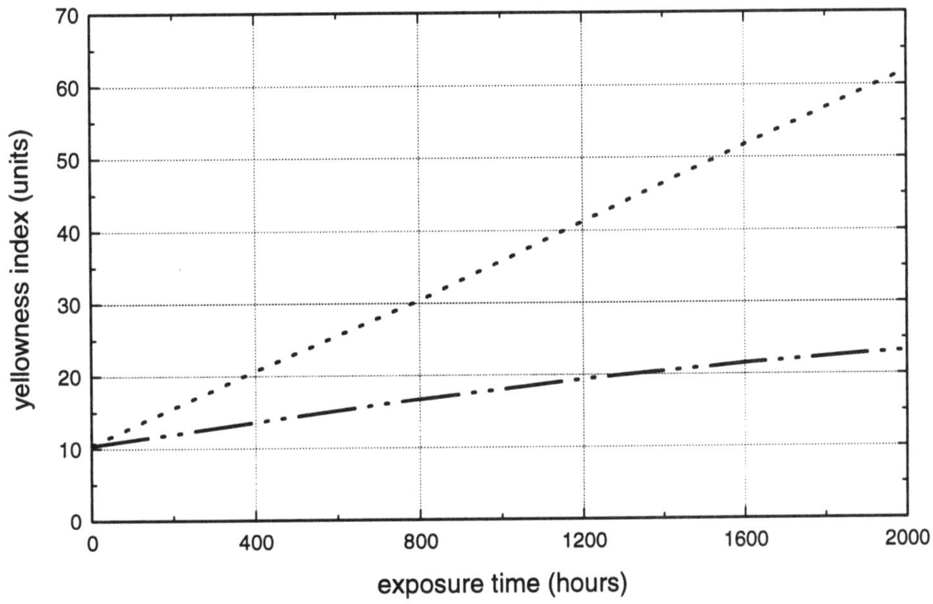

··············	BASF Elastollan 1185A-10 TPEU (black); 3.2 mm extruded sheets; cross-head speed: 50 cm/min
—··—··	BASF Elastollan 1185A-10 TPEU (UV stabilized; 2% UV Conc. 929); 3.2 mm extruded sheets; cross-head speed: 50 cm/min
Reference No.	130

GRAPH 211: Xenon Weatherometer Exposure Time vs. Tensile Strength of Thermoplastic Polyether Urethane Elastomer

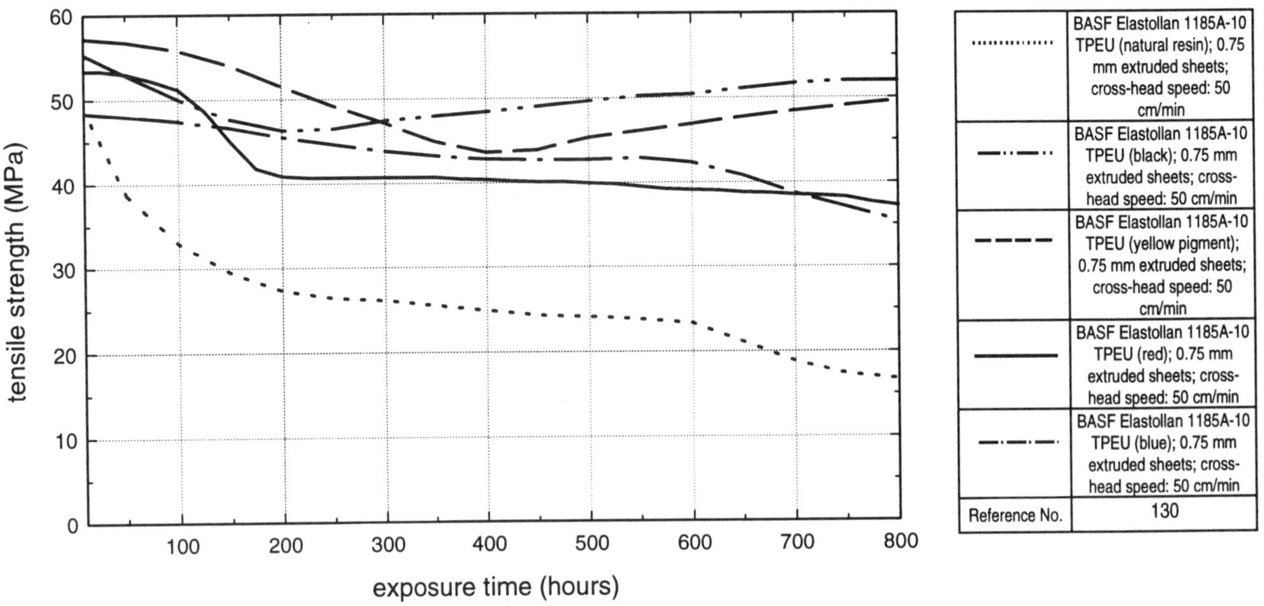

Chapter 67

Polyvinyl Chloride Polyol

TABLE 156: Outdoor Weathering, Accelerated Outdoor Weathering by EMMAQUA and Accelerated Weathering with a Xenon Arc Weatherometer of White Polyvinyl Chloride Polyol.

Material Family	Polyvinyl Chloride Polyol		
Features	white color		
Reference Number	155	155	155
EXPOSURE CONDITIONS			
Exposure Type	outdoor weathering	accelerated outdoor weathering	accelerated weathering
Exposure Location	Arizona	Arizona	
Exposure Country	USA	USA	
Exposure Apparatus		EMMAQUA	xenon arc weatherometer
Exposure Time (days)	365	365	214
Exposure Note	water spray added		
PROPERTIES RETAINED (%)			
Tensile Strength	98.0	72.4	69.1
Elongation	90.2	57.9	45.5
SURFACE AND APPEARANCE			
ΔE Color Change	5.9	17.4	12.5

TABLE 157: Outdoor Weathering in Arizona and Florida of Flexible, White Polyvinyl Chloride Polyol.

Material Family	Polyvinyl Chloride Polyol				
Features	white color				
Reference Number	155	155	155	155	155
EXPOSURE CONDITIONS					
Exposure Type	outdoor weathering				
Exposure Location	Arizona	Arizona	Arizona	Arizona	Florida
Exposure Country	USA	USA	USA	USA	USA
Exposure Time (days)	183	365	730	1461	730
PROPERTIES RETAINED (%)					
Tensile Strength					99 {ac}
Elongation					77 {ac}
CHANGE IN PHYSICAL CHARACTERISTICS					
Hardness Change (units)					2 {ae}
SURFACE AND APPEARANCE					
ΔE Color Change	5.12 (samples washed)	5.92 (samples washed)	20.11 (samples washed)	15.67 (samples washed)	

TABLE 158: Accelerated Weathering in an Atlas Weatherometer of Geon Company Geon Polyvinyl Chloride Polyol.

Material Family	Polyvinyl Chloride Polyol							
Material Supplier/ Grade	Geon Company Geon 83741		Geon Company Geon 83718		Geon Company Geon 83794		Geon Company Geon 83332	
Features	profile extrusion grade, weatherable		profile extrusion grade, weatherable		profile extrusion grade, weatherable		low temperature grade	
Reference Number	193	193	193	193	193	193	193	193
EXPOSURE CONDITIONS								
Exposure Type	accelerated weathering							
Exposure Test Method	ASTM E42-57 type E equipment							
Exposure Apparatus	Atlas weatherometer							
Exposure Time (days)	41.7	83.3	41.7	83.3	41.7	83.3	41.7	83.3
PROPERTY VALUES AFTER EXPOSURE								
100% Modulus (MPa)	5.0	5.0	6.9	6.9	12.4	11.0	4.4	5.4
Tensile Strength (MPa)	6.9	6.2	8.9	8.3	13.8	12.4	8.3	8.9
Elongation (%)	230	200	235	210	180	200	300	280
Shore A Hardness (units)	70	72	80	80	90	88	71	73
PROPERTIES RETAINED (%)								
Tensile Strength	65	58	72	67	91	82	86	93
Elongation	74	65	78	70	60	67	89	78

Chapter 68

Nitrile Thermoplastic Elastomer

Outdoor Weather Resistance

Goodyear: Chemigum TPE 2175 (features: black color, 75 Shore A hardness, 3.2 mm thick; manufacturing method: compression molding)

Chemigum TPE displayed excellent resistance to outdoor weathering. After two years' exposure in Florida, both direct and under glass, Chemigum did not exhibit severe chalking, exudation, cracks or unsightly blemishes on the surface.

Reference: *Florida Weathering Of Chemigum TPE,* supplier technical report (TPE 06-0292/498900-2/92) - Goodyear Chemicals, 1992.

TABLE 159: Outdoor Weathering in Florida of Goodyear Chemigum Nitrile Thermoplastic Elastomer.

Material Family	Nitrile Thermoplastic Elastomer	
Material Supplier/ Grade	Goodyear Chemigum TPE 2175	
Features	black color, medium gloss	
Manufacturing Method	compression molding	
Reference Number	175	175

MATERIAL CHARACTERISTICS

Shore A hardness	75	75
Sample Thickness (mm)	3.2	3.2

EXPOSURE CONDITIONS

Exposure Type	outdoor weathering	
Exposure Location	Florida	Florida
Exposure Country	USA	USA
Exposure Note	direct exposure; 5° south	under glass exposure; 5° south
Exposure Time (days)	730	730

PROPERTIES RETAINED (%)

100% Modulus	97	97
Tensile Strength	81	78
Elongation @ Break	84	80

CHANGE IN PHYSICAL CHARACTERISTICS

Shore A Hardness Change (units)	0	0

SURFACE AND APPEARANCE

Gloss @ 60° Retained (%)	12.6 {g}	30.9 {g}
ΔE Color	2.28 {h}	1.46 {h}
Visual Observation (as weathered)	slight chalking	no blemish
Visual Observation (washed)	no chalking, no cracks	clean, no cracks

Chapter 69

Butyl Rubber and Bromoisobutylene Isoprene Rubber

Weather Resistance

Exxon: Bromoisobutylene Isoprene Rubber (BIIR)

Carbon black filled bromobutyl compounds have high resistance to weathering. Light colored compounds should be compounded to minimize degradation by ultraviolet light. The following techniques are suggested for maximum weather resistance in mineral-filler compounds.

1) obtain a high state of cure.

2) use low levels of a high quality paraffinic plasticizer.

3) include ultraviolet light absorbers such as titanium dioxide.

4) depending upon the application, use up to 10 phr of paraffin wax to protect the surface.

5) avoid clays, to the extent possible, especially hard clays. Calcium carbonates, talcs, and silicas generally perform better.

Reference: *Bromobutyl Rubber Optimizing Key Properties,* supplier marketing literature - Exxon Chemicals.

Exxon: BROMO XP-50

Isobutylene backbone polymers are generally susceptible to attack by ultra-violet radiation. In light-colored compounds, this results in chain scission and the development of surface tack with high dirt retention. (Carbon black-filled isobutylene polymer compounds have good resistance to UV attack and weathering in general due to the masking and UV absorption effects of the black.) Experience with light-colored Bromo XP-50 stocks shows that the new polymer is less sensitive to UV attack than are other isobutylene polymers by virtue of the attached pendant aromatic rings. In fact, crosslinking reactions begin to predominate over chain scission above about 10% p-methylstyrene content.

Depending on the grade of Bromo XP-50 used and the characteristics of the other compound ingredients, it may still be desirable to compound light colored stocks to minimize potential effects of attack by ultraviolet light. The following compounding techniques are suggested:

1) Obtain a high state of cure.

2) Use a low content of a high quality paraffinic plasticizer

3) Include ultraviolet light absorbers such as titanium dioxide.

4) Use 25 phr or more of masking fillers such as talc.

5) To the extent possible, avoid clays, especially hard clays. Calcium carbonates and silicas generally perform better.

Reference: *BROMO XP-50 Optimizing Key Properties,* supplier marketing literature - Exxon Chemicals.

Exxon: Butyl Rubber

For weather resistance, as in rubber sheeting for roofs and water management application, the least unsaturated butyl is advantageously used.

Reference: Fusco, James V., Hous, Pierre, *Butyl And Halobutyl Rubbers,* reference book - Exxon Chemicals, 1987.

Ozone Resistance

Exxon: Bromoisobutylene Isoprene Rubber (BIIR)

Bromobutyl shows good resistance to attack by ozone due to its high saturation. It is superior to general purpose rubbers. High loadings of fillers and plasticizers are detrimental to ozone resistance, especially high aromatic oils even at moderate levels. Amine, resin, thiourea and alkylphenol disulfide cures of bromobutyl yield vulcanizates with high ozone resistance.

Reference: *Bromobutyl Rubber Optimizing Key Properties,* supplier marketing literature - Exxon Chemicals.

Exxon: BROMO XP-50

Because of the absence of polymer chain unsaturation, compounds containing only Bromo XP-50 polymers are inherently resistant to ozone attack. The ozone resistance of blend compounds will be a function of the amount of Bromo XP-50 in the stock. The Bromo XP-50 should be used at a minimum of 35 or 40% of the total polymer. Filler and plasticizer levels will be important in blend stocks, as will be the ability of the cure system to achieve covulcanization of the different polymer phases.

Reference: *BROMO XP-50 Optimizing Key Properties,* supplier marketing literature - Exxon Chemicals.

Exxon: Butyl Rubber

The low level of chemical unsaturation in the polymer chain produces an elastomer with greatly improved resistance to ozone when compared to polydiene rubbers. Butyl with the lowest level of unsaturation (Exxon Butyl 065) produces high levels of ozone resistance, which are also influenced by the type and concentration of vulcanizate crosslinks. For maximum ozone resistance, as in electrical insulation, the least unsaturated butyl is advantageously used.

Reference: Fusco, James V., Hous, Pierre, *Butyl And Halobutyl Rubbers,* reference book - Exxon Chemicals, 1987.

Chapter 70

Chlorosulfonated Polyethylene Rubber

Outdoor Weather Resistance

DuPont: Hypalon 20; Hypalon 30; Hypalon 40; Hypalon 45 (features: highly crystalline); **Hypalon 48**

Hypalon 40 is the most weather resistant type, and although the differences are small, they are enough to suggest that Hypalon 40 be used for most products.

Ten year test results on vulcanizates of Hypalon 48 do indicate that it sheds dirt better than Hypalon 40.

Hypalon 20 dissolves in solvent to give much lower solution viscosities (or higher solids content at equivalent viscosity). It is, therefore, suitable for the preparation of weather resistant flexible films from solution e.g., spread coating. Hypalon 30 also offers low solution viscosities but is a much stiffer material suitable for solution coatings over rigid substrates.

Hypalon 45, a more crystalling type, has sufficient strength to be used in the uncured state. Uncured compounds of Hypalon 45 actually may benefit from exposure to the weather by reason of gradual cross-linking promoted by UV exposure and moisture. Long-term exposure data demonstrates that, when properly compounded, Hypalon 45 has excellent weather resistance.

Color Pigments Type and amount of pigments are critical because Hypalon requires protection from ultraviolet light. Unpigmented compounds of Hypalon darken and craze after 6-12 months of direct exposure to sunlight.

Curing Systems In non-black stocks, magnesia/pentaerythritol is preferred except when maximum water resistance is required. In such a case, tribasic lead maleate is satisfactory. If flexibility is important, no more than five parts of magnesia should be used. Vulcanizates containing high amounts of magnesia stiffen on exposure to weather.

In black stocks, a litharge curing system is the most satisfactory.

Fillers Calcium carbonate is preferred as it chalks less than other fillers.

Plasticizers Aromatic and naphthenic oils discolor light colored stocks; esters and chlorinated hydrocarbons do not. Fungus nutrients such as safflower oil or other natural products should be avoided

Reference: Dupuis, I. C., Cumberland, D. W., *Compounding 'Hypalon' For Weather Resistance,* supplier technical report (E-23070-1 HP-515.1) - Du Pont Company, 1987.

Effect of Color Pigments on Weatherability

DuPont: Hypalon 20; Hypalon 30; Hypalon 40; Hypalon 45 (features: highly crystalline); **Hypalon 48**

The type and level of color pigment are the most important factors in determining the weather resistance of compounds of Hypalon. It is desirable to select only those color pigments which show a high degree of opacity to ultraviolet radiation. This restricts the penetration of the ultraviolet radiation and subsequent deterioration to the outermost surface where it is observed mainly as a loss of gloss rather than crazing. In addition, a pigment which protects the surface of a compound of Hypalon from deterioration during outdoor exposure is of little value unless it also maintains its color.

The retention of color by vulcanizates of Hypalon on exposure to weather varies considerably with the pigment used since all color pigments do not afford the same degree of color stability. It is important to remember that pigments differ in their color stability and opacity. Yellows and oranges, for example, are lower in tinctorial strength and durability than blues and greens. These differences are usually accentuated in pastels where orange and yellow fade much more rapidly than blue or green when used at the same ratio of color to titanium dioxide. (Blends of color pigment and titanium dioxide are frequently used for purposes of economy and/or color match.) Although adequate in mass tones, orange and yellow should not be used in very light tints where color stability is important.

All pigments do not provide equal protection. They can generally be classified based on the weight of pigment, in parts per hundred (phr) relative to the resin used, required to achieve a reasonable level of lightfastness and ultraviolet absorption. The study included pigments specifically recommended for use in compounds of Hypalon. The recommended pigments are divided into four groups ranging from the most to the least efficient on a weight basis. The most efficient are those where the minimum suggested amount is three parts per hundred of Hypalon. Compounds containing three parts have not crazed after ten years exposure to direct sunlight in Delaware. The next group - mass tones of orange and yellow - is somewhat less efficient, requiring 6 phr. They have been exposed to direct Delaware sunlight for fifteen years without objectionable color change or crazing, but pastels of these colors are much less permanent. The third group consists of red iron oxide. Iron oxide is very stable but is less brilliant than organic pigments. Because of its low tinctorial strength the minimum suggested amount is 10 parts. The last group contains titanium oxide, which is the only pigment satisfactory for use in Hypalon. The most chalk-resistant rutile grades should be used unless a self-cleaning surface is desired; the suggested minimum is 35 parts.

Color Pigments Recommended for Use in Hypalon

Color	Type	Number	Manufacturer	Remarks	Suggested Alternative
Group I - Minimum Recommended Amount - 3 phr					
Blue	Phthalocyanine	BT-284-D	DuPont		
			DuPont		BT-391-D
			DuPont		BT-449-D
			DuPont		BT-383-D
Green	Phthalocyanine	GT-674-D	DuPont	Blue shade	
			DuPont	Yellow shade	
			DuPont	Yellow shade - offered as a replacement for GT-751-D - no long term exposure experience	
Black	Carbon Black	SRF	Various		
Group II - Minimum Recommended Amount - 6 phr					
Red	Quinacridone	RT-742-D	DuPont	Blue shade	
			DuPont		
Violet	Quinacridone	RT-899-D	DuPont	No long term exposure experience	RT-790-D
Orange	Molybdate	YE-637-D	DuPont	Discolors in the presence of sulfur	
			DuPont	Heat stabilized pigment	KO-786-D
			DuPont	Heat stabilized pigment	KO-789-D
Yellow	Chrome	Y-469-D	DuPont	Discolors in the presence of sulfur	
			DuPont	Bleeds, particularly in tints	YT-808-D
Group III - Minimum Recommended Amount - 10 phr					
Red	Iron Oxide	RO-3097	Pfizer	Relatively dull	
Group IV - Minimum Recommended Amount - 35 phr					
White	Titanium Dioxide	R-960	DuPont	Extremely low chalk rate	
		OR-450	American Cyanamid	Moderate chalk rate - superior for dead white compound - self-cleaning after 10 years	
		R-902	DuPont	Moderate chalk rate - self-cleaning after 5 years	

Reference: Dupuis, I. C., Cumberland, D. W., *Compounding 'Hypalon' For Weather Resistance,* supplier technical report (E-23070-1 HP-515.1) - Du Pont Company, 1987.

Effect of Curing Systems on Weatherability

DuPont: Hypalon 20; Hypalon 30; Hypalon 40; Hypalon 45 (features: highly crystalline); **Hypalon 48**

The choice of curing system depends primarily upon the requirements for color and water resistance. There are four curing systems from which to choose: magnesia, magnesia/pentaerythritol, litharge, and tribasic lead maleate. Compounds designed to be used uncured can be formulated using Hypalon 45 as the base polymer.

Non Black (Colored) Compounds If maximum water resistance is not required, the compound may be cured with magnesia alone or with magnesia/pentaerythritol combination. Compounds cured with either system have excellent colorability and they retain their color and smooth surface during outdoor exposure. Of the two systems, the magnesia/pentaerythritol is generally the more useful. Compounds cured with 20 parts of magnesia have better abrasion resistance and retain color slightly better, but they are sometimes scorchy and they stiffen noticeably upon exposure to weather. (Note: If flexibility is important, magnesia content should not exceed five parts.) Compounds cured with magnesia/pentaerythritol generally exhibit greater processing safety and retain their flexibility on exposure to weather. Their color retention, if not quite the equal of an all magnesia-cured compound is nevertheless excellent.

If maximum water resistance is required in a light-colored compound, the most suitable curing agents are basic lead complexes such as tribasic lead maleate. The lead complex releases lead oxide slowly and thus allows the production of brightly colored products. It is essential, though, to select accelerators that release a minimum of free sulfur; mercaptobenzothiazole (MBT) is very effective when activated with hydrogenated wood rosin. Compounds cured with these systems have been exposed for ten years in Florida. They show good retention of surface smoothness, color and mechanical properties. In the "fadeometer," lead complex-cured compounds may discolor but only very slight discoloration has been observed in actual outdoor exposure. Although there is little apparent difference between 25 and 40 parts of lead complex, the results of other tests indicate, that for satisfactory resistance to crazing, a minimum of 40 parts of lead complex should be used.

Litharge is the recommended curing agent. It is necessary if low water swell is required and is preferred even if water resistance is not required. Litharge-cured compounds are safe processing and have good physical properties. In fifteen-year exposure tests they show trace crazing and still retain flexibility.

The magnesia/pentaerythritol combination may be used if a black, lead-free compound is required without maximum water resistance. Compounds cured with magnesia/pentaerythritol are resistant to crazing and they retain their strength and flexibility when aged outdoors.

Hypalon compounds retain much of their elongation during exposure to weather. Note also that the rate of elongation loss decreases with time - most loss occurs in the lightly loaded compound. Compounds containing higher levels of black are more stable.

Reference: Dupuis, I. C., Cumberland, D. W., *Compounding 'Hypalon' For Weather Resistance,* supplier technical report (E-23070-1 HP-515.1) - Du Pont Company, 1987.

Effect of Fillers on Weatherability

DuPont: Hypalon 20; Hypalon 30; Hypalon 40; Hypalon 45 (features: highly crystalline); **Hypalon 48**

Several types of filler are suitable for use in Hypalon. Whiting is preferred from the viewpoint of weatherability alone; however, it is necessary in many cases to use other extenders for specific properties (e.g., water resistance) not obtainable with whiting. Below is a summary of the performance of several commonly used fillers:

Whiting (calcium carbonate) Vulcanizates containing 50 parts show trace crazing and moderate chalking after ten years of direct sun exposure in Delaware. White vulcanizates containing 200 parts have been exposed for 18 months in Florida without surface deterioration. After fifteen years, masstone green vulcanizates containing 150 parts whiting have begun to chalk slightly and trace crazing is visible. Fine particle size precipitated calcium carbonate is entirely suitable for use in Hypalon and often gives better physical properties. Since Hypalon is not entirely dependent on pigment reinforcement, economic factors usually favor the use of ground whiting.

Clay (aluminum silicate) Yellow vulcanizates containing 50 parts of clay have shown excellent weather resistance.

Talc (magnesium silicate) Fading and chalking in crease with increasing amounts in colored compounds. Approximately 50 parts would be maximum loading in white or light colors and 75 parts in black compounds.

Silica (silicon dioxide) Silica is not suggested as a primary filler as silica-filled vulcanizates have shown visible crazing after only one year in Florida.

Reference: Dupuis, I. C., Cumberland, D. W., *Compounding 'Hypalon' For Weather Resistance,* supplier technical report (E-23070-1 HP-515.1) - Du Pont Company, 1987.

Effect of Plasticizers on Weatherability

DuPont: Hypalon 20; Hypalon 30; Hypalon 40; Hypalon 45 (features: highly crystalline); Hypalon 48

Plasticizers have little influence on the craze resistance of vulcanizates of Hypalon when used in moderate amounts (5 to 25 parts). No data are available on the durability of more highly plasticized formulations.

Plasticizers do have an influence on color stability. Aromatic and naphthenic oils discolor and should be avoided in light colored stocks. Esters and chlorinated hydrocarbons show excellent color stability. Paraffinic oils are also very stable, but their low order of compatibility with Hypalon limits their usefulness.

Hypalon itself does not nourish mold or fungus but some plasticizers do and they must be avoided in order to prevent fungus growth during exposure to weather. Due to the large number of plasticizers on the market, it is only possible to generalize on their fungus resistance. It has been found that phthalate and phosphate esters, chlorinated hydrocarbons, and some paraffin oils are satisfactory in this respect. Vegetable-based oils and other known fungus nutrients should be avoided.

Reference: Dupuis, I. C., Cumberland, D. W., *Compounding 'Hypalon' For Weather Resistance,* supplier technical report (E-23070-1 HP-515.1) - Du Pont Company, 1987.

TABLE 160: Outdoor Weathering, Accelerated Outdoor Weathering, and Accelerated Weathering of DuPont Hypalon Chlorosulfonated Polyethylene Rubber.

Material Family	Chlorosulfonated Polyethylene Rubber		
Material Supplier/ Grade	DuPont Hypalon	DuPont Hypalon	DuPont Hypalon
Features	black color	black color	black color
Reference Number	157	157	157
EXPOSURE CONDITIONS			
Exposure Type	outdoor weathering	accelerated outdoor weathering	accelerated weathering
Exposure Location	Arizona	Arizona	
Exposure Country	USA	USA	
Exposure Apparatus		EMMAQUA	xenon arc weatherometer
Exposure Time (days)	365	365	214
Exposure Note	water spray added		
PROPERTIES RETAINED (%)			
Tensile Strength	87.6	80.5	86.8
Elongation	83.7	47.9	49.5
SURFACE AND APPEARANCE			
ΔE Color Change	7.8	9.1	9.8

TABLE 161: Outdoor Weathering in Arizona of DuPont Hypalon Chlorosulfonated Polyethylene Rubber.

Material Family	Chlorosulfonated Polyethylene Rubber			
Material Supplier/ Grade	DuPont Hypalon 40			
Features	black color			
Reference Number	119	119	119	119

MATERIAL CHARACTERISTICS

Sample Thickness (mm)	0.76	0.76	0.76	0.76
Shore A hardness	76	76	76	76

MATERIAL COMPOSITION

Hypalon 40	100 phr	100 phr	100 phr	100 phr
polyethylene	3 phr	3 phr	3 phr	3 phr
DPTTS	2 phr	2 phr	2 phr	2 phr
litharge (90% paste)	25 phr	25 phr	25 phr	25 phr
Thiofide (Monsanto)	0.5 phr	0.5 phr	0.5 phr	0.5 phr
aromatic process oil	40 phr	40 phr	40 phr	40 phr
N-774 black	90 phr	90 phr	90 phr	90 phr

EXPOSURE CONDITIONS

Exposure Type	outdoor weathering			
Exposure Location	Phoenix, Arizona	Phoenix, Arizona	Phoenix, Arizona	Phoenix, Arizona
Exposure Country	USA	USA	USA	USA
Exposure Time (days)	182	365	182	365
Exposure Note	5° south latitude	5° south latitude	5° south latitude; water spray added	5° south latitude; water spray added
Test Lab	DSET Laboratories	DSET Laboratories	DSET Laboratories	DSET Laboratories

PROPERTIES RETAINED (%)

100% Modulus	150	104	138	121
Tensile Strength	113	85	108	87
Elongation @ Break	87	90	87	83

SURFACE AND APPEARANCE

ΔE Color Change	10.6 {b}	10 {b}	6.8 {b}	8 {b}

TABLE 162: Outdoor Weathering in Florida and Delaware of Wire Cable Compound DuPont Hypalon Chlorosulfonated Polyethylene Rubber.

Material Family	Chlorosulfonated Polyethylene Rubber		
Material Supplier/ Grade	DuPont Hypalon 40		
Product Form	1.2 mm insulation on No. 12 AWG aluminum		
Cure	60 seconds in 1.6 MPa steam		
Curing System	litharge	litharge	magnesia
Features	wire and cable compound, black color	wire and cable compound, black color	wire and cable compound, non-black color
Reference Number	139	139	139

MATERIAL COMPOSITION

Hypalon 40	100 phr	100 phr	100 phr
Chlorowax 40			25 phr
Heliozone	2 phr	2 phr	
Kenflex A	10 phr	10 phr	
litharge	25 phr	25 phr	
magnesia			4 phr
MBTS (2-mercaptobenzothiazole disulfide)	1 phr	1 phr	1 phr
NBC	3 phr	3 phr	
petrolatum	3 phr	3 phr	3 phr
stearic acid			2 phr
Tetrone A	2 phr	2 phr	1.5 phr
paraffin			3 phr
aromatic process oil	17 phr	17 phr	
TiO_2			35 phr
whiting			25 phr
FEF carbon black	15 phr	15 phr	
hard clay	60 phr	60 phr	100 phr

EXPOSURE CONDITIONS

Exposure Type	outdoor weathering		
Exposure Location	Delaware	Florida	Delaware
Exposure Country	USA	USA	USA
Exposure Time (days)	7304	7304	2191

PROPERTIES RETAINED (%)

200% Modulus	138	144	
Tensile Strength	124	133	110
Elongation @ Break	83	75	82

TABLE 163: Outdoor Weathering in Florida, Texas and California of Green Hose Cover Compound DuPont Hypalon Chlorosulfonated Polyethylene Rubber.

Material Family	Chlorosulfonated Polyethylene Rubber					
Material Supplier/ Grade	DuPont Hypalon 40					
Features	green color, hose cover compound					
Sample Designation	sample A	sample B	sample A	sample B	sample A	sample B
Reference Number	139	139	139	139	139	139
MATERIAL COMPOSITION						
Hypalon 40	100 phr	100 phr	100 phr	100 phr	100 phr	100 phr
A-C polyethylene 617A	3 phr	3 phr	3 phr	3 phr	3 phr	3 phr
cis-4-polybutadiene 1203	3 phr	3 phr	3 phr	3 phr	3 phr	3 phr
diocytl phthalate	20 phr	20 phr	20 phr	20 phr	20 phr	20 phr
magnesia	5 phr	5 phr	5 phr	5 phr	5 phr	5 phr
Pentaerythritol 200	3 phr	3 phr	3 phr	3 phr	3 phr	3 phr
TMTD (tetramethylthiuram disulfide)	2 phr	2 phr	2 phr	2 phr	2 phr	2 phr
paraffin	3 phr	3 phr	3 phr	3 phr	3 phr	3 phr
whiting	100 phr	100 phr	100 phr	100 phr	100 phr	100 phr
phthalocyanine green	3 phr	3 phr	3 phr	3 phr	3 phr	3 phr
sulfur	1 phr	1 phr	1 phr	1 phr	1 phr	1 phr
EXPOSURE CONDITIONS						
Exposure Type	outdoor weathering					
Exposure Location	Florida		Texas		California	
Exposure Note	tested after actual use		tested after actual use		tested after actual use	
Exposure Time (days)	1825		2555		2190	
SURFACE AND APPEARANCE						
Chalking	good (tight chalk) to very good	very good to like original	like original	like original	like original	like original
Crazing	very poor to poor	fair to good	poor	good	good	good
RESULTS OF EXPOSURE						
Mildew Resistance	good	very good	good	like original	like original	like original

TABLE 164: Outdoor Weathering in Florida of White DuPont Hypalon Chlorosulfonated Polyethylene Rubber.

Material Family	Chlorosulfonated Polyethylene Rubber				
Material Supplier/ Grade	DuPont Hypalon 40				
Features	white color				
Cure	30 minutes at 153°C				
Curing System	magnesia	magnesia / pentaerythriol	magnesia / pentaerythriol	tribasic lead maleate	tribasic lead maleate
Reference Number	139	139	139	139	139
MATERIAL COMPOSITION					
Hypalon 40	100 phr	100 phr	100 phr	100 phr	100 phr
Maglite D	20 phr	2 phr	5 phr		
MBT				2 phr	2 phr
Pentaerythritol 200		3 phr	3 phr		
Staybelite				2.5 phr	2.5 phr
Tetrone A	2 phr	2 phr	2 phr		
Tri-Mal				25 phr	40 phr
Zalba Special				2 phr	
Atomite	50 phr	50 phr	50 phr	50 phr	50 phr
TI-Pure R-960	50 phr	50 phr	50 phr	50 phr	50 phr
EXPOSURE CONDITIONS					
Exposure Type	outdoor weathering				
Exposure Location	Florida				
Exposure Country	USA				
Exposure Time (days)	3652	3652	3652	3652	3652
PROPERTIES RETAINED (%)					
100% Modulus	126.7	71.4	59.5	141.9	127
Tensile Strength	73.8	64.6	76.7	86.5	76.1
Elongation @ Break	51	71	84	49	45
CHANGE IN PHYSICAL CHARACTERISTICS					
Shore A Hardness Change (units)	12	4	1	4	-2
SURFACE AND APPEARANCE					
Chalking	moderate	moderate	moderate	considerable	considerable
Crazing	trace, under 20X magnification	trace, under 20X magnification	trace, under 20X magnification	trace, under 20X magnification	trace, under 20X magnification
Gloss	lost	lost	lost	lost	lost

TABLE 165: Outdoor Weathering in Delaware of DuPont Hypalon Chlorosulfonated Polyethylene Rubber.

Material Family	Chlorosulfonated Polyethylene Rubber	
Material Supplier/ Grade	DuPont Hypalon 20	
Features	black color	
Cure	30 minutes at 153°C	
Curing System	litharge	
Reference Number	139	139

MATERIAL COMPOSITION

Hypalon 20	100 phr	100 phr
sublimed litharge	20 phr	20 phr
MBTS (2-mercaptobenzothiazole disulfide)	0.5 phr	0.5 phr
Staybelite	2.5 phr	2.5 phr
Sundex 53 (replaced by Sundex 790)		35 phr
Tetrone A	0.75 phr	0.75 phr
SRF carbon black	10 phr	75 phr

EXPOSURE CONDITIONS

Exposure Type	outdoor weathering	
Exposure Location	Delaware	Delaware
Exposure Country	USA	USA
Exposure Time (days)	5479	5479

PROPERTIES RETAINED (%)

100% Modulus	120	155.9
Tensile Strength	66.7	129.7
Elongation @ Break	51	75

CHANGE IN PHYSICAL CHARACTERISTICS

Shore A Hardness Change (units)	4	8

SURFACE AND APPEARANCE

Chalking	slight	slight
Crazing	trace under 20X magnification	trace under 20X magnification
Gloss	lost	lost

TABLE 166: Outdoor Weathering in Panama of Pond Liner Formulation DuPont Hypalon Chlorosulfonated Polyethylene Rubber.

Material Family	Chlorosulfonated Polyethylene Rubber	
Material Supplier/ Grade	DuPont Hypalon 45	
Features	pond liner formulation, white color, uncured	pond liner formulation, black color, uncured
Reference Number	139	139

MATERIAL COMPOSITION

Hypalon 45	100 phr	100 phr
Black Pearls #2	0.015 phr	
Carbowax 4000	0.5 phr	0.5 phr
Ionol	2 phr	2 phr
Kemstrene amide SU	0.6 phr	0.6 phr
Maglite D	4 phr	4 phr
Atomite whiting	50 phr	
Ti-Pure R-610 (replaced by Ti-Pure R-960)	50 phr	
MT carbon black		100 phr

EXPOSURE CONDITIONS

Exposure Type	outdoor weathering	
Exposure Country	Panama	Panama
Exposure Note	45° south	45° south
Exposure Time (days)	3652	3652

PROPERTIES RETAINED (%)

100% Modulus	187.5	248.5
300% Modulus	147.1	
Tensile Strength	89.8	172.9
Elongation @ Break	45.8	38.9

SURFACE AND APPEARANCE

Chalking	excellent rating	por rating
Crazing	small checks at 10X magnification	no crazing

TABLE 167: Accelerated Outdoor Weathering by EMMA and EMMAQUA of DuPont Hypalon Chlorosulfonated Polyethylene Rubber.

Material Family	Chlorosulfonated Polyethylene Rubber			
Material Supplier/ Grade	DuPont Hypalon 40			
Features	black color	black color	black color	black color
Reference Number	119	119	119	119
MATERIAL CHARACTERISTICS				
Sample Thickness (mm)	0.76	0.76	0.76	0.76
Shore A hardness	76	76	76	76
MATERIAL COMPOSITION				
Hypalon 40	100 phr	100 phr	100 phr	100 phr
polyethylene	3 phr	3 phr	3 phr	3 phr
DPTTS	2 phr	2 phr	2 phr	2 phr
litharge (90% paste)	25 phr	25 phr	25 phr	25 phr
Thiofide (Monsanto)	0.5 phr	0.5 phr	0.5 phr	0.5 phr
aromatic process oil	40 phr	40 phr	40 phr	40 phr
N-774 black	90 phr	90 phr	90 phr	90 phr
EXPOSURE CONDITIONS				
Exposure Type	accelerated outdoor weathering			
Exposure Location	Phoenix, Arizona	Phoenix, Arizona	Phoenix, Arizona	Phoenix, Arizona
Exposure Country	USA	USA	USA	USA
Exposure Apparatus	EMMA	EMMA	EMMAQUA	EMMAQUA
Exposure Time (days)	182	365	182	365
Exposure Note	correlates to approximately 2.5 years actual Florida aging	correlates to approximately 5 years actual Florida aging	correlates to approximately 2.5 years actual Florida aging	correlates to approximately 5 years actual Florida aging
PROPERTIES RETAINED (%)				
100% Modulus	188	183		220
Tensile Strength	101	99	108	82
Elongation @ Break	61	59	62	47
SURFACE AND APPEARANCE				
ΔE Color Change	8.4 {b}	8.4 {b}	8.5 {b}	9.1 {b}

TABLE 168: Accelerated Weathering in a Xenon Arc Weatherometer of DuPont Hypalon Chlorosulfonated Polyethylene Rubber.

Material Family	Chlorosulfonated Polyethylene Rubber	
Material Supplier/ Grade	DuPont Hypalon 40	
Features	black color	
Reference Number	119	119

MATERIAL CHARACTERISTICS

Sample Thickness (mm)	0.76	0.76
Shore A hardness	76	76

MATERIAL COMPOSITION

Hypalon 40	100 phr	100 phr
polyethylene	3 phr	3 phr
DPTTS	2 phr	2 phr
litharge (90% paste)	25 phr	25 phr
Thiofide (Monsanto)	0.5 phr	0.5 phr
aromatic process oil	40 phr	40 phr
N-774 black	90 phr	90 phr

EXPOSURE CONDITIONS

Exposure Type	accelerated weathering	
Exposure Test Method	GM, TM30-2	GM, TM30-2
Exposure Apparatus	xenon arc weatherometer	xenon arc weatherometer
Exposure Time (days)	125	208
Exposure Temperature (°C)	89	89
Exposure Note	3.8 hrs in light at 50% RH, 1 hr in dark at 100% RH	3.8 hrs in light at 50% RH, 1 hr in dark at 100% RH

PROPERTIES RETAINED (%)

100% Modulus	220	
Tensile Strength	114	88
Elongation @ Break	54	50

SURFACE AND APPEARANCE

ΔE Color	8 {b}	9.8 {b}

GRAPH 212: Delaware Outdoor Weathering Exposure Time vs. vs. Elongation at Break of Chlorosulfonated Polyethylene Rubber

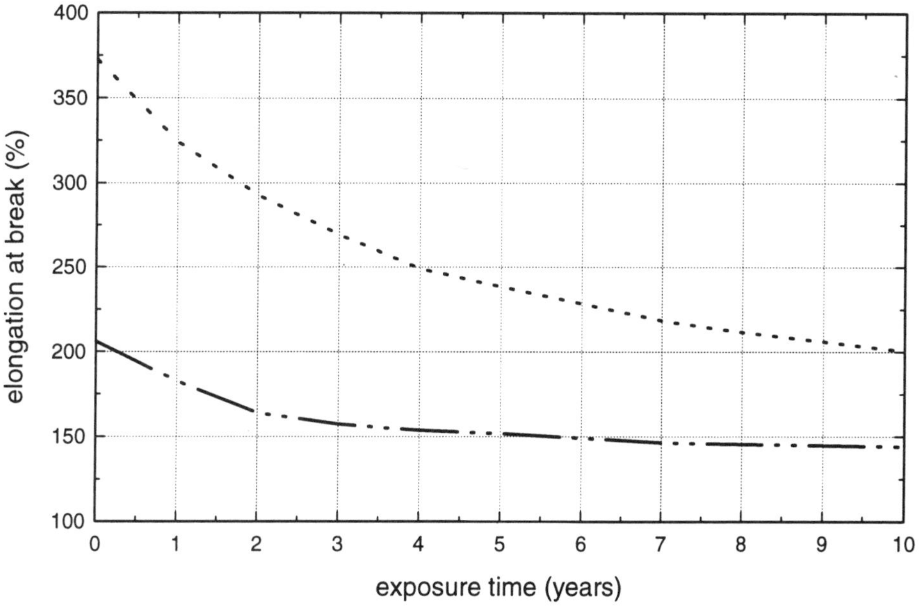

....	DuPont Hypalon 20 CSM (aliphatic PU; litharge cure; 100 phr Hypalon 20, 2.5 phr Staybelite, 20 phr sublimed litharge, 10 phr SRF black, 0.5 phr MBTS, 0.75 phr Tetrone A)
—··—··	DuPont Hypalon 20 CSM (aliphatic PU; litharge cure; 100 phr Hypalon 20, 2.5 phr Staybelite, 20 phr sublimed litharge, 75 phr SRF black, 35 phr Sundex 53, 0.5 phr MBTS, 0.75 phr Tetrone A)
Reference No.	139

Chapter 71

Ethylene Propylene Copolymer

UV Resistance

Ausimont: Dutral-CO

All ethylene propylene elastomers are sensitive to light and ultraviolet rays.

If the vulcanized part is black, the carbon black it contains acts as an absorber, protecting items exposed to atmospheric agents and light for decades.

In the case of light colored vulcanized items, it is advisable to attenuate the phenomenon with:

1) high molecular weight Dutral
2) high purity paraffinic oils
3) zinc oxide, 15-20 phr
4) rutile titanium dioxide
5) phthalocyanine-based pigments.

The various UV stabilizers currently used with plastomeric polymers are not particularly effective.

Reference: *Dutral And The Automotive Industry,* supplier marketing literature - Montedison Specialty Chemicals Ausimont.

GRAPH 213: **Xenon Arc Weatherometer Exposure Time vs. Carbonyl Formation - Area Normalized of Ethylene Propylene Copolymer**

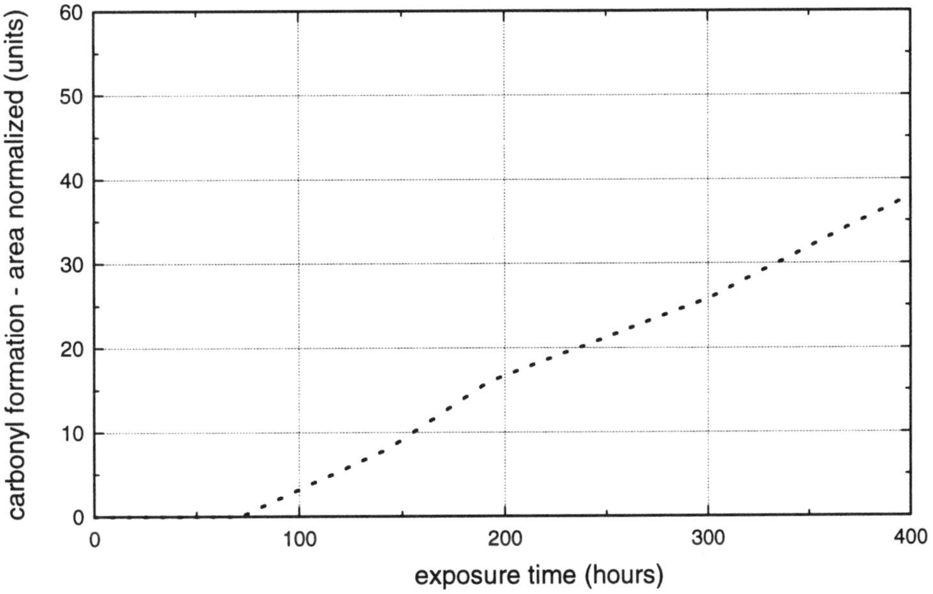

....	EPM (25 Mooney (ML 1+4 @ 125°C))
Reference No.	204

GRAPH 214: Xenon Arc Weatherometer Exposure Time vs. Decrease in Molecular Weight of Ethylene Propylene Copolymer

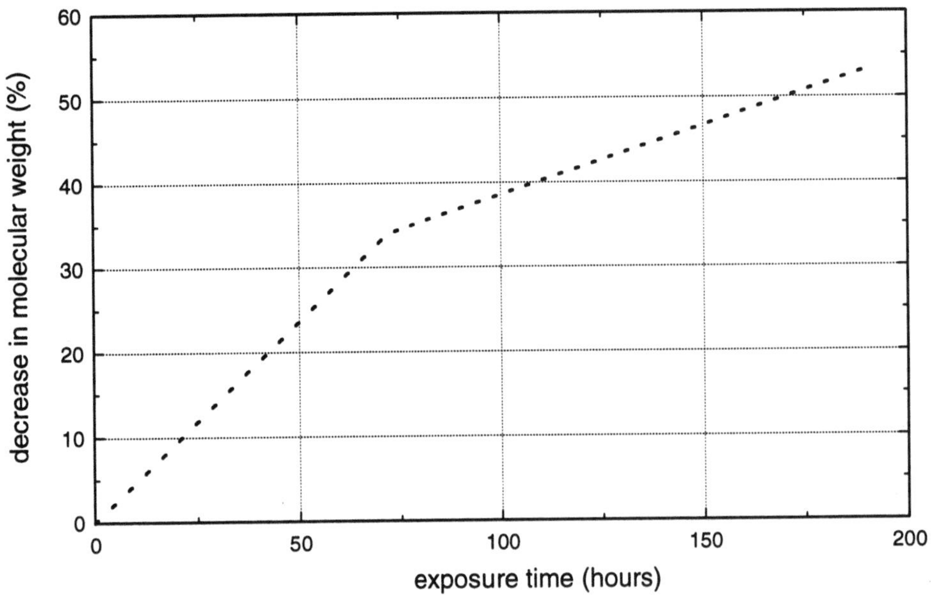

....	EPM (25 Mooney (ML 1+4 @ 125°C))
Reference No.	204

Chapter 72

Ethylene Propylene Diene Methylene Terpolymer

UV Resistance

Ausimont: Dutral-TER

All ethylene propylene elastomers are sensitive to light and ultraviolet rays. If the vulcanized part is black, the carbon black it contains acts as an absorber, protecting items exposed to atmospheric agents and light for decades.

In the case of light colored vulcanized items, it is advisable to attenuate the phenomenon with:

1) high molecular weight Dutral
2) high purity paraffinic oils
3) zinc oxide, 15-20 phr
4) rutile titanium dioxide
5) phthalocyanine-based pigments.

The various UV stabilizers currently used with plastomeric polymers are not particularly effective.

Reference: *Dutral And The Automotive Industry,* supplier marketing literature - Montedison Specialty Chemicals Ausimont.

Outdoor Weather Resistance

DuPont: Nordel

Nordel hydrocarbon rubber can be compounded to give economical, highly weather resistant vulcanizates. After 20 years of continuous outdoor exposure in Florida (45° South), properly compounded mineral-filled vulcanizates show slight checking and moderate chalking with an erosion rate of about 0.5 mils/year. After 20 years in Florida (45° South), general purpose black-loaded compounds retain good elastomeric properties, show some chalking but with erosion rates of less than 0.05 mils/year. Slight surface checking is visible at 10X magnification.

Maximum weather resistance can be obtained by using one or more of the following compounding techniques:

1. Adding pigments or fillers that give protection against UV light;
2. Using process oils that have a low aromaticity;
3. Adding zinc oxide beyond that recommended for curing;
4. Incorporating an antioxidant.

<u>Special Compounding When the Vulcanizate Will Be Weathered with a Static or Dynamic Strain</u>

General purpose black vulcanizates from EPDM, while exhibiting outstanding weather resistance under normal exposure conditions, may show accelerated development of surface crazing if they are exposed while subject to a continuous static or dynamic strain. Surface crazing of Nordel hydrocarbon rubber vulcanizates, under these conditions, can be minimized by:

- using a fine particle size black (e.g. HAF black),
- using a process oil with a low level of aromaticity (preferably less than 1% aromatics)
- using ca. 1 phr of an antioxidant. Neozone D is best, but a hindred bis-phenol antioxidant should be used if staining is a problem.

Special Compounding for Tensile Strength Retention

Mineral-filled vulcanizates from Nordel hydrocarbon rubber, like those from most other elastomers, tend to lose tensile strength when exposed to natural weathering. This loss can either be markedly reduced by compounding with 5 phr of Hypalon 45 synthetic rubber or completely eliminated by heat treating the stock with quinone dioxime. Quinone dioxime causes marked discoloration during cure, which limits its use to dark-colored vulcanizates.

Reference: Baseden, G. A., *Compounding Nordel Hydrocarbon Rubber For Good Weathering Resistance,* supplier technical report (E-88779 / 5/87 118 545/A) - Du Pont Company, 1987.

Exxon: Vistalon 5600 (features: black color, 76 Shore A hardness; material compostion: 100 phr Vistalon 5600, 100 phr N-550 carbon black, 100 phr N-774 black, 100 phr naphthenic process oil, 5 phr zinc oxide, 2 phr stearic acid, 1 phr Flectol H (Monsanto), 2 phr sulfur, 1.5 phr Thiotax (Monsanto), 0.8 phr TDEDC, 0.8 phr DPTTS, 0.8 phr Thiurad (Monsanto))

EPDM was exposed to conventional Arizona aging by DSET Laboratories, Inc. (Phoenix, AZ) at a 5° tilt from the horizontal. This tilt is preferable to 0°, as it allows for some drainage and dirt wash off during rains. Direct exposures are intended for materials which will be used outdoors and subjected to all elements of weather. The exposure period consisted of 6, 12, 24 and 48 months. EPDM shows a ΔE color change of less than 3 during the total aging cycle. Over the 48 month period EPDM experienced a significant increase in hardness and maintained reasonably good tensile strength. Although EPDM retained tensile strength, its ability to be elongated decreased with exposure time.

EPDM was also exposed to conventional Arizona aging with spray. This exposure method is the same as conventional aging, except a water spray is used to induce moisture weathering conditions. The introduction of moisture plays an important role in improving both the relevancy and reproducibility of the weathering test results. Spray nozzles are mounted above the face of the rack at points distributed to insure uniform wetting of the entire exposure area. Distilled water is sprayed for 4 hours preceding sunrise to soak the samples, and then twenty times during the day in 15 second bursts. The purpose of the wetting is twofold. First, the introduction of water in the otherwise arid climate induces and accelerates some degradation modes which do not occur as rapidly, if at all, without moisture. Second, a thermal shock causes a reduction in specimen surface temperatures as much as 14°C (25°F). This results in physical stresses which accelerate the degradation process. EPDM discolored within 6 months and continued to increase in ΔE with exposure. The hardness of EPDM increases significantly with exposure time. During the exposure period the material maintained tensile strength. The elongation of EPDM decreased significantly with exposure.

EPDM was also exposed to conventional Florida aging. This test method is a real time exposure by DSET Laboratories, Inc. (Homestead, FL) at a 5° tilt from the horizontal. Since this location has a much higher average humidity this exposure is harsher on some materials. Testing on EPDM was not accomplished over the total exposure period. Samples at 48 months had deteriorated past the point of meaningful testing. The data shows significant color change with exposure time. The material did sustain its tensile strength with exposure.

Conventional Florida aging with spray was used to test EPDM. This exposure method is the same as the conventional aging, except water spray is used to induce moisture weathering conditions. Moisture is introduced the same in Florida as it is in Arizona. EPDM showed significant color and hardness changes during the 48 months of exposure, but retained good tensile strength. The specimens had a continuing decrease in elongation with exposure time.

Reference: *Weathering Of Santoprene Thermoplastic Rubber Black Ultraviolet Grades,* supplier technical report (TCD00592) - Advanced Elastomer Systems, 1992.

EPDM (features: weatherable, black color, 65 Shore A hardness, 3.2 mm thick; manufacturing method: compression molding)

EPDM rubber did not weather well when exposed under glass. The surface was covered with mildew like growth. Unsightly blemishes remained visible even after washing.

Reference: *Florida Weathering Of Chemigum TPE,* supplier technical report (TPE 06-0292/498900-2/92) - Goodyear Chemicals, 1992.

Accelerated Outdoor Weathering Resistance

Exxon: Vistalon 5600 (features: black color, 76 Shore A hardness; material compostion: 100 phr Vistalon 5600, 100 phr N-550 carbon black, 100 phr N-774 black, 100 phr naphthenic process oil, 5 phr zinc oxide, 2 phr stearic acid, 1 phr Flectol H (Monsanto), 2 phr sulfur, 1.5 phr Thiotax (Monsanto), 0.8 phr TDEDC, 0.8 phr DPTTS, 0.8 phr Thiurad (Monsanto))

Outdoor accelerated exposure testing was done using equatorial mount with mirrors for acceleration (EMMA) on EPDM. This test method uses natural sunlight and special reflecting mirrors to concentrate the sunlight to the intensity of about eight suns. A blower which directs air over and under the samples is used to cool the specimens. This limits the increase in surface temperatures of most materials to 10 °C (18 °F) above the maximum surface temperature that is reached by identically mounted samples exposed to direct sunlight at the same time and locations without concentration. DSET Laboratories designed and maintains this equipment. The exposure period was a total of 6 and 12 months, which has been correlated to about 2-1/2 and 5 years of actual aging in a Florida environment. While EPDM retained its tensile strength, it showed a significant increase in hardness and a significant deterioration in elongation during the exposure period.

Specimens were also exposed to an equatorial mount with mirrors for acceleration plus water (EMMAQUA). This is an accelerated weathering test method which uses the same apparatus as the EMMA, except water spray is used to induce moisture weathering conditions. EPDM showed a major change in color and hardness along with significant decreases in elongation and tensile strength with exposure time.

Reference: *Weathering Of Santoprene Thermoplastic Rubber Black Ultraviolet Grades,* supplier technical report (TCD00592) - Advanced Elastomer Systems, 1992.

Accelerated Artificial Weathering Resistance

Exxon: Vistalon 5600 (features: black color, 76 Shore A hardness; material compostion: 100 phr Vistalon 5600, 100 phr N-550 carbon black, 100 phr N-774 black, 100 phr naphthenic process oil, 5 phr zinc oxide, 2 phr stearic acid, 1 phr Flectol H (Monsanto), 2 phr sulfur, 1.5 phr Thiotax (Monsanto), 0.8 phr TDEDC, 0.8 phr DPTTS, 0.8 phr Thiurad (Monsanto))

Under xenon arc weatherometer testing (to General Motors standard TM30-2 and SAE J 1885) EPDM displays a significant change in both color and hardness with exposure. Retention of tensile strength is good. However, elongation retention is poor.

Reference: *Weathering Of Santoprene Thermoplastic Rubber Black Ultraviolet Grades,* supplier technical report (TCD00592) - Advanced Elastomer Systems, 1992.

Effect of Carbon Black on Weatherability

DuPont: Nordel 1070 (cure: 20 minutes @ 160°C; features: black color; material compostion: 100 phr Nordel 1070, 1 phr stearic acid, 5 phr zinc oxide, 80 phr paraffinic oil, 1.5 phr zinc dimethyl dithiocarbonate, 1.5 phr tetramethyl thiuram disulfide, 1 phr zinc mercaptobenzothiazole, 2 phr sulfur, 80 phr FEF carbon black)
Nordel

Vulcanizates of Nordel hydrocarbon rubber have outstanding weather resistance when they contain at least 5 phr of a furnace black. In practical black-loaded vulcanizates, the concentrations of black and oil have no significant bearing on weather resistance within the ranges of 20 to 150 parts of black and 20 to 100 parts of oil. For example, weathering produces no more change in a vulcanizate loaded with 150 phr of black and 100 phr of oil than in one containing only 20 phr of each. Low concentrations (e.g. 5 phr) of furnace black provide very good weather protection in mineral-filled vulcanizates; compounds of this type are of particular interest in coverings for wires and cables.

Reference: Baseden, G. A., *Compounding Nordel Hydrocarbon Rubber For Good Weathering Resistance,* supplier technical report (E-88779 / 5/87 118 545/A) - Du Pont Company, 1987.

Effect of Color Pigments on Weatherability

DuPont: Nordel 1070 (cure: 20 minutes @ 160°C; features: white color; material compostion: 100 phr Nordel 1070, 1 phr stearic acid, 20 phr zinc oxide, 80 phr hard clay, 80 phr paraffinic oil, 1.5 phr zinc dimethyl dithiocarbonate, 1.5 phr tetramethyl thiuram disulfide, 1 phr zinc mercaptobenzothiazole, 2 phr sulfur, 1 phr phenolic antioxidant, 35 phr Ti-Pure R-610 (replaced by Ti-Pure R-960)); **Nordel**

All non-black vulcanizates should contain titanium dioxide and/or opaque colored pigments since conventional mineral fillers, by themselves, provide little protection against UV radiation. Combinations of titanium dioxide and colored pigments can be used to produce weather resistant vulcanizates with bright colors or pastel shades.

Following are proportions of individual pigments and pigment blends which provide optimum protection for mineral-filled vulcanizates of Nordel hydrocarbon rubber.

EFFECTIVE UV SCREENING AGENTS IN MINERAL FILLED VULCANIZATES

Pigment	Minimum Amount of Pigment Alone, Needed For Good UV Screening (phr)	Suggested Combination of Pigment and Titanium Dioxide For Bright Color And Good Weathering (phr)	
		Pigment	Titanium Dioxide
Furnace Black	5	-	-
Ti-Pure R-960 (DuPont)	25	-	-
Phthalocyanine Green	5*	2	20
Chrome Green	9	-	-
Phthalocyanine Blue	8*	1	25
Chrome Yellow	9	5 to 10	20
Iron Oxide (Yellow or Red)	10	-	20
Quinacridone Red	8*	2	-
Pyrazalone Red	8	-	

Remarks
Chrome yellow fades with clay loading. Super-Multiflex[1] loading gives better color stability. Pyrazalone Red gives a bright red color, but this fades with a clay loading; Super Multiflex gives better color stability. This pigment can stain painted surfaces by migration from the vulcanizate.

[1]With Super-Multifex loading, 5 phr Hypalon 45 in the compound enhances physical properties.
*Amount shown gives very dark colors.

Mineral-filled vulcanizates show better weather resistance at low to medium oil and filler loadings than at high extensions. Highly loaded vulcanizates (e.g. those containing 60 volumes of Nordel) tend to develop a brittle surface layer on weathering, causing fine cracks to appear on the surface if the vulcanizate is stretched or bent. This is not apparent when moderate loading levels are used, such as 30 volumes of filler per hundred volumes of Nordel.

Reference: Baseden, G. A., *Compounding Nordel Hydrocarbon Rubber For Good Weathering Resistance,* supplier technical report (E-88779 / 5/87 118 545/A) - Du Pont Company, 1987.

Effect of Curing Systems on Weatherability

DuPont: Nordel

Good weather resistance can be obtained with sulfur or peroxide cures.

Reference: Baseden, G. A., *Compounding Nordel Hydrocarbon Rubber For Good Weathering Resistance,* supplier technical report (E-88779 / 5/87 118 545/A) - Du Pont Company, 1987.

Effect of Plasticizers on Weatherability

DuPont: Nordel 1070; Nordel

Vulcanizates of Nordel containing at least 20 phr of a reinforcing carbon black are so weather resistant that in most applications the type of oil used is unimportant. Differences among oils do become significant, however, with very long (e.g. 20 years) outdoor exposure, or if outdoor exposure involves flexing or static strain of 20% or more; in such cases, naphthenic or paraffinic oils having a minimum aromatic content should be used, and 5% aromatic carbon atoms should be considered a maximum for these exposure conditions.

Process oils with a minimum aromatic content should be used in all mineral-filled compounds for best weather resistance as well as good color stability during cure and weathering. If good color stability is essential, the oil - whether paraffinic or naphthenic - should be selected on the basis of minimum aromatic content.

Reference: Baseden, G. A., *Compounding Nordel Hydrocarbon Rubber For Good Weathering Resistance,* supplier technical report (E-88779 / 5/87 118 545/A) - Du Pont Company, 1987.

Ozone Resistance

Exxon: Vistalon 5600 (material compostion: 100 phr Vistalon 5600, 100 phr N-550 carbon black, 100 phr N-774 black, 100 phr naphthenic process oil, 5 phr zinc oxide, 2 phr stearic acid, 1 phr Flectol H (Monsanto), 2 phr sulfur, 1.5 phr Thiotax (Monsanto), 0.8 phr TDEDC, 0.8 phr DPTTS, 0.8 phr Thiurad (Monsanto))

The results of both ASTM D1171 and ASTM D518 ozone testing indicate that EPDM rubber exhibits outstanding resistance to ozone.

Reference: *Ozone Resistance Of Santoprene Rubber,* supplier technical report (TCD01787) - Advanced Elastomer Systems, 1986.

TABLE 169: Outdoor Weathering and Accelerated Outdoor Weathering of White, Randomly Selected, Unstrained Ethylene Propylene Diene Methylene Terpolymer.

Material Family	Ethylene Propylene Diene Methylene Terpolymer						
Features	white color						
Reference Number	129	129	129	129	129	129	129
MATERIAL CHARACTERISTICS							
sample note	randomly selected samples from different manufacturers				randomly selected samples from different manufacturers		
EXPOSURE CONDITIONS							
Exposure Type	accelerated outdoor weathering				outdoor weathering		
Exposure Location	New River, Arizona				Homestead, Florida		
Exposure Country	USA				USA		
Exposure Apparatus	EMMA						
Exposure Apparatus Note	Fresnel concentrator with 10 mirrors						
Exposure Note	water spray at night only; specimens mounted on kiln dried white oak boards; unstrained specimens				45° angle south; specimens mounted on kiln dried white oak boards; unstrained specimens		
Exposure Time (days)	19	44	61	132	182	365	730
Total Irradiation (MJ/m^2)	4000	8000	12000	20000	4290	6276	12505
UV Irradiation (MJ/m^2)	110	242	394	462	144	282	562
PROPERTIES RETAINED (%)							
Tensile Strength	96.3	96.3	88.8	88.8	102	81.4	76.2
Elongation @ Break	104	104	98	96	104	106	98

TABLE 170: Outdoor Weathering and Accelerated Outdoor Weathering of Black, Weather Resistant, Unstrained Ethylene Propylene Diene Methylene Terpolymer.

Material Family	Ethylene Propylene Diene Methylene Terpolymer						
Features	black color; weatherable grade						
Reference Number	129	129	129	129	129	129	129
MATERIAL CHARACTERISTICS							
sample note	known formulation with 20 year history of proven weather performance				known formulation with 20 year history of proven weather performance		
EXPOSURE CONDITIONS							
Exposure Type	accelerated outdoor weathering				outdoor weathering		
Exposure Location	New River, Arizona				Homestead, Florida		
Exposure Country	USA				USA		
Exposure Apparatus	EMMA						
Exposure Apparatus Note	Fresnel concentrator with 10 mirrors						
Exposure Note	water spray at night only; specimens mounted on kiln dried white oak boards; unstrained specimens				45° angle south; specimens mounted on kiln dried white oak boards; unstrained specimens		
Exposure Time (days)	19	44	61	132	182	365	730
Total Irradiation (MJ/m^2)	4000	8000	12000	20000	4290	6276	12505
UV Irradiation (MJ/m^2)	110	242	394	462	144	282	562
PROPERTIES RETAINED (%)							
Tensile Strength	106.5 {ac}	107.1 {ac}	109.5 {ac}	108.9 {ac}	103.7 {ac}	106.5 {ac}	109.5 {ac}
Elongation @ Break	100 {ac}	100 {ac}	100 {ac}	86.7 {ac}	90 {ac}	95.6 {ac}	82.2 {ac}

TABLE 171: Outdoor Weathering and Accelerated Outdoor Weathering of Black, Randomly Selected, Unstrained Ethylene Propylene Diene Methylene Terpolymer.

Material Family	Ethylene Propylene Diene Methylene Terpolymer						
Features	black color						
Reference Number	129	129	129	129	129	129	129
MATERIAL CHARACTERISTICS							
sample note	randomly selected samples from different manufacturers				randomly selected samples from different manufacturers		
EXPOSURE CONDITIONS							
Exposure Type	accelerated outdoor weathering				outdoor weathering		
Exposure Location	New River, Arizona				Homestead, Florida		
Exposure Country	USA				USA		
Exposure Apparatus	EMMA						
Exposure Apparatus Note	Fresnel concentrator with 10 mirrors						
Exposure Note	water spray at night only; specimens mounted on kiln dried white oak boards; unstrained specimens				45° angle south; specimens mounted on kiln dried white oak boards; unstrained specimens		
Exposure Time (days)	19	44	61	132	182	365	730
Total Irradiation (MJ/m^2)	4000	8000	12000	20000	4290	6276	12505
UV Irradiation (MJ/m^2)	110	242	394	462	144	282	562
PROPERTIES RETAINED (%)							
Tensile Strength	96	92.7	97	97	97.8	99.2	100.3
Elongation @ Break	105.8	96.2	92.3	83.7	98.1	92.3	82.7

TABLE 172: Outdoor Weathering, Accelerated Outdoor Weathering by EMMAQUA and Accelerated Weathering with a Xenon Arc Weatherometer of Black Exxon Vistalon Ethylene Propylene Diene Methylene Terpolymer.

Material Family	Ethylene Propylene Diene Methylene Terpolymer		
Features	black color	black color	black color
Reference Number	157	157	157

EXPOSURE CONDITIONS

Exposure Type	outdoor weathering	accelerated outdoor weathering	accelerated weathering
Exposure Location	Arizona	Arizona	
Exposure Country	USA	USA	
Exposure Apparatus		EMMAQUA	xenon arc weatherometer
Exposure Time (days)	365	365	216
Exposure Note	water spray added		

PROPERTIES RETAINED (%)

Tensile Strength	98.2	90.9	87.9
Elongation	66.8	34.4	20.8

SURFACE AND APPEARANCE

ΔE Color Change	6.5	10.5	8.6

TABLE 173: Outdoor Weathering in Arizona With and Without Water Spray Added of Black Exxon Vistalon Ethylene Propylene Diene Methylene Terpolymer.

Material Family	Ethylene Propylene Diene Methylene Terpolymer							
Material Supplier/ Grade	Exxon Vistalon 5600							
Features	black color; sheet							
Reference Number	120	120	120	120	120	120	120	120
MATERIAL CHARACTERISTICS								
Shore A hardness	76	76	76	76	76	76	76	76
Sample Thickness (mm)	0.76	0.76	0.76	0.76	0.76	0.76	0.76	0.76
MATERIAL COMPOSITION								
Vistalon 5600	100 phr	100 phr	100 phr	100 phr	100 phr	100 phr	100 phr	100 phr
DPTTS	0.8 phr	0.8 phr	0.8 phr	0.8 phr	0.8 phr	0.8 phr	0.8 phr	0.8 phr
Flectol H (Monsanto)	1 phr	1 phr	1 phr	1 phr	1 phr	1 phr	1 phr	1 phr
stearic acid	2 phr	2 phr	2 phr	2 phr	2 phr	2 phr	2 phr	2 phr
TDEDC	0.8 phr	0.8 phr	0.8 phr	0.8 phr	0.8 phr	0.8 phr	0.8 phr	0.8 phr
Thiotax (Monsanto)	1.5 phr	1.5 phr	1.5 phr	1.5 phr	1.5 phr	1.5 phr	1.5 phr	1.5 phr
Thiurad (Monsanto)	0.8 phr	0.8 phr	0.8 phr	0.8 phr	0.8 phr	0.8 phr	0.8 phr	0.8 phr
naphthenic process oil	100 phr	100 phr	100 phr	100 phr	100 phr	100 phr	100 phr	100 phr
sulfur	2 phr	2 phr	2 phr	2 phr	2 phr	2 phr	2 phr	2 phr
zinc oxide	5 phr	5 phr	5 phr	5 phr	5 phr	5 phr	5 phr	5 phr
N-550 carbon black	100 phr	100 phr	100 phr	100 phr	100 phr	100 phr	100 phr	100 phr
N-774 black	100 phr	100 phr	100 phr	100 phr	100 phr	100 phr	100 phr	100 phr
EXPOSURE CONDITIONS								
Exposure Type	outdoor weathering				outdoor weathering			
Exposure Location	Phoenix, Arizona				Phoenix, Arizona			
Exposure Country	USA				USA			
Test Lab	DSET Laboratories				DSET Laboratories			
Exposure Note	5° angle				5° angle; with water spray (distilled water sprayed for 4 hours preceeding sunrise and then 20 times during the day in 15 second bursts)			
Exposure Time (days)	182	365	730	1461	182	365	730	1461
PROPERTIES RETAINED (%)								
Tensile Strength	97 {ac}	100 {ac}	111 {ac}	99 {ac}	94 {ac}	98 {ac}	109 {ac}	87 {ac}
Elongation	65 {ac}	67 {ac}	48 {ac}	47 {ac}	60 {ac}	67 {ac}	49 {ac}	52 {ac}
CHANGE IN PHYSICAL CHARACTERISTICS								
Hardness Change (units)	6 {ae}	6 {ae}	11 {ae}	11 {ae}	6 {ae}	6 {ae}	11 {ae}	14 {ae}
SURFACE AND APPEARANCE								
ΔE Color Change	7 {ab}	9.9 {ab}	5.4 {ab}	10.5 {ab}	7.5 {ab}	6.5 {ab}	9.2 {ab}	8.7 {ab}

TABLE 174: Outdoor Weathering in Arizona of Black Exxon Vistalon Ethylene Propylene Diene Methylene Terpolymer.

Material Family	Ethylene Propylene Diene Methylene Terpolymer			
Material Supplier/ Grade	Exxon Vistalon 5600			
Features	black color			
Reference Number	119	119	119	119

MATERIAL CHARACTERISTICS

Sample Thickness (mm)	0.76	0.76	0.76	0.76
Shore A hardness	76	76	76	76

MATERIAL COMPOSITION

Vistalon 5600	100 phr	100 phr	100 phr	100 phr
DPTTS	0.8 phr	0.8 phr	0.8 phr	0.8 phr
Flectol H (Monsanto)	1 phr	1 phr	1 phr	1 phr
stearic acid	2 phr	2 phr	2 phr	2 phr
TDEDC	0.8 phr	0.8 phr	0.8 phr	0.8 phr
Thiotax (Monsanto)	1.5 phr	1.5 phr	1.5 phr	1.5 phr
Thiurad (Monsanto)	0.8 phr	0.8 phr	0.8 phr	0.8 phr
naphthenic process oil	100 phr	100 phr	100 phr	100 phr
sulfur	2 phr	2 phr	2 phr	2 phr
zinc oxide	5 phr	5 phr	5 phr	5 phr
N-550 carbon black	100 phr	100 phr	100 phr	100 phr
N-774 black	100 phr	100 phr	100 phr	100 phr

EXPOSURE CONDITIONS

Exposure Type	outdoor weathering			
Exposure Location	Phoenix, Arizona	Phoenix, Arizona	Phoenix, Arizona	Phoenix, Arizona
Exposure Country	USA	USA	USA	USA
Exposure Time (days)	182	365	182	365
Exposure Note	5° south latitude	5° south latitude	5° south latitude; water spray added	5° south latitude; water spray added
Test Lab	DSET Laboratories	DSET Laboratories	DSET Laboratories	DSET Laboratories

PROPERTIES RETAINED (%)

100% Modulus	132	134	138	134
Tensile Strength	98	100	95	99
Elongation @ Break	66	68	60	66

SURFACE AND APPEARANCE

ΔE Color Change	7 {b}	10.1 {b}	7.4 {b}	6.5 {b}

TABLE 175: Outdoor Weathering in Florida and Accelerated Outdoor Weathering by EMMA of Black, Weather Resistant, Strained Ethylene Propylene Diene Methylene Terpolymer.

Material Family	Ethylene Propylene Diene Methylene Terpolymer						
Features	black color; weatherable grade						
Reference Number	129	129	129	129	129	129	129
MATERIAL CHARACTERISTICS							
Sample Note	known formulation with 20 year history of proven weather performance				known formulation with 20 year history of proven weather performance		
EXPOSURE CONDITIONS							
Exposure Type	accelerated outdoor weathering				outdoor weathering		
Exposure Location	New River, Arizona				Homestead, Florida		
Exposure Country	USA				USA		
Exposure Apparatus	EMMA						
Exposure Apparatus Note	Fresnel concentrator with 10 mirrors						
Exposure Note	water spray at night only; specimens mounted on kiln dried white oak boards; specimens strained @ 50%				45° angle south; specimens mounted on kiln dried white oak boards; specimens strained @ 50%		
Exposure Time (days)	19	44	61	132	182	365	730
Total Irradiation (MJ/m²)	4000	8000	12000	20000	4290	6276	12505
UV Irradiation (MJ/m²)	110	242	394	462	144	282	562
PROPERTIES RETAINED (%)							
Tensile Strength	117.5 {ac}	112.6 {ac}	125.5 {ac}	120.6 {ac}	116.3 {ac}	116.9 {ac}	104 {ac}
Elongation @ Break	91.1 {ac}	70 {ac}	67.8 {ac}	72.2 {ac}	72.2 {ac}	65.6 {ac}	46.7 {ac}

TABLE 176: Outdoor Weathering in Florida With and Without Water Spray Added of Black Exxon Vistalon Ethylene Propylene Diene Methylene Terpolymer.

Material Family	Ethylene Propylene Diene Methylene Terpolymer							
Material Supplier/ Grade	Exxon Vistalon 5600							
Features	black color, sheet							
Reference Number	120	120	120	120	120	120	120	120
MATERIAL CHARACTERISTICS								
Shore A hardness	76	76	76	76	76	76	76	76
Sample Thickness (mm)	0.76	0.76	0.76	0.76	0.76	0.76	0.76	0.76
MATERIAL COMPOSITION								
Vistalon 5600	100 phr	100 phr	100 phr	100 phr	100 phr	100 phr	100 phr	100 phr
DPTTS	0.8 phr	0.8 phr	0.8 phr	0.8 phr	0.8 phr	0.8 phr	0.8 phr	0.8 phr
Flectol H (Monsanto)	1 phr	1 phr	1 phr	1 phr	1 phr	1 phr	1 phr	1 phr
stearic acid	2 phr	2 phr	2 phr	2 phr	2 phr	2 phr	2 phr	2 phr
TDEDC	0.8 phr	0.8 phr	0.8 phr	0.8 phr	0.8 phr	0.8 phr	0.8 phr	0.8 phr
Thiotax (Monsanto)	1.5 phr	1.5 phr	1.5 phr	1.5 phr	1.5 phr	1.5 phr	1.5 phr	1.5 phr
Thiurad (Monsanto)	0.8 phr	0.8 phr	0.8 phr	0.8 phr	0.8 phr	0.8 phr	0.8 phr	0.8 phr
naphthenic process oil	100 phr	100 phr	100 phr	100 phr	100 phr	100 phr	100 phr	100 phr
sulfur	2 phr	2 phr	2 phr	2 phr	2 phr	2 phr	2 phr	2 phr
zinc oxide	5 phr	5 phr	5 phr	5 phr	5 phr	5 phr	5 phr	5 phr
N-550 carbon black	100 phr	100 phr	100 phr	100 phr	100 phr	100 phr	100 phr	100 phr
N-774 black	100 phr	100 phr	100 phr	100 phr	100 phr	100 phr	100 phr	100 phr
EXPOSURE CONDITIONS								
Exposure Type	outdoor weathering				outdoor weathering			
Exposure Location	Homestead, Florida				Homestead, Florida			
Exposure Country	USA				USA			
Test Lab	DSET Laboratories				DSET Laboratories			
Exposure Note	5° angle				5° angle; with water spray (distilled water sprayed for 4 hours preceeding sunrise and then 20 times during the day in 15 second bursts)			
Exposure Time (days)	182	365	730	1461	182	365	730	1461
PROPERTIES RETAINED (%)								
Tensile Strength	90 {ac}	94 {ac}	110{ac}	{ac}	97 {ac}	{ac}	109 {ac}	95 {ac}
Elongation	72 {ac}	71 {ac}	62 {ac}	{ac}	77 {ac}	{ac}	67 {ac}	61 {ac}
CHANGE IN PHYSICAL CHARACTERISTICS								
Hardness Change (units)	-1 {ae}	1 {ae}	9 {ae}	{ae}	13 {ae}	{ae}	6 {ae}	8 {ae}
SURFACE AND APPEARANCE								
ΔE Color Change	5.3 {ab}	5.8 {ab}	{ab}	{ab}	4.5 {ab}	{ab}	{ab}	{ab}

TABLE 177: Outdoor Weathering in Florida of Weatherable Ethylene Propylene Diene Methylene Terpolymer.

Material Family	Ethylene Propylene Diene Methylene Terpolymer	
Features	black color, weatherable, medium gloss	
Manufacturing Method	compression molding	
Reference Number	175	175

MATERIAL CHARACTERISTICS

Shore A hardness	65	65
Sample Thickness (mm)	3.2	3.2

EXPOSURE CONDITIONS

Exposure Type	outdoor weathering	
Exposure Location	Florida	Florida
Exposure Country	USA	USA
Exposure Note	direct exposure; 5° south	under glass exposure; 5° south
Exposure Time (days)	730	730

PROPERTIES RETAINED (%)

100% Modulus	115	132
Tensile Strength	101	103
Elongation @ Break	86	67

CHANGE IN PHYSICAL CHARACTERISTICS

Shore A Hardness Change (units)	8	13

SURFACE AND APPEARANCE

Gloss @ 60° Retained (%)	57 {g}	14.3 {g}
ΔE Color	0.8 {h}	0.5 {h}
Visual Observation (as weathered)	slight chalking	severe mildew
Visual Observation (washed)	no chalking, no cracks	spotted, no cracks

TABLE 178: Outdoor Weathering in Florida and Accelerated Outdoor Weathering by EMMA of Black, Randomly Selected, Strained Ethylene Propylene Diene Methylene Terpolymer.

Material Family	Ethylene Propylene Diene Methylene Terpolymer						
Features	black color						
Reference Number	129	129	129	129	129	129	129
MATERIAL CHARACTERISTICS							
Sample Note	randomly selected samples from different manufacturers				randomly selected samples from different manufacturers		
EXPOSURE CONDITIONS							
Exposure Type	accelerated outdoor weathering				outdoor weathering		
Exposure Location	New River, Arizona				Homestead, Florida		
Exposure Country	USA				USA		
Exposure Apparatus	EMMA						
Exposure Apparatus Note	Fresnel concentrator with 10 mirrors						
Exposure Note	water spray at night only; specimens mounted on kiln dried white oak boards; specimens strained @ 50%				45° angle south; specimens mounted on kiln dried white oak boards; specimens strained @ 50%		
Exposure Time (days)	19	44	61	132	182	365	730
Total Irradiation (MJ/m^2)	4000	8000	12000	20000	4290	6276	12505
UV Irradiation (MJ/m^2)	110	242	394	462	144	282	562
PROPERTIES RETAINED (%)							
Tensile Strength	110	107.8	117.5	115.4	112.1	110.5	105.7
Elongation @ Break	88.5	81.7	76.9	75	83.7	76	59.6

TABLE 179: Outdoor Weathering in Florida and Accelerated Outdoor Weathering by EMMA of White, Randomly Selected, Strained Ethylene Propylene Diene Methylene Terpolymer.

Material Family	Ethylene Propylene Diene Methylene Terpolymer						
Features	white color						
Reference Number	129	129	129	129	129	129	129

MATERIAL CHARACTERISTICS

Sample Note	randomly selected samples from different manufacturers				randomly selected samples from different manufacturers		

EXPOSURE CONDITIONS

Exposure Type	accelerated outdoor weathering				outdoor weathering		
Exposure Location	New River, Arizona				Homestead, Florida		
Exposure Country	USA				USA		
Exposure Apparatus	EMMA						
Exposure Apparatus Note	Fresnel concentrator with 10 mirrors						
Exposure Note	water spray at night only; specimens mounted on kiln dried white oak boards; specimens strained @ 50%				45° angle south; specimens mounted on kiln dried white oak boards; specimens strained @ 50%		
Exposure Time (days)	19	44	61	132	182	365	730
Total Irradiation (MJ/m^2)	4000	8000	12000	20000	4290	6276	12505
UV Irradiation (MJ/m^2)	110	242	394	462	144	282	562

PROPERTIES RETAINED (%)

Tensile Strength	112.9	107.2	108.9	105.4	121.8	102	92.3
Elongation @ Break	92	90	82	73	82	80	68

TABLE 180: Outdoor Weathering in Florida of Black Ethylene Propylene Diene Methylene Terpolymer.

Material Family	Ethylene Propylene Diene Methylene Terpolymer				
Features	black color, commercially available grade				
Reference Number	155	155	155	155	155
EXPOSURE CONDITIONS					
Exposure Type	outdoor weathering				
Exposure Location	Arizona	Arizona	Arizona	Arizona	Florida
Exposure Country	USA	USA	USA	USA	USA
Exposure Time (days)	183	365	730	1461	730
PROPERTIES RETAINED (%)					
Tensile Strength					148 {ac}
Elongation					44 {ac}
CHANGE IN PHYSICAL CHARACTERISTICS					
Hardness Change (units)					24 {ae}
SURFACE AND APPEARANCE					
ΔE Color Change	7.02 (samples washed)	9.9 (samples washed)	5.4 (samples washed)	10.53 (samples washed)	

TABLE 181: Accelerated Outdoor Weathering by EMMA and EMMAQUA and Accelerated Weathering in a Xenon Arc Weatherometer of Black Exxon Vistalon Ethylene Propylene Diene Methylene Terpolymer.

Material Family	Ethylene Propylene Diene Methylene Terpolymer							
Material Supplier/ Grade	Exxon Vistalon 5600							
Features	black color							
Reference Number	120	120	120	120	120	120	120	120
MATERIAL CHARACTERISTICS								
Shore A hardness	76	76	76	76	76	76	76	76
Sample Thickness (mm)	0.76	0.76	0.76	0.76	0.76	0.76	0.76	0.76
MATERIAL COMPOSITION								
Vistalon 5600	100 phr	100 phr	100 phr	100 phr	100 phr	100 phr	100 phr	100 phr
DPTTS	0.8 phr	0.8 phr	0.8 phr	0.8 phr	0.8 phr	0.8 phr	0.8 phr	0.8 phr
Flectol H (Monsanto)	1 phr	1 phr	1 phr	1 phr	1 phr	1 phr	1 phr	1 phr
stearic acid	2 phr	2 phr	2 phr	2 phr	2 phr	2 phr	2 phr	2 phr
TDEDC	0.8 phr	0.8 phr	0.8 phr	0.8 phr	0.8 phr	0.8 phr	0.8 phr	0.8 phr
Thiotax (Monsanto)	1.5 phr	1.5 phr	1.5 phr	1.5 phr	1.5 phr	1.5 phr	1.5 phr	1.5 phr
Thiurad (Monsanto)	0.8 phr	0.8 phr	0.8 phr	0.8 phr	0.8 phr	0.8 phr	0.8 phr	0.8 phr
naphthenic process oil	100 phr	100 phr	100 phr	100 phr	100 phr	100 phr	100 phr	100 phr
sulfur	2 phr	2 phr	2 phr	2 phr	2 phr	2 phr	2 phr	2 phr
zinc oxide	5 phr	5 phr	5 phr	5 phr	5 phr	5 phr	5 phr	5 phr
N-550 carbon black	100 phr	100 phr	100 phr	100 phr	100 phr	100 phr	100 phr	100 phr
N-774 black	100 phr	100 phr	100 phr	100 phr	100 phr	100 phr	100 phr	100 phr
EXPOSURE CONDITIONS								
Exposure Type	accelerated weathering		accelerated outdoor weathering			accelerated outdoor weathering		
Exposure Apparatus	xenon arc weatherometer		EMMA			EMMAQUA		
Exposure Note	temperature: 89°C; 3.8 hrs in light at 50% RH, 1 hr in dark at 100% RH; test method: SAE J1885; GM TM30-2							
Exposure Time (days)	125	208	182	365	730	182	365	730
Energy @ 340nm (kJ/m²)	4703	7838						
Energy @ <380nm (MJ/m²)	403	672						
PROPERTIES RETAINED (%)								
Tensile Strength	85 {ac}	88 {ac}	93 {ac}	89 {ac}	83 {ac}	91 {ac}	91 {ac}	69 {ac}
Elongation	23 {ac}	21 {ac}	38 {ac}	28 {ac}	17 {ac}	47 {ac}	34 {ac}	14 {ac}
CHANGE IN PHYSICAL CHARACTERISTICS								
Hardness Change (units)	16 {ae}	16 {ae}	-1 {ae}	12 {ae}	19 {ae}	15 {ae}	13 {ae}	22 {ae}
SURFACE AND APPEARANCE								
ΔE Color Change	5.3 {ab}	8.6 {ab}	7 {ab}	9.7 {ab}	9.8 {ab}	7.6 {ab}	10.5 {ab}	8.7 {ab}

TABLE 182: Accelerated Outdoor Weathering in Arizona by EMMA and EMMAQUA of Black Exxon Vistalon Ethylene Propylene Diene Methylene Terpolymer.

Material Family	Ethylene Propylene Diene Methylene Terpolymer			
Material Supplier/ Grade	Exxon Vistalon 5600			
Features	black color			
Reference Number	119	119	119	119

MATERIAL CHARACTERISTICS

Sample Thickness (mm)	0.76	0.76	0.76	0.76
Shore A hardness	76	76	76	76

MATERIAL COMPOSITION

Vistalon 5600	100 phr	100 phr	100 phr	100 phr
DPTTS	0.8 phr	0.8 phr	0.8 phr	0.8 phr
Flectol H (Monsanto)	1 phr	1 phr	1 phr	1 phr
stearic acid	2 phr	2 phr	2 phr	2 phr
TDEDC	0.8 phr	0.8 phr	0.8 phr	0.8 phr
Thiotax (Monsanto)	1.5 phr	1.5 phr	1.5 phr	1.5 phr
Thiurad (Monsanto)	0.8 phr	0.8 phr	0.8 phr	0.8 phr
naphthenic process oil	100 phr	100 phr	100 phr	100 phr
sulfur	2 phr	2 phr	2 phr	2 phr
zinc oxide	5 phr	5 phr	5 phr	5 phr
N-550 carbon black	100 phr	100 phr	100 phr	100 phr
N-774 black	100 phr	100 phr	100 phr	100 phr

EXPOSURE CONDITIONS

Exposure Type	accelerated outdoor weathering			
Exposure Location	Phoenix, Arizona	Phoenix, Arizona	Phoenix, Arizona	Phoenix, Arizona
Exposure Country	USA	USA	USA	USA
Exposure Apparatus	EMMA	EMMA	EMMAQUA	EMMAQUA
Exposure Time (days)	182	365	182	365
Exposure Note	correlates to approximately 2.5 years actual Florida aging	correlates to approximately 5 years actual Florida aging	correlates to approximately 2.5 years actual Florida aging	correlates to approximately 5 years actual Florida aging

PROPERTIES RETAINED (%)

100% Modulus				156
Tensile Strength	95	92	92	92
Elongation @ Break	38	29	47	35

SURFACE AND APPEARANCE

ΔE Color Change	7.1 {b}	7.1 {b}	7.8{b}	10.6{b}

TABLE 183: Accelerated Weathering in a UVCON and a Xenon Arc Weatherometer of White, Randomly Selected, Strained Ethylene Propylene Diene Methylene Terpolymer.

Material Family	Ethylene Propylene Diene Methylene Terpolymer									
Features	white color									
Reference Number	129	129	129	129	129	129	129	129	129	129
MATERIAL CHARACTERISTICS										
Sample Note	randomly selected samples from different manufacturers					randomly selected samples from different manufacturers				
EXPOSURE CONDITIONS										
Exposure Type	accelerated weathering					accelerated weathering				
Exposure Apparatus	xenon arc weatherometer					UV-CON				
Exposure Apparatus Note	filter type: borosilicate inner and outer; spray nozel: F-80; exposure: 0.35 W/m² @ 340 nm					lamp type: fluorescent UVB-313 (UVB-B)				
Exposure Cycle Note	690 minutes light, 30 minutes light and deionized water spray					20 hours UV at 80°C, 4 hours condensation at 50°C				
Exposure Test Method	ASTM G26					ASTM G53				
Exposure Note	specimens mounted on exterior grade plywood covered with aluminum foil; specimen rotation every 250 hours; specimens strained @ 50%					specimens mounted on exterior grade plywood covered with aluminum foil; specimens strained @ 50%				
Exposure Relative Humidity (%)	45-55	45-55	45-55	45-55	45-55					
Exposure Temperature (°C)	77-83 (black panel temp.)	77-83 (black panel temp.)	77-83 (black panel temp.)	77-83 (black panel temp.)	77-83 (black panel temp.)					
Exposure Time (days)	20.8	41.7	83.3	125	166.6	20.8	41.7	83.3	125	166.6
Total Irradiation (MJ/m²)	711 (calculated)	1422	2844	4266	5688	53 (calculated)	106	213	319	425
UV Irradiation (MJ/m²)	72 (calculated)	144	288	432	576	52 (calculated)	103	206	310	413
PROPERTIES RETAINED (%)										
Tensile Strength	105.4	102	96.9	10.9 (sample broke)	(sample broke)	105.4	86.5	85.4	64.8 (sample broke)	21.8 (sample broke)
Elongation @ Break	91	80	79	32 (sample broke)	(sample broke)	99	87	90	48 (sample broke)	16 (sample broke)

TABLE 184: Accelerated Weathering in a UVCON and a Xenon Arc Weatherometer of White, Randomly Selected, Unstrained Ethylene Propylene Diene Methylene Terpolymer.

Material Family	Ethylene Propylene Diene Methylene Terpolymer									
Features	white color									
Reference Number	129	129	129	129	129	129	129	129	129	129

MATERIAL CHARACTERISTICS

Sample Note	randomly selected samples from different manufacturers					randomly selected samples from different manufacturers				

EXPOSURE CONDITIONS

Exposure Type	accelerated weathering					accelerated weathering				
Exposure Apparatus	xenon arc weatherometer					UV-CON				
Exposure Apparatus Note	filter type: borosilicate inner and outer; spray nozel: F-80; exposure: 0.35 W/m² @ 340 nm					lamp type: fluorescent UVB-313 (UVB-B)				
Exposure Cycle Note	690 minutes light, 30 minutes light and deionized water spray					20 hours UV at 80°C, 4 hours condensation at 50°C				
Exposure Test Method	ASTM G26					ASTM G53				
Exposure Note	specimens mounted on exterior grade plywood covered with aluminum foil; specimen rotation every 250 hours; unstrained specimens					specimens mounted on exterior grade plywood covered with aluminum foil; unstrained specimens				
Exposure Relative Humidity (%)	45-55	45-55	45-55	45-55	45-55					
Exposure Temperature (°C)	77-83 (black panel temp.)	77-83 (black panel temp.)	77-83 (black panel temp.)	77-83 (black panel temp.)	77-83 (black panel temp.)					
Exposure Time (days)	20.8	41.7	83.3	125	166.6	20.8	41.7	83.3	125	166.6
Total Irradiation (MJ/m²)	711 (calculated)	1422	2844	4266	5688	53 (calculated)	106	213	319	425
UV Irradiation (MJ/m²)	72 (calculated)	144	288	432	576	52 (calculated)	103	206	310	413

PROPERTIES RETAINED (%)

Tensile Strength	91.7	86.5	76.8	64.2	29.8	92.8	84.8	86.5	48.1	31 (wide spread in test results)
Elongation @ Break	101	106	100	102	86	100	88	98	62	28 (wide spread in test results)

TABLE 185: Accelerated Weathering in a UVCON and a Xenon Arc Weatherometer of Black, Weather Resistant, Strained Ethylene Propylene Diene Methylene Terpolymer.

Material Family	Ethylene Propylene Diene Methylene Terpolymer									
Features	black color; weatherable grade									
Reference Number	129	129	129	129	129	129	129	129	129	129
MATERIAL CHARACTERISTICS										
Sample Note	known formulation with 20 year history of proven weather performance					known formulation with 20 year history of proven weather performance				
EXPOSURE CONDITIONS										
Exposure Type	accelerated weathering					accelerated weathering				
Exposure Apparatus	xenon arc weatherometer					UV-CON				
Exposure Apparatus Note	filter type: borosilicate inner and outer; spray nozel: F-80; exposure: 0.35 W/m² @ 340 nm					lamp type: fluorescent UVB-313 (UVB-B)				
Exposure Cycle Note	690 minutes light, 30 minutes light and deionized water spray					20 hours UV at 80°C, 4 hours condensation at 50°C				
Exposure Test Method	ASTM G26					ASTM G53				
Exposure Note	specimens mounted on exterior grade plywood covered with aluminum foil; specimen rotation every 250 hours; specimens strained @ 50%					specimens mounted on exterior grade plywood covered with aluminum foil; specimens strained @ 50%				
Exposure Relative Humidity (%)	45-55	45-55	45-55	45-55	45-55					
Exposure Temperature (°C)	77-83 (black panel temp.)	77-83 (black panel temp.)	77-83 (black panel temp.)	77-83 (black panel temp.)	77-83 (black panel temp.)					
Exposure Time (days)	20.8	41.7	83.3	125	166.6	20.8	41.7	83.3	125	166.6
Total Irradiation (MJ/m²)	711 (calculated)	1422	2844	4266	5688	53 (calculated)	106	213	319	425
UV Irradiation (MJ/m²)	72 (calculated)	144	288	432	576	52 (calculated)	103	206	310	413
PROPERTIES RETAINED (%)										
Tensile Strength	133.2 {ac}	136 {ac}	128 {ac}	94.8 {ac}	94.8 {ac}	120 {ac}	114.5 {ac}	118.2 {ac}	112 {ac}	108.3 {ac}
Elongation @ Break	77.8 {ac}	67.8 {ac}	46.7 {ac}	36.7 {ac}	37.8 {ac}	88.9 {ac}	72.2 {ac}	66.7 {ac}	36.7 {ac}	37.8 {ac}

TABLE 186: **Accelerated Weathering in a UVCON and a Xenon Arc Weatherometer of Black, Randomly Selected, Unstrained Ethylene Propylene Diene Methylene Terpolymer.**

Material Family	Ethylene Propylene Diene Methylene Terpolymer									
Features	black color									
Reference Number	129	129	129	129	129	129	129	129	129	129
MATERIAL CHARACTERISTICS										
Sample Note	randomly selected samples from different manufacturers					randomly selected samples from different manufacturers				
EXPOSURE CONDITIONS										
Exposure Type	accelerated weathering					accelerated weathering				
Exposure Apparatus	xenon arc weatherometer					UV-CON				
Exposure Apparatus Note	filter type: borosilicate inner and outer; spray nozel: F-80; exposure: 0.35 W/m^2 @ 340 nm					lamp type: fluorescent UVB-313 (UVB-B)				
Exposure Cycle Note	690 minutes light, 30 minutes light and deionized water spray					20 hours UV at 80°C, 4 hours condensation at 50°C				
Exposure Test Method	ASTM G26					ASTM G53				
Exposure Note	specimens mounted on exterior grade plywood covered with aluminum foil; specimen rotation every 250 hours; unstrained specimens					specimens mounted on exterior grade plywood covered with aluminum foil; unstrained specimens				
Exposure Relative Humidity (%)	45-55	45-55	45-55	45-55	45-55					
Exposure Temperature (°C)	77-83 (black panel temp.)	77-83 (black panel temp.)	77-83 (black panel temp.)	77-83 (black panel temp.)	77-83 (black panel temp.)					
Exposure Time (days)	20.8	41.7	83.3	125	166.6	20.8	41.7	83.3	125	166.6
Total Irradiation (MJ/m^2)	711 (calculated)	1422	2844	4266	5688	53 (calculated)	106	213	319	425
UV Irradiation (MJ/m^2)	72 (calculated)	144	288	432	576	52 (calculated)	103	206	310	413
PROPERTIES RETAINED (%)										
Tensile Strength	97	99.2	97	90.1	78.7	95.2	99.7	105.7	90	79.8
Elongation @ Break	83.7	76	66.3	58.7	62.5	74	80.8	69.2	55.8	48

TABLE 187: Accelerated Weathering in a Xenon Arc Weatherometer of Black Exxon Vistalon Ethylene Propylene Diene Methylene Terpolymer.

Material Family	Ethylene Propylene Diene Methylene Terpolymer	
Material Supplier/ Grade	Exxon Vistalon 5600	
Features	black color	
Reference Number	119	119

MATERIAL CHARACTERISTICS

Sample Thickness (mm)	0.76	0.76
Shore A hardness	76	76

MATERIAL COMPOSITION

Vistalon 5600	100 phr	100 phr
DPTTS	0.8 phr	0.8 phr
Flectol H (Monsanto)	1 phr	1 phr
stearic acid	2 phr	2 phr
TDEDC	0.8 phr	0.8 phr
Thiotax (Monsanto)	1.5 phr	1.5 phr
Thiurad (Monsanto)	0.8 phr	0.8 phr
naphthenic process oil	100 phr	100 phr
sulfur	2 phr	2 phr
zinc oxide	5 phr	5 phr
N-550 carbon black	100 phr	100 phr
N-774 black	100 phr	100 phr

EXPOSURE CONDITIONS

Exposure Type	accelerated weathering	
Exposure Test Method	GM TM30-2	GM TM30-2
Exposure Apparatus	xenon arc weatherometer	xenon arc weatherometer
Exposure Time (days)	125	208
Exposure Temperature (°C)	89	89
Exposure Note	3.8 hrs in light at 50% RH, 1 hr in dark at 100% RH	3.8 hrs in light at 50% RH, 1 hr in dark at 100% RH

PROPERTIES RETAINED (%)

Tensile Strength	86	88
Elongation @ Break	23	21

SURFACE AND APPEARANCE

ΔE Color Change	5.3 {b}	8.8 {b}

TABLE 188: Accelerated Weathering in a UVCON and a Xenon Arc Weatherometer of Black, Weather Resistant, Unstrained Ethylene Propylene Diene Methylene Terpolymer.

Material Family	Ethylene Propylene Diene Methylene Terpolymer									
Features	black color; weatherable grade									
Reference Number	129	129	129	129	129	129	129	129	129	129
MATERIAL CHARACTERISTICS										
Sample Note	known formulation with 20 year history of proven weather performance					known formulation with 20 year history of proven weather performance				
EXPOSURE CONDITIONS										
Exposure Type	accelerated weathering					accelerated weathering				
Exposure Apparatus	xenon arc weatherometer					UV-CON				
Exposure Apparatus Note	filter type: borosilicate inner and outer; spray nozel: F-80; exposure: 0.35 W/m² @ 340 nm					lamp type: fluorescent UVB-313 (UVB-B)				
Exposure Cycle Note	690 minutes light, 30 minutes light and deionized water spray					20 hours UV at 80°C, 4 hours condensation at 50°C				
Exposure Test Method	ASTM G26					ASTM G53				
Exposure Note	specimens mounted on exterior grade plywood covered with aluminum foil; specimen rotation every 250 hours; unstrained specimens					specimens mounted on exterior grade plywood covered with aluminum foil; unstrained specimens				
Exposure Relative Humidity (%)	45-55	45-55	45-55	45-55	45-55					
Exposure Temperature (°C)	77-83 (black panel temp.)	77-83 (black panel temp.)	77-83 (black panel temp.)	77-83 (black panel temp.)	77-83 (black panel temp.)					
Exposure Time (days)	20.8	41.7	83.3	125	166.6	20.8	41.7	83.3	125	166.6
Total Irradiation (MJ/m²)	711 (calculated)	1422	2844	4266	5688	53 (calculated)	106	213	319	425
UV Irradiation (MJ/m²)	72 (calculated)	144	288	432	576	52 (calculated)	103	206	310	413
PROPERTIES RETAINED (%)										
Tensile Strength	104 {ac}	100.3 {ac}	105.8 {ac}	89.8 {ac}	83.7 {ac}	103.4 {ac}	103.4 {ac}	105.5 {ac}	93.5 {ac}	85.5 {ac}
Elongation @ Break	83.3 {ac}	75.6 {ac}	74.4 {ac}	48.9 {ac}	40 {ac}	88.9 {ac}	82.2 {ac}	71.1 {ac}	48.9 {ac}	42.2 {ac}

TABLE 189: Accelerated Weathering in a UVCON and a Xenon Arc Weatherometer of Black, Randomly Selected, Strained Ethylene Propylene Diene Methylene Terpolymer.

Material Family	Ethylene Propylene Diene Methylene Terpolymer									
Features	black color									
Reference Number	129	129	129	129	129	129	129	129	129	129

MATERIAL CHARACTERISTICS

Sample Note	randomly selected samples from different manufacturers					randomly selected samples from different manufacturers				

EXPOSURE CONDITIONS

Exposure Type	accelerated weathering					accelerated weathering				
Exposure Apparatus	xenon arc weatherometer					UV-CON				
Exposure Apparatus Note	filter type: borosilicate inner and outer; spray nozel: F-80; exposure: 0.35 W/m^2 @ 340 nm					lamp type: fluorescent UVB-313 (UVB-B)				
Exposure Cycle Note	690 minutes light, 30 minutes light and deionized water spray					20 hours UV at 80°C, 4 hours condensation at 50°C				
Exposure Test Method	ASTM G26					ASTM G53				
Exposure Note	specimens mounted on exterior grade plywood covered with aluminum foil; specimen rotation every 250 hours; specimens strained @ 50%					specimens mounted on exterior grade plywood covered with aluminum foil; specimens strained @ 50%				
Exposure Relative Humidity (%)	45-55	45-55	45-55	45-55	45-55					
Exposure Temperature (°C)	77-83 (black panel temp.)	77-83 (black panel temp.)	77-83 (black panel temp.)	77-83 (black panel temp.)	77-83 (black panel temp.)					
Exposure Time (days)	20.8	41.7	83.3	125	166.6	20.8	41.7	83.3	125	166.6
Total Irradiation (MJ/m^2)	711 (calculated)	1422	2844	4266	5688	53 (calculated)	106	213	319	425
UV Irradiation (MJ/m^2)	72 (calculated)	144	288	432	576	52 (calculated)	103	206	310	413

PROPERTIES RETAINED (%)

Tensile Strength	118.6	119.1	114.3	101.9	105.1	114	116.2	112.9	110	98
Elongation @ Break	71.7	64.5	55.8	52.9	49	73.1	76	61.5	51.9	44.2

TABLE 190: Ozone Resistance of Exxon Vistalon Ethylene Propylene Diene Methylene Terpolymer.

Material Family	Ethylene Propylene Diene Methylene Terpolymer		
Material Supplier/ Grade	Exxon Vistalon 5600	Exxon Vistalon 5600	
Reference Number	158	158	155
MATERIAL COMPOSITION			
Vistalon 5600	100 phr	100 phr	
DPTTS	0.8 phr	0.8 phr	
Flectol H (Monsanto)	1 phr	1 phr	
stearic acid	2 phr	2 phr	
TDEDC	0.8 phr	0.8 phr	
Thiotax (Monsanto)	1.5 phr	1.5 phr	
Thiurad (Monsanto)	0.8 phr	0.8 phr	
naphthenic process oil	100 phr	100 phr	
sulfur	2 phr	2 phr	
zinc oxide	5 phr	5 phr	
N-550 carbon black	100 phr	100 phr	
N-774 black	100 phr	100 phr	
EXPOSURE CONDITIONS			
Exposure Type	ozone	ozone	ozone
Exposure Test Method	ASTM D1171	ASTM D518	ASTM D1171
Exposure Apparatus	Monsanto's ozone chamber	Monsanto's ozone chamber	
Exposure Specimen Note	extruded triangle profile mounted over a mandrel	bent loop specimen	
Exposure Temperature (°C)			40
Exposure Time (hours)	336	336	100
Exposure Note	168 hours @ 400 pphm, 168 hours @ 500 pphm	168 hours @ 400 pphm, 168 hours @ 500 pphm	300 pphm ozone
SURFACE AND APPEARANCE			
Time to Initial Crack (hours)	no cracking	no cracking	
Shell Rating (0-10 scale, 0 represents total cracking, 10 represents no cracking)	10	10	
RESULTS OF EXPOSURE			
Ozone Resistance			good to excellent

TABLE 191: Ozone Resistance of Ethylene Propylene Diene Methylene Terpolymer.

Material Family	Ethylene Propylene Diene Methylene Terpolymer		
Cure	10 min. @ 175°C		
Reference Number	204	204	204
MATERIAL COMPOSITION			
oil modified EPDM	70 phr	70 phr	70 phr
EPDM	50 phr	50 phr	50 phr
dicumyl peroxide (40%)	10 phr	10 phr	10 phr
magnesium oxide	10 phr	10 phr	10 phr
zinc stearate	0.5 phr	0.5 phr	0.5 phr
paraffinic plasticizer	15 phr	15 phr	15 phr
antioxidant	4 phr	4 phr	4 phr
zinc oxide	5 phr	5 phr	5 phr
N-550 carbon black	85 phr	85 phr	85 phr
co-agent	1 phr	1 phr	1 phr
EXPOSURE CONDITIONS			
Exposure Type	ozone		
Exposure Test Method	ASTM D1149		
Exposure Note	100 pphm ozone; samples at 20% strain		
Exposure Temperature (°C)	40	40	40
Exposure Time (hours)	24	48	72
SURFACE AND APPEARANCE			
Number of Cracks	0	0	0

GRAPH 215: Xenon Arc Weatherometer Exposure Time vs. Carbonyl Formation - Area Normalized of Ethylene Propylene Diene Methylene Terpolymer

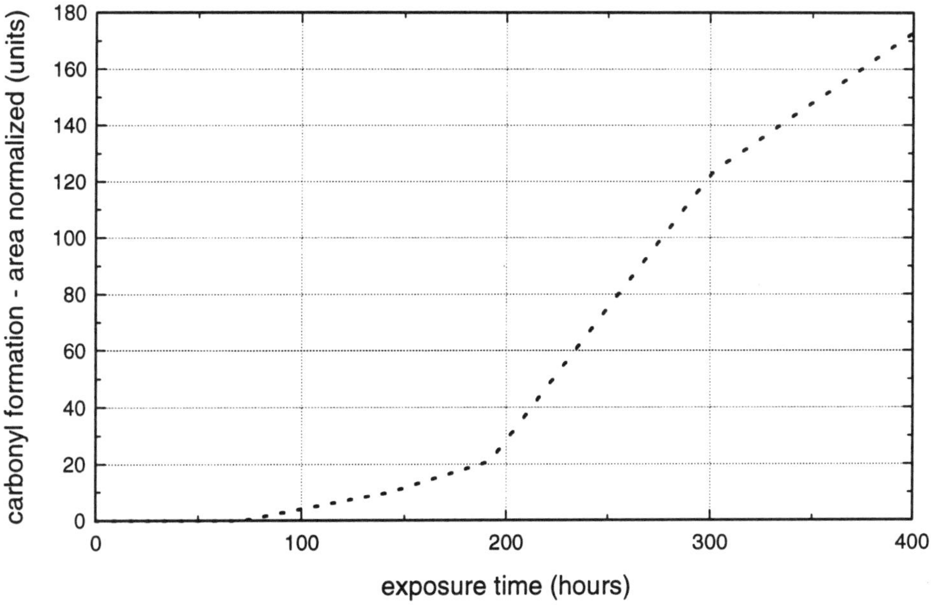

···············	EPDM (26 Mooney (ML 1+4 @ 125°C))
Reference No.	204

GRAPH 216: Xenon Arc Weatherometer Exposure Time vs. Decrease in Molecular Weight of Ethylene Propylene Diene Methylene Terpolymer

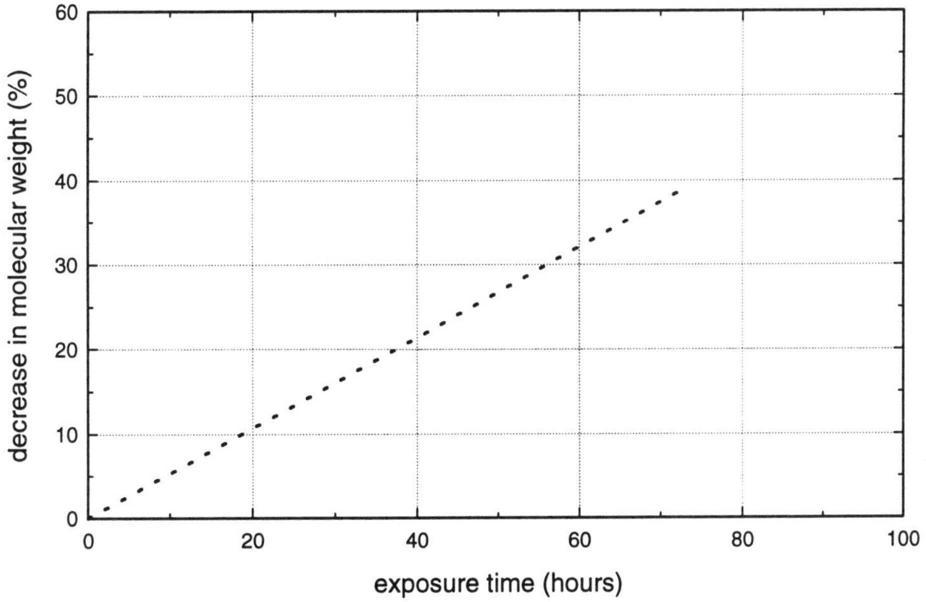

···············	EPDM (26 Mooney (ML 1+4 @ 125°C)); crosslinking occurred after further exposure
Reference No.	204

Chapter 73

Fluoroelastomer

Outdoor Weather Resistance

Montedison: Tecnoflon

Tecnoflon exhibits excellent weathering resistance to sunlight.

Reference: *Tecnoflon,* supplier marketing literature - Montefluos.

Ozone Resistance

Montedison: Tecnoflon

Because of its chemically saturated structure, Tecnoflon exhibits excellent weathering resistance to sunlight and especially to ozone.

Cured items, bent at an angle of 180°, show no cracks after several hundred hours of exposure to ozone concentrations as high as 100 ppm. This is particularly important when considering that standard tests (e.g. in the automotive industry) require 0.5 ppm ozone concentrations.

Reference: *Tecnoflon,* supplier marketing literature - Montefluos.

Chapter 74

Neoprene Rubber

Outdoor Weather Resistance

DuPont: Neoprene W (features: black color, 76 Shore A hardness; material compostion: 100 phr Neoprene W, 1 phr stearic acid, 4 phr magnesium oxide, 2 phr Flectol ODP (Monsanto), 70 phr N-774 black, 25 phr N-330 black, 25 phr process oil, 5 phr zinc oxide, 1 phr sulfur, 0.75 phr monothiruad (Monsanto), 0.75 phr DOTG)

Chloroprene was exposed to conventional Arizona aging by DSET Laboratories, Inc. (Phoenix, AZ) at a 5° tilt from the horizontal. This tilt is preferable to 0° as it allows for some drainage and dirt wash off during rains. Direct exposures are intended for materials which will be used outdoors and subjected to all elements of weather. The exposure period consisted of 6, 12, 24 and 48 months. Chloroprene showed a ΔE color change of less than 3 during the total aging cycle. Over the 48 month period chloroprene experienced a significant increase in hardness and maintained reasonably good tensile strength. Although chloroprene retained tensile strength, its ability to be elongated decreased with exposure time.

Chloroprene was also exposed to conventional Arizona aging with spray. This exposure method is the same as conventional aging, except a water spray is used to induce moisture weathering conditions. The introduction of moisture plays an important role in improving both the relevancy and reproducibility of the weathering test results. Spray nozzles are mounted above the face of the rack at points distributed to insure uniform wetting of the entire exposure area. Distilled water is sprayed for 4 hours preceding sunrise to soak the samples, and then twenty times during the day in 15 second bursts. The purpose of the wetting is twofold. First, the introduction of water in the otherwise arid climate induces and accelerates some degradation modes which do not occur as rapidly, if at all, without moisture. Second, a thermal shock causes a reduction in specimen surface temperatures as much as 14 °C (25 °F). This results in physical stresses which accelerate the degradation process. Chloroprene discolored within 6 months and continued to increase in ΔE with exposure. The hardness of chloroprene increased significantly while the elongation decreased significantly with exposure time. During the exposure period the material maintained tensile strength.

Chloroprene was also exposed to conventional Florida aging. This test method is a real time exposure by DSET Laboratories, Inc. (Homestead, FL) at a 5° tilt from the horizontal. Since this location has a much higher average humidity this exposure is harsher on some materials. Testing on chloroprene was not accomplished over the total exposure period. Samples at 48 months had deteriorated past the point of meaningful testing. The data shows significant color change with exposure time. Chloroprene showed a major decrease in both tensile strength and elongation retention almost immediately after exposure.

Conventional Florida aging with spray was used to test chloroprene. This exposure method is the same as the conventional aging, except water spray is used to induce moisture weathering conditions. Moisture is introduced the same in Florida as it is in Arizona. Chloroprene showed significant color, hardness and tensile strength changes during the 48 months of exposure. The specimens also had a continuing decrease in elongation with exposure time.

Reference: *Weathering Of Santoprene Thermoplastic Rubber Black Ultraviolet Grades,* supplier technical report (TCD00592) - Advanced Elastomer Systems, 1992.

Accelerated Outdoor Weathering Resistance

DuPont: Neoprene W (features: black color, 76 Shore A hardness; material compostion: 100 phr Neoprene W, 1 phr stearic acid, 4 phr magnesium oxide, 2 phr Flectol ODP (Monsanto), 70 phr N-774 black, 25 phr N-330 black, 25 phr process oil, 5 phr zinc oxide, 1 phr sulfur, 0.75 phr monothiruad (Monsanto), 0.75 phr DOTG)

Outdoor accelerated exposure testing was done using equatorial mount with mirrors for acceleration (EMMA) on chloroprene. This test method uses natural sunlight and special reflecting mirrors to concentrate the sunlight to the intensity of about eight suns. A blower which directs air over and under the samples is used to cool the specimens. This limits the increase in surface temperatures of most materials to 10 °C (18 °F) above the maximum surface temperature that is reached by identically mounted samples exposed to direct sunlight at the same time and locations without concentration. DSET Laboratories designed and maintains this equipment. The exposure period was a total of 6 and 12 months, which has been correlated to about 2-1/2 and 5 years of actual aging in a Florida environment. While chloroprene retained its tensile strength, it showed a significant increase in hardness and a significant deterioration in elongation during the exposure period.

Specimens were also exposed to an equatorial mount with mirrors for acceleration plus water (EMMAQUA). This is an accelerated weathering test method which uses the same apparatus as the EMMA, except water spray is used to induce moisture weathering conditions. Chloroprene specimens at 24 months were deteriorated past the ability to obtain reasonable data. Prior to that point there were large color changes, significant loss in tensile strength and a major decrease in elongation with exposure time.

Reference: *Weathering Of Santoprene Thermoplastic Rubber Black Ultraviolet Grades,* supplier technical report (TCD00592) - Advanced Elastomer Systems, 1992.

Accelerated Artificial Weathering Resistance

DuPont: Neoprene W (features: black color, 76 Shore A hardness; material compostion: 100 phr Neoprene W, 1 phr stearic acid, 4 phr magnesium oxide, 2 phr Flectol ODP (Monsanto), 70 phr N-774 black, 25 phr N-330 black, 25 phr process oil, 5 phr zinc oxide, 1 phr sulfur, 0.75 phr monothiruad (Monsanto), 0.75 phr DOTG)

Under xenon arc weatherometer testing (to General Motors standard TM30-2 and SAE J 1885) chloroprene displays a significant change in both color and hardness with exposure. Retention of tensile strength is good. However, elongation retention is poor.

Reference: *Weathering Of Santoprene Thermoplastic Rubber Black Ultraviolet Grades,* supplier technical report (TCD00592) - Advanced Elastomer Systems, 1992.

Ozone Resistance

DuPont: Neoprene W (material compostion: 100 phr Neoprene W, 1 phr stearic acid, 4 phr magnesium oxide, 2 phr Flectol ODP (Monsanto), 70 phr N-774 black, 25 phr N-330 black, 25 phr process oil, 5 phr zinc oxide, 1 phr sulfur, 0.75 phr monothiruad (Monsanto), 0.75 phr DOTG)

Using test method ASTM D1171, neoprene cracked within eight hours. Using test method ASTM D518, neoprene rubber cracked within three hours. Of all rubbers tested in this report neoprene was the only to exhibit cracking and thus, poor ozone resistance.

Reference: *Ozone Resistance Of Santoprene Rubber,* supplier technical report (TCD01787) - Advanced Elastomer Systems, 1986.

TABLE 192: **Outdoor Weathering, Accelerated Outdoor Weathering by EMMAQUA and Accelerated Weathering with a Xenon Arc Weatherometer of Black DuPont Neoprene W Neoprene Rubber.**

Material Family	Neoprene Rubber		
Features	black color	black color	black color
Reference Number	157	157	157

EXPOSURE CONDITIONS

Exposure Type	outdoor weathering	accelerated outdoor weathering	accelerated weathering
Exposure Location	Arizona	Arizona	
Exposure Country	USA	USA	
Exposure Apparatus		EMMAQUA	xenon arc weatherometer
Exposure Time (days)	365	365	216
Exposure Note	water spray added		

PROPERTIES RETAINED (%)

Tensile Strength	79.5	60.1	70.4
Elongation	77.9	28.2	15.3

SURFACE AND APPEARANCE

ΔE Color Change	7.7	9.5	9.2

TABLE 193: **Outdoor Weathering in Arizona and Florida of Black Neoprene Rubber.**

Material Family	Neoprene Rubber				
Features	black color, commercially available grade				
Reference Number	155	155	155	155	155

EXPOSURE CONDITIONS

Exposure Type	outdoor weathering				
Exposure Location	Arizona	Arizona	Arizona	Arizona	Florida
Exposure Country	USA	USA	USA	USA	USA
Exposure Time (days)	183	365	730	1461	730

PROPERTIES RETAINED (%)

Tensile Strength					71 {ac}
Elongation					70 {ac}

CHANGE IN PHYSICAL CHARACTERISTICS

Hardness Change (units)					4 {ae}

SURFACE AND APPEARANCE

ΔE Color Change	9.6 (samples washed)	10.18 (samples washed)	3.2 (samples washed)	11.0 (samples washed)	

TABLE 194: Outdoor Weathering in Arizona of Black DuPont Neoprene W Neoprene Rubber.

Material Family	Neoprene Rubber			
Material Supplier/ Grade	DuPont Neoprene W			
Features	black color			
Reference Number	119	119	119	119
MATERIAL CHARACTERISTICS				
Sample Thickness (mm)	0.76	0.76	0.76	0.76
Shore A hardness	74	74	74	74
MATERIAL COMPOSITION				
Neoprene W	100 phr	100 phr	100 phr	100 phr
DOTG	0.75 phr	0.75 phr	0.75 phr	0.75 phr
Flectol ODP (Monsanto)	2 phr	2 phr	2 phr	2 phr
magnesium oxide	4 phr	4 phr	4 phr	4 phr
monothiruad (Monsanto)	0.75 phr	0.75 phr	0.75 phr	0.75 phr
stearic acid	1 phr	1 phr	1 phr	1 phr
process oil	25 phr	25 phr	25 phr	25 phr
sulfur	1 phr	1 phr	1 phr	1 phr
zinc oxide	5 phr	5 phr	5 phr	5 phr
N-330 black	25 phr	25 phr	25 phr	25 phr
N-774 black	70 phr	70 phr	70 phr	70 phr
EXPOSURE CONDITIONS				
Exposure Type	outdoor weathering			
Exposure Location	Phoenix, Arizona	Phoenix, Arizona	Phoenix, Arizona	Phoenix, Arizona
Exposure Country	USA	USA	USA	USA
Exposure Time (days)	182	365	182	365
Exposure Note	5° south latitude	5° south latitude	5° south latitude; water spray added	5° south latitude; water spray added
Test Lab	DSET Laboratories	DSET Laboratories	DSET Laboratories	DSET Laboratories
PROPERTIES RETAINED (%)				
100% Modulus	139	111	140	125
Tensile Strength	84	71	97	81
Elongation @ Break	70	78	80	79
SURFACE AND APPEARANCE				
ΔE Color Change	9.8 {b}	10.3 {b}	6.1 {b}	7.9 {b}

TABLE 195: Outdoor Weathering in Arizona With and Without Water Spray Added of Black DuPont Neoprene W Neoprene Rubber.

Material Family	Neoprene Rubber							
Material Supplier/ Grade	DuPont Neoprene W							
Features	black color; sheet							
Reference Number	120	120	120	120	120	120	120	120
MATERIAL CHARACTERISTICS								
Shore A hardness	76	76	76	76	76	76	76	76
Sample Thickness (mm)	0.76	0.76	0.76	0.76	0.76	0.76	0.76	0.76
MATERIAL COMPOSITION								
Neoprene W	100 phr	100 phr	100 phr	100 phr	100 phr	100 phr	100 phr	100 phr
DOTG	0.75 phr	0.75 phr	0.75 phr	0.75 phr	0.75 phr	0.75 phr	0.75 phr	0.75 phr
Flectol ODP (Monsanto)	2 phr	2 phr	2 phr	2 phr	2 phr	2 phr	2 phr	2 phr
magnesium oxide	4 phr	4 phr	4 phr	4 phr	4 phr	4 phr	4 phr	4 phr
monothiruad (Monsanto)	0.75 phr	0.75 phr	0.75 phr	0.75 phr	0.75 phr	0.75 phr	0.75 phr	0.75 phr
stearic acid	1 phr	1 phr	1 phr	1 phr	1 phr	1 phr	1 phr	1 phr
process oil	25 phr	25 phr	25 phr	25 phr	25 phr	25 phr	25 phr	25 phr
sulfur	1 phr	1 phr	1 phr	1 phr	1 phr	1 phr	1 phr	1 phr
zinc oxide	5 phr	5 phr	5 phr	5 phr	5 phr	5 phr	5 phr	5 phr
N-330 black	25 phr	25 phr	25 phr	25 phr	25 phr	25 phr	25 phr	25 phr
N-774 black	70 phr	70 phr	70 phr	70 phr	70 phr	70 phr	70 phr	70 phr
EXPOSURE CONDITIONS								
Exposure Type	outdoor weathering				outdoor weathering			
Exposure Location	Phoenix, Arizona				Phoenix, Arizona			
Exposure Country	USA				USA			
Test Lab	DSET Laboratories				DSET Laboratories			
Exposure Note	5° angle				5° angle; with water spray (distilled water sprayed for 4 hours preceeding sunrise and then 20 times during the day in 15 second bursts)			
Exposure Time (days)	182	365	730	1461	182	365	730	1461
PROPERTIES RETAINED (%)								
Tensile Strength	83 {ac}	71 {ac}	68 {ac}	67 {ac}	96 {ac}	80 {ac}	107 {ac}	53 {ac}
Elongation	69 {ac}	77 {ac}	58 {ac}	42 {ac}	80 {ac}	78 {ac}	67 {ac}	25 {ac}
CHANGE IN PHYSICAL CHARACTERISTICS								
Hardness Change (units)	8 {ae}	10 {ae}	8 {ae}	12 {ae}	6 {ae}	8 {ae}	4 {ae}	15 {ae}
SURFACE AND APPEARANCE								
ΔE Color Change	9.6 {ab}	10.2 {ab}	3.2 {ab}	11 {ab}	6 {ab}	7.7 {ab}	6.1 {ab}	9.3 {ab}

TABLE 196: Outdoor Weathering in Florida With and Without Water Spray Added of Black DuPont Neoprene W Neoprene Rubber.

Material Family	Neoprene Rubber							
Material Supplier/ Grade	DuPont Neoprene W							
Features	black color; sheet							
Reference Number	120	120	120	120	120	120	120	120
MATERIAL CHARACTERISTICS								
Shore A hardness	76	76	76	76	76	76	76	76
Sample Thickness (mm)	0.76	0.76	0.76	0.76	0.76	0.76	0.76	0.76
MATERIAL COMPOSITION								
Neoprene W	100 phr	100 phr	100 phr	100 phr	100 phr	100 phr	100 phr	100 phr
DOTG	0.75 phr	0.75 phr	0.75 phr	0.75 phr	0.75 phr	0.75 phr	0.75 phr	0.75 phr
Flectol ODP (Monsanto)	2 phr	2 phr	2 phr	2 phr	2 phr	2 phr	2 phr	2 phr
magnesium oxide	4 phr	4 phr	4 phr	4 phr	4 phr	4 phr	4 phr	4 phr
monothiruad (Monsanto)	0.75 phr	0.75 phr	0.75 phr	0.75 phr	0.75 phr	0.75 phr	0.75 phr	0.75 phr
stearic acid	1 phr	1 phr	1 phr	1 phr	1 phr	1 phr	1 phr	1 phr
process oil	25 phr	25 phr	25 phr	25 phr	25 phr	25 phr	25 phr	25 phr
sulfur	1 phr	1 phr	1 phr	1 phr	1 phr	1 phr	1 phr	1 phr
zinc oxide	5 phr	5 phr	5 phr	5 phr	5 phr	5 phr	5 phr	5 phr
N-330 black	25 phr	25 phr	25 phr	25 phr	25 phr	25 phr	25 phr	25 phr
N-774 black	70 phr	70 phr	70 phr	70 phr	70 phr	70 phr	70 phr	70 phr
EXPOSURE CONDITIONS								
Exposure Type	outdoor weathering				outdoor weathering			
Exposure Location	Homestead, Florida				Homestead, Florida			
Exposure Country	USA				USA			
Test Lab	DSET Laboratories				DSET Laboratories			
Exposure Note	5° angle				5° angle; with water spray (distilled water sprayed for 4 hours preceeding sunrise and then 20 times during the day in 15 second bursts)			
Exposure Time (days)	182	365	730	1461	182	365	730	1461
PROPERTIES RETAINED (%)								
Tensile Strength	34 {ac}	40 {ac}	71 {ac}		85 {ac}		81 {ac}	42 {ac}
Elongation	46 {ac}	48 {ac}	70 {ac}		98 {ac}		72 {ac}	37 {ac}
CHANGE IN PHYSICAL CHARACTERISTICS								
Hardness Change (units)	-1 {ae}	1 {ae}	4 {ae}		2 {ae}		6 {ae}	12 {ae}
SURFACE AND APPEARANCE								
ΔE Color Change	8.3 {ab}				9.5 {ab}			

TABLE 197: Accelerated Outdoor Weathering by EMMA and EMMAQUA and Accelerated Weathering in a Xenon Arc Weatherometer of Black DuPont Neoprene W Neoprene Rubber.

Material Family	Neoprene Rubber							
Material Supplier/ Grade	DuPont Neoprene W							
Features	black color; sheet							
Reference Number	120	120	120	120	120	120	120	120
MATERIAL CHARACTERISTICS								
Shore A hardness	74	74	74	74	74	74	74	74
Sample Thickness (mm)	0.76	0.76	0.76	0.76	0.76	0.76	0.76	0.76
MATERIAL COMPOSITION								
Hypalon 40	100 phr	100 phr	100 phr	100 phr	100 phr	100 phr	100 phr	100 phr
polyethylene	3 phr	3 phr	3 phr	3 phr	3 phr	3 phr	3 phr	3 phr
DPTTS	2 phr	2 phr	2 phr	2 phr	2 phr	2 phr	2 phr	2 phr
litharge (90% paste)	25 phr	25 phr	25 phr	25 phr	25 phr	25 phr	25 phr	25 phr
Thiofide (Monsanto)	0.5 phr	0.5 phr	0.5 phr	0.5 phr	0.5 phr	0.5 phr	0.5 phr	0.5 phr
aromatic process oil	40 phr	40 phr	40 phr	40 phr	40 phr	40 phr	40 phr	40 phr
N-774 black	90 phr	90 phr	90 phr	90 phr	90 phr	90 phr	90 phr	90 phr
EXPOSURE CONDITIONS								
Exposure Type	accelerated weathering		accelerated outdoor weathering			accelerated outdoor weathering		
Exposure Apparatus	xenon arc weatherometer		EMMA			EMMAQUA		
Exposure Note	temperature: 89°C; 3.8 hrs in light at 50% RH, 1 hr in dark at 100% RH; test method: SAE J1885; GM TM30-2							
Exposure Time (days)	125	208	182	365	730	182	365	730
Energy @ 340nm (kJ/m²)	4703	7838						
Energy @ <380nm (MJ/m²)	403	672						
PROPERTIES RETAINED (%)								
Tensile Strength	71 {ac}	70 {ac}	73 {ac}	73 {ac}	83 {ac}	71 {ac}	60 {ac}	
Elongation	19 {ac}	15 {ac}	41 {ac}	20 {ac}	14 {ac}	28 {ac}	29 {ac}	
CHANGE IN PHYSICAL CHARACTERISTICS								
Hardness Change (units)	16 {ae}	16 {ae}	13 {ae}	8 {ae}	18 {ae}	14 {ae}	12 {ae}	
SURFACE AND APPEARANCE								
ΔE Color Change	6.1 {ab}	9.2 {ab}	7.1 {ab}	10.8 {ab}	9.3 {ab}	8.1 {ab}	9.5 {ab}	

TABLE 198: Accelerated Outdoor Weathering in Arizona by EMMA and EMMAQUA of Black DuPont Neoprene W Neoprene Rubber.

Material Family	Neoprene Rubber			
Material Supplier/ Grade	DuPont Neoprene W			
Features	black color			
Reference Number	119	119	119	119
MATERIAL CHARACTERISTICS				
Sample Thickness (mm)	0.76	0.76	0.76	0.76
Shore A hardness	74	74	74	74
MATERIAL COMPOSITION				
Neoprene W	100 phr	100 phr	100 phr	100 phr
DOTG	0.75 phr	0.75 phr	0.75 phr	0.75 phr
Flectol ODP (Monsanto)	2 phr	2 phr	2 phr	2 phr
magnesium oxide	4 phr	4 phr	4 phr	4 phr
monothiruad (Monsanto)	0.75 phr	0.75 phr	0.75 phr	0.75 phr
stearic acid	1 phr	1 phr	1 phr	1 phr
process oil	25 phr	25 phr	25 phr	25 phr
sulfur	1 phr	1 phr	1 phr	1 phr
zinc oxide	5 phr	5 phr	5 phr	5 phr
N-330 black	25 phr	25 phr	25 phr	25 phr
N-774 black	70 phr	70 phr	70 phr	70 phr
EXPOSURE CONDITIONS				
Exposure Type	accelerated outdoor weathering			
Exposure Location	Phoenix, Arizona	Phoenix, Arizona	Phoenix, Arizona	Phoenix, Arizona
Exposure Country	USA	USA	USA	USA
Exposure Apparatus	EMMA	EMMA	EMMAQUA	EMMAQUA
Exposure Time (days)	182	365	182	365
Exposure Note	correlates to approximately 2.5 years actual Florida aging	correlates to approximately 5 years actual Florida aging	correlates to approximately 2.5 years actual Florida aging	correlates to approximately 5 years actual Florida aging
PROPERTIES RETAINED (%)				
Tensile Strength	74	74	72	60
Elongation @ Break	41	32	28	29
SURFACE AND APPEARANCE				
ΔE Color Change	7.2 {b}	7.2 {b}	8.1{b}	9.7 {b}

TABLE 199: Accelerated Weathering in a Xenon Arc Weatherometer of Black DuPont Neoprene W Neoprene Rubber.

Material Family	Neoprene Rubber	
Material Supplier/ Grade	DuPont Neoprene W	
Features	black color	
Reference Number	119	119

MATERIAL CHARACTERISTICS

Sample Thickness (mm)	0.76	0.76
Shore A hardness	74	74

MATERIAL COMPOSITION

Neoprene W	100 phr	100 phr
DOTG	0.75 phr	0.75 phr
Flectol ODP (Monsanto)	2 phr	2 phr
magnesium oxide	4 phr	4 phr
monothiruad (Monsanto)	0.75 phr	0.75 phr
stearic acid	1 phr	1 phr
process oil	25 phr	25 phr
sulfur	1 phr	1 phr
zinc oxide	5 phr	5 phr
N-330 black	25 phr	25 phr
N-774 black	70 phr	70 phr

EXPOSURE CONDITIONS

Exposure Type	accelerated weathering	
Exposure Test Method	GM, TM30-2	GM, TM30-2
Exposure Apparatus	xenon arc weatherometer	xenon arc weatherometer
Exposure Time (days)	125	208
Exposure Temperature (°C)	89	89
Exposure Note	3.8 hrs in light at 50% RH, 1 hr in dark at 100% RH	3.8 hrs in light at 50% RH, 1 hr in dark at 100% RH

PROPERTIES RETAINED (%)

Tensile Strength	71	70
Elongation @ Break	21	17

SURFACE AND APPEARANCE

ΔE Color	6 {b}	9.2 {b}

TABLE 200: Ozone Resistance of DuPont Neoprene Rubber.

Material Family	Neoprene Rubber		
Material Supplier/ Grade	DuPont Neoprene W	DuPont Neoprene W	
Reference Number	158	158	155
MATERIAL COMPOSITION			
Neoprene W	100 phr	100 phr	
DOTG	0.75 phr	0.75 phr	
Flectol ODP (Monsanto)	2 phr	2 phr	
magnesium oxide	4 phr	4 phr	
monothiruad (Monsanto)	0.75 phr	0.75 phr	
stearic acid	1 phr	1 phr	
process oil	25 phr	25 phr	
sulfur	1 phr	1 phr	
zinc oxide	5 phr	5 phr	
N-330 black	25 phr	25 phr	
N-774 black	70 phr	70 phr	
EXPOSURE CONDITIONS			
Exposure Type	ozone	ozone	ozone
Exposure Test Method	ASTM D1171	ASTM D518	ASTM D1171
Exposure Apparatus	Monsanto's ozone chamber	Monsanto's ozone chamber	
Exposure Specimen Note	extruded triangle profile mounted over a mandrel	bent loop specimen	
Exposure Temperature (°C)			40
Exposure Time (hours)	336	336	100
Exposure Note	168 hours @ 400 pphm, 168 hours @ 500 pphm	168 hours @ 400 pphm, 168 hours @ 500 pphm	300 pphm ozone
SURFACE AND APPEARANCE			
Time to Initial Crack (hours)	<8	<3	
Shell Rating (0-10 scale, 0 represents total cracking, 10 represents no cracking)	1	0	
RESULTS OF EXPOSURE			
Ozone Resistance			poor

Chapter 75

Polybutadiene

TABLE 201: **Ozone Resistance of Japan Synthetic Rubber JSR BR Polybutadiene Rubber.**

Material Family	Polybutadiene Rubber												
Material Supplier/ Grade	Japan Synthetic Rubber Company JSR BR 01												
Cure	145°C for 30 minutes												
Reference Number	216	216	216	216	216	216	216	216	216	216	216	216	216
MATERIAL COMPOSITION													
JSR BR 01 (cis 1,4-polybutadiene)	100 phr	100 phr	100 phr	100 phr	70 phr	70 phr	70 phr	70 phr	60 phr	60 phr	60 phr	60 phr	50 phr
JSR RB 820 (syndiotactic 1,2-polybutadiene)					30 phr	30 phr	30 phr	30 phr	40 phr	40 phr	40 phr	40 phr	50 phr
stearic acid	2 phr	2 phr	2 phr	2 phr	2 phr	2 phr	2 phr	2 phr	2 phr	2 phr	2 phr	2 phr	2 phr
process oil	5 phr	5 phr	5 phr	5 phr	5 phr	5 phr	5 phr	5 phr	5 phr	5 phr	5 phr	5 phr	5 phr
antioxidant	1 phr	1 phr	1 phr	1 phr	1 phr	1 phr	1 phr	1 phr	1 phr	1 phr	1 phr	1 phr	1 phr
sulfur	1.75 phr	1.75 phr	1.75 phr	1.75 phr	1.75 phr	1.75 phr	1.75 phr	1.75 phr	1.75 phr	1.75 phr	1.75 phr	1.75 phr	1.75 phr
zinc oxide	3 phr	3 phr	3 phr	3 phr	3 phr	3 phr	3 phr	3 phr	3 phr	3 phr	3 phr	3 phr	3 phr
HAF carbon black	50 phr	50 phr	50 phr	50 phr	50 phr	50 phr	50 phr	50 phr	50 phr	50 phr	50 phr	50 phr	50 phr
curing accelerator	0.8 phr	0.8 phr	0.8 phr	0.8 phr	0.8 phr	0.8 phr	0.8 phr	0.8 phr	0.8 phr	0.8 phr	0.8 phr	0.8 phr	0.8 phr
EXPOSURE CONDITIONS													
Exposure Type	ozone												
Exposure Note	50 pphm ozone; sample at 20% strain												
Exposure Temperature (°C)	40	40	40	40	40	40	40	40	40	40	40	40	40
Exposure Time (hours)	3	10	48	96	3	10	48	96	3	24	48	96	96
RESULTS OF EXPOSURE													
Number of Cracks	innumerable	innumerable	innumerable	innumerable	many	many	many	many	a few	a few	a few	a few	a few
Size of Cracks	can be detected with naked eye	relatively large cracks of <1 mm depth	large cracks of 1-3 mm depth	cracks of 3 mm or more	can be detected with naked eye	relatively large cracks of <1 mm depth	large cracks of 1-3 mm depth	cracks of 3 mm or more	cannot be detected with naked eye but visible through 10x magnifying glass	can be detected with naked eye	large cracks of 1-3 mm depth	cracks of 3 mm or more	cannot be detected with naked eye but visible through 10x magnifying glass

Chapter 76

Polyisoprene Rubber

Ozone Resistance

Goodyear: Natsyn

Like most unsaturated polymers, Natsyn synthetic rubber does not resist degradation from ultraviolet radiation and ozone unless special compounding techniques are employed. In a recipe similar to a truck tire tread compound, two antiozonants are compared alone and in combination with Sunolite 240 wax. Agerite Resin D at 10 phr is included because it is a way to provide ozone protection and avoid the staining characteristics of most antiozonants.

In general, the alkyl-aryl p-phenylenediamines, of which Wingstay 300 is a representative type, give the best ozone resistance to Natsyn and natural rubber compounds. Dialkyl p-phenylenediamines, such as Santoflex 77, are also effective but less persistent. When a dialkyl p-phenylenediamines is used a secondary antiozonant, e.g. Wingstay 100, will extend the longevity of protection. Wingstay 100 should not be used in excess of 0.75-1.00 phr because of its limited solubility in Natsyn.

It is frequently advantageous to use special waxes in combination with the antiozonant. Wax is especially suitable for static applications because it blooms to the surface of the compound and becomes a protective coating. However, wax has a significant adverse effect on ozone resistance in dynamic applications and so a good antiozonant is required for maximum protection.

Polymer blending is another way of improving the aging resistance of Natsyn. Ethylene-propylene terpolymers (EPDM) are inherently immune to attack by ozone and oxygen and, when properly blended, are capable of extending this protection to Natsyn. In addition, EPDM polymers do not stain nor do they disappear through volatilization, and thus these blends are suitable for use in light colored stocks.

In order to achieve the maximum benefit from the EPDM, proper mixing is essential. Natsyn breaks down rapidly on a mill, but EPDM polymers do not. A better dispersion of the two polymers will result if both are close in Mooney viscosity. Experience had shown that ozone resistance is greatly enhanced if the polymers are well blended before other compounding materials are added. Care should be taken to keep the mixing temperature below 149°C in the Banbury to prevent degradation of the Natsyn. Otherwise normal mixing procedures are satisfactory.

Reference: *Natsyn Polyisoprene Rubber,* supplier design guide (700-821-980-540) - Goodyear Chemicals, 1988.

TABLE 202: Ozone Resistance of Goodyear Natsyn Polyisoprene Rubber.

Material Family	Polyisoprene Rubber							
Material Supplier/ Grade	Goodyear Natsyn 2200							
Features	control recipe							
Ingredients Added To Control Recipe		Wingstay 300	Santoflex 77	Agerite Resin D	wax	Wingstay 300 plus wax	Santoflex 77 plus wax	Agerite Resin D plus wax
Reference Number	174	174	174	174	174	174	174	174

MATERIAL COMPOSITION

Natsyn 2200	100 phr	100 phr	100 phr	100 phr	100 phr	100 phr	100 phr	100 phr
Agerite Resin D (antioxidant - RT Vanderbilt)				10 phr				10 phr
Amax (accelerator - RT Vanderbilt)	1 phr	1 phr	1 phr	1 phr	1 phr	1 phr	1 phr	1 phr
Santoflex 77 (antiozonant - Monsanto)			3 phr				3 phr	
stearic acid	2 phr	2 phr	2 phr	2 phr	2 phr	2 phr	2 phr	2 phr
Sunolite 240 Wax (Sun Oil Co.)					2 phr	2 phr	2 phr	2 phr
Wingstay 100 (antiozonant - Goodyear)			1 phr					
Wingstay 29 (antioxidant - Goodyear)	1 phr	1 phr	1 phr	1 phr	1 phr	1 phr	1 phr	1 phr
Wingstay 300 (antiozonant - Goodyear)		4 phr				4 phr		
naphthenic process oil	10 phr	10 phr	10 phr	10 phr	10 phr	10 phr	10 phr	10 phr
sulfur	2 phr	2 phr	2 phr	2 phr	2 phr	2 phr	2 phr	2 phr
zinc oxide	3 phr	3 phr	3 phr	3 phr	3 phr	3 phr	3 phr	3 phr
N-330 black	45 phr	45 phr	45 phr	45 phr	45 phr	45 phr	45 phr	45 phr

EXPOSURE CONDITIONS

Exposure Type	ozone															
Exposure Test Method	Annulus Test															
Exposure Specimen Note	die cut from an ASTM test sheet															
Exposure Note	50 pphm ozone															
Pre Exposure Conditioning		48 hrs. @ 100°C		48 hrs. @ 100°C		48 hrs. @ 100°C		48 hrs. @ 100°C		48 hrs. @ 100°C		48 hrs. @ 100°C		48 hrs. @ 100°C		48 hrs. @ 100°C
Exposure Time (hours)	56	48	56	48	56	48	56	48	56	48	56	48	56	48	56	48
Exposure Temperature (°C)	38	38	38	38	38	38	38	38	38	38	38	38	38	38	38	38
RESULTS OF EXPOSURE																
Minimum Elongation @ Which Cracks Occur (%)	5	5	15	10	25	5	5	15	5	20	50	70	70	30	50	40

TABLE 203: Ozone Resistance of Goodyear Natsyn Polyisoprene Rubber.

Material Family	Polyisoprene Rubber							
Material Supplier/ Grade	Goodyear Natsyn 2200							
Features	control recipe							
Ingredients Added To Control Recipe		Wingstay 300	Santoflex 77	Agerite Resin D	wax	Wingstay 300 plus wax	Santoflex 77 plus wax	Agerite Resin D plus wax
Reference Number	174	174	174	174	174	174	174	174

MATERIAL COMPOSITION

Natsyn 2200	100 phr	100 phr	100 phr	100 phr	100 phr	100 phr	100 phr	100 phr
Agerite Resin D (antioxidant - RT Vanderbilt)				10 phr				10 phr
Amax (accelerator - RT Vanderbilt)	1 phr	1 phr	1 phr	1 phr	1 phr	1 phr	1 phr	1 phr
Santoflex 77 (antiozonant - Monsanto)			3 phr				3 phr	
stearic acid	2 phr	2 phr	2 phr	2 phr	2 phr	2 phr	2 phr	2 phr
Sunolite 240 Wax (Sun Oil Co.)					2 phr	2 phr	2 phr	2 phr
Wingstay 100 (antiozonant - Goodyear)			1 phr					
Wingstay 29 (antioxidant - Goodyear)	1 phr	1 phr	1 phr	1 phr	1 phr	1 phr	1 phr	1 phr
Wingstay 300 (antiozonant - Goodyear)		4 phr				4 phr		
naphthenic process oil	10 phr	10 phr	10 phr	10 phr	10 phr	10 phr	10 phr	10 phr
sulfur	2 phr	2 phr	2 phr	2 phr	2 phr	2 phr	2 phr	2 phr
zinc oxide	3 phr	3 phr	3 phr	3 phr	3 phr	3 phr	3 phr	3 phr
N-330 black	45 phr	45 phr	45 phr	45 phr	45 phr	45 phr	45 phr	45 phr

EXPOSURE CONDITIONS

Exposure Type	ozone															
Exposure Test Method	ASTM D1171 Loop Ozone Test															
Exposure Specimen Note	bent loop with triangular cross section															
Exposure Note	50 pphm ozone; specimens stretched to 23% elongation and relaxed to 0 at rate of 30 cycles per minute															
Pre Exposure Conditioning		48 hrs. @ 100°C		48 hrs. @ 100°C		48 hrs. @ 100°C		48 hrs. @ 100°C		48 hrs. @ 100°C		48 hrs. @ 100°C		48 hrs. @ 100°C		48 hrs. @ 100°C
Exposure Time (hours)	72	48	72	48	72	48	72	48	72	48	72	48	72	48	72	48
Exposure Temperature (°C)	38	38	38	38	38	38	38	38	38	38	38	38	38	38	38	38

RESULTS OF EXPOSURE

ASTM D1171 Rating (0 to 3 scale with 0 as most resistant)	3	2	3	2	3	2	3	2	3	3	1	0	1	2	1	0

TABLE 204: Ozone Resistance of Goodyear Natsyn Polyisoprene Rubber.

Material Family	Polyisoprene Rubber							
Material Supplier/ Grade	Goodyear Natsyn 2200							
Features	control recipe							
Ingredients Added To Control Recipe		Wingstay 300	Santoflex 77	Agerite Resin D	wax	Wingstay 300 plus wax	Santoflex 77 plus wax	Agerite Resin D plus wax
Reference Number	174	174	174	174	174	174	174	174

MATERIAL COMPOSITION

Natsyn 2200	100 phr	100 phr	100 phr	100 phr	100 phr	100 phr	100 phr	100 phr
Agerite Resin D (antioxidant - RT Vanderbilt)				10 phr				10 phr
Amax (accelerator - RT Vanderbilt)	1 phr	1 phr	1 phr	1 phr	1 phr	1 phr	1 phr	1 phr
Santoflex 77 (antiozonant - Monsanto)			3 phr				3 phr	
stearic acid	2 phr	2 phr	2 phr	2 phr	2 phr	2 phr	2 phr	2 phr
Sunolite 240 Wax (Sun Oil Co.)					2 phr	2 phr	2 phr	2 phr
Wingstay 100 (antiozonant - Goodyear)			1 phr					
Wingstay 29 (antioxidant - Goodyear)	1 phr	1 phr	1 phr	1 phr	1 phr	1 phr	1 phr	1 phr
Wingstay 300 (antiozonant - Goodyear)		4 phr				4 phr		
naphthenic process oil	10 phr	10 phr	10 phr	10 phr	10 phr	10 phr	10 phr	10 phr
sulfur	2 phr	2 phr	2 phr	2 phr	2 phr	2 phr	2 phr	2 phr
zinc oxide	3 phr	3 phr	3 phr	3 phr	3 phr	3 phr	3 phr	3 phr
N-330 black	45 phr	45 phr	45 phr	45 phr	45 phr	45 phr	45 phr	45 phr

EXPOSURE CONDITIONS

Exposure Type	ozone															
Exposure Test Method	Static Strip Test															
Exposure Specimen Note	die cut from an ASTM test sheet															
Exposure Note	50 pphm ozone; specimens stretched 15%															
Pre Exposure Conditioning		48 hrs. @ 100°C		48 hrs. @ 100°C		48 hrs. @ 100°C		48 hrs. @ 100°C		48 hrs. @ 100°C		48 hrs. @ 100°C		48 hrs. @ 100°C		48 hrs. @ 100°C
Exposure Time (hours)	72	48	72	48	72	48	72	48	72	48	72	48	72	48	72	48
Exposure Temperature (°C)	38	38	38	38	38	38	38	38	38	38	38	38	38	38	38	38

RESULTS OF EXPOSURE

# Quarters Showing Cracks {e}	4	4	1	1	2	2	3	3	2	1	0	0	0	0	0	0
Size of Cracks (mm) {e}	3.8-5.1	3.8-5.1	>6.35	>6.35	5.1-6.4	5.1-6.4	>6.35	>6.35	5.1-6.4	>6.35	0	0	0	0	0	0

TABLE 205: Ozone Resistance of Goodyear Natsyn Polyisoprene Rubber.

Material Family	Polyisoprene Rubber							
Material Supplier/ Grade	Goodyear Natsyn 2200							
Features	control recipe							
Ingredients Added To Control Recipe		Wingstay 300	Santoflex 77	Agerite Resin D	wax	Wingstay 300 plus wax	Santoflex 77 plus wax	Agerite Resin D plus wax
Reference Number	174	174	174	174	174	174	174	174

MATERIAL COMPOSITION

Natsyn 2200	100 phr	100 phr	100 phr	100 phr	100 phr	100 phr	100 phr	100 phr
Agerite Resin D (antioxidant - RT Vanderbilt)				10 phr				10 phr
Amax (accelerator - RT Vanderbilt)	1 phr	1 phr	1 phr	1 phr	1 phr	1 phr	1 phr	1 phr
Santoflex 77 (antiozonant - Monsanto)			3 phr				3 phr	
stearic acid	2 phr	2 phr	2 phr	2 phr	2 phr	2 phr	2 phr	2 phr
Sunolite 240 Wax (Sun Oil Co.)					2 phr	2 phr	2 phr	2 phr
Wingstay 100 (antiozonant - Goodyear)			1 phr					
Wingstay 29 (antioxidant - Goodyear)	1 phr	1 phr	1 phr	1 phr	1 phr	1 phr	1 phr	1 phr
Wingstay 300 (antiozonant - Goodyear)		4 phr				4 phr		
naphthenic process oil	10 phr	10 phr	10 phr	10 phr	10 phr	10 phr	10 phr	10 phr
sulfur	2 phr	2 phr	2 phr	2 phr	2 phr	2 phr	2 phr	2 phr
zinc oxide	3 phr	3 phr	3 phr	3 phr	3 phr	3 phr	3 phr	3 phr
N-330 black	45 phr	45 phr	45 phr	45 phr	45 phr	45 phr	45 phr	45 phr

EXPOSURE CONDITIONS

Exposure Type	ozone															
Exposure Test Method	Kinetic Stretch Test															
Exposure Specimen Note	die cut from an ASTM test sheet															
Exposure Note	50 pphm ozone; specimens stretched to 23% elongation and relaxed to 0 at rate of 30 cycles per minute															
Pre Exposure Conditioning		48 hrs. @ 100°C		48 hrs. @ 100°C		48 hrs. @ 100°C		48 hrs. @ 100°C		48 hrs. @ 100°C		48 hrs. @ 100°C		48 hrs. @ 100°C		48 hrs. @ 100°C
Exposure Time (hours)	72	48	72	48	72	48	72	48	72	48	72	48	72	48	72	48
Exposure Temperature (°C)	38	38	38	38	38	38	38	38	38	38	38	38	38	38	38	38

RESULTS OF EXPOSURE

# Quarters Showing Cracks {e}	4	4	4	4	4	4	4	4	4	4	4	4	4	4	4	4
Size of Cracks (mm) {e}	1.02-1.5	0.5-0.76	0.25-0.5	0.25-0.5	0.25-0.5	0.25-0.5	1.5-2.03	0.76-1.0	>6.35	0.5-0.76	0.5-0.76	0.5-0.76	0.25-0.5	0.76-1.0	0.76-1.0	0.76-1.0

Chapter 77

Polyurethane

Accelerated Artificial Weathering Resistance

American Cyanamid: TMXDI (META) (chemical type: aliphatic isocyanate (meta-tetramethylxylxylene diisocyanate)); **Polyurethane** (chemical type: $H_{12}MDI$)

The weathering of polyurethanes based on TMXDI (META) aliphatic isocyanate parallels that of other aliphatic isocyanates in that surface quality and gloss retention are excellent. Both TMXDI (META) and an H_{12} MDI systems for urethane roof topcoats were exposed to a Xenon arc weatherometer. QUV values parallel the Xenon data reported., The weatherability of TMXDI (META) aliphatic isocyanate exceeds that observed for the H_{12} MDI system.

Although the molecule of TMXDI (META) aliphatic isocyanate is built around an aromatic ring, the isocyanate functionalities are not conjugated and are aliphatic. Thus, the quinoid structures which give rise to the poor weatherability of true aromatics, such as toluene diisocyanate (TDI), are not possible. In addition, the presence of methyl groups in the place of benzylic hydrogens further enhances the UV stability.

Reference: *Cyanamid TMXDI (META) Aliphatic Isocyanate,* supplier marketing literature (90-4-849 3K 5/90 - UPT-061) - American Cyanamid Company Urethane Chemicals, 1990.

TABLE 206: Accelerated Weathering in a Xenon Arc Weatherometer of Polyurethane Rubber.

Material Family	Polyurethane Rubber			
Material Supplier/ Grade	American Cyanmid TMXDI (META)			
Features	original gloss of approximately 35, roof coating formulation		original gloss of approximately 35	
Reference Number	148	148	148	148
MATERIAL COMPOSITION				
chemical type	aliphatic isocyanate (meta-tetramethylxylxylene diisocyanate)		$H_{12}MDI$	
EXPOSURE CONDITIONS				
Exposure Type	accelerated weathering		accelerated weathering	
Exposure Apparatus	xenon arc weatherometer		xenon arc weatherometer	
Exposure Time (days)	41.6	83.3	41.6	83.3
SURFACE AND APPEARANCE				
Gloss Retained (%)	100	100	61	24

GRAPH 217: Florida Outdoor Weathering Exposure Time vs. Delta b Color Scale of Polyurethane

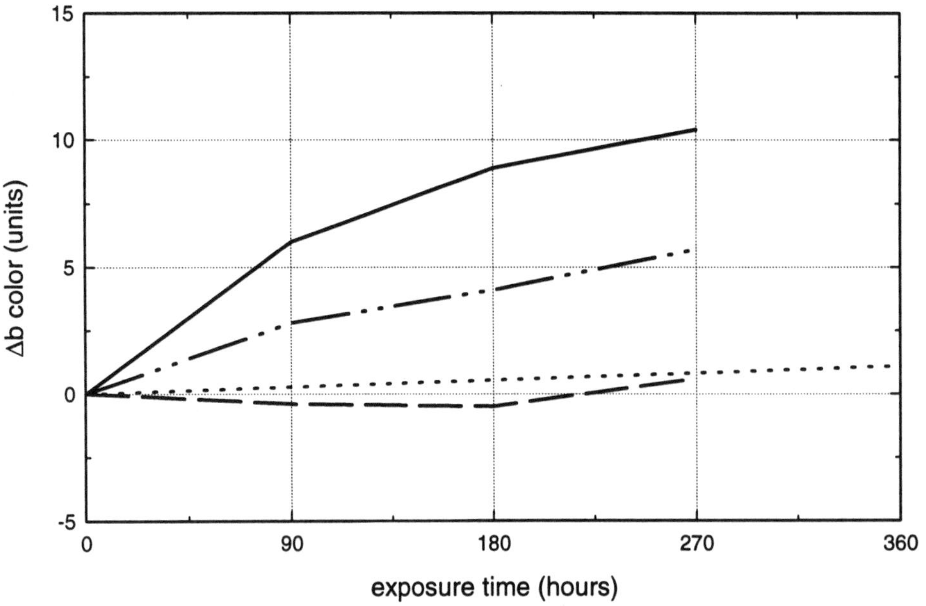

··············	Recticel Colo-Fast HM 100 RIM PU (aliphatic PU; red, tan, brown, high modulus elastomer)
—··—··	PU (aromatic PU, MDI; red, tan, brown; rigid foam)
– – – –	Recticel Colo-Fast DF 230 RIM PU (aliphatic PU; red, tan, brown; rigid structural foam)
———	PU (aromatic PU, MDI; red, tan, brown, high modulus elastomer)
Reference No.	224

Chapter 78

Silicone

Weather Resistance

Shin-Etsu: RTV Silicone

Because these materials are 100% solids and contain no extenders or plasticizers, there is little or no shrinkage due to weathering. Because of their chemical make-up, their resistance to ozone and ultraviolet light is excellent.

Reference: *High Technology RTV Silicones One and Two Component,* supplier marketing literature (SE 5/89 10M) - Shin Etsu Silicones of America, 1989.

TABLE 207: Outdoor Weathering, Accelerated Outdoor Weathering by EMMAQUA, and Accelerated Weathering in a Xenon Arc Weatherometer of White Silicone Rubber.

Material Family	Silicone		
Features	white color	white color	white color
Reference Number	157	157	157
EXPOSURE CONDITIONS			
Exposure Type	outdoor weathering	accelerated outdoor weathering	accelerated weathering
Exposure Location	Arizona	Arizona	
Exposure Country	USA	USA	
Exposure Apparatus		EMMAQUA	xenon arc weatherometer
Exposure Time (days)	365	365	214
Exposure Note	water spray added		
PROPERTIES RETAINED (%)			
Tensile Strength	98.0	72.4	69.1
Elongation	106	55.8	91.6
SURFACE AND APPEARANCE			
ΔE Color Change	0.7	1.7	1.5

TABLE 208: Outdoor Weathering in Arizona and Florida of White Silicone Rubber.

Material Family	Silicone				
Features	white color, commercially available grade				
Reference Number	155	155	155	155	155
EXPOSURE CONDITIONS					
Exposure Type	outdoor weathering				
Exposure Location	Arizona	Arizona	Arizona	Arizona	Florida
Exposure Country	USA	USA	USA	USA	USA
Exposure Time (days)	183	365	730	1461	730
PROPERTIES RETAINED (%)					
Tensile Strength					85 {ac}
Elongation					87 {ac}
CHANGE IN PHYSICAL CHARACTERISTICS					
Hardness Change (units)					2 {ae}
SURFACE AND APPEARANCE					
ΔE Color Change	4.6 (samples washed)	0.68 (samples washed)	3.37 (samples washed)	3.6 (samples washed)	

Glossary of Terms

AATCC method 16 A method by the American Association of Textile Chemists and Colorists for the accelerated testing of the colorfastness of fabrics and yarns to light. Exposure conditions are: for option A - continuous borosilicate glass-filtered core/solid carbon-arc lamp irradiation, option C - continuous daylight carbon-arc lamp irradiation, option D - same as option A but alternating with dark periods, option E - continuous borosilicate/soda-lime glass-filtered water-cooled xenon-arc lamp irradiation, option F - same as option E but alternating with dark periods, option I - continuous soda-lime glass-filtered air-cooled xenon-arc lamp irradiation, option J - the same as option I but alternating with dark periods. The color change of white card-mounted specimens is evaluated by rating against AATCC gray scale, colorimetry in AATCC fading units or determination of delta E value. The exposure stages are determined with simultaneously exposed AATCC Blue Wool Standards or reference specimens. Also called AATCC method 16C.

AATCC method 16C See *AATCC method 16.*

AATCC method 111B A method by the American Association of Textile Chemists and Colorists for the weather resistance testing of fabrics and yarns. The specimens, secured in special frames, are exposed to natural sunlight and weather around the clock in uncovered cabinets with or without backing on racks facing the equator. The conditions of the test are recorded by measuring maximum and minimum temperatures and relative humidity values, hours of wetness for rain and rain plus dew, total and UV radiant energy. The weatherability of the specimens is assessed by comparing with standard samples and by measuring percent residual strength (breaking, tearing, or burst) and/or colorfastness.

AATCC method 111D A method by the American Association of Textile Chemists and Colorists for the testing of the resistance of fabrics and yarns to weather, excluding precipitation. The specimens, secured in special frames, are exposed to natural sunlight and weather around the clock in glass-covered ventilated cabinets with backing on sloped racks. The glass is 2.0-2.5 mm thick, absorbing radiation less than 310 nm. The conditions of the test are recorded by measuring maximum and minimum temperatures and relative humidity values, radiant energy. The weatherability of the specimens is assessed by comparing with standard samples and by measuring percent residual strength (breaking, tearing, or burst) and/or colorfastness.

AATCC method 169 A method by the American Association of Textile Chemists and Colorists for the accelerated testing of the resistance of fabrics and yarns to weather in an artificial weathering apparatus. The specimens are exposed to water- or air-cooled long arc xenon lamp irradiation on variously positioned racks at different radiant energy levels. In the semitropical climate exposure (option 1), the specimens are irradiated 90 minute and water-sprayed 30 minute at 77°C and relative humidity 70%. In the arid climate exposure (option 3), the specimens are irradiated only. In the Columbus, Ohio climate exposure (option 4), the specimens are irradiated at 102 minute and water-sprayed 18 minute at 63°C and relative humidity 50%. The weatherability of the specimens is assessed by comparing with standard samples and by measuring percent residual strength (breaking, tearing, or burst) and/or colorfastness.

ABS See *acrylonitrile butadiene styrene polymer.*

ABS nylon alloy See *acrylonitrile butadiene styrene polymer nylon alloy.*

ABS PC alloy See *acrylonitrile butadiene styrene polymer polycarbonate alloy.*

ABS resin See *acrylonitrile butadiene styrene polymer.*

accelerant See *accelerator.*

accelerated indoor colorfastness test See *accelerated indoor light colorfastness test.*

accelerated indoor light colorfastness test An indoor test that measures the resistance of a colored plastic to fading and/or discoloration on prolonged exposure to a common source of UV radiation such as sunlight, glass-filtered daylight or fluorescent lighting. The test is performed in controlled simulated environment using artificial light sources at high levels of radiance to reduce the test time. The sources used include xenon arc, carbon arc and fluorescent lamps. Also called accelerated indoor colorfastness test.

accelerator A chemical substance that accelerates chemical, photochemical, biochemical, etc. reaction or process, such as crosslinking or degradation of polymers, that is triggered and/or sustained by another substance, such as a curing agent or catalyst, or environmental factor, such as heat, radiation or a microorganism. Also called accelerant, promoter, cocatalyst.

acetal resins Thermoplastics prepared by polymerization of formaldehyde or its trioxane trimer. Acetals have high impact strength and stiffness, low friction coefficient and permeability, good dimensional stability and dielectric properties, and high fatigue strength and thermal stability. Acetals have poor acid and UV resistance and are flammable. Processed by injection and blow molding and extrusion. Used in mechanical parts such as gears and bearings, automotive components, appliances, and plumbing and electronic applications. Also called acetals.

acetals See *acetal resins.*

acrylate styrene acrylonitrile polymer Acrylic rubber-modified thermoplastic with high weatherability. ASA has good heat and chemical resistance, toughness, rigidity, and antistatic properties. Processed by extrusion, thermoforming, and molding. Used in construction, leisure, and automotive applications such as siding, exterior auto trim, and outdoor furniture. Also called ASA.

acrylic resins Thermoplastic polymers of alkyl acrylates such as methyl methacrylates. Acrylic resins have good optical clarity, weatherability, surface hardness, chemical resistance, rigidity, impact strength, and dimensional stability. They have poor solvent resistance, resistance to stress cracking, flexibility, and thermal stability. Processed by casting, extrusion, injection molding, and thermoforming. Used in transparent parts, auto trim, household items, light fixtures, and medical devices. Also called polyacrylates.

acrylonitrile butadiene styrene polymer ABS resins are thermoplastics comprised of a mixture of styrene-acrylonitrile copolymer (SAN) and SAN-grafted butadiene rubber. They have high impact resistance, toughness, rigidity and processability, but low dielectric strength, continuous service temperature, and elongation. Outdoor use requires protective coatings in some cases. Plating grades provide excellent adhesion to metals. Processed by extrusion, blow molding, thermoforming, calendaring and injection molding. Used in household appliances, tools, nonfood packaging, business machinery, interior automotive parts, extruded sheet, pipe and pipe fittings. Also called ABS, ABS resin, acrylonitrile-butadiene-styrene polymer.

acrylonitrile butadiene styrene polymer nylon alloy A thermoplastic processed by injection molding, with properties similar to ABS but higher elongation at yield. Also called ABS nylon alloy.

acrylonitrile butadiene styrene polymer polycarbonate alloy A thermoplastic processed by injection molding and extrusion, with properties similar to ABS. Used in automotive applications. Also called ABS PC alloy.

acrylonitrile copolymer A thermoplastic prepared by copolymerization of acrylonitrile with small amounts of other unsaturated monomers. Has good gas barrier properties and chemical resistance. Processed by extrusion, injection molding, and thermoforming. Used in food packaging.

acrylonitrile-butadiene-styrene polymer See *acrylonitrile butadiene styrene polymer.*

alcohols A class of hydroxy compounds in which a hydroxy group(s) is attached to a carbon chain or ring. Alcohols are produced synthetically from petroleum stock, e.g., by hydration of ethylene, or derived from natural products, e.g., by fermentation of grain. The alcohols are divided in the following groups: monohydric, dihydric, trihydric and polyhydric. Used in organic synthesis, as solvents, plasticizers, fuels, beverages, detergents, etc.

amorphous nylon Transparent aromatic polyamide thermoplastics.

anatase TiO_2 See *anatase titanium dioxide.*

anatase titanium dioxide One of the naturally occurring crystal forms of titanium dioxide. Used as a white or opacifying pigment in a wide range of materials including coatings and plastics. Anatase titanium dioxide has a lower refractive index and opacity than rutile titanium dioxide, another crystal form of this oxide. The pigment is non-migrating, heat resistant, chemically inert and lightfast. Also called anatase TiO_2.

annulus test An ozone resistance test for rubbers that involves a flat-ring specimen mounted as a band over a rack, stretched 0 to 100% and subjected to ozone attack in the test chamber. The specimen is evaluated by comparing to a calibrated template to determine the minimum elongation at which cracking occurred.

anthraquinone An aromatic compound comprising two benzene rings linked by two carbonyl (C=O) groups, $C_6H_4(CO)_2C_6H_4$. Combustible. Used as an intermediate in organic synthesis, mainly in the manufacture of anthraquinone dyes and pigments. One method of preparation is by condensation of 1,4-naphthaquinone with butadiene.

antioxidant A chemical substance capable of inhibiting oxidation or oxidative degradation of another substance such as plastic in which it is incorporated. Antioxidants act by terminating chain-propagating free radicals or by decomposing peroxides, formed during oxidation, into stable products. The first group of antioxidants include hindered phenols and amines; the second - sulfur compounds such as thiols.

Arizona aging An outdoor exposure test performed in Arizona, USA under climatic conditions characterized by high annual sun radiation and temperature to evaluate the weatherability of materials such as coatings or plastics. Special panels with specimens are exposed at a standard tilt angle for 6-48 months. The aging of the specimens is assessed visually and by measuring the changes in color, surface, mechanical (e.g., hardness and tensile strength) and other properties.

aromatic polyester estercarbonate A thermoplastic block copolymer of an aromatic polyester with polycarbonate. Has higher heat distortion temperature than regular polycarbonate.

aromatic polyesters Engineering thermoplastics prepared by polymerization of aromatic polyol with aromatic dicarboxylic anhydride. They are tough with somewhat low chemical resistance. Processed by injection and blow molding, extrusion, and thermoforming. Drying is required. Used in automotive housings and trim, electrical wire jacketing, printed circuit boards, and appliance enclosures.

artificial accelerated weathering An indoor test in which a material, such as plastic, is exposed to simulated outdoor conditions (sunlight, temperature changes, humidity, marine environment) harsher than normal to produce changes or degradation faster. The test is performed in a controlled environment such as climatic or weathering chamber with artificial light sources like xenon arc lamps. The weathering of the specimens is assessed visually and by measuring the changes in color, surface, mechanical (e.g., hardness and tensile strength) and other properties.

ASA See *acrylate styrene acrylonitrile polymer.*

Aspergillus flavus A specie of common mold belonging to the genus Aspergillus. Used alone or in artificial mixtures with other fungi to prepare culture for the testing of mildew resistance of materials such as plastics or fungicidal activity of antimildew agents or fungicides.

Aspergillus niger A specie of common mold belonging to the genus Aspergillus. Used alone or in artificial mixtures with other fungi to prepare culture for the testing of mildew resistance of materials such as plastics or fungicidal activity of antimildew agents or fungicides.

Aspergillus versicolor A specie of common mold belonging to the genus Aspergillus. Used alone or in artificial mixtures with other fungi to prepare culture for the testing of mildew resistance of materials such as plastics or fungicidal activity of antimildew agents or fungicides.

ASTM B117 An American Society for Testing of Materials (ASTM) standard method for salt spray (fog) testing of materials such as metallic and nonmetallic coatings on metal substrates. The specimens are exposed to a fine spray of a NaCl solution at about 35°C in a special chamber for an extended period of time. The results are assessed visually by checking for a specified extent of corrosion damage or by a measuring technique, such as impedance.

ASTM B368 An American Society for Testing of Materials (ASTM) standard method for copper-accelerated acetic acid-salt spray (fog) testing of copper-nickel-chromium or nickel-chromium coatings on metal and plastic substrates. The method sets forth test conditions for evaluating anticorrosive properties of coatings exposed to a fine spray of a $NaCl\text{-}CuCl_2\text{-}AcOH$ solution (pH 3.1-3.3) at about 49°C in a special chamber for 6-720 hours. The results are assessed by measuring the rate of corrosion, i.e., weight loss per unit area of the test panel.

ASTM C793 An American Society for Testing of Materials (ASTM) standard method for testing the effects of accelerated weathering on elastomeric joint sealants used in building construction. The sealants are spread on aluminum plates and exposed to 250 hours of ultraviolet radiation with intermittent water spray in a weathering chamber, followed by freezing at about -26°C for 24 hours with subsequent bending over a mandrel at a specified temperature. The results are evaluated by visual examination of the specimens for cracks.

ASTM D256 An American Society for Testing of Materials (ASTM) standard method for determination of the resistance to breakage by flexural shock of plastics and electrical insulating materials, as indicated by the energy extracted from standard pendulum-type hammers in breaking standard specimens with one pendulum swing. The hammers are mounted on standard machines of either Izod or Charpy type. **Note:** Impact properties determined include Izod or Charpy impact energy normalized per width of the specimen. Also called ASTM method D256-84. See also *impact energy*.

ASTM D279 An American Society for Testing of Materials (ASTM) standard method for determining the bleeding (migration) characteristics of dry pigments by direct solvent extraction of the pigment or by overstriping a film containing the pigment with a white coating and observing for the color migration from the base coat containing the pigment. During extraction, the pigment is shaken with toluene, filtered and filtrate observed for color. The degree of bleeding is rated from none to severe.

ASTM D395 An American Society for Testing of Materials (ASTM) standard methods for testing the capacity of rubber to recover from compressive stress in air or liquid media. The specimen is subjected to compression by a specified force for a definite time at a specified temperature. The difference between the original and the final specimen thickness or compression set is calculated as a percentage of the original thickness by measuring the final thickness 30 minute after stress removal.

ASTM D412 An American Society for Testing of Materials (ASTM) standard methods for determining tensile strength, tensile stress, ultimate elongation, tensile set and set after break of rubber at low, ambient and elevated temperatures using straight, dumbbell and cut-ring specimens.

ASTM D471 An American Society for Testing of Materials (ASTM) standard method for determining the resistance of nonporous rubber to hydrocarbon oils, fuels, service fluids and water. The specimens are immersed in fluids for 22-670 hours at -75 to 250°C, followed by measuring of the changes in mass, volume, tensile strength, elongation and hardness for solid specimens and the changes in breaking strength, burst strength, tear strength and adhesion for rubber-coated fabrics.

ASTM D570 An American Society for Testing of Materials (ASTM) standard method for determining relative rate of water absorption of immersed plastics. The test applies to all kinds of plastics: molded, cast, laminated, etc. The specimens are immersed for 2 to 24 hours or until saturation at ambient temperature, or for 1/2 to 2 hours in boiling water. The absorption is calculated as a percentage of weight gain.

ASTM D638 An American Society for Testing of Materials (ASTM) standard method for determination of tensile properties of unreinforced and reinforced plastics in the form of standard dumbbell-shaped specimens under defined conditions of pretreatment, temperature, humidity and testing machine speed. **Note:** Tensile properties determined include tensile stress (strength) at yield and at break, percentage elongation at yield or at break and modulus of elasticity. Also called ASTM method D638-84. See also *tensile strength*.

ASTM D638 An American Society for Testing of Materials (ASTM) standard method for determining tensile strength, elongation and modulus of elasticity of reinforced or unreinforced plastics in the form of sheet, plate, moldings, rigid tubes and rods. Five (I-V) types, depending on dimensions, of dumbbell-shaped specimens with thickness not exceeding 14 mm are specified. Specified speed of testing varies depending on the specimen type and plastic rigidity. Also called ASTM D638, type IV.

ASTM D638, type IV See *ASTM D638.*

ASTM D746 An American Society for Testing of Materials (ASTM) standard method for determining brittleness temperature of plastics and elastomers by impact. The brittleness temperature is the temperature, at which 50% of cantilever beam specimens fail on impact of a striking edge moving at a linear speed of 1.8-2.1 m/s and striking the specimen at a specified distance from the clamp. The temperature of the specimen is controlled by placing it in a heat-transfer medium, the temperature of which (usually subfreezing) is controlled by a thermocouple.

ASTM D750 An American Society for Testing of Materials (ASTM) standard method for the testing of rubber deterioration in carbon-arc or weathering apparatus. Vulcanized rubber specimens are exposed to the light of carbon arc lamp simulating the sunlight with or without strain and sprayed with water 18 minute every 102 minute. Testing may be carried out in the presence of ozone. The aging is evaluated after specified duration of exposure by determining the percentage decrease in tensile strength and elongation at break and by observing the extent of surface crazing and cracking.

ASTM D1003 An American Society for Testing of Materials (ASTM) standard method for measuring the haze and luminous transmittance of transparent plastics using hazemeter or spectrophotometer

ASTM D-1925-63T See *ASTM D1925.*

ASTM D1006 An American Society for Testing of Materials (ASTM) standard practice for conducting exterior exposure tests of house and trim paints on new wood. Painted testing panels (boards or plywood) are exposed for several years on vertical fences facing both north and south and visually examined for failures at prescribed intervals (1-6 months).

ASTM D1171 An American Society for Testing of Materials (ASTM) standard method for determining the resistance of rubber to outdoor weathering and to surface ozone cracking in a special chamber. The specimens with triangular cross sections are mounted strained around circular mandrels and exposed outdoors for a specified period of time or in the chamber for 72 hours at about 40°C and ozone partial pressure about 50 MPa. After the exposure the specimens are compared to the reference standards to evaluate the degree of cracking in terms of a rating. Also called ASTM D1171 Loop.

ASTM D1171 Loop See *ASTM D1171.*

ASTM D1435 An American Society for Testing of Materials (ASTM) standard practice for outdoor weathering of plastics. The specimens are mounted on special racks at different tilts and exposed outdoors for an extended period of time while the solar radiation, temperature, rainfall, etc. are measured. The weathering of the specimens is assessed visually and by measuring the changes in color, surface, mechanical (e.g., hardness and tensile strength) and other properties.

ASTM D1499 An American Society for Testing of Materials (ASTM) standard practice for operating light- and water-exposure apparatus for plastics. The specimens are exposed to light from a carbon arc lamp and intermittent water spray (18 minute every 2 hours) at about 63°C for 720 hours. The degradation of the specimens is assessed visually and by measuring the changes in surface, color, mechanical (e.g., hardness and tensile strength) and other properties.

ASTM D1708 An American Society for Testing of Materials (ASTM) standard method for determining tensile properties of plastics using microtensile specimens with maximum thickness 3.2 mm and minimum length 38.1 mm, including thin films. Tensile properties include yield strength, tensile strength, tensile strength at break, elongation at break, etc. determined per ASTM D638.

ASTM D1925 An American Society for Testing of Materials (ASTM) standard method for determining the yellowness index or its change for clear or white plastics exposed to daylight as measured by a spectrophotometer. Also called ASTM D-1925-63T.

ASTM D2240 An American Society for Testing of Materials (ASTM) standard method for determining the hardness of materials ranging from soft rubbers to some rigid plastics by measuring the penetration of a blunt (type A) or sharp (type D) indenter of a durometer at a specified force. The blunt indenter is used for softer materials and the sharp indenter - for more rigid materials.

ASTM D2565 An American Society for Testing of Materials (ASTM) standard practice for operating xenon arc-type light exposure apparatus with or without water spray exposure for plastics. The practice specifies the number and location of xenon arc lamps and other characteristics of the apparatus and the specimens. The xenon arc lamps used may be water- or air-cooled and are equipped with a proper filter to simulate sunlight. The test is conducted for 720 hours at about 63°C and at a specified intensity of radiation. The degradation of the specimens is assessed visually and by measuring the changes in surface, color, mechanical (e.g., hardness and tensile strength) and other properties.

ASTM D3679 An American Society for Testing of Materials (ASTM) standard specification for extruded single-wall rigid poly(vinyl chloride) siding that establishes its physical requirements (dimensions, weight, weatherability, impact resistance, expansion, shrinkage, and appearance), test methods for physical requirements and marking.

ASTM D3763 An American Society for Testing of Materials (ASTM) standard method for determination of the resistance of plastics, including films, to high-speed puncture over a broad range of test velocities using load and displacement sensors. **Note:** Puncture properties determined include maximum load, deflection to maximum load point, energy to maximum load point and total energy. Also called ASTM method D3763-86. See also *impact energy*.

ASTM D3841 An American Society for Testing of Materials (ASTM) standard specification for glass fiber-reinforced polyester construction panels. The specification covers classification, inspection, certification, dimensions, weight, appearance, light transmission, weatherability, expansion, impact resistance, flammability and load-deflection properties of panels and their methods of testing.

ASTM D4141 An American Society for Testing of Materials (ASTM) standard practice for conducting accelerated outdoor exposure tests for evaluating exterior durability of coatings applied to metal substrates. The degradation of coatings is accelerated by maximizing the temperature, using a heated (Procedure B) or unheated (Procedure A) black box panel, or by maximizing sunlight irradiation, using Fresnel reflector (Procedure C) panel. A black box panel is a coated metal panel mounted as a closure on a black box to simulate the conditions on the hoods, roofs and deck leads of automobiles parked in direct sunlight. The degradation of the specimens is evaluated by measuring loss of gloss, discoloration, checking, cracking, chalking and blistering. Also called ASTM D4141 - A&B.

ASTM D4141 - A&B See *ASTM D4141.*

ASTM D4275 An American Society for Testing of Materials (ASTM) standard method for determination of butylated hydroxytoluene in ethylene polymers and ethylene-vinyl acetate copolymers by solvent extraction followed by gas chromatographic analysis. Detection of butylated hydroxytoluene is achieved by flame ionization. Butylated hydroxytoluene is a stabilizer used in the manufacture of the ethylene polymers.

ASTM D4434 An American Society for Testing of Materials (ASTM) standard specification for poly(vinyl chloride) sheet roofing, used as a single-ply roof membrane. The material may be unreinforced or reinforced and may contain fibers or fabrics. The specification specifies types, dimensions, mechanical properties, weatherability, resistance to heat aging, appearance, and test methods. The mechanical properties tested include tensile strength and elongation at break, seam strength, tear resistance and tearing strength. The exposure tests include accelerated weathering, water exposure, xenon arc light exposure and fluorescent UV/condensation exposure. Also called ASTM DS D4434.

ASTM D4637 An American Society for Testing of Materials (ASTM) standard specification for unreinforced or fabric-reinforced vulcanized rubber sheet made from EPDM or chloroprene rubber and used as single-ply roof membranes. The specification specifies grades, dimensions, mechanical properties, weatherability, resistance to ozone and heat aging, appearance, and test methods. The mechanical properties tested include tensile strength, set and elongation, seam strength, tear resistance and tearing strength. The exposure tests include water absorption. Also called ASTM DS D4637.

ASTM D5071 An American Society for Testing of Materials (ASTM) standard practice for operating xenon arc-type light exposure apparatus with water spray exposure of photodegradable plastics. The practice specifies the type and irradiance capability of xenon arc lamps and other characteristics of the apparatus and the specimens. The xenon arc lamps used must have a proper filter to simulate sunlight. The test is conducted for specified time at about 63°C and a specified intensity of radiation. Alternating light and dark periods with moisture test programs are recommended. The degradation of the specimens is assessed visually and by measuring the changes in surface, color, mechanical (e.g., hardness and tensile strength) and other properties.

ASTM DS D4434 See *ASTM D4434.*

ASTM DS D4637 See *ASTM D4637.*

ASTM E42 See *ASTM G23.*

ASTM E42-57 See *ASTM G23.*

ASTM E313 An American Society for Testing of Materials (ASTM) standard method for determination of indexes of whiteness and yellowness of near-white, opaque materials such as textiles, paints and plastics. The whiteness or yellowness indexes are one-scale colorimetric attributes measured with a spectrophotometer or a colorimeter having green and blue source-filter-photodetector combination. Yellowness index is calculated as 100(1 - B/G), where B is blue light reflectance and G is daylight luminous reflectance of the specimen. The whiteness index is calculated as 4B - 3G. G and B are proportional to the light flux reflected by the specimen for the CIE Source C when viewed under specified geometric conditions by a receptor whose spectral response duplicates the luminosity function y and z, respectively.

ASTM E838 See *ASTM G90.*

ASTM E896 An American Society for Testing of Materials (ASTM) standard method for determination of photolysis rates, quantum yields and phototransformation products of materials that absorb light directly (without the presence of light sensitizers) in aqueous media, to estimate the environmental rates of photolysis, i.e., light-induced changes in the structure of a molecule. A 3-tier system of testing of increasing complexity is employed. The simplest, tier I test involves the measurement of the concentration of residual material in aqueous solution after up to 6 hours of sunlight exposure. In the tier II test, photolysis rate and rate constants are determined, and in the tier III test - phototransformation products. Also called ASTM E896-92.

ASTM E896-92 See *ASTM E896.*

ASTM G7 An American Society for Testing of Materials (ASTM) standard practice for atmospheric environmental exposure (weathering) testing of nonmetallic materials. The practice specifies test variables that are important to produce consistent results. These variables include exposure location; type, position and construction of specimen panel racks; instrumentation for determining climatological data such as relative humidity; and type and duration of exposure. The types of exposure include direct weathering, exposure behind glass, sheltered storage, undercover and warehouse exposure. The changes in specimens are evaluated by rating against standards.

ASTM G23 An American Society for Testing of Materials (ASTM) standard practice for operating light-exposure apparatus with or without water spray for determination of lightfastness of nonmetallic materials. The specimens are exposed to light from a carbon arc lamp with or without alternating periods of darkness and intermittent water spray at about 63°C for a specified extended period of time. The apparatus (weatherometers) used are as follows: type D - twin enclosed carbon-arc lamp apparatus with rotating specimen drum, type E - single open-flame sunshine carbon-arc lamp apparatus with rotating specimen rack, type H - single enclosed carbon-arc lamp apparatus with rotating specimen rack. The degradation of the specimens is assessed visually and by measuring the changes in color, surface, mechanical (e.g., hardness and tensile strength) and other properties. Also called ASTM E42, ASTM E42-57.

ASTM G26 An American Society for Testing of Materials (ASTM) standard practice for operating light-exposure apparatus with or without water spray for determination of lightfastness of nonmetallic materials. The specimens are exposed to light from a xenon arc lamp with or without alternating periods of darkness and intermittent water spray at about 63°C for a specified extended period of time. The apparatus (weatherometers) used are as follows: type A and B - water-cooled long-arc vertical xenon lamp with borosilicate glass filters, type C and D - single air-cooled xenon-arc lamp apparatus with IR optical filters, type E - triple xenon-arc lamp apparatus with IR optical filters. The degradation of the specimens is assessed visually and by measuring the changes in color, mechanical (e.g., impact and tensile strength) and other properties.

ASTM G53 An American Society for Testing of Materials (ASTM) standard practice for operating light- and water-exposure apparatus for determination of the resistance to deterioration of nonmetallic materials exposed to sunlight and water as rain and dew. The specimens are exposed alternately to light from a fluorescent UV lamp and to condensation in a repetitive cycle. Condensation is produced by exposing one surface of the specimen to a heated, saturated mixture of air and water vapor, while cooling the opposite surface. The degradation of the specimens is assessed visually and by measuring the changes in color, surface, mechanical (e.g., hardness and tensile strength) and other properties.

ASTM G85 An American Society for Testing of Materials (ASTM) standard practice for modified salt spray (fog) testing of materials such as metals, metallic coatings and nonmetallic coatings on metal substrates. The method sets forth the conditions for evaluating anticorrosive properties of materials exposed to a fine spray of a saline solution at about 35°C in a special chamber for an extended period of time. The test may be continuous or cyclic. The saline solution may be acetic acid-NaCl solution (pH 3.1-3.3), acidified seawater or SO_2-NaCl solution. The results are assessed visually by checking for a specified extent of corrosion damage or by a measuring technique, such as impedance.

ASTM G90 An American Society for Testing of Materials (ASTM) standard practice for performing accelerated outdoor weathering of nonmetallic materials using natural sunlight concentrated by Fresnel reflector without (type A) or with (type B) periodic water spray to simulate arid or humid climatic conditions. Fresnel reflector machine comprises a system of flat mirrors that follows the sun with the help of 2 photoreceptor cells, to maximize irradiation and temperature of specimen panels and accelerate weathering. Total solar radiation dose is reported. The degradation of the specimens is assessed visually and by measuring the changes in color, surface, mechanical (e.g., hardness and tensile strength) and other properties. Also called ASTM E838.

ASTM method D256-84 See *ASTM D256.*

ASTM method D3763-86 See *ASTM D3763.*

ASTM method D638-84 See *ASTM D638.*

Atlas Ci65 xenon arc weatherometer See *xenon arc weatherometer.*

Atlas fadeometer See *fadeometer.*

Atlas UV-CON See *fluorescent UV lamp-condensation apparatus.*

azo A prefix indicating an organic group of two nitrogen atoms linked by a double bond, -N=N-, or a class of chemical compounds containing this group, like azo dyes.

B

backed exposure rack A rack for holding specimens or specimen panels during exposure testing that is enclosed from the back to better control the effect of exposure on the exposed side of the specimen.

bending properties See *flexural properties.*

bending strength See *flexural strength.*

bending stress See *flexural stress.*

benzotriazoles A family of UV absorbers for plastics and rubbers, comprising derivatives of 2-(2'-hydroxyphenyl)benzotriazole. They offer strong intensity and broad UV absorption with fairly sharp wavelength cutoff close to the visible region. Higher alkyl derivatives are less volatile and therefore more suitable for higher temperature processing.

biodegradation Microorganism-induced degradation of the material that may involve a negative effect such as loss of performance and cracking of an underground pipe or a positive effect such as decomposition of material waste to simple chemical compounds. Usually, the microorganisms such as fungi induce biodegradation by generating the enzymes, proteins that catalyze degradation reactions. Also called microbiological attack.

bisphenol A polyester A thermoset unsaturated polyester based on bisphenol A and fumaric acid.

black panel temperature Temperature measured by sensors mounted on the black-coated stainless steel panel as specified in the ASTM G26. It is used as the standard reference to control an indoor light-exposure test temperature.

bleaching Complete loss of color of the material as a result of degradation or removal of colored substances present on its surface. Bleaching can be caused by chemical reactions, radiation, etc.

blistering The formation of bubbles on the surface of a nonmetallic coating or a plastic specimen or article as a result of air or other gases or evaporation of moisture or other volatiles trapped beneath. Blistering is often caused by improper application or excessive mixing of paints, heat and polymer degradation.

borosilicate outer filter A borosilicate glass outer filter of xenon-arc lamp that in combination with soda-lime or quartz inner filter selectively screens radiation output, especially in the short UV wavelength region, to simulate the window glass-filtered daylight and sunlight, respectively, in an accelerated light exposure testing apparatus.

breaking elongation See *elongation.*

bubbling The presence of bubbles of trapped air and/or volatile vapors in nonmetallic coating or plastic specimen or article. Bubbling is often caused by improper application or excessive mixing of paints or degassing.

C

C.I. Pigment Orange 20 See *cadmium orange.*

C.I. Pigment Red 108 See *cadmium red.*

C.I. Pigment Yellow 37:1 See *lithopone yellow.*

CA See *cellulose acetate.*

CAB See *cellulose acetate butyrate.*

cadmium orange An orange nonbleeding inorganic pigment based on cadmium sulfide and sulfoselenide (Color Index Number 77202), having high lightfastness and good heat and alkali resistance. Used in PVC and polyolefin plastics, paints and high-gloss baking enamels. Also called C.I. Pigment Orange 20.

cadmium red A red nonbleeding inorganic pigment based on cadmium sulfide and sulfoselenide (Color Index Number 77202), having high lightfastness and good heat and alkali resistance. Used in PVC and polyolefin plastics, paints and high-gloss baking enamels. Also called CP cadmium red, C.I. Pigment Red 108.

cadmium yellow A yellow nonbleeding inorganic pigment based on cadmium sulfide and sulfoselenide (Color Index Number 77202), having high lightfastness and good heat and alkali resistance. Used in PVC and polyolefin plastics, paints and high-gloss baking enamels.

carbon arc weatherometer An apparatus for accelerated indoor weatherability testing of materials such as plastics. Equipped with one to two enclosed or open-flame carbon arc lamps with borosilicate glass filters to simulate the sunlight and with a water spraying device. The lamps used in this apparatus have unnaturally high irradiance in the short wavelength region, especially at 390 nm. As a result it produces less realistic degradation than xenon arc apparatus. Most models allow controlling and monitoring temperature and humidity inside the apparatus and allow alternating dark and light cycles of exposure.

carbon black A black colloidal carbon filler made by the partial combustion or thermal cracking of natural gas, oil, or another hydrocarbon. There are several types of carbon black depending on the starting material and the method of manufacture. Each type of carbon black comes in several grades. Carbon black is widely used as a filler and pigment in rubbers and plastics. It reinforces, increases the resistance to UV light and reduces static charging.

cellulose acetate Thermoplastic esters of cellulose with acetic acid. Have good toughness, gloss, clarity, processability, stiffness, hardness, and dielectric properties, but poor chemical, fire and water resistance and compressive strength. Processed by injection and blow molding and extrusion. Used for appliance cases, steering wheels, pens, handles, containers, eyeglass frames, brushes, and sheeting. Also called CA.

cellulose acetate butyrate Thermoplastic mixed esters of cellulose with acetic and butyric acids. Have good toughness, gloss, clarity, processability, dimensional stability, weatherability, and dielectric properties, but poor chemical, fire and water resistance and compressive strength. Processed by injection and blow molding and extrusion. Used for appliance cases, steering wheels, pens, handles, containers, eyeglass frames, brushes, and sheeting. Also called CAB.

cellulose propionate Thermoplastic esters of cellulose with propionic acid. Have good toughness, gloss, clarity, processability, dimensional stability, weatherability, and dielectric properties, but poor chemical, fire and water resistance and compressive strength. Processed by injection and blow molding and extrusion. Used for appliance cases, steering wheels, pens, handles, containers, eyeglass frames, brushes, and sheeting. Also called CP.

cellulosic plastics Thermoplastic cellulose esters and ethers. Have good toughness, gloss, clarity, processability, and dielectric properties, but poor chemical, fire and water resistance and compressive strength. Processed by injection and blow molding and extrusion. Used for appliance cases, steering wheels, pens, handles, containers, eyeglass frames, brushes, and sheeting.

Chaetomium globosum A specie of common mold belonging to the genus Chaetomium. Used alone or in artificial mixtures with other fungi to prepare culture for the testing of mildew resistance of materials such as plastics or fungicidal activity of antimildew agents or fungicides.

chain scission Breaking of the chainlike molecule of a polymer as a result of chemical, photochemical, etc. reaction such as thermal degradation or photolysis.

chalking Formation of a dry, chalk-like, loose powder on or just beneath the surface of paint film or plastic caused by the exudation of a compounding ingredient such as pigment, often as a result of ingredient migration to the surface and surface degradation.

channel black Carbon black made by impingement of a natural gas flame against a metal plate or channel iron, from which a deposit is scraped. Used as a reinforcing filler in rubbers. Also called gas black.

checking A defect on the surface of a topcoat paint manifesting itself by slight breaks in the film so that underlying coats are visible. Some checks are so small that they are invisible without magnification. Also called surface checks.

chemical saturation Absence of double or triple bonds in a chain organic molecule such as that of most polymers, usually between carbon atoms. Saturation makes the molecule less reactive and polymers less susceptible to degradation and crosslinking. Also called chemically saturated structure.

chemical unsaturation Presence of double or triple bonds in a chain organic molecule such as that of some polymers, usually between carbon atoms. Unsaturation makes the molecule more reactive, especially in free-radical addition reactions such as addition polymerization, and polymers more susceptible to degradation, crosslinking and chemical modification. Also called polymer chain unsaturation.

chemically saturated structure See *chemical saturation.*

chlorendic polyester A chlorendic anhydride-based unsaturated polyester.

chlorinated polyvinyl chloride Thermoplastic produced by chlorination of polyvinyl chloride. Has increased glass transition temperature, chemical and fire resistance, rigidity, tensile strength, and weatherability as compared to PVC. Processed by extrusion, injection molding, casting, and calendering. Used for pipes, auto parts, waste disposal devices, and outdoor applications. Also called CPVC.

chloroethyl alcohol(2-) See *ethylene chlorohydrin.*

chlorohydrins Halohydrins with chlorine as a halogen atom. One of the most reactive of halohydrins. Dichlorohydrins are used in the preparation of epichlorohydrins, important monomers in the manufacture of epoxy resins. Most chlorohydrins are reactive colorless liquids, soluble in polar solvents such as alcohols. **Note:** Chlorohydrins are a class of organic compounds, not to be mixed with a specific member of this class, 1-chloropropane-2,3-diol sometimes called chlorohydrin.

chlorosulfonated polyethylene rubber Thermosetting elastomers containing 20- 40% chlorine. Have good weatherability and heat and chemical resistance. Used for hoses, tubes, sheets, footwear soles, and inflatable boats.

chrome green A green inorganic pigment consisting mainly of lead chromate and used in paints, rubber and plastics. Chrome green has good lightfastness, brightness, weatherability and chemical resistance.

Chromophtal green A green organometallic pigment based on chromium phthalocyanine.

Chromophtal red A red organometallic pigment based on chromium phthalocyanine.

Ci65 xenon arc weatherometer See *xenon arc weatherometer.*

coated molybdate See *coated molybdate orange pigment.*

coated molybdate orange pigment Solid solutions of lead chromate, lead molybdate and lead sulfate and used as dark orange to light red inorganic pigments for plastics. Coated with silica, these pigments exhibit high hiding power, brightness, lightfastness, thermal stability and resistance to bleeding. Also called coated molybdate.

cocatalyst See *accelerator.*

color The wavelength composition of light, specifically of the light reflected or emitted by the material and its visual appearance (red, blue, ect.). Also called hue, tint, coloration.

color change See *discoloration.*

color concentrate let down Reducing the intensity or depth of the color of a concentrated colored pigment dispersion (or paste) in a vehicle (water, binder or solvent) by the addition of a white, or sometimes colorless, pigment.

color difference The square root of the sum of the squares of the chromaticity difference and the lightness difference. Also called delta E, delta E color change.

color masking agent See *masking filler.*

coloration See *color.*

colorimeter An optical instrument for determining or matching colors. The sample's color is matched visually with the color resulting from superposition of the light that passes through 3 primary color filters adjusted to transmit a varying amount of light. Or, an optical instrument for determining concentration of a colored solution by comparing its absorbance in a certain wavelength region with that of a standard solution with known concentration. Also called Hunter color meter.

composite spore suspension A mixture of spores of different fungus species suspended in culture media and used for testing mildew resistance of materials such as paints and plastics and activity of antimildew agents and fungicides.

concentration units The units for measuring the content of a distinct material or substance in a medium other than this material or substance, such as solvent. **Note:** The concentration units are usually expressed in the units of mass or volume of substance per one unit of mass or volume of medium. When the units of substance and medium are the same, the percentage is often used.

conditioning Process of bringing the material or apparatus to a certain condition, e.g., moisture content or temperature, prior to further processing, treatment, etc. Also called conditioning cycle.

conditioning cycle See *conditioning.*

conventional aging Prolonged exposure of materials such as plastics to natural or artificial environmental conditions to produce degradation as in weatherability testing, without accelerating the process by using above normal temperature, irradiation, etc.

conventional aging with spray Prolonged exposure of materials such as plastics to natural or artificial environmental conditions, including water or salt spray, to produce degradation as in weatherability testing, without accelerating the process by using above normal temperature, irradiation, etc.

covulcanization Simultaneous vulcanization of a blend of two or more different rubbers to enhance their individual properties such as ozone resistance. Rubbers are often modified to improve covulcanization.

CP See *cellulose propionate.*

CP cadmium red See *cadmium red.*

CPVC See *chlorinated polyvinyl chloride.*

cracking Appearance of external and/or internal cracks in the material as a result of stress that exceeds the strength of the material. The stress can be external and/or internal and can be caused by a variety of adverse conditions: structural defects, impact, aging, corrosion, etc. or a combination of thereof. Also called cracks. See also *processing defects.*

cracks See *cracking.*

crazes See *crazing.*

crazing Appearance of thin cracks on the surface of the material or, sometimes, minute frost-like internal cracks, as a result of stress that exceeds the strength of the material. Impact, temperature changes, degragation ect. Also called crazes.

crosslinked polyethylene Polyethylene thermoplastics partially photochemically or chemically crosslinked. Have improved tensile strength, dielectric properties, and impact strength at low and elevated temperatures.

crosslinking Reaction of formation of covalent bonds between chain-like polymer molecules or between polymer molecules and low-molecular compounds such as carbon black fillers. As a result of crosslinking polymers, such as thermosetting resins, may become hard and infusible. Crosslinking is induced by heat, UV or electron-beam radiation, oxidation, etc. Crosslinking can be achieved ether between polymer molecules alone as in unsaturated polyesters or with the help of multifunctional crosslinking agents such as diamines that react with functional side groups of the polymers. Crosslinking can be catalysed by the presence of transition metal complexes, thiols and other compounds.

crystal polystyrene See *general purpose polystyrene.*

CTFE See *polychlorotrifluoroethylene.*

CTH-Glas Trac A sun-tracking, automotive glass-covered cabinet for accelerated outdoor weathering of automotive interior materials. The air temperature inside the cabinet is controlled and usually maintained at 70°C during daylight hours and 38°C during night. The relative humidity during night is controlled and usually maintained at 75%. Produced by Heraeus DSET Laboratories, Inc., Phoenix, Arizona.

cycle time See *processing time.*

cyclic compounds A broad class of organic compounds consisting of carbon rings that are saturated, partially unsaturated or aromatic, in which some carbon atoms may be replaced by other atoms such as oxygen, sulfur and nitrogen.NTE.

D

DAP See *diallyl phthalate resins.*

dark cycle In weathering and light exposure testing that simulates outdoor environment, a period when the specimen is not irradiated that alternates with the period of irradiation.

decoloration Complete or partial loss of color of the material as a result of degradation or removal of colored substances present in it. Also called decoloring.

decoloring See *decoloration.*

defects See *processing defects.*

deflection temperature under load See *heat deflection temperature.*

degradation Loss or undesirable change in the properties, such as color, of a material as a result of aging, chemical reaction, wear, exposure, etc. See also *stability.*

delta E See *color difference.*

delta E color change See *color difference.*

dew cycle In light and water exposure testing that simulates outdoor environment, a period when the specimen is exposed to condensation instead of radiation, that alternates with the period of irradiation. Condensation is produced by exposing one surface of the specimen to a heated, saturated mixture of air and water vapor, while cooling the opposite surface.

diallyl phthalate resins Thermosets supplied as diallyl phthalate prepolymer or monomer. Have high chemical, heat and water resistance, dimensional stability, and strength. Shrink during peroxide curing. Processed by injection, compression and transfer molding. Used in glass-reinforced tubing, auto parts, and electrical components. Also called DAP.

dihydric alcohols See *glycols.*

dihydroxy alcohols See *glycols.*

DIN 6167 A German Standards Institute (DIN) standard specifying conditions for determination of yellowness (yellowness index) of near-white or near-colorless materials such as plastics and nonmetallic coatings.

DIN 50031 A German Standards Institute (DIN) standard specifying conditions for salt spray testing of anticorrosive properties of materials such as metallic and nonmetallic coatings on metal substrates. The specimens are exposed to a fine spray of a NaCl (5 g/100 mL), acetic acid-NaCl (pH 3.1-3.4), or $CuCl_2$-acetic acid-NaCl (CASS test, pH 3.1-3.4) solution at about 35°C in a special chamber for 96 hours. The results are assessed by measuring the rate of corrosion, i.e., weight loss per unit area of the test panel.

DIN 53231 A German Standards Institute (DIN) standard specifying conditions for artificial weathering (with wetting) and aging (without wetting) of coatings by exposure to filtered xenon-arc lamp irradiation. The specimens are exposed to 550 W/m^3 average hourly irradiance at 290-800 nm wavelength with (method 1) or without (method 2) wetting and filtering of the light through a 3 mm-thick window glass at relative humidity 40-60%. For method 1, rain is simulated by immersion or spraying and condensation is simulated by spraying the back of the test panels with cold water. The wetting may be continuous or periodic with 102 or 17 minute dry periods. The degradation of the specimens is assessed visually and by measuring the changes in color, mechanical (e.g., impact and tensile strength) and other properties. Also called DIN 53231 method 1, DIN 53231 method 2.

DIN 53231 method 1 See *DIN 53231.*

DIN 53231 method 2 See *DIN 53231.*

DIN 53387 A German Standards Institute (DIN) standard specifying conditions for artificial weathering (with wetting) and aging (without wetting) of plastics and elastomers by exposure to filtered xenon-arc lamp irradiation. The specimens are exposed to 550 W/m^3 average hourly irradiance at 290-800 nm wavelength with (method 1) or without (method 2) wetting and filtering of the light through a 3 mm-thick window glass at relative humidity 40-60%. For method 1, rain is simulated by immersion or spraying and condensation is simulated by spraying the back of the test panels with cold water. The wetting may be continuous or periodic with 102 or 17 minute dry periods. The degradation of the specimens is assessed visually and by measuring the changes in color, mechanical (e.g., impact and tensile strength) and other properties. Also called DIN 53387 method 1, DIN 53387 method 2.

DIN 53387 method 1 See *DIN 53387.*

DIN 53387 method 2 See *DIN 53387.*

DIN 53388 A German Standards Institute (DIN) standard specifying conditions for the testing of resistance to degradation of plastics and elastomers exposed to window glass-filtered daylight. Also called ISO 877, DIN 53388 scale.

DIN 53388 scale See *DIN 53388.*

DIN 53453 A German Standards Institute (DIN) standard specifying conditions for the flexural impact testing of molded or laminated plastics. The bar specimens are either unnotched or notched on one side, mounted on two-point support and struck in the middle (on the unnotched side for notched specimens) by a hammer of the pendulum impact machine. Impact strength of the specimen is calculated relative to the cross-sectional area of the specimen as the energy required to break the specimen equal to the difference between the energy in the pendulum at the instant of impact and the energy remaining after complete fracture of the specimen. Also called DIN 53453 impact test.

DIN 53453 impact test See *DIN 53453.*

DIN 54001 A German Standards Institute (DIN) standard specifying conditions for the preparation and use of gray scale for assessing the change in color during accelerated testing of colorfastness of dyed and printed textiles. Also called ISO 105-A02.

DIN 54003 A German Standards Institute (DIN) standard specifying conditions for the accelerated testing of colorfastness of interior materials in motor vehicles to light by irradiation with glass-filtered xenon-arc lamp. Also called DIN 54003 (FAKRA), ISO 105-B06.

DIN 54003 (FAKRA) See *DIN 54003.*

DIN 54004 A German Standards Institute (DIN) standard specifying conditions for the accelerated testing of colorfastness of dyed and printed textiles to light by irradiation with xenon-arc fading lamp. Also called ISO 105-B02.

DIN 54071 A German Standards Institute (DIN) standard specifying conditions for the accelerated testing of colorfastness of dyed and printed textiles to weather by irradiation with xenon-arc lamp. Also called ISO 105-B04.

discoloration A change in color due to chemical or physical changes in the material. Also called color change.

disperse dyes Nonionic dyes insoluble in water and used mainly as fine aqueous dispersions in dying acetate, polyester and polyamide fibers. A large subclass of disperse dyes comprises low-molecular-weight aromatic azo compounds with amino, hydroxy and alkoxy groups that fix on fibers by forming Van der Waals and hydrogen bonds.

displacement Process of removing one object, e.g., a medium in an apparatus, or its part, and replacing it with another. Also called displacement cycle.

displacement cycle See *displacement.*

double carbon arc weatherometer An apparatus for accelerated indoor testing of weatherability of materials such as plastics. Equipped with carbon arc lamp having a combination of neutral solid and cored electrodes enclosed in a borosilicate glass filter to simulate the sunlight and with a water spraying device. Most models allow controlling and monitoring temperature and humidity inside the apparatus and allow alternating dark and light cycles of exposure.

drop dart impact See *falling weight impact energy.*

drop dart impact energy See *falling weight impact energy.*

drop dart impact strength See *falling weight impact energy.*

drop weight impact See *falling weight impact energy.*

drop weight impact energy See *falling weight impact energy.*

drop weight impact strength See *falling weight impact energy.*

DSET Heraeus DSET Laboratories, Inc. A Phoenix, Arizona company specializing in conventional and accelerated weatherability testing services and equipment.

E

ECTFE See *ethylene chlorotrifluoroethylene copolymer.*

elongation The increase in gauge length of a specimen in tension, measured at or after the fracture, depending on the viscoelastic properties of the material. **Note:** Elongation is usually expressed as a percentage of the original gauge length. Also called tensile elongation, elongation at break, ultimate elongation, breaking elongation, elongation at rupture. See also *tensile strain.*

elongation at break See *elongation.*

elongation at rupture See *elongation.*

EMAC See *ethylene methyl acrylate copolymer.*

embrittlement A reduction or loss of ductility or toughness in materials such as plastics resulting from chemical or physical damage.

EMMA See *equatorial mount with mirrors for acceleration.*

EMMAQUA See *equatorial mount with mirrors for acceleration and water spray.*

enclosed carbon arc See *enclosed carbon arc lamp.*

enclosed carbon arc lamp A light source for accelerated indoor weatherability testing of materials such as plastics that consists of a carbon arc enclosed in a borosilicate glass filter for short wavelengths to simulate the sunlight. The enclosed carbon arc lamps have unnaturally high irradiance in the short wavelength region, especially at 390 nm. As a result they produce less realistic degradation than xenon arc lamps. Also called enclosed carbon arc.

energy quencher A low-molecular-weight organic compound such as polycyclic aromatic compound that retards ionizing radiation-induced polymer degradation by scavenging or trapping part of excited-state energy of the polymer without undergoing significant chemical change due to the highly efficient decay of its own excited states. Also called energy quencher additives, energy scavenger.

energy quencher additives See *energy quencher.*

energy scavenger See *energy quencher.*

EPDM See *EPDM rubber.*

EPDM rubber Sulfur-vulcanizable thermosetting elastomers produced from ethylene, propylene, and a small amount of nonconjugated diene such as hexadiene. Have good weatherability and chemical and heat resistance. Used as impact modifiers and for weather stripping, auto parts, cable insulation, conveyor belts, hoses, and tubing. Also called EPDM.

epoxides Organic compounds containing three-membered cyclic group(s) in which two carbon atoms are linked with an oxygen atom as in an ether. This group is called an epoxy group and is quite reactive, allowing the use of epoxides as intermediates in preparation of certain fluorocarbons and cellulose derivatives and as monomers in preparation of epoxy resins. Also called epoxy compounds.

epoxies See *epoxy resins.*

epoxy compounds See *epoxides.*

epoxy resins Thermosetting polyethers containing crosslinkable glycidyl groups. Usually prepared by polymerization of bisphenol A and epichlorohydrin or reacting phenolic novolaks with epichlorohydrin. Can be made unsaturated by acrylation. Unmodified varieties are cured at room or elevated temperatures with polyamines or anhydrides. Bisphenol A epoxy resins have excellent adhesion and very low shrinkage during curing. Cured novolak epoxies have good UV stability and dielectric properties. Cured acrylated epoxies have high strength and chemical resistance. Processed by molding, casting, coating, and lamination. Used as protective coatings, adhesives, potting compounds, and binders in laminates and composites. Also called epoxies.

epoxyethane See *ethylene oxide.*

EPR See *ethylene propene rubber.*

equatorial mount with mirrors See *equatorial mount with mirrors for acceleration.*

equatorial mount with mirrors and H_2O spray See *equatorial mount with mirrors for acceleration and water spray.*

equatorial mount with mirrors for acceleration An accelerated outdoor weathering test method by Heraeus DSET Laboratories, Inc. for exterior materials with equatorial mount of specimen panels and reflector mirrors to increase solar irradiation. The equatorial mount means that the specimen panels are continuously maintained in the position facing and normal to the sun. The mirrors follow the sun for maximum reflection. Also called EMMA, equatorial mount with mirrors.

equatorial mount with mirrors for acceleration and water spray An accelerated outdoor weathering test method by Heraeus DSET Laboratories, Inc. for exterior materials with equatorial mount of specimen panels and reflector mirrors to increase solar irradiation and water spray to simulate humid climate. The equatorial mount means that the specimen panels are continuously maintained in the position facing and normal to the sun. The mirrors follow the sun for maximum reflection. Also called EMMAQUA, equatorial mount with mirrors and H_2O spray.

ETFE See *ethylene tetrafluoroethylene copolymer.*

ethanediol(1,2-) See *ethylene glycol.*

ethers A class of organic compounds in which an oxygen atom is interposed between two carbon atoms in a chain or a ring. Ethers are derived mainly by catalytic hydration of olefins. The lower molecular weight ethers are dangerous fire and explosion hazards. **Note:** Major types of ethers include aliphatic, cyclic and polymeric ethers.

ethylene acrylic rubber Copolymers of ethylene and acrylic esters. Have good toughness, low temperature properties, and resistance to heat, oil, and water. Used in auto and heavy equipment parts.

ethylene alcohol See *ethylene glycol.*

ethylene chlorohydrin (C_2H_5ClO) Ethylene chlorohydrin, $ClCH_2CH_2OH$, is a colorless liquid easily soluble in most organic liquids and water. It has an autoignition temperature of 450° C (797 °F) and is a moderate fire hazard. Derived by reaction of hydrochlorous acid with ethylene. It is a strong irritant, deadly via inhalation, skin absorption, etc. with TLV of 1 ppm in air. Penetrates through rubber gloves. Used as a solvent for cellulose derivatives, intermediate in organic synthesis (e.g., for ethylene oxide) and sprouting activator. **Note:** Hydrolysis of ethylene oxide during sterilization can result in the formation of ethylene chlorohydrin and its residual presence in sterilized goods. Also called 2-chloroethyl alcohol, glycol chlorohydrin. See also *chemical sterilization agent hydrolysis products.*

ethylene copolymers See *ethylene polymers.*

ethylene methyl acrylate copolymer Thermoplastic copolymers of ethylene with <40% methyl acrylate. Have good dielectric properties, toughness, thermal stability, stress crack resistance, and compatibility with other polyolefins. Transparency decreases with increasing content of acrylate. Processed by blow film extrusion and blow and injection molding. Used in heat-sealable films, disposable gloves, and packaging. Some grades are FDA-approved for food packaging. Also called EMAC.

ethylene polymers Ethylene polymers include ethylene homopolymers and copolymers with other unsaturated monomers, most importantly olefins such as propylene and polar substances such as vinyl acetate. The properties and uses of ethylene polymers depend on the molecular structure and weight. Also called ethylene copolymers.

ethylene propene rubber Stereospecific copolymers of ethylene with propylene. Used as impact modifiers for plastics. Also called EPR.

ethylene tetrafluoroethylene copolymer Thermoplastic alternating copolymer of ethylene and tetrafluoroethylene. Has good impact strength, abrasion and chemical resistance, weatherability, and dielectric properties. Processed by molding, extrusion, and powder coating. Used in tubing, cables, pump parts, and tower packing in a wide temperature range. Also called ETFE.

ethylene vinyl alcohol copolymer Thermoplastics prepared by hydrolysis of ethylene-vinyl acetate polymers. Have good barrier properties, mechanical strength, gloss, elasticity, weatherability, clarity, and abrasion resistance. Barrier properties and processibility improve with increasing content of ethylene due to lower absorption of moisture. Processed by extrusion, coating, blow and blow film molding, and thermoforming. Used as packaging films and container liners. Also called EVOH.

EVOH See *ethylene vinyl alcohol copolymer.*

extenders Relatively inexpensive resin, plasticizer or filler such as carbonate used to reduce cost and/or to improve processing of plastics, rubbers or nonmetallic coatings.

exterior rutile TiO_2 See *exterior rutile titanium dioxide.*

exterior rutile titanium dioxide Special grades of rutile titanium dioxide with increased weatherability that are used as a white or opacifying pigment in a wide range of exterior materials including coatings and plastics. Exterior grades of rutile are often chemically modified, e.g., with chromia, to increase their durability. Also called exterior rutile TiO_2.

F

F40 UVB See *QFS-40 lamp.*

fade-o-meter See *fadeometer.*

fadeometer A light exposure apparatus for accelerated testing of lightfastness of colored materials such as plastics or textiles. The apparatus is equipped with a light source that simulates sunlight but provides a more intense irradiation. Light sources used are carbon- or xenon-arc lamps. Among the manufactures of fadeometers is Atlas Electric Devices Co., Chicago, Illinois. Also called fade-o-meter, Atlas fadeometer.

falling dart impact See *falling weight impact energy.*

falling dart impact energy See *falling weight impact energy.*

falling dart impact strength See *falling weight impact energy.*

falling sand abrasion test A test for determining abrasion resistance of coatings by the amount of abrasive sand required to wear through a unit thickness of the coating, when the sand falls against it at a specified angle from a specified height through a guide tube. Also called falling sand test method.

falling sand test method See *falling sand abrasion test.*

falling weight impact See *falling weight impact energy.*

falling weight impact energy The mean energy of a free-falling dart or weight (tup) that will cause 50% failures after 50 tests to a directly or indirectly stricken specimen. The energy is calculated by multiplying dart mass, gravitational acceleration and drop height. Also called falling weight impact strength, falling weight impact, falling dart impact energy, falling dart impact strength, falling dart impact, drop dart impact energy, drop dart impact strength.

falling weight impact strength See *falling weight impact energy.*

FEP See *fluorinated ethylene propylene copolymer.*

fireproofing agent See *flame retardant.*

five-membered heterocyclic compounds A class of heterocyclic compounds containing rings that consist of five atoms.

five-membered heterocyclic nitrogen compounds A class of heterocyclic compounds containing rings that consist of five atoms, some of which is a nitrogen.

five-membered heterocyclic oxygen compounds A class of heterocyclic compounds containing rings that consist of five atoms, some of which is an oxygen.

flame retardant A substance that reduce the flammability of materials such as plastics or textiles in which it is incorporated. There are inorganic flame retardants such as antimony trioxide (Sb_2O_3) and organic flame retardants such as brominated polyols. The mechanisms of flame retardation vary depending on the nature of material and flame retardant. For example, some flame retardants yield a substantial volume of coke on burning, which prevents oxygen from reaching inside the material and blocks further combustion. Also called fireproofing agent, flame retardant chemical additives, ignition resistant chemical additives.

flame retardant chemical additives See *flame retardant.*

flaw See *processing defects.*

flexural properties Properties describing the reaction of physical systems to flexural stress and strain. Also called bending properties.

flexural strength The maximum stress in the extreme fiber of a specimen loaded to failure in bending. **Note:** Flexural strength is calculated as a function of load, support span and specimen geometry. Also called modulus of rupture in bending, modulus of rupture, bending strength.

flexural stress The maximum stress in the extreme fiber of a specimen in bending. **Note:** Flexural stress is calculated as a function of load at a given strain or at failure, support span and specimen geometry. Also called bending stress.

fluorescent sunlamp with dew See *fluorescent UV lamp-condensation apparatus.*

fluorescent UV lamp-condensation apparatus An apparatus for accelerated indoor weathering of materials such as plastics equipped with a fluorescent UV-A lamp like UVA-340 that produces an energy spectrum with a peak emission at 340 nm and simulates closely the short wavelength region of sun radiation, or with fluorescent UV-B lamp like UVB-313 with peak emission at 313 nm that provides a significantly higher UV radiation output for faster testing. The apparatus is also equipped with a condensation unit that supplies water vapor. The vapor condenses on the surface of the specimen cooled from behind to simulate the dew. Among the manufactures of fluorescent UV lamp-condensation apparatus is Atlas Electric Devices Co., Chicago, Illinois, and The Q-Panel Co., Cleveland, Ohio (Q-U-V accelerated weathering tester). Also called fluorescent sunlamp with dew, fluorescent UV-condensation apparatus, Atlas UV-CON, UV-CON, QUV, Q-U-V accelerated weathering tester.

fluorescent UV-condensation apparatus See *fluorescent UV lamp-condensation apparatus.*

fluorinated ethylene propylene copolymer Thermoplastic copolymer of tetrafluoroethylene and hexafluoropropylene. Has decreased tensile strength and wear and creep resistance, but good weatherability, dielectric properties, fire and chemical resistance, and friction. Decomposes above 204° C (400 °F), releasing toxic products. Processed by molding, extrusion, and powder coating. Used in chemical apparatus liners, pipes, containers, bearings, films, coatings, and cables. Also called FEP.

fluoro rubber See *fluoroelastomers.*

fluoroelastomers Fluorine-containing synthetic rubber with good chemical and heat resistance. Used in underhood applications such as fuel lines, oil and coolant seals, and fuel pumps, and as a flow additive for polyolefins. Also called fluoro rubber.

fluoroplastics See *fluoropolymers.*

fluoropolymers Polymers prepared from unsaturated fluorine-containing hydrocarbons. Have good chemical resistance, weatherability, thermal stability, antiadhesive properties and low friction and flammability, but low creep resistance and strength and poor processibility. The properties vary with the fluorine content. Processed by extrusion and molding. Used as liners in chemical apparatus, in bearings, films, coatings, and containers. Also called fluoroplastics.

fluorosilicones Polymers with chains of alternating silicon and oxygen atoms and trifluoropropyl pendant groups. Most are rubbers.

FMQ See *methylfluorosilicones.*

Fourier-transform infrared spectrometry A spectroscopic technique in which all wavelengths in the infrared region (750-1E+06 nm) simultaneously used to irradiate the sample for a short period of time, and the absorption spectrum is found by mathematical manipulation of the Fourier transform (a periodic function) obtained. Also called FTIR analysis.

FS-40 See *QFS-40 lamp.*

FS-40 (UV-B) lamps See *QFS-40 lamp.*

FS-40 lamp See *QFS-40 lamp.*

FTIR analysis See *Fourier-transform infrared spectrometry.*

fungus resistance See *mildew resistance.*

furnace black The most common type of carbon black made by burning vaporized heavy oil fractions in a furnace with 50% of the air required for complete combustion. It comes in high abrasion, fast extrusion, high modulus, general purpose, semireinforcing, conducting, high elongation, reinforcing and fast-extruding grades among others. Furnace black is widely used as a filler and pigment in rubbers and plastics. It reinforces, increases the resistance to UV light and reduces static charging.

G

gas black See *channel black.*

general purpose polystyrene General purpose polystyrene is an amorphous thermoplastic prepared by homopolymerization of styrene. It has good tensile and flexural strengths, high light transmission and adequate resistance to water, detergents and inorganic chemicals. It is attached by hydrocarbons and has a relatively low impact resistance. Processed by injection molding and foam extrusion. Used to manufacture containers, health care items such as pipettes, kitchen and bathroom housewares, stereo and camera parts and foam sheets for food packaging. Also called crystal polystyrene.

gloss The ratio of the light specularly reflected from a surface of material such as plastic or nonmetallic coating, to the total light reflected. The gloss is measured at a specified angle of incidence of light, e.g., 60°. It usually decreases as a result of weathering.

glycol modified polycyclohexylenedimethylene terephthalate Thermoplastic polyester prepared from glycol, cyclohexylenedimethanol, and terephthalic acid. Has good impact strength and other mechanical properties, chemical resistance, and clarity. Processed by injection molding and extrusion. Can be blended with polycarbonate. Also called PCTG.

glycols Aliphatic alcohols with two hydroxy groups attached to a carbon chain. Can be produced by oxidation of alkenes followed by hydration. Also called dihydric alcohols, dihydroxy alcohols.

gray scale rating Evaluating light-induced changes in the color of materials such as plastics, rubbers and nonmetallic coatings, by comparing to gray scale. Gray scale is a series of achromatic tones (usually ten) having varying proportions of white and black, to give a full range of grays between white and black.

H

halogen compounds A class of organic compounds containing halogen atoms such as chlorine. A simple example is halocarbons but many other subclasses with various functional groups and of different molecular structure exist as well., organic compounds

halohydrins Halogen compounds that contain a halogen atom(s) and a hydroxy (OH) group(s) attached to a carbon chain or ring. Can be prepared by reaction of halogens with alkenes in the presence of water or by reaction of halogens with triols. Halohydrins can be easily dehydrochlorinated in the presence of a base to give an epoxy compound.

HALS See *hindered amine light stabilizer.*

hard clays Sedimentary rocks composed mainly of fine clay mineral material without natural plasticity, or any compacted or indurated clay.

haze The percentage of transmitted light which, in passing through a plastic specimen, deviates from the incident beam via forward scattering more that 2.5° on average (ASTM D883).

HDPE See *high density polyethylene.*

HDT See *heat deflection temperature.*

heat deflection point See *heat deflection temperature.*

heat deflection temperature The temperature at which a material specimen (standard bar) is deflected by a certain degree under specified load. Also called heat distortion temperature, heat distortion point, heat deflection point, deflection temperature under load, tensile heat distortion temperature, HDT.

heat distortion point See *heat deflection temperature.*

heat distortion temperature See *heat deflection temperature.*

heterocyclic compounds A class of cyclic compounds containing rings with some carbon atoms replaced by other atoms such as oxygen, sulfur and nitrogen.

hiding power The capacity of a coating material such as paint and, by extension, of the pigment in it to render invisible or cover up a surface on which it is applied as a film. For paints, hiding power is often expressed in gallons per square foot. Also called opacity.

high density polyethylene A linear polyethylene with density 0.94-0.97 g/cm^3. Has good toughness at low temperatures, chemical resistance, and dielectric properties and high softening temperature, but poor weatherability. Processed by extrusion, blow and injection molding, and powder coating. Used in houseware, containers, food packaging, liners, cable insulation, pipes, bottles, and toys. Also called HDPE.

high impact polystyrene See *impact polystyrene.*

high molecular weight low density polyethylene Thermoplastic with improved abrasion and stress crack resistance and impact strength, but poor processibility and reduced tensile strength. Also called HMWLDPE.

hindered amine light stabilizer Amines, such as piperidine derivatives, with bulky, sterically hindered molecular structure. These light stabilizers photooxidize readily to nitroxyl radicals that neutralize, via recombination, alkyl radicals formed during photodegradation of polymers such as polyolefins and therefore retard this process. Also called HALS.

HIPS See *impact polystyrene.*

HMWLDPE See *high molecular weight low density polyethylene.*

H.P.U.V. A test used to simulate the effect of fluoroescent lighting and filtered sunlight. The H.P.U.V. test uses two lamps simultaneously, a cool white fluoroescent lamp and a filtered sunlamp.

hue See *color.*

Hunter color meter See *colorimeter.*

hydrophilic starch surface See *hydrophilic surface.*

hydrophilic surface Surface of a hydrophilic substance that has a strong ability to bind, adsorb or absorb water; a surface that is readily wettable with water. Hydrophilic substances include carbohydrates such as starch. Also called hydrophilic starch surface.

hydroxy compounds A broad class of organic compounds that contain a hydroxy (OH) group(s) that is not part of another functional group such as carboxylic group. Also called hydroxyl-containing compounds.

hydroxybenzophenone See *2-hydroxybenzophenone.*

hydroxybenzophenone (2-) An aromatic ketone, $C_6H_5COC_6H_4OH$, used as UV absorber in plastics. A solid at room temperature, insoluble in water but soluble in alcohols. Also called hydroxybenzophenone.

hydroxyl-containing compounds See *hydroxy compounds.*

I

ignition resistant chemical additives See *flame retardant.*

impact energy The energy required to break a specimen, equal to the difference between the energy in the striking member of the impact apparatus at the instant of impact and the energy remaining after complete fracture of the specimen. Also called impact strength. See also *ASTM D256, ASTM D3763.*

impact polystyrene Impact polystyrene is a thermoplastic produced by polymerizing styrene dissolved in butadiene rubber. Impact polystyrene has good dimensional stability, high rigidity and good low temperature impact strength, but poor barrier properties, grease resistance and heat resistance. Processed by extrusion, injection molding, thermoforming and structural foam molding. Used in food packaging, kitchen housewares, toys, small appliances, personal care items and audio products. Also called IPS, high impact polystyrene, HIPS, impact PS.

impact property tests Names and designations of the methods for impact testing of materials. Also called impact tests. See also *impact toughness*.

impact PS See *impact polystyrene*.

impact strength The energy required to break a specimen, equal to the difference between the energy in the striking member of the impact apparatus at the instant of impact with the specimen and the energy remaining after complete fracture of the specimen.

impact strength See *impact energy*.

impact tests See *impact property tests*.

impact toughness Property of a material indicating its ability to absorb energy of a high-speed impact by plastic deformation rather than crack or fracture. See also *impact property tests*.

inoculum A small amount of medium containing microorganisms from a pure culture which is used to start a new culture or to introduce microorganisms into a specimen.

ionomers Thermoplastics containing a relatively small amount of pendant ionized acid groups. Have good flexibility and impact strength in a wide temperature range, puncture and chemical resistance, adhesion, and dielectric properties, but poor weatherability, fire resistance, and thermal stability. Processed by injection, blow and rotational molding, blown film extrusion, and extrusion coating. Used in food packaging, auto bumpers, sporting goods, and foam sheets.

IPS See *impact polystyrene*.

iron oxide A dark red powder, Fe_2O_3, widely used, especially as a heat-stable, anticorrosive pigment in coatings. Produced synthetically or from iron ores.

irradience The amount of radiant power per unit area of irradiated surface at a point in time. A measure of radiation exposure, it is often expressed in the units of watt per square meter (W/m^2). Also called radiant flux density.

ISO 105-A02 See *DIN 54001*.

ISO 105-B02 See *DIN 54004*.

ISO 105-B04 See *DIN 54071*.

ISO 105-B06 See *DIN 54003*.

ISO 877 See *DIN 53388*

ISO 4665 part 2 An international standard describing the methods of outdoor exposure of vulcanized rubber to assess its resistance to weathering and ozone cracking under atmospheric conditions with or without glass cover. The specimens are mounted on a sloped rack facing the equator in direct sun, normally without backing, for up to 6 years. Strain is applied for the ozone cracking resistance test. Solar radiation is measured by a photoreceptor or by Blue Wool Standards. Other climatic conditions are recorded as well. The deterioration of the specimen is assessed visually and by measuring the change in color or other properties.

ISO 4665 part 3 An international standard describing the methods of exposure to the artificial daylight from xenon-arc lamp during accelerated light resistance testing of vulcanized rubber. The lamp is equipped with a filter to reduce short wavelength emission and is installed in an enclosure. The test is carried out at black panel temperature about 55 °C, and relative humidity of about 65% without water spray or with intermittent water spray. The deterioration of the specimen is assessed visually and by measuring the change in color or other properties. Radiation dosage is determined by a photoreceptor, Blue Wool Standards (ISO 105-B01) and the gray scale (ISO 105-A02), or other physical standards.

ISO 4892 An international standard describing the methods of exposure to artificial light from xenon-arc, enclosed carbon-arc or open-flame carbon-arc lamp during accelerated light resistance testing of plastics and textiles. The lamp is equipped with a filter to reduce short wavelength emission and is installed in an enclosure. The test is carried out at black panel temperature about 45-63°C, and relative humidity of about 35-90% without water spray or with intermittent water spray. The deterioration of the specimen is assessed visually and by measuring the change in color or other properties. Radiation dosage is determined by a photoreceptor, Blue Wool Standards (ISO 105-B01) and the gray scale (ISO 105-A02), or other physical standards. Also called ISO 4892/2 method A, ISO 4892/2 method B.

ISO 4892/2 method A See *ISO 4892*.

ISO 4892/2 method B See *ISO 4892*.

isophthalate polyester An unsaturated polyester based on isophthalic acid.

Izod See *Izod impact energy*.

Izod impact See *Izod impact energy*.

Izod impact energy The energy required to break a specimen equal to the difference between the energy in the striking member of the Izod-type impact apparatus at the instant of impact and the energy remaining after complete fracture of the specimen. Also called Izod impact, Izod impact strength, Izod.

Izod impact strength See *Izod impact energy*.

J

J See *joule*.

joule A unit of energy in SI system that is equal to the work done when the point of application of a force of one newton (N) is displaced through distance of one meter (m) in the direction of the force. The dimension of joule is N m. Also called J.

K

kinetic strip test An ozone resistance test for rubbers that involves a strip-shaped specimen stretched to 23% and relax to 0 at a rate of 30 cycles per minute, while subjected to ozone attack in the test chamber. The results of the test are reported with 2 digits separated with a virgule. The number before the virgule indicates the number of quarters of the test strip which showed the cracks. The number after the virgule indicates the size of the cracks in length perpendicular to the length of the strip.

L

langley A unit of total solar radiation that is equal to one calorie of heat energy per square centimeter of irradiated surface.

LCP See *liquid crystal polymers*.

LDPE See *low density polyethylene*.

light cycle In weathering and light exposure testing that simulates outdoor environment, a period when the specimen is irradiated that alternates with the period of darkness.

light stability See *lightfastness.*

light transmission See *transmittance.*

lightfastness The resistance of material to deterioration as evident by a change in color, performance, mechanical properties, etc. as a result of exposure to sunlight or artificial light source. Also called light stability.

linear low density polyethylene Linear polyethylenes with density 0.91-0.94 g/cm^3. Has better tensile, tear, and impact strength and crack resistance properties, but poorer haze and gloss than branched low-density polyethylene. Processed by extrusion at increased pressure and higher melt temperatures compared to branched low-density polyethylene, and by molding. Used to manufacture film, sheet, pipe, electrical insulation, liners, bags and food wraps. Also called LLDPE, LLDPE resin.

linear polyethylenes Linear polyethylenes are polyolefins with linear carbon chains. They are prepared by copolymerization of ethylene with small amounts of higher alfa-olefins such as 1-butene. Linear polyethylenes are stiff, tough and have good resistance to environmental cracking and low temperatures. Processed by extrusion and molding. Used to manufacture film, bags, containers, liners, profiles and pipe.

liquid crystal polymers Thermoplastic aromatic copolyesters with highly ordered structure. Have good tensile and flexural properties at high temperatures, chemical, radiation and fire resistance, and weatherability. Processed by sintering and injection molding. Used to substitute ceramics and metals in electrical components, electronics, chemical apparatus, and aerospace and auto parts. Also called LCP.

lithopone red A weather-resistant inorganic red pigment containing cadmium sulfoselenide, zinc sulfide, barium sulfate and zinc oxide, used in plastics.

lithopone yellow A weather-resistant inorganic yellow pigment (Color Index Number 77199:1) containing cadmium sulfide, zinc sulfide, barium sulfate and zinc oxide, used in plastics. Also called C.I. Pigment Yellow 37:1.

LLDPE See *linear low density polyethylene.*

LLDPE resin See *linear low density polyethylene.*

low density polyethylene A branched-chain thermoplastic with density 0.91-0.94 g/cm^3. Has good impact strength, flexibility, transparency, chemical resistance, dielectric properties, and low water permeability and brittleness temperature, but poor heat, stress cracking and fire resistance and weatherability. Processed by extrusion coating, injection and blow molding, and film extrusion. Can be crosslinked. Used in packaging and shrink films, toys, bottle caps, cable insulation, and coatings. Also called LDPE.

luminous transmittance See *transmittance.*

M

macroscopic properties See *thermodynamic properties.*

magnesia See *magnesium oxide.*

magnesium oxide A white powder, MgO, produced by calcining magnesium carbonate or hydroxide in several grades (technical, fused, rubber, etc.). Used as filler, thickening agent in polyesters and inorganic rubber accelerator. Also called magnesia.

masking filler A filler or pigment with low hiding power added in small amounts to clear plastics to mask their natural tint, e.g., blue pigments are added to mask the yellow tint. Also called color masking agent.

masstone green vulcanizates Vulcanized rubber containing no other pigments but a green one.

matte surface A dull or low-gloss surface that is more prone to light scattering than reflection.

MBT See *2-mercaptobenzothiazole.*

mechanical properties Properties describing the reaction of physical systems to stress and strain.

melamine resins Thermosetting resins prepared by condensation of formaldehyde with melamine. Have good hardness, scratch and fire resistance, clarity, colorability, rigidity, dielectric properties, and tensile strength, but poor impact strength. Molding grades are filled. Processed by compression, transfer, and injection molding, impregnation, and coating. Used in cosmetic containers, appliances, tableware, electrical insulators, furniture laminates, adhesives, and coatings.

mercaptobenzothiazole (2-) A nitrogen- and sulfur-containing polyheterocyclic organic thiol used as vulcanization accelerator for rubber. Requires zinc oxide as an activator. Its vulcanizates have a good aging resistance. A yellowish powder with distinctive odor. Combustible. Also called MBT.

mercury cadmium red An inorganic red pigment containing mercury and cadmium sulfides; used mainly in rubber; has good light and heat resistance.

methylfluorosilicones Silicone rubbers containing pendant fluorine and methyl groups. Have good chemical and heat resistance. Used in gasoline lines, gaskets, and seals. Also called FMQ.

methylphenylsilicones Silicone rubbers containing pendant phenyl and methyl groups. Have good resistance to heat, oxidation, and radiation, and compatibility with plastics.

methylsilicone Silicone rubbers containing pendant methyl groups. Have good heat and oxidation resistance. Used in electrical insulation and coatings. Also called MQ.

methylvinylfluorosilicone Silicone rubbers containing pendant vinyl, methyl, and fluorine groups. Can be additionally crosslinked via vinyl groups. Have good resistance to petroleum products at elevated temperatures.

methylvinylsilicone Silicone rubbers containing pendant methyl and vinyl groups. Can be additionally crosslinked via vinyl groups. vulcanized to high degrees of crosslinking. Used in sealants, adhesives, coatings, cables, gaskets, tubing, and electrical tape.

microbiological attack See *biodegradation.*

micron A unit of length equal to 1E-6 meter. Its symbol is Greek small letter mu or mum.

microtensile specimen A small specimen as specified in ASTM D1708 for determining tensile properties of plastics. It has maximum thickness 3.2 mm and minimum length 38.1 mm. Tensile properties determined with this specimen include yield strength, tensile strength, tensile strength at break and elongation at break.

migration A mass-transfer process in which the matter moves from one place to another usually in a slow and spontaneous fashion. In plastics and coatings, migration of pigments, fillers, plasticizers and other ingredients via diffusion or floating to the surface or through interface to other materials results in various defects called blooming, chalking, bronzing, flooding, bleeding, etc.

mildew resistance Ability of a material such as plastic or nonmetallic coating to resist fungus growth and deterioration caused by fungi such as common mold, including polymer degradation and discoloration. Also called fungus resistance.

mineral salt medium A corrosive medium such as aqueous solution, containing mineral or inorganic salt such as sodium chloride (NaCl). Used in material testing, especially of anticorrosive properties.

modified polyphenylene ether Thermoplastic polyphenylene ether alloys with impact polystyrene. Have good impact strength, resistance to heat and fire, but poor resistance to solvents. Processed by injection and structural foam molding and extrusion. Used in auto parts, appliances, and telecommunication devices. Also called MPE, MPO, modified polyphenylene oxide.

modified polyphenylene oxide See *modified polyphenylene ether.*

modulus of rupture See *flexural strength.*

modulus of rupture in bending See *flexural strength.*

molding defects Structural and other defects in material caused inadvertently during molding by using wrong tooling, process parameters, ingredients, pa Also called molding flaw. See also *design, etc. Usually preventable.*

molding flaw See *molding defects.*

molecular weight The sum of the atomic weights of all atoms in a molecule. Also called MW.

molecular weight distribution The relative amounts of polymeric molecules of different weights in a specimen. **Note:** The molecular weight distribution can be expressed in terms of the ratio between weight- and number-average molecular weights. Also called polydispersity, MWD, molecular weight ratio.

molecular weight ratio See *molecular weight distribution.*

Monastral Blue A blue copper phthalocyanine pigment with excellent light stability and high stability to vulcanization and aging; non-bleeding. Used in paints, rubbers, plastics such as PVC and textiles such as rayon. Since it has a low hiding power small amounts of it (2-10 ppm) are added to clear plastics to neutralize slightly yellow tint.

MPE See *modified polyphenylene ether.*

MPO See *modified polyphenylene ether.*

MQ See *methylsilicone.*

mulch film A film, usually dark colored PVC film, used instead of mulch in agriculture, e.g., to prevent fruit rotting and runners and weed growth in cultivation of strawberrys.

MW See *molecular weight.*

MWD See *molecular weight distribution.*

N

nanometer A unit of length equal to 1E-9 meter. Often used to denote the wavelength of radiation, especially in UV and visible spectral region. Also called nm.

neoprene rubber Polychloroprene rubbers with good resistance to petroleum products, heat, and ozone, weatherability, and toughness.

nickel complex light stabilizer Light stabilizers for plastics comprising nickel complexes, such nickel acetylacetonate, dithiolate, or pyridylbenzimidazole complexes. Their stabilization mechanism differs but most act as UV absorbers.

nitrile rubber Rubbers prepared by free-radical polymerization of acrylonitrile with butadiene. Have good resistance to petroleum products, heat, and abrasion. Used in fuel hoses, shoe soles, gaskets, oil seals, and adhesives.

nitroarylamine A class of aromatic amines containing benzene ring(s) with nitro (NO^2 group substituent(s), such as nitroaniline ($O_2NC_6H_4NH_2$). Used as organic intermediates (e.g., in dye synthesis) and antioxidants in propellants and plastics.

nm See *nanometer.*

nonelastomeric thermoplastic polyurethanes See *rigid thermoplastic polyurethanes.*

nonelastomeric thermosetting polyurethane Curable mixtures of isocyanate prepolymers or monomers. Have good abrasion resistance and low-temperature stability, but poor heat, fire, and solvent resistance and weatherability. Processed by reaction injection and structural foam molding, casting, potting, encapsulation, and coating. Used in heat insulation, auto panels and trim, and housings for electronic devices.

notch effect The effect of the presence of specimen notch or its geometry on the outcome of a test such as an impact strength test of plastics. Notching results in local stresses and accelerates failure in both static and cycling testing (mechanical, ozone cracking, etc.).

notched Izod See *notched Izod impact energy.*

notched Izod impact See *notched Izod impact energy.*

notched Izod impact energy The energy required to break a notched specimen equal to the difference between the energy in the striking member of the Izod-type impact apparatus at the instant of impact and the energy remaining after complete fracture of the specimen. **Note:** Energy depends on geometry (e.g., width, depth, shape) of the notch, on the cross-sectional area of the specimen and on the place of impact (on the side of the notch or on the opposite side). In some tests notch is made on both sides of the specimen Also called notched Izod impact strength, notched Izod impact, notched Izod.

notched Izod impact strength See *notched Izod impact energy.*

nylon Thermoplastic polyamides often prepared by ring-opening polymerization of lactam. Have good resistance to most chemicals, abrasion, and creep, good impact and tensile strengths, barrier properties, and low friction, but poor resistance to moisture and light. Have high mold shrinkage. Processed by injection, blow, and rotational molding, extrusion, and powder coating. Used in fibers, auto parts, electrical devices, gears, pumps, appliance housings, cable jacketing, pipes, and films.

nylon 6 Thermoplastic polymer of caprolactam. Has good weldability and mechanical properties but rapidly picks up moisture which results in strength losses. Processed by injection, blow, and rotational molding and extrusion. Used in fibers, tire cord, and machine parts.

nylon 11 Thermoplastic polymer of 11-aminoundecanoic acid having good impact strength, hardness, abrasion resistance, processability, and dimensional stability. Processed by powder coating, rotational molding, extrusion, and injection molding.

Used in electric insulation, tubing, profiles, bearings, and coatings.

nylon 12 Thermoplastic polymer of lauric lactam having good impact strength, hardness, abrasion resistance, and dimensional stability. Processed by powder coating, rotational molding, extrusion, and injection molding. Used in sporting goods and auto parts.

nylon 46 Thermoplastic copolymer of 2-pyrrolidone and caprolactam.

nylon 66 Thermoplastic polymer of adipic acid and hexamethylenediamine having good tensile strength, elasticity, toughness, heat resistance, abrasion resistance, and solvent resistance but low weatherability and color resistance. Processed by injection molding and extrusion. Used in fibers, bearings, gears, rollers, and wire jackets.

nylon 610 Thermoplastic polymer of hexamethylenediamine and sebacic acid having decreased melting point and water absorption and good retention of mechanical properties. Processed by injection molding and extrusion. Used in fibers and machine parts.

nylon 612 Thermoplastic polymer of 1,12-dodecanedioic acid and hexamethylenediamine having good dimensional stability, low moisture absorption, and good retention of mechanical properties. Processed by injection molding and extrusion. Used in wire jackets, cable sheath, packaging film, fibers, bushings, and housings.

nylon 666 Thermoplastic polymer of adipic acid, caprolactam, and hexamethylenediamine having good strength, toughness, abrasion and fatigue resistance, and low friction but high moisture absorption and low dimensional stability. Processed by injection molding and extrusion. Used in electrical devices and auto and mechanical parts.

nylon MXD6 Thermoplastic polymer of m-xylyleneadipamide having good flexural strength and chemical resistance but decreased tensile strength.

O

olefin resins See *polyolefins.*

olefinic resins See *polyolefins.*

olefinic thermoplastic elastomers Blends of EPDM or EP rubbers with polypropylene or polyethylene, optionally crosslinked. Have low density, good dielectric and mechanical properties, and processibility but low oil resistance and high flammability. Processed by extrusion, injection and blow molding, thermoforming, and calendering. Used in auto parts, construction, wire jackets, and sporting goods. Also called TPO.

opacity See *hiding power.*

optical characteristics See *optical properties.*

optical properties The effects of a material or medium on light or other electromagnetic radiation passing through it, such as absorption, reflection, etc. Also called optical characteristics, optical property.

optical property See *optical properties.*

optical transmittance See *transmittance.*

organic compounds Chemical compounds based on carbon chains and rings and also containing hydrogen that can be entirely or partially substituted with oxygen, nitrogen and other elements. Also called organic substances.

organic compounds See *halogen compounds.*

organic substances See *organic compounds.*

outgassing rate See *degassing rate.*

oxazolines Heterocyclic compounds containing five-membered rings in which one carbon is replaced with an oxygen atom and another with a nitrogen atom. Oxazolines are colorless liquids soluble in organic solvents and water. Used as intermediates, e.g., in synthesis of surfactants.

ozone An allotropic form of oxygen, O_3. Unstable gas formed naturally, in air by lightening or in stratosphere by the UV portion of solar radiation, or formed as a result of combustion of fossil fuels, i.e., in exhaust gases from automobiles. O3 is an active oxidizing agent that accelerates deterioration of rubber.

P

PA See *polyamides.*

PABM See *polyaminobismaleimide resins.*

paraffinic plasticizer Plasticizers for plastics comprising liquid or solid long-chain alkanes or paraffins (saturated linear or branched hydrocarbons).

parts per hundred A relative unit of concentration, parts of one substance per 100 parts of another. Parts can be measured by weight, volume, count or any other suitable unit of measure. Used often to denote composition of a blend or mixture, such as plastic, in terms of the parts of a minor ingredient, such as plasticizer, per 100 parts of a major, such as resin. Also called phr.

parts per hundred million A relative unit of concentration, parts of one substance per 100 million parts of another. Parts can be measured by weight, volume, count or any other suitable unit of measure. Used often to denote very small concentration of a substance, such as impurity or toxin, in a medium, such as air. Also called pphm.

PBI See *polybenzimidazoles.*

PBT See *polybutylene terephthalate.*

PC See *polycarbonates.*

PCT See *polycyclohexylenedimethylene terephthalate.*

PCTG See *glycol modified polycyclohexylenedimethylene terephthalate.*

PE copolymer See *polyethylene copolymer.*

PEEK See *polyetheretherketone.*

PEI See *polyetherimides.*

PEK See *polyetherketone.*

pendant aromatic rings Aromatic (conjugated unsaturated rings such as those of benzene, C_6H_6) rings attached to the main chain of a polymer molecule.

Penicillium funiculosum A specie of common mold belonging to the genus Penicillium. Used alone or in artificial mixtures with other fungi to prepare culture for the testing of mildew resistance of materials such as plastics or fungicidal activity of antimildew agents or fungicides.

pentaerythritol A polyol, $C(CH_2OH)_4$, prepared by reaction of acetaldehyde with an excess formaldehyde in alkaline medium. Used as plasticizer and as monomer in alkyd resins.

percent light transmittance See *transmittance.*

perfluoroalkoxy resins Thermoplastic polymers of perfluoroalkoxyethylenes having good creep, heat, and chemical resistance and processibility but low compressive and tensile strengths. Processed by molding, extrusion, rotational molding, and powder coating. Used in films, coatings, pipes, containers, and chemical apparatus linings. Also called PFA.

PES See *polyethersulfone.*

PET See *polyethylene terephthalate.*

PETG See *polycyclohexylenedimethylene ethylene terephthalate.*

PFA See *perfluoroalkoxy resins.*

phase transition See *phase transition properties.*

phase transition point The temperature at which a phase transition occurs in a physical system such as material. **Note:** An example of phase transition is glass transition. Also called phase transition temperature, transition point, transition temperature.

phase transition properties Properties of physical systems such as materials associated with their transition from one phase to another, e.g., from liquid to solid phase. Also called phase transition.

phase transition temperature See *phase transition point.*

phenolic resins Thermoset polymers of phenols with excess or deficiency of aldehydes, mainly formaldehyde, to give resole or novolak resins, respectively. Heat-cured resins have good dielectric properties, hardness, thermal stability, rigidity, and compressive strength but poor chemical resistance and dark color. Processed by coating, potting, compression, transfer, or injection molding and extrusion. Used in coatings, adhesives, potting compounds, handles, electrical devices, and auto parts.

photo bleaching See *photochemical bleaching.*

photo-Fries rearrangement See *photochemical Fries rearrangement.*

photochemical bleaching Complete loss of color of the material as a result of photodegradation of colored substances present in its surface layer. Also called photo bleaching.

photochemical degradation Degradation as a result of light-induced reactions such as photolysis. Also called photodegradation.

photochemical Fries rearrangement Rearrangement of phenolic esters to o- and/or p-phenolic ketones induced by light. Also called photo-Fries rearrangement.

photodegradation See *photochemical degradation.*

photooxidation Oxidation of a substance such as polymer, initiated by light, especially UV portion of it. An important part of polymer photodegradation. Usually proceeds via formation of peroxides, which readily decompose to highly reactive free radicals. Inhibited or retarded in polymers by antioxidants and light stabilizers.

phr See *parts per hundred.*

phthalocyanine A nitrogen-containing heterocyclic organic compound, $(C_6H_4C_2N)_2(C_6H_4C_2NH)_2N_4$, belonging to the group of benzoporphyrins and comprising 4 isoindole groups jointed by 4 nitrogen atoms. Readily forms salt complexes with copper, chromium, iron, etc., that are important green and blue dyes and pigments. These pigments have high light and chemical stability. Used in coatings, plastics and textiles.

PI See *polyimides.*

plasticizer A substance incorporated into a material such as plastic or rubber to increase its softness, processability and flexibility via solvent or lubricating action or by lowering its molecular weight. Plasticizers can lower melt viscosity, improve flow and increase low-temperature resilience of material. Most plasticizers are nonvolatile organic liquids or low-melting-point solids, such as dioctyl phthalate or stearic acid. They have to be non-bleeding, nontoxic and compatible with material. Sometimes plasticizers play a dual role as stabilizers or crosslinkers.

plastics See *polymers.*

PMMA See *polymethyl methacrylate.*

PMP See *polymethylpentene.*

polyacrylates See *acrylic resins.*

polyallomer Crystalline thermoplastic block copolymers of ethylene, propylene, and other olefins. Have good impact strength and flex life and low density.

polyamide thermoplastic elastomers Copolymers containing soft polyether and hard polyamide blocks having good chemical, abrasion, and heat resistance, impact strength, and tensile properties. Processed by extrusion and injection and blow molding. Used in sporting goods, auto parts, and electrical devices. Also called polyamide TPE.

polyamide TPE See *polyamide thermoplastic elastomers.*

polyamides Thermoplastic aromatic or aliphatic polymers of dicarboxylic acids and diamines, of amino acids, or of lactams. Have good mechanical properties, chemical resistance, and antifriction properties. Processed by extrusion and molding. Used in fibers and molded parts. Also called PA.

polyaminobismaleimide resins Thermoset polymers of aromatic diamines and bismaleimides having good flow and thermochemical properties and flame and radiation resistance. Processed by casting and compression molding. Used in aircraft parts and electrical devices. Also called PABM.

polyarylamides Thermoplastic crystalline polymers of aromatic diamines and aromatic dicarboxylic anhydrides having good heat, fire, and chemical resistance, property retention at high temperatures, dielectric and mechanical properties, and stiffness but poor light resistance and processibility. Processed by solution casting, molding, and extrusion. Used in films, fibers, and molded parts.

polyarylsulfone Thermoplastic aromatic polyether-polysulfone having good heat, fire, and chemical resistance, impact strength, resistance to environmental stress cracking, dielectric properties, and rigidity. Processed by injection and compression molding and extrusion. Used in circuit boards, lamp housings, piping, and auto parts.

polybenzimidazoles Mainly polymers of 3,3',4,4'-tetraminonbiphenyl(diaminobenzidine) and diphenyl isophthalate. Have good heat, fire, and chemical resistance. Used as coatings and fibers in aerospace and other high-temperature applications. Also called PBI.

polybutylene terephthalate Thermoplastic polymer of dimethyl terephthalate and butanediol having good tensile strength, dielectric properties, and chemical and water resistance, but poor impact strength and heat resistance. Processed by injection and blow molding, extrusion, and thermoforming. Used in auto body parts, electrical devices, appliances, and housings. Also called PBT.

polycarbodiimide Polymers containing -N=C=N- linkages in the main chain, typically formed by catalyzed polycondensation of polyisocyanates. They are used to prepare open-celled foams with superior thermal stability. Sterically hindered polycarbodiimides are used as hydrolytic stabilizers for polyester-based urethane elastomers.

polycarbonate See *polycarbonates.*

polycarbonate polyester alloys High-performance thermoplastics processed by injection and blow molding. Used in auto parts.

polycarbonate resins See *polycarbonates.*

polycarbonates Polycarbonates are thermoplastics prepared by either phosgenation of dihydric aromatic alcohols such as bisphenol A or by transesterification of these alcohols with carbonates, e.g., diphenyl carbonate. Polycarbonates consist of chains with repeating carbonyldioxy groups and can be aliphatic or aromatic. They have very good mechanical properties, especially impact strength, low moisture absorption and good thermal and oxidative stability. They are self-extinguishing and some grades are transparent. Polycarbonates have relatively low chemical resistance and resistance to stress cracking. Processed by injection and blow molding, extrusion, thermoforming at relatively high processing temperatures. Used in telephone parts, dentures, business machine housings, safety equipment, nonstaining dinnerware, food packaging, etc. Also called polycarbonate, PC, polycarbonate resins.

polychlorotrifluoroethylene Thermoplastic polymer of chlorotrifluoroethylene having good transparency, barrier properties, tensile strength, and creep resistance, modest dielectric properties and solvent resistance, and poor processibility. Processed by extrusion, injection and compression molding, and coating. Used in chemical apparatus, low-temperature seals, films, and internal lubricants. Also called CTFE.

polycyclohexylenedimethylene ethylene terephthalate Thermoplastic polymer of cyclohexylenedimethylenediol, ethylene glycol, and terephthalic acid. Has good clarity, stiffness, hardness, and low-temperature toughness. Processed by injection and blow molding and extrusion. Used in containers for cosmetics and foods, packaging film, medical devices, machine guards, and toys. Also called PETG.

polycyclohexylenedimethylene terephthalate Thermoplastic polymer of cyclohexylenedimethylenediol and terephthalic acid having good heat resistance. Processed by molding and extrusion. Also called PCT.

polydispersity See *molecular weight distribution.*

polyester resins See *polyesters.*

polyester thermoplastic elastomers Copolymers containing soft polyether and hard polyester blocks having good dielectric strength, chemical and creep resistance, dynamic performance, appearance, and retention of properties in a wide temperature range but poor light resistance. Processed by injection, blow, and rotational molding, extrusion casting, and film blowing. Used in electrical insulation, medical products, auto parts, and business equipment. Also called polyester TPE.

polyester TPE See *polyester thermoplastic elastomers.*

polyesters A broad class of polymers usually made by condensation of a diol with dicarboxylic acid or anhydride. Polyesters consist of chains with repeating carbonyloxy group and can be aliphatic or aromatic. There are thermosetting polyesters, such as alkyd resins and unsaturated polyesters, and thermoplastic polyesters such as PET. The properties, processing methods and applications of polyesters vary widely. Also called polyester resins.

polyetheretherketone Semi-crystalline thermoplastic aromatic polymer having good chemical, heat, fire, and radiation resistance, toughness, rigidity, bearing strength, and processibility. Processed by injection molding, spinning, cold forming, and extrusion. Used in fibers, films, auto engine parts, aerospace composites, and electrical insulation. Also called PEEK.

polyetherimides Thermoplastic cyclized polymers of aromatic diether dianhydrides and aromatic diamine. Have good chemical, creep, and heat resistance and dielectric properties. Processed by extrusion, thermoforming, and compression, injection, and blow molding. Used in auto parts, jet engines, surgical instruments, industrial apparatus, food packaging, cookware, and computer disks. Also called PEI.

polyetherketone Thermoplastic having good heat and chemical resistance. thermal stability. Used in advanced composites, wire coating, filters, integrated circuit boards, and bearings. Also called PEK.

polyethersulfone Thermoplastic aromatic polymer having good heat and fire resistance, transparency, dielectric properties, dimensional stability, rigidity, and toughness, but poor solvent and stress cracking resistance, processibility, and weatherability. Processed by injection, blow, and compression molding and extrusion. Used in high temperature applications electrical devices, medical devices, housings, and aircraft and auto parts. Also called PES.

polyethylene copolymer Thermoplastics polymers of ethylene with other olefins such as propylene. Processed by molding and extrusion. Also called PE copolymer.

polyethylene terephthalate Thermoplastic polymer of ethylene glycol with terephthalic acid. Has good hardness, wear and chemical resistance, dimensional stability, and dielectric properties. High-crystallinity grades have good tensile strength and heat resistance. Processed by extrusion and injection and blow molding. Used in fibers, food packaging (films, bottles, trays), magnetic tapes, and photo films. Also called PET.

polyimides Thermoplastic aromatic cyclized polymers of trimellitic anhydride and aromatic diamine. Have good tensile strength, dimensional stability, dielectric and barrier properties, and creep, impact, heat, and fire resistance, but poor processibility. Processed by compression and injection molding, powder sintering, film casting, and solution coating. Thermoset uncyclized polymers are heat curable and have good processability. Processed by transfer and injection molding, lamination, and coating. Used in jet engines, compressors, sealing coatings, auto parts, and business machines. Also called PI.

polymer chain unsaturation See *chemical unsaturation.*

polymers Polymers are high-molecular-weight organic or inorganic compounds the molecules of which comprise linear, branched, crosslinked or otherwise shaped chains of repeating molecular groups. Synthetic polymers are prepared by polymerization of one or more monomers. The monomers are low-molecular-weight substances with one or more reactive bonds or functional groups. Also called resins, plastics.

polymethyl methacrylate Thermoplastic polymer of methyl methacrylate having good transparency, weatherability, impact strength, and dielectric properties. Processed by compression and injection molding, casting, and extrusion. Used in lenses, sheets, airplane canopies, signs, and lighting fixtures. Also called PMMA.

polymethylpentene Thermoplastic polymer of 4-methyl-1-pentene having low density, good transparency, rigidity, dielectric and tensile properties, and heat and chemical resistance. Processed by injection and blow molding and extrusion. Used in laboratory ware, coated paper, light fixtures, auto parts, and electrical insulation. Also called PMP.

polyolefin resins See *polyolefins.*

polyolefins Polyolefins are a broad class of hydrocarbon-chain elastomers or thermoplastics usually prepared by addition (co)polymerization of alkenes such as ethylene. There are branched and linear polyolefins and some are chemically or physically modified. Unmodified polyolefins have relatively low thermal stability and a nonporous, nonpolar surface with poor adhesive properties. Processed by extrusion, injection molding, blow molding and rotational molding. Polyolefins are used more and have more applications than any other polymers. Also called olefinic resins, olefin resins, polyolefin resins.

polyphenylene ether nylon alloys Thermoplastics having improved heat and chemical resistance and toughness. Processed by molding and extrusion. Used in auto body parts.

polyphenylene sulfide High-performance engineering thermoplastic having good chemical, water, fire, and radiation resistance, dimensional stability, and dielectric properties, but decreased impact strength and poor processibility. Processed by injection, compression, and transfer molding and extrusion. Used in hydraulic components, bearings, electronic parts, appliances, and auto parts. Also called PPS.

polyphenylene sulfide sulfone Thermoplastic having good heat, fire, creep, and chemical resistance and dielectric properties. Processed by injection molding. Used in electrical devices. Also called PPSS.

polyphthalamide Thermoplastic polymer of aromatic diamine and phthalic anhydride. Has good heat, chemical, and fire resistance, impact strength, retention of properties at high temperatures, dielectric properties, and stiffness, but decreased light resistance and poor processibility. Processed by solution casting, molding, and extrusion. Used in films, fibers, and molded parts. Also called PPA.

polypropylene Thermoplastic polymer of propylene having low density and good flexibility and resistance to chemicals, abrasion, moisture, and stress cracking, but decreased dimensional stability, mechanical strength, and light, fire, and heat resistance. Processed by injection molding, spinning, and extrusion. Used in fibers and films for adhesive tapes and packaging. Also called PP.

polystyrene Polystyrenes are thermoplastics produced by polymerization of styrene with or without modification (e.g., by copolymerization or blending) to make impact resistant or expandable grades. They have good rigidity, high dimensional stability, low moisture absorption, optical clarity, high gloss and good dielectric properties. Unmodified polystyrenes have poor impact strength and resistance to solvents, heat and UV radiation. Processed by injection molding, extrusion, compression molding, and foam molding. Used widely in medical devices, housewares, food packaging, electronics and foam insulation. Also called polystyrenes, PS, polystyrol.

polystyrenes See *polystyrene.*

polystyrol See *polystyrene.*

polysulfones Thermoplastics, often aromatic and with ether linkages, having good heat, fire, and creep resistance, dielectric properties, transparency, but poor weatherability, processibility, and stress cracking resistance. Processed by injection, compression, and blow molding and extrusion. Used in appliances, electronic devices, auto parts, and electric insulators. Also called PSO.

polytetrafluoroethylene Thermoplastic polymer of tetrafluoroethylene having good dielectric properties, chemical, heat, abrasion, and fire resistance, antiadhesive properties, impact strength, and weatherability, but decreased strength, processibility, barrier properties, and creep resistance. Processed by sinter molding and powder coating. Used in nonstick coatings, chemical apparatus, electrical devices, bearings, and containers. Also called PTFE.

polyurethane resins See *polyurethanes.*

polyurethanes Polyurethanes (PUs) are a broad class of polymers consisting of chains with a repeating urethane group, prepared by condensation of polyisocyanates with polyols, e.g., polyester or polyether diols. PUs may be thermoplastic or thermosetting, elastomeric or rigid, cellular or solid, and offer a wide range of properties depending on composition and molecular structure. Many PUs have high abrasion resistance, good retention of properties at low temperatures and good foamability. Some have poor heat resistance, weatherability and resistance to solvents. PUs are flammable and can release toxic substances. Thermoplastic PUs are not crosslinked and are processed by injection molding and extrusion. Thermosetting PUs can be cured at relatively low temperatures and give foams with good heat insulating properties. They are processed by reaction injection molding, rigid and flexible foam methods, casting and coating. PUs are used in load bearing rollers and wheels, acoustic clamping materials, sporting goods, seals and gaskets, heat insulation, potting and encapsulation. Also called PUR, PU, urethane polymers, urethane resins, urethanes, polyurethane resins.

polyvinyl chloride Thermoplastic polymer of vinyl chloride, available in rigid and flexible forms. Has good dimensional stability, fire resistance, and weatherability, but decreased heat and solvent resistance and high density. Processed by injection and blow molding, calendering, extrusion, and powder coating. Used in films, fabric coatings, wire insulation, toys, bottles, and pipes. Also called PVC.

polyvinyl fluoride Crystalline thermoplastic polymer of vinyl fluoride having good toughness, flexibility, weatherability, and low-temperature and abrasion resistance. Processed by film techniques. Used in packaging, glazing, and electrical devices. Also called PVF.

polyvinylidene chloride Stereoregular thermoplastic polymer of vinylidene chloride having good abrasion and chemical resistance and barrier properties. Processed by molding and extrusion. Used in food packaging films, bag liners, pipes, upholstery, fibers, and coatings. Also called PVDC.

polyvinylidene fluoride Thermoplastic polymer of vinylidene fluoride having good strength, processibility, wear, fire, solvent, and creep resistance, and weatherability, but decreased dielectric properties and heat resistance. Processed by extrusion, injection and transfer molding, and powder coating. Used in electrical insulation, pipes, chemical apparatus, coatings, films, containers, and fibers. Also called PVDF.

PP See *polypropylene.*

PPA See *polyphthalamide.*

pphm See *parts per hundred million.*

ppm A unit for measuring small concentrations of material or substance as the number of its parts (arbitrary quantity) per million parts of medium consisting of another material or substance.

PPS See *polyphenylene sulfide.*

PPSS See *polyphenylene sulfide sulfone.*

prevulcanization See *scorching.*

process characteristics See *processing parameters.*

process conditions See *processing parameters.*

process media See *processing agents.*

process parameters See *processing parameters.*

process pressure See *processing pressure.*

process rate See *processing rate.*

process speed See *processing rate.*

process time See *processing time.*

process velocity See *processing rate.*

processing additives See *processing agents.*

processing agents Agents or media used in the manufacture, preparation and treatment of a material or article to improve its processing or properties. The agents often become a part of the material. Also called process media, processing aids, processing additives.

processing aids See *processing agents.*

processing defects Structural and other defects in material or article caused inadvertently during manufacturing, preparation and treatment processes by using wrong tooling, process parameters, ingredients, part design, etc. Usually preventable. Also called processing flaw, defects, flaw. See also *cracking.*

processing flaw See *processing defects.*

processing methods Method names and designations for material or article manufacturing, preparation and treatment processes. **Note:** Both common and standardized names are used. Also called processing procedures.

processing parameters Measurable parameters such as temperature prescribed or maintained during material or article manufacture, preparation and treatment processes. Also called process characteristics, process conditions, process parameters.

processing pressure Pressure maintained in an apparatus during material or article manufacture, preparation and treatment processes. Also called process pressure. See also *pressure.*

processing procedures See *processing methods.*

processing rate Speed of the process in manufacture, preparation and treatment of a material or article. It usually denotes the change in a process parameter per unit of time or the throughput speed of material in a unit of weight, volume, etc. per unit of time. Also called process speed, process velocity, process rate.

processing time Time required for the completion of a process in the manufacture, preparation and treatment of a material or article. Also called process time, cycle time. See also *time.*

promoter See *accelerator.*

PS See *polystyrene.*

PSO See *polysulfones.*

PTFE See *polytetrafluoroethylene.*

PU See *polyurethanes.*

PUR See *polyurethanes.*

PVC See *polyvinyl chloride.*

PVDC See *polyvinylidene chloride.*

PVDF See *polyvinylidene fluoride.*

PVF See *polyvinyl fluoride.*

Q

Q-U-V accelerated weathering tester See *fluorescent UV lamp-condensation apparatus.*

QFS-40 lamp A fluorescent UV-B lamp with peak emission at 313 nm that provides high UV radiation output for accelerated indoor lightfastness and weatherability testing of materials such as plastics, nonmetallic coatings and textiles. The lamp does not match closely sunlight spectrum in the short wavelength region. Also called FS-40 (UV-B) lamps, FS-40, F40 UVB, F40-UVB, FS-40 lamp.

quartz inner filter An inner filter from quartz glass in xenon-arc lamp that in combination with soda-lime or borosilicate glass outer filter selectively screens radiation output, especially in the short UV wavelength region, to simulate window glass-filtered daylight or sunlight, respectively, in an accelerated light exposure testing apparatus.

quinacridone red A light-fast, chemically resistant organic red pigment used in paints, inks and plastics. Properties of quinacridone pigments are similar to those of phthalocyanine pigments.

QUV See *fluorescent UV lamp-condensation apparatus.*

R

Ra See *roughness average.*

radiant flux density See *irradience.*

reaction injection molding system Liquid compositions, mostly polyurethane-based, of thermosetting resins, prepolymers, monomers, or their mixtures. Have good processibility, dimensional stability, and flexibility. Processed by foam molding with in-mold curing at high temperatures. Used in auto parts and office furniture. Also called RIM.

relative humidity The ratio of the actual vapor pressure of the air to the saturation vapor pressure. Also called RH.

relative viscosity The ratio of solution viscosity to the viscosity of the solvent.

resins See *polymers.*

resorcinol modified phenolic resins Thermosetting polymers of phenol, formaldehyde, and resorcinol having good heat and creep resistance and dimensional stability.

RH See *relative humidity.*

rigid thermoplastic polyurethanes Rigid thermoplastic polyurethanes are not chemically crosslinked. They have high abrasion resistance, good retention of properties at low temperatures, but poor heat resistance, weatherability and resistance to solvents. Rigid thermoplastic polyurethanes are flammable and can release toxic substances. Processed by injection molding and extrusion. Also called rigid thermoplastic urethanes, nonelastomeric thermoplastic polyurethanes.

rigid thermoplastic urethanes See *rigid thermoplastic polyurethanes.*

RIM See *reaction injection molding system.*

roughness average A height parameter of surface roughness equal to the average absolute deviation of surface profile from the mean line, calculated as the integrated area of peaks and valleys above and below the mean line, respectively, divided by the length of this line. Also called Ra.

rutile TiO_2 See *rutile titanium oxide.*

rutile titanium oxide One of the naturally occurring crystal forms of titanium dioxide. Used as a white or opacifying pigment in a wide range of materials including coatings and plastics. Rutile titanium dioxide has a higher refractive index and opacity than anatase titanium dioxide, another crystal form of this oxide. The pigment is non-migrating, heat resistant, chemically inert and lightfast. Also called rutile TiO_2.

S

SAE J576 A Society of Automotive Engineers (SAE) recommended practice for evaluating the suitability of plastics intended for molded optical parts, such as lenses and reflectors of motor vehicle lighting devices. The suitability is determined by the extent of change of optical properties after outdoor conventional weathering (Arizona, Florida). The properties determined after exposure are luminous transmittance, chromaticity coordinates, haze and appearance.

SAE J1545 A Society of Automotive Engineers (SAE) recommended practice for instrumented color difference measurement against reference standards for exterior finishes such as topcoat paints, interior textiles and colored exterior and interior hard trim used in motor vehicles. The color is measured with a spectrophotometer or colorimeter that meet specified requirements. The color difference is determined using lightness, chroma and hue difference scales.

SAE J1885 A Society of Automotive Engineers (SAE) recommended practice for accelerated exposure of automotive interior trim components to determine their colorfastness using a water-cooled xenon-arc lamp apparatus. The lamp is equipped with quartz inner and borosilicate outer filters. The amount of heat, relative humidity and irradiance are controlled to simulate the extreme conditions that may exist inside a motor vehicle. Alternating irradiation is used with 3.8 hour light cycle (black panel temperature 89°C, relative humidity 50%) and 1.0 hour dark cycle (38 °C, 95%). The fading of the specimens is evaluated visually using Gray Scale or instrumentally by measuring color difference values.

SAE J1960 A Society of Automotive Engineers (SAE) standard test method for accelerated exposure of automotive exterior materials to determine their colorfastness using a water-cooled xenon-arc lamp apparatus. The lamp is equipped with quartz inner and borosilicate outer filters. The amount of heat, moisture (humidity, condensation or rain) and irradiance are controlled to simulate the extreme conditions that may exist outside a motor vehicle. Alternating irradiation is used with 2.0 hour light cycle (black panel temperature 70°C, relative humidity 50%), including 20 minute with water spray, and 1.O hour dark cycle (38°C, 95%) with condensation. The fading of the specimens is evaluated visually using Gray Scale or instrumentally by measuring color difference values.

SAE J1961 A Society of Automotive Engineers (SAE) standard test method for accelerated outdoor exposure of automotive exterior materials using a solar Fresnel-reflector apparatus to simulate extreme environmental conditions encountered outside a vehicle due to sunlight, heat and moisture (as humidity, condensation or rain). The flat Fresnel mirrors of the apparatus focus direct sunlight onto an air-cooled specimen area. The apparatus can be either backed or unbacked and is equipped with a water spray device. Spraying, when used, is done during the night only for 3 minutes at a time with 12 minute dry intervals. The report includes exposure time and radiant exposure.

SAE J1976 A Society of Automotive Engineers (SAE) standard test method for outdoor weathering of exterior automotive materials such as coatings for determination of their weatherability. The method specifies the exposure racks, black boxes and instrumentation. In Procedure A, test racks with or without backing are positioned at a fixed angle of 5 deg from the horizontal facing due south. In Procedure B, specimens are exposed in an similarly positioned unheated black. The conditions of the test are recorded by measuring maximum and minimum temperatures and relative humidity values, hours of wetness, total and UV radiant energy.

SAE J2020 A Society of Automotive Engineers (SAE) standard test method for accelerated exposure of automotive exterior components using a fluorescent UV lamp and condensation apparatus to simulate extreme environmental conditions on the outside of an automobile due to sunlight, heat, humidity, etc. to predict the performance of exterior materials such as topcoat paint. The condensation in the apparatus is achieved by evaporation of water from a heated pan and exposure of the back sides of the specimens to the cooling effect of ambient air. The specimens are irradiated 8 hours at 70°C, alternating with 4-hour condensation exposure at 50°C.

SAE J2212 A Society of Automotive Engineers (SAE) recommended practice for accelerated exposure of automotive interior trim components to determine their colorfastness using an air-cooled xenon-arc lamp apparatus. The lamp is equipped with Suprax 1/3 filter system. The amount of heat, relative humidity and irradiance are controlled to simulate the extreme conditions that may exist inside a motor vehicle. Alternating irradiation (irradiance 80 W/m^2 at 300-400 nm wavelength) is used with light cycle (chamber temperature 62°C, relative humidity 50%) and dark cycle (38°C, 95%). The fading of the specimens is evaluated visually using Gray Scale or instrumentally by measuring color difference values.

SAE J2230 A Society of Automotive Engineers (SAE) standard test method for accelerated exposure of automotive interior trim materials using outdoor under-glass sun-tracking temperature and humidity apparatus in which the temperature is controlled in a 24-hour cycle and the humidity is controlled during dark (night) portion of the cycle. The test is designed to simulate extreme environmental conditions encountered inside a vehicle due to sunlight, heat and humidity to determine the colorfastness of interior materials such as textiles. The specimen cabinet is covered with 3-mm thick tempered safety glass, maintained facing direct sunlight and equipped with heaters, humidifiers, UV radiometers, sensors and controller. The temperature is maintained at 70°C during the day and 38°C (at 75% relative humidity) during the night. The fading of the specimens is evaluated visually using Gray Scale or instrumentally by measuring color difference values.

SAN See *styrene acrylonitrile copolymer.*

SAN copolymer See *styrene acrylonitrile copolymer.*

SAN resin See *styrene acrylonitrile copolymer.*

scorch See *scorching.*

scorching Premature vulcanization of rubber during processing, e.g., on a calender. Resistance of rubber to scorching is tested by heating it while subjecting to shear, e.g., in Mooney viscometer, for a certain period of time. Also called scorch, prevulcanization.

service life The period of time required for the specified properties of the material to deteriorate under normal use conditions to the minimum allowable level with material retaining its overall usability.

shelf life Time during which a physical system, such as a material, retains its storage stability under specified conditions. Also called storage life.

short wavelength cutoff Selectively filtering the radiation from artificial light sources to cut it off in the short UV wavelength region below approximately 300 nm to simulate sunlight. Also called solar cutoff.

shortwave ultraviolet radiation On earth's surface, electromagnetic radiation in the 315-280 nm wavelength region (UV-B) of the solar spectrum. In outer space, radiation in the 280-100 nm wavelength region (UV-C). Also called shortwave UV.

shortwave UV See *shortwave ultraviolet radiation.*

silicone There are rigid thermoplastic and liquid silicones and silicone rubbers consisting of alternating silicone and oxygen atom chains with organic pendant groups, prepared by hydrolytic polymcondensation of chlorosilanes, followed by crosslinking. Silicone rubbers have good adhesion, flexibility, dielectric properties, weatherability, barrier properties, and heat and fire resistance, but decreased strength. Rigid silicones have good flexibility, weatherability, soil repelling properties, dimensional stability, but poor solvent resistance. Processed by coating, casting, and injection compression, and transfer molding. Used in coatings, electronic devises, diaphragms, medical products, adhesives, and sealants. Also called siloxane.

siloxane See *silicone.*

silver streaking See *silver streaks.*

silver streaks Scars or surface defects on injection moldings caused by the high velocity injection of a stream of molten material into the mold ahead of the normally advancing material front and its premature solidification. Also similar appearance defects resulting from exposure or stress. Also called silver streaking, splay marks.

SMA See *styrene maleic anhydride copolymer.*

SMA PTB alloy See *styrene maleic anhydride copolymer PBT alloy.*

softening point Temperature at which the material changes from rigid to soft or exhibits a sudden and substantial decrease in hardness. Also called softening temperature, softening range.

softening range See *softening point.*

softening temperature See *softening point.*

solar cutoff See *short wavelength cutoff.*

solar radiation Electromagnetic radiation with wavelengths ranging from 1E-09 cm to 30 km emitted by sun. The intensity of solar radiation in the short UV wavelength region of spectrum changes from outer space to the earth's surface because of the absorption of the UV light below approximately 295 nm by the ozone layer of the atmosphere.

splay marks See *silver streaks.*

stability The ability of a physical system, such as a material, to resist a change or degradation under exposure to outside forces, including mechanical force, heat and weather. See also *degradation.*

starch A polysaccharide, consisting of amylose and amylopectin, found in plants such as potatoes. Gels in water. Used in adhesives, textile sizes, thickeners and in manufacture of biodegradable polymers such as polyesters. The grades include technical and edible.

starch modified low density polyethylene Biodegradable thermoplastic starch-grafted low-density polyethylene.

starch modified polypropylene Biodegradable thermoplastic starch-grafted polypropylene.

starch modified polyurethane Biodegradable thermoplastic starch-grafted polyurethane.

static strip test An ozone resistance test for rubbers that involves a strip-shaped specimen mounted as a test board, stretched to 15% and subjected to ozone attack in the test chamber. The results of the test are reported with 2 digits separated with a virgule. The number before the virgule indicates the number of quarters of the test strip which showed the cracks. The number after the virgule indicates the size of the cracks in length perpendicular to the length of the strip.

storage life See *shelf life.*

storage stability The resistance of a physical system, such as a material, to decomposition, deterioration of properties or any type of degradation in storage under specified conditions.

strain The per unit change, due to force, in the size or shape of a body referred to its original size or shape. **Note:** Strain is nondimensional but is often expressed in unit of length per unit of length or percent.

stress cracking Appearance of external and/or internal cracks in the material as a result of stress that is lower than its short-term strength.

stress pattern Distribution of applied or residual stress in a specimen, usually throughout its bulk. Applied stress is a stress induced by an outside force, e.g., by loading. Residual stress or stress memory may be a result of processing or exposure. The stress pattern can be made visible in transparent materials by polarized light.

styrene acrylonitrile copolymer SAN resins are thermoplastic copolymers of about 70% styrene and 30% acrylonitrile with higher strength, rigidity and chemical resistance than polystyrene. Characterized by transparency, high heat deflection properties, excellent gloss, hardness and dimensional stability. Have low continuous service temperature and impact strength. Processed by injection molding, extrusion, injection-blow molding and compression molding. Used in appliances, housewares, instrument lenses for automobiles, medical devices, and electronics. Also called styrene-acrylonitrile copolymer, SAN, SAN resin, SAN copolymer.

styrene butadiene block copolymer Thermoplastic amorphous block polymer of butadiene and styrene having good impact strength, rigidity, gloss, compatibility with other styrenic resins, water resistance, and processibility. Used in food and display containers, toys, and shrink wrap.

styrene butadiene copolymer Thermoplastic polymers of butadiene and >50% styrene having good transparency, toughness, and processibility. Processed by extrusion, injection and blow molding, and thermoforming. Used in film wraps, disposable packaging, medical devices, toys, display racks, and office supplies.

styrene maleic anhydride copolymer Thermoplastic copolymer of styrene with maleic anhydride having good thermal stability and adhesion, but decreased chemical and light resistance. Processed by injection and foam molding and extrusion. Used in auto parts, appliances, door panels, pumps, and business machines. Also called SMA.

styrene maleic anhydride copolymer PBT alloy Thermoplastic alloy of styrene maleic anhydride copolymer and polybutylene terephthalate having improved dimensional stability and tensile strength. Processed by injection molding. Also called SMA PTB alloy.

styrene plastics See *styrenic resins.*

styrene polymers See *styrenic resins.*

styrene resins See *styrenic resins.*

styrene-acrylonitrile copolymer See *styrene acrylonitrile copolymer.*

styrenic resins Styrenic resins are thermoplastics prepared by free-radical polymerization of styrene alone or with other unsaturated monomers. The properties of styrenic resins vary widely with molecular structure, attaining the high performance level of engineering plastics. Processed by blow and injection molding, extrusion, thermoforming, film techniques and structural foam molding. Used heavily for the manufacture of automotive parts, household goods, packaging, films, tools, containers and pipes. Also called styrene resins, styrene polymers, styrene plastics.

styrenic thermoplastic elastomers Linear or branched copolymers containing polystyrene end blocks and elastomer (e.g., isoprene rubber) middle blocks. Have a wide range of hardnesses, tensile strength, and elongation, and good low-temperature flexibility, dielectric properties, and hydrolytic stability. Processed by injection and blow molding and extrusion. Used in coatings, sealants, impact modifiers, shoe soles, medical devices, tubing, electrical insulation, and auto parts. Also called TES.

sunshine carbon lamp Open-flame carbon-arc lamp equipped with glass filters (e.g., Corex D) to better match the sunlight. Sunshine lamp emits more radiant energy in the short UV (260-320 nm) wavelength region than sun on the earth's surface and therefore can produce unrealistic light exposure results. However, radiation of sunshine lamp is more realistic than that of the enclosed carbon-arc lamp.

Suntest See *Suntest CPS.*

Suntest CPS The Suntest CPS (Controlled Power System) is an accelerated weathering chamber of table-top size with automated control and monitoring of temperature and irradiance from an air-cooled xenon-arc lamp with 3 filter systems: Max UV (high output in short UV wavelength region), Suprax (best sunlight match) and window glass. Cycling of light and dark periods and water immersion module are available. Produced by Heraeus DSET Laboratories, Inc., Phoenix, Arizona. Also called Suntest.

Super-Maq A large apparatus for accelerated outdoor weathering equipped with sun-tracking Fresnel mirror-reflecting solar concentrator and water spray. Super-Maq allows testing of the complete components. Produced by Heraeus DSET Laboratories, Inc., Phoenix, Arizona.

superficial surface oxidation Oxidation of the material surface that is relatively insignificant and is restricted to the thin surface layer of the material.

surface checks See *checking.*

surface roughness Relatively fine spaced surface irregularities, the heights, widths and directions of which establish the predominant surface pattern.

surface tack Stickiness of a surface of a material such as wet paint when touched.

syndiotactic A polymer molecule in which pendant groups and atoms attached to the main chain are arranged in a symmetrical and recurring fashion relative to it in a single plane.

synergistic effect The boosting effect of one substance on the property of another so that the total effect of both substances in a mixture is greater than the sum of the effects of each substance individually, such as synergistic effect of zinc bis(dibutyldithiocarbamate) on the UV absorption by zinc oxide.

T

tautomeric Pertaining to tautomerism, i.e., isomerism in which migration of a hydrogen atom results in two or more structures, called tautomers that are in equilibrium. For example, enol and keto tautomers of acetoacetate.

temperature Property which determines the direction of heat flow between objects. **Note:** The heat flows from the object with higher temperature to that with lower.

tensile elongation See *elongation.*

tensile heat distortion temperature See *heat deflection temperature.*

tensile properties Properties describing the reaction of physical systems to tensile stress and strain. See also *tensile property tests.*

tensile property tests Names and designations of the methods for tensile testing of materials. Also called tensile tests. See also *tensile properties.*

tensile strain The relative length deformation exhibited by a specimen in tension. See also *elongation.*

tensile strength The maximum tensile stress that a specimen can sustain in a test carried to failure. **Note:** The maximum stress can be measured at or after the failure or reached before the fracture, depending on the viscoelastic behavior of the material. Also called tensile ultimate strength, ultimate tensile strength, UTS, tensile strength at break, ultimate tensile stress. See also *ASTM D638.*

tensile strength at break See *tensile strength.*

tensile stress The stress is perpendicular and directed to the opposite plane on which the forces act.

tensile tests See *tensile property tests.*

tensile ultimate strength See *tensile strength.*

terephthalate polyester Thermoset unsaturated polymer of terephthalic anhydride.

TES See *styrenic thermoplastic elastomers.*

test methods Names and designations of material test methods. Also called testing methods.

test variables Terms related to the testing of materials such as test method names.

testing methods See *test methods.*

tetrafluoroethylene propylene copolymer Thermosetting elastomeric polymer of tetrafluoroethylene and propylene having good chemical and heat resistance and flexibility. Used in auto parts.

thermal properties Properties related to the effects of heat on physical systems such as materials and heat transport. The effects of heat include the effects on structure, geometry, performance, aging, stress-strain behavior, etc.

thermal stability The resistance of a physical system, such as a material, to decomposition, deterioration of properties or any type of degradation in storage under specified conditions.

thermodynamic properties A quantity that is either an attribute of the entire system or is a function of position, which is continuous and does not vary rapidly over microscopic distances, except possibility for abrupt changes at boundaries between phases of the system. Also called macroscopic properties.

thermoplastic polyesters A class of polyesters that can be repeatedly made soft and pliable on heating and hard (flexible or rigid) on subsequent cooling.

thermoplastic polyurethanes A class of polyurethanes including rigid and elastomeric polymers that can be repeatedly made soft and pliable on heating and hard (flexible or rigid) on subsequent cooling. Also called thermoplastic urethanes, TPUR, TPU.

thermoplastic urethanes See *thermoplastic polyurethanes.*

three-membered heterocyclic compounds A class of heterocyclic compounds containing rings that consist of three atoms.

three-membered heterocyclic oxygen compounds A class of heterocyclic compounds containing rings that consist of three atoms, one or two of which is an oxygen.

time One of basic dimensions of the universe designating the duration and order of events at a given place. See also *processing time*.

tinctorial strength Measure of the effectiveness with which a unit quantity of a pigment or colorant to change the color of a material. Also called tint strength.

tint See *color.*

tint strength See *tinctorial strength.*

titanium dioxide A white pigment and opacifying agent, TiO_2, with the greatest hiding power. Exists in two crystal forms: rutile, with a higher refractive index and opacity, and anatase, with a lower refractive index and opacity. Manufactured in bulk by sulfation process from the mineral ilmenite, or by chlorination process from the mineral rutile. The pigment is non-migrating, heat resistant, chemically inert and lightfast. Used widely in paints, rubber, plastics, paper, synthetic fibers, cosmetics, enamel frits, floor coverings, etc.

total solar irradience The amount of radiant power of sunlight integrated over all its wavelengths per unit area of irradiated surface at a point in time. A measure of radiation exposure, it is often expressed in the units of watt per square meter (W/m^2).

toughness Property of a material indicating its ability to absorb energy by plastic deformation rather than crack or fracture.

TPO See *olefinic thermoplastic elastomers.*

TPU See *thermoplastic polyurethanes.*

TPU See *urethane thermoplastic elastomers.*

TPUR See *thermoplastic polyurethanes.*

transition point See *phase transition point.*

transition temperature See *phase transition point.*

transmittance The ratio of the light intensity transmitted by a body to the incident light intensity. Also called percent light transmittance, light transmission, luminous transmittance, optical transmittance, transmittancy, transparency, transparence.

transmittancy See *transmittance.*

transparence See *transmittance.*

transparency See *transmittance.*

transparent pigment Pigments such as some organic pigments, having low hiding power.

tribasic lead maleate A salt of maleic acid, highly effective as heat stabilizer for polymeric materials. Limited to use in applications where toxicity and lack of clarity can be tolerated.

turbidity The cloudiness in a liquid caused by a suspension of colloidal liquid droplets or fine solids.

U

UHMWPE See *ultrahigh molecular weight polyethylene.*

ultimate elongation See *elongation.*

ultimate tensile strength See *tensile strength.*

ultimate tensile stress See *tensile strength.*

ultrahigh molecular weight polyethylene Thermoplastic linear polymer of ethylene with molecular weight in the millions. Has good wear and chemical resistance, toughness, and antifriction properties, but poor processibility. Processed by compression molding and ram extrusion. Used in bearings, gears, and sliding surfaces. Also called UHMWPE.

ultramarine blue An inorganic blue pigment with good alkali and heat resistance, low hiding power, poor acid resistance and weatherability. Prepared by heating a mixture of sulfur, clay, alkali and a reducing agent. Used in coatings, inks, rubber, laundry blues. In low concentration can neutralize yellow tint of white or clear materials.

ultraviolet filter Glass filters that selectively transmit (UV-bandpass filters) or block (longpass filters) UV light. Also called UV filters.

ultraviolet light See *ultraviolet radiation.*

ultraviolet light See *ultraviolet radiation.*

ultraviolet radiation Electromagnetic radiation in the wavelength range 13-400 nm below the short-wavelength limit of the visible light. **Note:** UV light comprises a significant portion of the natural sun light. Also called ultraviolet light, UV light, UV radiation.

ultraviolet radiation Electromagnetic radiation in the 13-400 nm wavelength region. Below the short wavelenght limit of visiable light. Sun is the main natural source of UV radiation on the earth. Artificial sources are many, including fluorescent UV lamps. UV radiation causes polymer photodegradation and other chemical reactions. Also called ultraviolet light, UV, UV light, UV radiation.

ultraviolet wavelength Any wavelength in the 13-400 nm wavelength region of electromagnetic radiation. Also called UV wavelength.

unbacked exposure rack A rack for holding specimens or specimen panels during exposure testing that is not enclosed from the back.

units See *units of measurement.*

units of measurement Systematic and non-systematic units for measuring physical quantities, including metric and US pound-inch systems. Also called units.

unstable isotopes See *radioisotopes.*

urea resins Thermosetting polymers of formaldehyde and urea having good clarity, colorability, scratch, fire, and solvent resistance, rigidity, dielectric properties, and tensile strength, but decreased impact strength and chemical, heat, and moisture resistance. Must be filled for molding. Processed by compression and injection molding, impregnation, and coating. Used in cosmetic containers, housings, tableware, electrical insulators, countertop laminates, adhesives, and coatings.

urethane polymers See *polyurethanes.*

urethane resins See *polyurethanes.*

urethane thermoplastic elastomers Block polyether or polyester polyurethanes containing soft and hard segments. Have good tensile strength, elongation, adhesion, and a broad hardness and service temperature ranges, but decreased moisture resistance and processibility. Processed by extrusion, injection molding, film blowing, and coating. Used in tubing, packaging film, adhesives, medical devices, conveyor belts, auto parts, and cable jackets. Also called TPU.

urethanes See *polyurethanes.*

UTS See *tensile strength.*

UV See *ultraviolet radiation.*

UV absorber A low-molecular-weight organic compound such as hydroxybenzophenone derivatives that is capable of absorbing significant amount of radiant energy in the ultraviolet wavelength region, thus protecting the material such as plastic in which it is incorporated from the damaging (degrading) effect of the energy. The absorbed energy is dissipated by UV absorber without significant chemical change via tautomerism of hydrogen bonds. Also called UV stabilizer.

UV filters See *ultraviolet filter.*

UV light See *ultraviolet radiation.*

UV light See *ultraviolet radiation.*

UV radiation See *ultraviolet radiation.*

UV radiation See *ultraviolet radiation.*

UV stabilizer See *UV absorber.*

UV wavelength See *ultraviolet wavelength.*

UV-A radiation A portion of ultraviolet radiation in the 315-400 nm wavelength range. Causes polymer damage.

UV-B radiation A portion of ultraviolet radiation in the 280-315 nm wavelength range. Includes the shortest wavelengths of sunlight found at the earth's surface. Causes severe polymer damage; absorbed by window glass.

UV-C radiation A portion of ultraviolet radiation in the 100-280 nm wavelength range. A part of sunlight spectra found only in outer space because of the absoption by the earth's atmosphere. Germicidal.

UV-CON See *fluorescent UV lamp-condensation apparatus.*

UVA-340 lamp A fluorescent lamp with peak emission at 340 nm that provides high UV-A radiation output for accelerated indoor lightfastness and weatherability testing of materials such as plastics, nonmetallic coatings and textiles. The lamp matches closely sunlight spectrum in the UV wavelength region.

UVB-313 lamp A fluorescent lamp with peak emission at 313 nm that provides high UV-B radiation output for accelerated indoor lightfastness and weatherability testing of materials such as plastics, nonmetallic coatings and textiles. The lamp does not match closely sunlight spectrum in the short UV wavelength region. Used for testing of automotive materials. Its output is more stable and higher than that of the FS-40 lamp.

V

veneer In rubber industry, a thin film applied on a rubber article to protect it against oxygen and ozone attack, act as a migration barrier or for decorative purposes.

Vicat softening point The temperature at which a flat-ended needle of prescribed geometry will penetrate a thermoplastic specimen to a certain depth under a specified load using a uniform rate of temperature rise. **Note:** Vicat softening point is determined according to ASTM D1525 test for thermoplastics such as polyethylene which have no definite melting point. Also called Vicat softening temperature.

Vicat softening temperature See *Vicat softening point.*

vinyl ester resins Thermosetting acrylated epoxy resins containing styrene reactive diluent. Cured by catalyzed polymerization of vinyl groups and crosslinking of hydroxy groups at room or elevated temperatures. Have good chemical, solvent, and heat resistance, toughness, and flexibility, but shrink during cure. Processed by filament winding, transfer molding, pultrusion, coating, and lamination. Used in structural composites, coatings, sheet molding compounds, and chemical apparatus.

vinyl resins Thermoplastics polymers of vinyl compounds such as vinyl chloride or vinyl acetate. Have good weatherability, barrier properties, and flexibility, but decreased solvent and heat resistance. Processed by molding, extrusion, and coating. Used in films and packaging.

vinyl thermoplastic elastomers Vinyl resin alloys having good fire and aging resistance, flexibility, dielectric properties, and toughness. Processed by extrusion. Used in cable jackets and wire insulation.

vinylidene fluoride hexafluoropropylene copolymer Thermoplastic polymer of vinylidene fluoride and hexafluoropropylene having good antistick, dielectric, and antifriction properties and chemical and heat resistance, but decreased mechanical strength and creep resistance and poor processibility. Processed by molding, extrusion, and coating. Used in chemical apparatus, containers, films, and coatings.

vinylidene fluoride hexafluoropropylene tetrafluoroethylene terpolymer Thermosetting elastomeric polymer of vinylidene fluoride, hexafluoropropylene, and tetrafluoroethylene having good chemical and heat resistance and flexibility. Used in auto parts.

vulcanizate Rubber that had been irreversibly transformed from predominantly plastic to predominantly elastic material by vulcanization (chemical curing or crosslinking) using heat, vulcanization agents, accelerants, etc.

vulcanizate crosslinks Chemical bonds formed between polymeric chains in rubber as a result of vulcanization.

W

warpage See *warping.*

warping Dimensional distortion or deviation from the intended shape of a plastic or rubber article as a result of nonuniform internal stress, e.g., caused by uneven heat shrinkage. Also called warpage.

water swell Expansion of material volume as a result of water absorption.

weatherometer An apparatus for accelerated indoor testing of weatherability of materials such as plastics. Equipped with carbon- or xenon-arc lamps having glass filters to simulate the sunlight and with a water spraying device. Most models allow controlling and monitoring temperature and humidity inside the apparatus and allow alternating dark and light cycles of exposure.

weight The gravitational force with which the earth attracts a body.

wet bulb depression The difference between the temperatures shown by the wet and dry thermometers of a psychrometer, an instrument for measuring the content of moisture (humidity) in the air.

whiting A finely divided form of calcium carbonate ($CaCO_3$) obtained by milling high-calcium limestone, marble, shell or chemically precipitated calcium carbonate. Used as an extender filler in plastics and rubbers.

X

xenon arc lamp An inert gas xenon-filled quartz tube with two electrodes to produce an electric arc discharge that emits radiation in the 200-1200 nm wavelength region. Can be air- or water-cooled. Equipped with glass filters to selectively block short wavelength UV light to simulate sunlight. Usually there are 2 filters, inner and outer, one of which is from borosilicate glass. In the water-cooled lamps, cooling demineralized water flows between the inner and outer filters. Used in the apparatus for accelerated indoor testing of weatherability and lightfastness of materials, these lamps produce more realistic degradation than carbon arc lamps. Also called xenon lamp.

xenon arc weatherometer An apparatus for accelerated indoor testing of weatherability of materials such as plastics. Equipped with one to three water- or air-cooled xenon arc lamps with borosilicate glass filters to simulate the sunlight and with a water spraying device. Produces more realistic degradation that carbon arc weatherometers. Most models allow controlling and monitoring temperature and humidity inside the apparatus and allow alternating dark and light cycles of exposure. Among the manufactures of xenon arc weatherometers is Atlas Electric Devices Co., Chicago, Illinois. Atlas Ci65 model has a two-tier inclined specimen rack and the xenon-arc lamps located vertically at the central axis of the racks. Also called Ci65 xenon arc weatherometer, Atlas Ci65 xenon arc weatherometer.

xenon lamp See *xenon arc lamp.*

Xenotest 1200 A computerized chamber for accelerated weatherability testing of materials manufactured by Heraeus DSET Laboratories, Inc., Pheonix, Arizona. Equipped with 3 air-cooled xenon-arc lamps, optical filter system for selectively blocking UV light, rotating specimen holders, sample spray system, specimen back side cooling for dew simulation, rain water heating system and control systems for irradiance and temperature, speed and humidity of inside air.

Y

yellowing Developing of yellow color in near-white or near-transparent materials such as plastics or coatings as a result of degradation on exposure to light, heat aging, weathering, etc. Usually is measured in terms of yellowness index.

yellowness index A measure of the tendency of materials such as plastics to become yellow as a result of: long-term exposure to light, irradiation, etc.

Z

zinc oxide An amorphous white powder, ZnO, used as a pigment in plastics and coatings, as an activator of rubber vulcanization accelerators and as reinforcing filler. Having the greatest UV light absorbing power of all commercially available pigment, it can act as a UV stabilizer, especially in synergistic mixtures with organics such as zinc bis(dibutyldithiocarbamate).

Graph Index

GRAPH: 01 Outdoor Weathering Exposure Time vs. Yellowness Index of ABS 5

GRAPH: 02 Arizona Outdoor Weathering Exposure Time vs. Drop Dart Impact Strength of ABS 6

GRAPH: 03 Arizona Outdoor Weathering Exposure Time vs. Elongation of ABS 6

GRAPH: 04 Arizona Outdoor Weathering Exposure Time vs. Tensile Strength at Yield of ABS 7

GRAPH: 05 Arizona Outdoor Weathering Exposure vs. Delta E Color Change of ABS 7

GRAPH: 06 Arizona, Florida and Ohio Outdoor Weathering Exposure Time vs. Drop Dart Impact Strength of ABS 8

GRAPH: 07 Florida Outdoor Weathering Exposure Time vs. Drop Dart Impact Strength of ABS 8

GRAPH: 08 Florida Outdoor Weathering Exposure Time vs. Drop Weight Impact of ABS 9

GRAPH: 09 Florida Outdoor Weathering Exposure vs. Delta E Color Change of ABS 9

GRAPH: 10 Florida Weathering Exposure Time vs. Chip Impact Strength of ABS 10

GRAPH: 11 Florida Weathering Exposure Time vs. Chip Impact Strength of ABS 10

GRAPH: 12 Ohio Outdoor Weathering Exposure Time vs. Delta E Color Change of ABS 11

GRAPH: 13 Ohio Outdoor Weathering Exposure Time vs. Drop Dart Impact Strength of ABS 11

GRAPH: 14 Okinawa, Japan Outdoor Weathering Exposure Time vs. Delta E Color Change of ABS 12

GRAPH: 15 Okinawa, Japan Outdoor Weathering Exposure Time vs. Dynstat Impact Strength Retained of ABS 12

GRAPH: 16 Okinawa, Japan Outdoor Weathering Exposure Time vs. Elongation at Break Retained of ABS 13

GRAPH: 17 Okinawa, Japan Outdoor Weathering Exposure Time vs. Gloss Retained of ABS 13

GRAPH: 18 West Virginia Outdoor Weathering Exposure Time vs. Falling Dart Impact of ABS 14

GRAPH: 19 West Virginia Outdoor Weathering Exposure Time vs. Falling Dart Impact of ABS 14

GRAPH: 20 West Virginia Outdoor Weathering Exposure Time vs. Falling Dart Impact of ABS 15

GRAPH: 21 West Virginia Outdoor Weathering Exposure Time vs. Flexural Modulus Retained of ABS 15

GRAPH: 22 West Virginia Outdoor Weathering Exposure Time vs. Flexural Strength of ABS 16

GRAPH: 23 West Virginia Outdoor Weathering Exposure Time vs. Flexural Strength Retained of ABS 16

GRAPH: 24 West Virginia Outdoor Weathering Exposure Time vs. Izod Impact Strength Retained of ABS 17

GRAPH: 25 West Virginia Outdoor Weathering Exposure Time vs. Tensile Strength Retained of ABS 17

GRAPH: 26 Sunshine Weatherometer Exposure Time vs. Dynstat Impact Strength Retained of ABS 18

GRAPH: 27 Sunshine Weatherometer Exposure Time vs. Elongation at Break Retained of ABS 18

GRAPH: 28 Sunshine Weatherometer Exposure Time vs. Gloss Retained of ABS 19

GRAPH: 29 Weatherometer Exposure Time vs. Impact Strength of ABS 19

GRAPH: 30 Xenotest 1200 Exposure Time vs. Impact Strength of ABS 20

GRAPH: 31 Accelerated Indoor UV Exposure Time vs. Delta E Color Change of ABS 20

GRAPH: 32 Outdoor Exposure Time vs. Impact Strength Retained of Acetal Copolymer 24

GRAPH: 33 New Jersey and Arizona Outdoor Exposure Time vs. Melt Index of Acetal Copolymer 25

GRAPH: 34 New Jersey and Arizona Outdoor Exposure Time vs. Tensile Impact of Acetal Copolymer 25

GRAPH: 35 New Jersey and Arizona Outdoor Exposure Time vs. Tensile Strength at Yield of Acetal Copolymer 26

GRAPH: 36 New Jersey Outdoor Exposure Time vs. Tensile Strength at Yield of Acetal Copolymer 26

GRAPH: 37 QUV Exposure Time vs. Delta E Color Change of Acetal Copolymer 27

GRAPH: 38 Sunshine Weatherometer Exposure Time vs. Discoloration of Acetal Copolymer 27

GRAPH: 39 Sunshine Weatherometer Exposure Time vs. Elongation Retained of Acetal Copolymer 28

GRAPH: 40 Sunshine Weatherometer Exposure Time vs. Tensile Strength Retained of Acetal Copolymer 28

GRAPH: 41 Xenon Arc Weatherometer Exposure Time vs. Relative Gloss of Acetal Copolymer 29

GRAPH: 42 Outdoor Weathering Exposure Time vs. Yellowness Index of ASA 33

GRAPH: 43 Okinawa, Japan Outdoor Weathering Exposure Time vs. Delta E Color Change of ASA 33

GRAPH: 44 Okinawa, Japan Outdoor Weathering Exposure Time vs. Dynstat Impact Strength Retained of ASA 34

GRAPH: 45 Okinawa, Japan Outdoor Weathering Exposure Time vs. Elongation at Break Retained of ASA 34

GRAPH: 46 Okinawa, Japan Outdoor Weathering Exposure Time vs. Gloss Retained of ASA 35

GRAPH: 47 Sunshine Weatherometer Exposure Time vs. Dynastat Impact Strength Retained of ASA 35

GRAPH: 48 Sunshine Weatherometer Exposure Time vs. Dynstat Impact Strength Retained of ASA 36

GRAPH: 49 Sunshine Weatherometer Exposure Time vs. Elongation at Break Retained of ASA 36

GRAPH: 50 Sunshine Weatherometer Exposure Time vs. Gloss Retained of ASA 37

GRAPH: 51 Sunshine Weatherometer Exposure Time vs. Gloss Retained of ASA 37

GRAPH: 52 Weatherometer Exposure Time vs. Impact Strength of ASA 38

GRAPH: 53 Xenotest 1200 Exposure Time vs. Impact Strength of ASA 38

GRAPH: 54 Florence, Kentucky Outside Weathering Exposure Time vs. Yellowness Index of Acrylic Resin 48

GRAPH: 55 EMMAQUA Accelerated Weathering Exposure Time vs. Light Transmission of Acrylic Resin 48

GRAPH: 56 EMMAQUA Accelerated Weathering Exposure Time vs. Yellowness Index of Acrylic Resin 49

GRAPH: 57 EMMAQUA Arizona Accelerated Weathering Exposure Time vs. Yellowness Index of Acrylic Resin 49

GRAPH: 58 Twin Carbon Arc Weatherometer Exposure Time vs. Yellowness Index of Acrylic Resin 50

GRAPH: 59 Fadeometer Exposure Time vs. Yellowness Index of Acrylic Resin 50

GRAPH: 60 UV-CON Accelerated Weathering Exposure Time vs. Yellowness Index of Acrylic Resin 51

GRAPH: 61 Fadeometer Exposure Time vs. Yellowness Index of Acrylic Copolymer 53

GRAPH: 62 UV-CON Accelerated Weathering Exposure Time vs. Yellowness Index of Acrylic Copolymer 54

GRAPH: 63 Florence, Kentucky Outside Weathering Exposure Time vs. Yellowness Index of Cellulose Acetate Butyrate 58

GRAPH: 64 Arizona Outdoor Weathering Exposure Time vs. Elongation at Break of Cellulose Acetate Butyrate 58

GRAPH: 65 Arizona Outdoor Weathering Exposure Time vs. Tensile Strength at Break of Cellulose Acetate Butyrate 59

GRAPH: 66 Kingsport, Tennessee Outdoor Weathering Exposure Time vs. Plaque Impact Strength of Cellulose Acetate Butyrate 59

GRAPH: 67 EMMAQUA Arizona Accelerated Weathering Exposure Time vs. Yellowness Index of Cellulose Acetate Butyrate 60

GRAPH: 68 Twin Carbon Arc Weatherometer Exposure Time vs. Yellowness Index of Cellulose Acetate Butyrate 60

GRAPH: 69 Weatherometer Exposure Time vs. Elongation Retained of Polychlorotrifluoroethylene 67

GRAPH: 70 Weatherometer Exposure Time vs. Elongation Retained of Polychlorotrifluoroethylene 68

GRAPH: 71 Weatherometer Exposure Time vs. Tensile Strength Retained of Polychlorotrifluoroethylene 68

GRAPH: 72 Weatherometer Exposure Time vs. Tensile Strength Retained of Polychlorotrifluoroethylene 69

GRAPH: 73 Accelerated Indoor UV Exposure Time vs. Delta E Color Change of Modified Polyphenylene Oxide 83

GRAPH: 74 Arizona Outdoor Weathering Exposure Time vs. Drop Dart Impact Strength of Modified Polyphenylene Oxide 83

GRAPH: 75 Arizona Outdoor Weathering Exposure Time vs. Elongation of Modified Polyphenylene Oxide 84

GRAPH: 76 Arizona Outdoor Weathering Exposure Time vs. Tensile Strength at Yield of Modified Polyphenylene Oxide 84

GRAPH: 77 Arizona Outdoor Weathering Exposure vs. Delta E Color Change of Modified Polyphenylene Oxide 85

GRAPH: 78 Ohio Outdoor Weathering Exposure vs. Delta E Color Change of Modified Polyphenylene Oxide 85

GRAPH: 79 Ohio Outdoor Weathering Exposure vs. Drop Dart Impact Strength of Modified Polyphenylene Oxide 86

GRAPH: 80 Weatherometer Exposure Time vs. Delta E Color Change of Nylon 12 88

GRAPH: 81 Weatherometer Exposure Time vs. Delta E Color Change of Nylon 12 88

GRAPH: 82 Weatherometer Exposure Time vs. Tensile Impact Strength of Nylon 12 89

GRAPH: 83 Outdoor Exposure Time vs. Elongation at Break of Nylon 6 93

GRAPH: 84 Outdoor Exposure Time vs. Flexural Modulus of Nylon 6 94

GRAPH: 85 Outdoor Exposure Time vs. Notched Izod Impact Strength of Nylon 6 94

GRAPH: 86 Outdoor Exposure Time vs. Tensile Strength of Nylon 6 95

GRAPH: 87 Hiratsuka, Japan Outdoor Exposure Time vs. Breaking Stress in Flexure of Nylon 6 95

GRAPH: 88 Hiratsuka, Japan Outdoor Exposure Time vs. Flexural Modulus of Nylon 6 96

GRAPH: 89 Hiratsuka, Japan Outdoor Exposure Time vs. Notched Izod Impact Strength of Nylon 6 96

GRAPH: 90 Hiratsuka, Japan Outdoor Exposure Time vs. Percent Weight Change of Nylon 6 97

GRAPH: 91 Hiratsuka, Japan Outdoor Weathering Exposure Time vs. Flexural Strength of Nylon 6 97

GRAPH: 92 Hiratsuka, Japan Outdoor Weathering Exposure Time vs. Tensile Strength of Nylon 6 98

GRAPH: 93 Sunshine Weatherometer Exposure Time vs. Elongation of Nylon 6 98

GRAPH: 94 Sunshine Weatherometer Exposure Time vs. Tensile Strength of Nylon 6 99

GRAPH: 95 Outdoor Exposure Time vs. Elongation at Break of Nylon 66 112

GRAPH: 96 Outdoor Exposure Time vs. Flexural Modulus of Nylon 66 112

GRAPH: 97 Outdoor Exposure Time vs. Notched Izod Impact Strength of Nylon 66 113
GRAPH: 98 Outdoor Exposure Time vs. Tensile Strength of Nylon 66 113
GRAPH: 99 Florida Outdoor Weathering Exposure Time vs. Tensile Strength of Nylon 66 114
GRAPH: 100 Hiratsuka, Japan Outdoor Exposure Time vs. Breaking Stress in Flexure of Nylon 66 114
GRAPH: 101 Hiratsuka, Japan Outdoor Exposure Time vs. Flexural Modulus of Nylon 66 115
GRAPH: 102 Hiratsuka, Japan Outdoor Exposure Time vs. Notched Izod Impact Strength of Nylon 66 115
GRAPH: 103 Hiratsuka, Japan Outdoor Exposure Time vs. Percent Weight Change of Nylon 66 116
GRAPH: 104 Hiratsuka, Japan Outdoor Weathering Exposure Time vs. Flexural Strength of Nylon 66 116
GRAPH: 105 Hiratsuka, Japan Outdoor Weathering Exposure Time vs. Tensile Strength of Nylon 66 117
GRAPH: 106 X-W Weatherometer Exposure Time vs. Tensile Strength of Nylon 66 117
GRAPH: 107 Hiratsuka, Japan Outdoor Exposure Time vs. Flexural Modulus of Nylon MXD6 121
GRAPH: 108 Hiratsuka, Japan Outdoor Exposure Time vs. Notched Izod Impact Strength of Nylon MXD6 122
GRAPH: 109 Hiratsuka, Japan Outdoor Weathering Exposure Time vs. Flexural Strength of Nylon MXD6 122
GRAPH: 110 Hiratsuka, Japan Outdoor Weathering Exposure Time vs. Tensile Strength of Nylon MXD6 123
GRAPH: 111 Sunshine Weatherometer Exposure Time vs. Elongation of Nylon MXD6 123
GRAPH: 112 Sunshine Weatherometer Exposure Time vs. Tensile Strength of Nylon MXD6 124
GRAPH: 113 Hiratsuka, Japan Outdoor Exposure Time vs. Breaking Stress in Flexure of Polyarylamide 126
GRAPH: 114 Hiratsuka, Japan Outdoor Exposure Time vs. Flexural Modulus of Polyarylamide 126
GRAPH: 115 Hiratsuka, Japan Outdoor Exposure Time vs. Notched Izod Impact Strength of Polyarylamide 127
GRAPH: 116 Hiratsuka, Japan Outdoor Exposure Time vs. Percent Weight Change of Polyarylamide 127
GRAPH: 117 Florence, Kentucky Outside Weathering Exposure Time vs. Yellowness Index of Polycarbonate 139
GRAPH: 118 Arizona Outdoor Weathering Exposure Time vs. Drop Dart Impact Strength of Polycarbonate 139
GRAPH: 119 Arizona Outdoor Weathering Exposure Time vs. Elongation of Polycarbonate 140
GRAPH: 120 Arizona Outdoor Weathering Exposure Time vs. Tensile Strength at Yield of Polycarbonate 140
GRAPH: 121 Arizona Outdoor Weathering Exposure Time vs. Delta E Color Change of Polycarbonate 141
GRAPH: 122 Florida Outdoor Weathering Exposure Time vs. Drop Dart Impact Strength of Polycarbonate 141
GRAPH: 123 Florida Outdoor Weathering Exposure vs. Delta E Color Change of Polycarbonate 142
GRAPH: 124 Ohio Outdoor Weathering Exposure vs. Delta E Color Change of Polycarbonate 142
GRAPH: 125 Ohio Outdoor Weathering Exposure vs. Drop Dart Impact Strength of Polycarbonate 143
GRAPH: 126 EMMAQUA Accelerated Weathering Exposure Time vs. Light Transmission of Polycarbonate 143
GRAPH: 127 EMMAQUA Accelerated Weathering Exposure Time vs. Yellowness Index of Polycarbonate 144
GRAPH: 128 EMMAQUA Arizona Accelerated Weathering Exposure Time vs. Yellowness Index of Polycarbonate 144
GRAPH: 129 Carbon Arc XW Weatherometer Exposure Time vs. Haze of Polycarbonate 145

GRAPH: 130 Twin Carbon Arc Weatherometer Exposure Time vs. Yellowness Index of Polycarbonate 145

GRAPH: 131 Accelerated Indoor UV Exposure Time vs. Delta E Color Change of Polycarbonate 146

GRAPH: 134 Florida and Arizona Outdoor Weathering Exposure Time vs. Notched Izod Impact Strength of Polybutylene Terephthalate 148

GRAPH: 135 Florida and Arizona Outdoor Weathering Exposure Time vs. Tensile Strength of Polybutylene Terephthalate 148

GRAPH: 136 Hiratsuka, Japan Outdoor Exposure Time vs. Breaking Stress in Flexure of Polybutylene Terephthalate 149

GRAPH: 137 Hiratsuka, Japan Outdoor Exposure Time vs. Flexural Modulus of Polybutylene Terephthalate 149

GRAPH: 138 Hiratsuka, Japan Outdoor Exposure Time vs. Notched Izod Impact Strength of Polybutylene Terephthalate 150

GRAPH: 139 Hiratsuka, Japan Outdoor Exposure Time vs. Percent Weight Change of Polybutylene Terephthalate 150

GRAPH: 140 Sunshine Carbon Arc Weatherometer Exposure Time vs. Impact Strength of Polybutylene Terephthalate 151

GRAPH: 141 Sunshine Weatherometer Exposure Time vs. Elongation of Polybutylene Terephthalate 151

GRAPH: 142 Sunshine Weatherometer Exposure Time vs. Tensile Strength of Polybutylene Terephthalate 152

GRAPH: 143 Weatherometer Exposure Time vs. Tensile Strength Retained of Polybutylene Terephthalate 152

GRAPH: 144 Sunshine Weatherometer Exposure Time vs. Elongation of Polyethylene Terephthalate 156

GRAPH: 145 Sunshine Weatherometer Exposure Time vs. Tensile Strength of Polyethylene Terephthalate 157

GRAPH: 146 Carbon Arc XW Weatherometer Exposure Time vs. Haze of Polyarylate 166

GRAPH: 147 Florida Outdoor Weathering Exposure Time vs. Ultimate Elongation of Polyimide 168

GRAPH: 148 Atlas Weatherometer Exposure Time vs. Ultimate Elongation of Polyimide 168

GRAPH: 149 Sunshine Weatherometer Exposure Time vs. Elongation Retained of Polyimide 169

GRAPH: 150 Sunshine Weatherometer Exposure Time vs. Flexural Strength Retained of Polyimide 169

GRAPH: 151 Sunshine Weatherometer Exposure Time vs. Tensile Strength Retained of Polyimide 170

GRAPH: 152 UV-CON Exposure Time vs. Flexural Strength Retained of Polyimide 170

GRAPH: 153 Weatherometer Exposure Time vs. Elongation of Polyamideimide 171

GRAPH: 154 Weatherometer Exposure Time vs. Tensile Strength of Polyamideimide 172

GRAPH: 155 Xenon Arc Weatherometer Exposure Time vs. Tensile Strength of Polyetherimide 173

GRAPH: 156 Xenon Weatherometer Exposure Time vs. Elongation Retained of Low Density Polyethylene 173

GRAPH: 157 Composting Exposure Time vs. Elongation Retained of Low Density Polyethylene 183

GRAPH: 158 Arizona Outdoor Weathering Exposure Time vs. Tensile Strength of High Density Polyethylene 196

GRAPH: 159 Weatherometer Exposure Time vs. Tensile Strength of High Density Polyethylene 196

GRAPH: 160 Weatherometer Exposure Time vs. Tensile Strength of High Density Polyethylene 197

GRAPH: 161 Weatherometer Exposure Time vs. Tensile Strength of High Density Polyethylene 197

GRAPH: 162 Weatherometer Exposure Time vs. Tensile Strength of High Density Polyethylene 198

GRAPH: 163 Weatherometer Exposure Time vs. Tensile Strength of High Density Polyethylene 198

GRAPH: 164 Weatherometer Exposure Time vs. Tensile Strength of High Density Polyethylene 199

GRAPH: 165 Weatherometer Exposure Time vs. Tensile Strength of High Density Polyethylene 199

GRAPH: 166 Weatherometer Exposure Time vs. Tensile Strength of High Density Polyethylene 200

GRAPH: 167 Weatherometer Exposure Time vs. Tensile Strength of High Density Polyethylene 200

GRAPH: 168 Weatherometer Exposure Time vs. Tensile Strength of High Density Polyethylene 201

GRAPH: 169 Outdoor Exposure Time vs. Chip Impact Strength of Polypropylene Copolymer 211

GRAPH: 170 Outdoor Exposure Time vs. Delta E Color Change of Polypropylene Copolymer 211

GRAPH: 171 Outdoor Exposure Time vs. Flexural Strength of Polypropylene Copolymer 212

GRAPH: 172 Outdoor Exposure Time vs. Tangent Modulus of Polypropylene Copolymer 212

GRAPH: 173 Outdoor Exposure Time vs. Tensile Strength of Polypropylene Copolymer 213

GRAPH: 174 Weatherometer Exposure Time vs. Izod Impact Strength Retained of Polymethylpentene 215

GRAPH: 175 Fadeometer Exposure Time vs. Yellowness Index of General Purpose Polystyrene 220

GRAPH: 176 Fluorescent Lamp Exposure Time vs. Yellowness Index of General Purpose Polystyrene 221

GRAPH: 177 Fadeometer Exposure Time vs. Yellowness Index of Impact Polystyrene 224

GRAPH: 178 Xenon Arc Weatherometer Exposure Time vs. Tensile Strength of Polysulfone 226

GRAPH: 179 Xenon Arc Weatherometer Exposure Time vs. Tensile Strength of Polyethersulphone 227

GRAPH: 180 Arizona Outdoor Weathering Exposure Time vs. Yellowness Index of Styrene Acrylonitrile Copolymer 230

GRAPH: 181 UV-CON Accelerated Weathering Exposure Time vs. Yellowness Index of Styrene Acrylonitrile Copolymer 230

GRAPH: 182 Florida Outdoor Weathering Exposure Time vs. Drop Weight Impact Retained of Olefin Modified Styrene Acrylonitrile Copolymer 233

GRAPH: 183 Florida Weathering Exposure Time vs. Chip Impact Strength of Olefin Modified Styrene Acrylonitrile Copolymer 233

GRAPH: 184 Arizona Outdoor Weathering Exposure Time vs. Drop Dart Impact Strength of Polyvinyl Chloride 250

GRAPH: 185 Arizona Outdoor Weathering Exposure Time vs. Elongation of Polyvinyl Chloride 250

GRAPH: 186 Arizona Outdoor Weathering Exposure Time vs. Elongation of Polyvinyl Chloride 251

GRAPH: 187 Arizona Outdoor Weathering Exposure Time vs. Tensile Strength at Yield of Polyvinyl Chloride 251

GRAPH: 188 Arizona Outdoor Weathering Exposure Time vs. Delta E Color Change of Polyvinyl Chloride 252

GRAPH: 189 Arizona, Florida and Ohio Outdoor Weathering Exposure Time vs. Drop Dart Impact Strength of Polyvinyl Chloride 252

GRAPH: 190 Florida Outdoor Weathering Exposure Time vs. Drop Dart Impact Strength of Polyvinyl Chloride 253

GRAPH: 191 Florida Outdoor Weathering Exposure Time vs. Drop Weight Impact Retained of Polyvinyl Chloride 253

GRAPH: 192 Florida Outdoor Weathering Exposure Time vs. Delta E Color Change of Polyvinyl Chloride 254

GRAPH: 193 Ohio Outdoor Weathering Exposure vs. Delta E Color Change of Polyvinyl Chloride 254

GRAPH: 194 Ohio Outdoor Weathering Exposure vs. Drop Dart Impact Strength of Polyvinyl Chloride 255

GRAPH: 195 Weatherometer Exposure Time vs. Tensile Strength at Yield of Polyvinyl Chloride 255

GRAPH: 196 Florida Outdoor Weathering Exposure Time vs. Drop Weight Impact Retained of Chlorinated Polyvinyl Chloride 257

GRAPH: 197 Accelerated Indoor UV Exposure Time vs. Delta E Color Change of ABS Polyvinyl Chloride Alloy 206

GRAPH: 197a Xenon Weatherometer Exposure Time vs. Elongation Retained of Low Density Polyethylene 265

GRAPH: 198 Composting Exposure Time vs. Elongation Retained of Low Density Polyethylene 265

GRAPH: 199 Burial Time vs. Starch Granules Digested of Starch Modified Polyethylene Alloy 267

GRAPH: 200 Florence, Kentucky Outside Weathering Exposure Time vs. Yellowness Index of Polyester 275

GRAPH: 201 EMMAQUA Arizona Accelerated Weathering Exposure Time vs. Yellowness Index of Polyester 275

GRAPH: 202 Twin Carbon Arc Weatherometer Exposure Time vs. Yellowness Index of Polyester 276

GRAPH: 203 Florida Outdoor Weathering Exposure Time vs. Delta b Color Scale of Polyurethane 280

GRAPH: 204 QUV Exposure Time vs. Gloss Retained of Polyurethane Reaction Injection Molding System 281

GRAPH: 205 Sunshine Carbon Arc Weatherometer Exposure Time vs. Gloss Retained of Polyurethane Reaction Injection Molding System 281

GRAPH: 205a Xenon Arc Weatherometer Exposure Time vs. Carbonyl Formation - Area Normalized of Olefinic Thermoplastic Elastomer .. 329

GRAPH: 206 Xenon Arc Weatherometer Exposure Time vs. Decrease in Molecular Weight of Olefinic Thermoplastic Elastomer 329

GRAPH: 207 QUV Exposure Time vs. Elongation of Urethane Thermoplastic Elastomer 350

GRAPH: 208 QUV Exposure Time vs. Tensile Strength of Urethane Thermoplastic Elastomer 350

GRAPH: 209 QUV Exposure Time vs. Yellowness Index of Urethane Thermoplastic Elastomer 351

GRAPH: 210 QUV Exposure Time vs. Yellowness Index of Thermoplastic Polyether Urethane Elastomer 351

GRAPH: 211 Xenon Weatherometer Exposure Time vs. Tensile Strength of Thermoplastic Polyether Urethane Elastomer 352

GRAPH: 212 Delaware Outdoor Weathering Exposure Time vs. vs. Elongation at Break of Chlorosulfonated Polyethylene Rubber 373

GRAPH: 213 Xenon Arc Weatherometer Exposure Time vs. Carbonyl Formation - Area Normalized of Ethylene Propylene Copolymer 375

GRAPH: 214 Xenon Arc Weatherometer Exposure Time vs. Decrease in Molecular Weight of Ethylene Propylene Copolymer 376

GRAPH: 215 Xenon Arc Weatherometer Exposure Time vs. Carbonyl Formation - Area Normalized of Ethylene Propylene Diene Methylene Terpolymer 405

GRAPH: 216 Xenon Arc Weatherometer Exposure Time vs. Decrease in Molecular Weight of Ethylene Propylene Diene Methylene Terpolymer 405

GRAPH: 217 Florida Outdoor Weathering Exposure Time vs. Delta b Color Scale of Polyurethane 428

Table Index

TABLE: 01 Outdoor Weathering in Florida of White ABS. 3

TABLE: 02 Outdoor Weathering in Ludwigshafen, Germany of ABS. 3

TABLE: 03 Accelerated Indoor Exposure by HPUV of General Electric Cycolac ABS. 4

TABLE: 04 Accelerated Indoor Expsoure to Fluorescent Light of General Electric Cycolac ABS. 4

TABLE: 05 Accelerated Indoor Exposure to Fluorescent Light of General Electric Cycolac ABS. 5

TABLE: 06 Outdoor Weathering in Arizona of UV Stabilized DuPont Delrin Acetal Resin. 22

TABLE: 07 Outdoor Weathering in Ludwigshafen, Germany of BASF AG Luran Acrylate Styrene Acrylonitrile Polymer. 32

TABLE: 08 Outdoor Weathering in Arizona of Aristech Acrylic Resin. 41

TABLE: 09 Outdoor Weathering in Arizona of Cyro Acrylite Plus Acrylic Resin. 42

TABLE: 10 Outdoor Weathering in Florida of ICI Perspex Acrylic Resin. 42

TABLE: 11 Outdoor Weathering in Florida of ICI Perspex Acrylic Resin. 43

TABLE: 12 Outdoor Weathering in Florida of ICI Perspex Acrylic Resin. 44

TABLE: 13 Outdoor Weathering in Kentucky and Accelerated Outdoor Weathering by EMMAQUA of Acrylic Resin. 45

TABLE: 14 Accelerated Weathering in a Xenon Arc Weatherometer of Aristech Acrylic Resin. 46

TABLE: 15 Accelerated Weathering in a Carbon Arc Weatherometer of Aristech Acrylic and DuPont Lucite Acrylic Resin. 47

TABLE: 16 Outdoor Weathering in Kentucky, Accelerated Outdoor Weathering by EMMAQUA and Accelerated Weathering in a Carbon Arc Weatherometer of Uvex Cellulose Acetate Butyrate. 57

TABLE: 17 Accelerated Weathering in a Xenon Arc Weatherometer of Ausimont Halar Ethylene Chlorotrifluoroethylene Copolymer. 61

TABLE: 18 Accelerated Weathering in a Weatherometer of Ausimont Hyflon Ethylene Tetrafluoroethylene Copolymer. 64

TABLE: 19 Accelerated Weathering in a Weatherometer of Ausimont Hyflon Ethylene Tetrafluoroethylene Copolymer. 64

TABLE: 20 Accelerated Weathering in a Weatherometer of DuPont Tefzel Ethylene Tetrafluoroethylene Copolymer. 65

TABLE: 21 Outdoor Weathering of Atochem Kynar Polyvinylidene Fluoride. 72

TABLE: 22 Accelerated Weathering in a Weatherometer of Polyvinylidene Fluoride. 73

TABLE: 23 Outdoor Weathering in Florida and Arizona of DuPont Surlyn Ionomer. 76

TABLE: 24 Effect of Pigments, UV Stabilizers and Antioxidants on the Accelerated Weathering in an Atlas Weatherometer of Zinc Ion Type DuPont Surlyn Ionomer. 77

TABLE: 25 Effect of Pigments, UV Stabilizers and Antioxidants on the Accelerated Weathering in an Atlas Weatherometer of Zinc Ion Type DuPont Surlyn Ionomer. 78

TABLE: 26 Effect of Pigments, UV Stabilizers and Antioxidants on the Accelerated Weathering in an Atlas Weatherometer of Sodium Ion Type DuPont Surlyn Ionomer. 79

TABLE: 27 Accelerated Weathering in a QUV of Zinc Ion Type DuPont Surlyn Ionomer. 80

TABLE: 28 Outdoor Weathering in Arizona, Florida and New York of General Electric Noryl Modified Polyphenylene Oxide. 82

TABLE: 29	Outdoor Weathering in Florida of Allied Signal Capron Nylon 6.	92
TABLE: 30	Outdoor Weathering in California and Pennsylvania of Nylon 6.	93
TABLE: 31	Outdoor Weathering in California and Pennsylvania of Glass Reinforced Nylon 610.	102
TABLE: 32	Outdoor Weathering in Arizona of DuPont Zytel Nylon 66.	105
TABLE: 33	Outdoor Weathering in Florida and Accelerated Weathering in XW Weatherometer of Mineral Filled DuPont Minlon Nylon 66.	106
TABLE: 34	Outdoor Weathering in Florida of DuPont Zytel Nylon 66.	107
TABLE: 35	Outdoor Weathering in California and Pennsylvania of Glass Reinforced Nylon 66.	108
TABLE: 36	Outdoor Weathering in Delaware of DuPont Zytel Nylon 66.	109
TABLE: 37	Accelerated Weathering in an XW Weatherometer of DuPont Zytel Nylon 66.	110
TABLE: 38	Accelerated Weathering in an XW Weatherometer of DuPont Zytel Nylon 66.	111
TABLE: 39	Outdoor Weathering in Arizona of Dow Calibre Polycarbonate.	133
TABLE: 40	Outdoor Weathering in California and Pennsylvania of Glass Reinforced Polycarbonate.	134
TABLE: 41	Outdoor Weathering in Pennsylvania of Miles Makrolon Polycarbonate.	135
TABLE: 42	Outdoor Weathering in Pennsylvania of Miles Makrolon Polycarbonate.	136
TABLE: 43	Outdoor Weathering in Kentucky, Accelerated Outdoor Weathering by EMMAQUA and Accelerated Weathering in a Carbon Arc Weatherometer of General Electric Lexan Polycarbonate.	137
TABLE: 44	Accelerated Weathering in an XW Weatherometer of General Electric Lexan Polycarbonate.	138
TABLE: 45	Accelerated Indoor Exposure by HPUV of General Electric Lexan Polycarbonate.	138
TABLE: 46	Outdoor Weathering in Arizona of DuPont Rynite Polyethylene Terephthalate.	154
TABLE: 47	Outdoor Weathering in Arizona of DuPont Rynite Polyethylene Terephthalate.	154
TABLE: 48	Outdoor Weathering in Florida of DuPont Rynite Polyethylene Terephthalate.	155
TABLE: 49	Outdoor Weathering in Florida of DuPont Rynite Polyethylene Terephthalate.	155
TABLE: 50	Accelerated Outdoor Weathering in Arizona by EMMA and EMMAQUA of DuPont Rynite Polyethylene Terephthalate.	156
TABLE: 51	Accelerated Weathering with a Xenon Arc Lamp of Hoechst AG Vectra Liquid Crystal Polymer.	163
TABLE: 52	Accelerated Weathering in an XW Weatherometer of DuPont Ardel Polyarylate.	165
TABLE: 53	Outdoor Weathering in the United Kingdom of ICI Victrex Polyetheretherketone.	175
TABLE: 54	Effect of Pigments on Outdoor Weathering in the United Kingdom of ICI Victrex USA Victrex Polyetheretherketone.	176
TABLE: 55	Effect of UV Stabilizers and UV Absorbers on Outdoor Weathering of Polyethylene Greenhouse Film.	180
TABLE: 56	Effect of UV Stabilizer Amount on Outdoor Weathering of Polyethylene Greenhouse Film.	180
TABLE: 57	Outdoor Weathering in California and Pennsylvania of Glass Reinforced Polyethyelene.	181
TABLE: 58	Effect of Carbon Black Type on Accelerated Outdoor Weathering by EMMA of Phillips Marlex High Density Polyethylene.	189
TABLE: 59	Effect of Color Dispersion on Accelerated Weathering in a Weatherometer of Phillips Marlex High Density Polyethylene with 0.5% CP Cadmium Red Pigment.	190
TABLE: 60	Effect of Pigments on Accelerated Weathering in a Weatherometer of Phillips Marlex High Density Polyethylene.	191

TABLE: 61	Effect of UV Stabilizers on Accelerated Weathering in a Weatherometer of Phillips Marlex High Density Polyethylene with 2% Cadmium Yellow Pigment.	192
TABLE: 62	Effect of Yellow Pigments on Accelerated Weathering in a Weatherometer of Phillips Marlex High Density Polyethylene.	193
TABLE: 63	Effect of UV Stabilizers on Accelerated Weathering in a Weatherometer of Phillips Marlex High Density Polyethylene with 2% Titanium Dioxide.	194
TABLE: 64	Effect of Antioxidants and UV Absorber on Accelerated Weathering in a Xenon Weatherometer of Green High Density Polyethylene.	195
TABLE: 65	Fungus Resistance of Phillips Marlex High Density Polyethylene.	195
TABLE: 66	Effect of UV Stabilizers on Accelerated Weathering in a Xenon Arc Weatherometer of Ethylene Vinyl Acetate Polyethylene Copolymer Greenhouse Film.	206
TABLE: 67	Effect of Antioxidants on Outdoor Weathering in Florida and Puerto Rico of Polypropylene.	207
TABLE: 68	Outdoor Weathering in California and Pennsylvania of Glass Reinforced Polypropylene.	208
TABLE: 69	Effect of Stabilizers and Antioxidants on Outdoor Weathering in Puerto Rico of Polypropylene.	209
TABLE: 70	Effect of ECC International Microcal Calcium Carbonate on Accelerated Weathering in QUV of Polypropylene.	210
TABLE: 71	Accelerated Weathering in an Atlas Weatherometer of Phillips Ryton Polyphenylene Sulfide.	218
TABLE: 72	Outdoor Weathering in California and Pennsylvania of Glass Reinforced General Purpose Polystyrene.	220
TABLE: 73	Outdoor Weathering in California and Pennsylvania of Glass Reinforced Polysulfone.	225
TABLE: 74	Outdoor Weathering in Arizona of Dow Chemical Tyril Styrene Acrylonitrile Copolymer.	229
TABLE: 75	Outdoor Weathering in Florida of Dow Chemical Rovel Olefin Modified Styrene Acrylonitrile Copolymer.	231
TABLE: 76	Outdoor Weathering in Florida of White Dow Chemical Rovel Olefin Modified Styrene Acrylonitrile Copolymer.	232
TABLE: 77	Outdoor Weathering in Arizona, Florida and Ohio of Gold Geon Company Geon Polyvinyl Chloride.	241
TABLE: 78	Outdoor Weathering in Arizona, Florida and Ohio of Yellow Geon Company Geon Polyvinyl Chloride.	241
TABLE: 79	Outdoor Weathering in Arizona, Florida and Ohio of White Geon Company Geon Polyvinyl Chloride.	242
TABLE: 80	Outdoor Weathering in Arizona, Florida and Ohio of Olive Geon Company Geon Polyvinyl Chloride.	242
TABLE: 81	Outdoor Weathering in Arizona, Florida and Ohio of White Geon Company Geon Polyvinyl Chloride.	243
TABLE: 82	Outdoor Weathering in Arizona, Florida and Ohio of Tan Geon Company Geon Polyvinyl Chloride.	243
TABLE: 83	Outdoor Weathering in Arizona, Florida and Ohio of Red Geon Company Geon Polyvinyl Chloride.	244
TABLE: 84	Outdoor Weathering in Arizona, Florida and Ohio of Green Geon Company Geon Polyvinyl Chloride.	244
TABLE: 85	Outdoor Weathering in Arizona, Florida and Ohio of Grey Geon Company Geon Polyvinyl Chloride.	245
TABLE: 86	Outdoor Weathering in Arizona, Florida and Ohio of Brown Geon Company Geon Polyvinyl Chloride.	245
TABLE: 87	Outdoor Weathering in Arizona, Florida and Ohio of Ivory Geon Company Geon Polyvinyl Chloride.	246
TABLE: 88	Outdoor Weathering in Arizona, Florida and Ohio of Brown Geon Company Geon Polyvinyl Chloride.	246
TABLE: 89	Outdoor Weathering in Arizona, Florida and Ohio of Blue Geon Company Geon Polyvinyl Chloride.	247
TABLE: 90	Outdoor Weathering in Arizona, Florida and Ohio of Black Geon Company Geon Polyvinyl Chloride.	247
TABLE: 91	Outdoor Weathering in Florida of Polyvinyl Chloride.	248

TABLE: 92 Outdoor Weathering in California and Pennsylvania of Glass Reinforced Polyvinyl Chloride. 248

TABLE: 93 Accelerated Indoor Exposure by HPUV of Geon Company Geon Polyvinyl Chloride. 249

TABLE: 94 Accelerated Indoor Exposure by HPUV of Indoor UV STABLE: , White Geon Company Geon Polyvinyl Chloride. 249

TABLE: 95 Accelerated Weathering in a QUV of Novatec Novaloy 9000 ABS Polyvinyl Chloride Alloy. 259

TABLE: 96 Accelerated Indoor Exposure by HPUV and Xeneon Arc Weatherometer of Dow Chemical Pulse Polycarbonate ABS Alloy. 263

TABLE: 97 Accelerated Indoor Exposure by HPUV of General Electric Cycoloy Polycarbonate ABS Alloy. 263

TABLE: 98 Soil Burial and Fungus Resistance of Biodegradable Novamont Mater-Bi Starch Synthetic Resin Alloy. 270

TABLE: 99 Outdoor Weathering in New Jersey of Filled and Reinforced Diallyl Phthalate Resin. 271

TABLE: 100 Outdoor Weathering in Kentucky, Accelerated Outdoor Weathering by EMMAQUA and Accelerated Weathering in a Carbon Arc Weatherometer of Kalwall Sunlite Acrylic Coated Polyester. 274

TABLE: 101 Accelerated Weathering in a Xenon Arc Weatherometer of Recticel Colo-Fast Polyurethane Reaction Injection Molding System. 278

TABLE: 102 Accelerated Weathering in a Xenon Arc Weatherometer of Recticel Colo-Fast Polyurethane Reaction Injection Molding System. 279

TABLE: 103 Accelerated Weathering in a Fadeometer of Recticel Colo-Fast Polyurethane Reaction Injection Molding System. 280

TABLE: 104 Outdoor Weathering, Accelerated Outdoor Weathering by EMMAQUA and Accelerated Weathering with a Xenon Arc Weatherometer of Black Advanced Elastomer Systems Santoprene Olefinic Thermoplastic Elastomer. 290

TABLE: 105 Outdoor Weathering, Accelerated Outdoor Weathering by EMMAQUA and Accelerated Weathering in a Xenon Arc Weatherometer of Black, UV Stabilized Advanced Elastomer Systems Santoprene Olefinic Thermoplastic Elastomer. 291

TABLE: 106 Outdoor Weathering in Arizona With and Without Water Spray of Black, General Purpose Advanced Elastomer Systems Santoprene Olefinic Thermoplastic Elastomer. 292

TABLE: 107 Outdoor Weathering in Arizona and Accelerated Outdoor Weathering by EMMAQUA of Colorable Advanced Elastomer Systems Olefinic Thermoplastic Elastomer. 293

TABLE: 108 Outdoor Weathering in Arizona and Florida of Black Advanced Elastomer Systems Santoprene Olefinic Thermoplastic Elastomer. 294

TABLE: 109 Outdoor Weathering in Arizona and Florida of Colored Advanced Elastomer Systems Santoprene Olefinic Thermoplastic Elastomer. 295

TABLE: 110 Outdoor Weathering in Arizona of Black Advanced Elastomer Systems Santoprene Olefinic Thermoplastic Elastomer. 296

TABLE: 111 Outdoor Weathering in Arizona With and Without Water Spray of Black, General Purpose Advanced Elastomer Systems Santoprene Olefinic Thermoplastic Elastomer. 297

TABLE: 112 Outdoor Weathering in Arizona With and Without Water Spray of Black, UV Stabilized Advanced Elastomer Systems Santoprene Olefinic Thermoplastic Elastomer. 298

TABLE: 113 Outdoor Weathering in Arizona With and Without Water Spray of Black, UV Stabilized Advanced Elastomer Systems Santoprene Olefinic Thermoplastic Elastomer. 299

TABLE: 114 Outdoor Weathering in Arizona With and Without Water Spray of Black, UV Stabilized Advanced Elastomer Systems Santoprene Olefinic Thermoplastic Elastomer. 300

TABLE: 115 Outdoor Weathering in Arizona with Water Spray Added of Black Advanced Elastomer Systems Olefinic Thermoplastic Elastomer. 301

TABLE: 116 Outdoor Weathering in Arizona With and Without Water Spray of Advanced Elastomer Systems Santoprene Olefinic Thermoplastic Elastomer. 302

TABLE: 117 Outdoor Weathering in Florida of Black Advanced Elastomer Systems Olefinic Thermoplastic Elastomer. 303

TABLE: 118 Outdoor Weathering in Florida With and Without Water Spray of Black, General Purpose Advanced Elastomer Systems Santoprene Olefinic Thermoplastic Elastomer. 304

TABLE: 119 Outdoor Weathering in Florida With and Without Water Spray of Black, General Purpose Advanced Elastomer Systems Santoprene Olefinic Thermoplastic Elastomer. 305

TABLE: 120 Outdoor Weathering in Florida With and Without Water Spray of Black, UV Stabilized Advanced Elastomer Systems Santoprene Olefinic Thermoplastic Elastomer. 306

TABLE: 121 Outdoor Weathering in Florida With and Without Water Spray of Black, UV Stabilized Advanced Elastomer Systems Santoprene Olefinic Thermoplastic Elastomer. 307

TABLE: 122 Outdoor Weathering in Florida With and Without Water Spray of Black, UV Stabilized Advanced Elastomer Systems Santoprene Olefinic Thermoplastic Elastomer. 308

TABLE: 123 Outdoor Weathering in Florida With and Without Water Spray of Advanced Elastomer Systems Santoprene Olefinic Thermoplastic Elastomer. 309

TABLE: 124 Accelerated Outdoor Weathering by EMMA and EMMAQUA and Accelerated Weathering in a Xenon Arc Weatherometer of Black, UV Stabilized Advanced Elastomer Systems Santoprene Olefinic Thermoplastic Elastomer. 310

TABLE: 125 Accelerated Outdoor Weathering by EMMA and EMMAQUABlack, General Purpose Advanced Elastomer Systems Santoprene Olefinic Thermoplastic Elastomer. 311

TABLE: 126 Accelerated Outdoor Weathering by EMMAQUA of Black Advanced Elastomer Systems Santoprene Olefinic Thermoplastic Elastomer. 312

TABLE: 127 Accelerated Outdoor Weathering in Arizona by EMMA and EMMAQUA of Black Advanced Elastomer Systems Santoprene Olefinic Thermoplastic Elastomer. 313

TABLE: 128 Accelerated Outdoor Weathering by EMMA and EMMAQUA and Accelerated Weathering in a Xenon Arc Weatherometer of Black, UV Stabilized Advanced Elastomer Systems Santoprene Olefinic Thermoplastic Elastomer. 314

TABLE: 129 Accelerated Weathering in a Xenon Arc Weatherometer of Black Advanced Elastomer Systems Santoprene Olefinic Thermoplastic Elastomer. 315

TABLE: 130 Accelerated Weathering in a Xenon Arc Weatherometer of Advanced Elastomer Systems Santoprene Olefinic Thermoplastic Elastomer. 316

TABLE: 131 Accelerated Weathering in a Xenon Arc Weatherometer of Black, General Purpose Advanced Elastomer Systems Santoprene Olefinic Thermoplastic Elastomer. 317

TABLE: 132 Accelerated Weathering in a Xenon Arc Weatherometer of Black, UV Stabilized Advanced Elastomer Systems Santoprene Olefinic Thermoplastic Elastomer. 318

TABLE: 133 Accelerated Weathering in a Xenon Arc Weatherometer of Black, UV Stabilized Advanced Elastomer Systems Santoprene Olefinic Thermoplastic Elastomer. 319

TABLE: 134 Accelerated Weathering in a Xenon Arc Weatherometer of Black, UV Stabilized Advanced Elastomer Systems Santoprene Olefinic Thermoplastic Elastomer. 320

TABLE: 135 Accelerated Weathering in a Xenon Arc Weatherometer of Black, UV Stabilized Advanced Elastomer Systems Santoprene Olefinic Thermoplastic Elastomer. 321

TABLE: 136 Accelerated Weathering in a Xenon Arc Weatherometer of Black, UV Stabilized Advanced Elastomer Systems Santoprene Olefinic Thermoplastic Elastomer. 322

TABLE: 137 Accelerated Indoor Exposure in UVCON of Evode Plastics Forprene Olefinic Thermoplastic Elastomer. 323

TABLE: 138 Accelerated Indoor Exposure in Xenon Lamp of Evode Plastics Forprene Olefinic Thermoplastic Elastomer. 324

TABLE: 139 Ozone Resistance of Evode Plastics Forprene Olefinic Thermoplastic Elastomer. 324

TABLE: 140 Ozone Resistance of Advanced Elastomer Systems Santoprene Olefinic Thermoplastic Elastomer. 325

TABLE: 141 Ozone Resistance of Dow Chemical Engage Olefinic Thermoplastic Elastomer. 326

TABLE: 142	Ozone Resistance of Black Advanced Elastomer Systems Santoprene Olefinic Thermoplastic Elastomer.	327
TABLE: 143	Ozone Resistance of Advanced Elastomer Systems Santoprene, Technor Apex Telcar and BP Chemicals TPR Olefinic Thermoplastic Elastomer.	328
TABLE: 144	Outdoor Weathering in Florida of DuPont Hytrel Polyester Thermoplastic Elastomer.	333
TABLE: 145	Effect of Carbon Black on Outdoor Exposure in Florida of DuPont Hytrel Polyester Thermoplastic Elastomer.	334
TABLE: 146	Effect of Carbon Black Level on Outdoor Weathering in Florida of DuPont Hytrel Polyester Thermoplastic Elastomer.	335
TABLE: 147	Effect of Carbon Black and Film Thickness on Outdoor Weathering in Florida of DuPont Hytrel Polyester Thermoplastic Elastomer.	336
TABLE: 148	Effect of Carbon Black on Accelerated Weathering in a Weatherometer of DuPont Hytrel Polyester Thermoplastic Elastomer.	337
TABLE: 149	Effect of Carbon Black Level on Accelerated Weathering in a Weatherometer of DuPont Hytrel Polyester Thermoplastic Elastomer.	338
TABLE: 150	Effect of Carbon Black on Accelerated Weathering in a Weatherometer of DuPont Hytrel Polyester Thermoplastic Elastomer.	339
TABLE: 151	Soil Burial and Fungus Resistance of DuPont Hytrel Polyester Thermoplastic Elastomer.	340
TABLE: 152	Ozone Resistance of Shell Chemcial Kraton Styrenic Thermoplastic Elastomer.	343
TABLE: 153	Accelerated Weathering in a Fadeometer of BF Goodrich Estane Urethane Thermoplastic Elastomer.	347
TABLE: 154	Accelerated Weathering in a Fadeometer and a QUV of BF Goodrich Estane Urethane Thermoplastic Elastomer.	348
TABLE: 155	Accelerated Weathering in a Fadeometer of BF Goodrich Estane Urethane Thermoplastic Elastomer.	349
TABLE: 156	Outdoor Weathering, Accelerated Outdoor Weathering by EMMAQUA and Accelerated Weathering with a Xenon Arc Weatherometer of White Polyvinyl Chloride Polyol.	353
TABLE: 157	Outdoor Weathering in Arizona and Florida of Flexible, White Polyvinyl Chloride Polyol.	354
TABLE: 158	Accelerated Weathering in an Atlas Weatherometer of Geon Company Geon Polyvinyl Chloride Polyol.	355
TABLE: 159	Outdoor Weathering in Florida of Goodyear Chemigum Nitrile Thermoplastic Elastomer.	358
TABLE: 160	Outdoor Weathering, Accelerated Outdoor Weathering, and Accelerated Weathering of DuPont Hypalon Chlorosulfonated Polyethylene Rubber.	364
TABLE: 161	Outdoor Weathering in Arizona of DuPont Hypalon Chlorosulfonated Polyethylene Rubber.	365
TABLE: 162	Outdoor Weathering in Florida and Delaware of Wire Cable Compound DuPont Hypalon Chlorosulfonated Polyethylene Rubber.	366
TABLE: 163	Outdoor Weathering in Florida, Texas and California of Green Hose Cover Compound DuPont Hypalon Chlorosulfonated Polyethylene Rubber.	367
TABLE: 164	Outdoor Weathering in Florida of White DuPont Hypalon Chlorosulfonated Polyethylene Rubber.	368
TABLE: 165	Outdoor Weathering in Delaware of DuPont Hypalon Chlorosulfonated Polyethylene Rubber.	369
TABLE: 166	Outdoor Weathering in Panama of Pond Liner Formulation DuPont Hypalon Chlorosulfonated Polyethylene Rubber.	370
TABLE: 167	Accelerated Outdoor Weathering by EMMA and EMMAQUA of DuPont Hypalon Chlorosulfonated Polyethylene Rubber.	371
TABLE: 168	Accelerated Weathering in a Xenon Arc Weatherometer of DuPont Hypalon Chlorosulfonated Polyethylene Rubber.	372
TABLE: 169	Outdoor Weathering and Accelerated Outdoor Weathering of White, Randomly Selected, Unstrained Ethylene Propylene Diene Methylene Terpolymer.	382

TABLE: 170 Outdoor Weathering and Accelerated Outdoor Weathering of Black, Weather Resistant, Unstrained Ethylene Propylene Diene Methylene Terpolymer. 383

TABLE: 171 Outdoor Weathering and Accelerated Outdoor Weathering of Black, Randomly Selected, Unstrained Ethylene Propylene Diene Methylene Terpolymer. 384

TABLE: 172 Outdoor Weathering, Accelerated Outdoor Weathering by EMMAQUA and Accelerated Weathering with a Xenon Arc Weatherometer of Black Exxon Vistalon Ethylene Propylene Diene Methylene Terpolymer. 385

TABLE: 173 Outdoor Weathering in Arizona With and Without Water Spray Added of Black Exxon Vistalon Ethylene Propylene Diene Methylene Terpolymer. 386

TABLE: 174 Outdoor Weathering in Arizona of Black Exxon Vistalon Ethylene Propylene Diene Methylene Terpolymer. 387

TABLE: 175 Outdoor Weathering in Florida and Accelerated Outdoor Weathering by EMMA of Black, Weather Resistant, Strained Ethylene Propylene Diene Methylene Terpolymer. 388

TABLE: 176 Outdoor Weathering in Florida With and Without Water Spray Added of Black Exxon Vistalon Ethylene Propylene Diene Methylene Terpolymer. 389

TABLE: 177 Outdoor Weathering in Florida of Weatherable Ethylene Propylene Diene Methylene Terpolymer. 390

TABLE: 178 Outdoor Weathering in Florida and Accelerated Outdoor Weathering by EMMA of Black, Randomly Selected, Strained Ethylene Propylene Diene Methylene Terpolymer. 391

TABLE: 179 Outdoor Weathering in Florida and Accelerated Outdoor Weathering by EMMA of White, Randomly Selected, Strained Ethylene Propylene Diene Methylene Terpolymer. 392

TABLE: 180 Outdoor Weathering in Florida of Black Ethylene Propylene Diene Methylene Terpolymer. 393

TABLE: 181 Accelerated Outdoor Weathering by EMMA and EMMAQUA and Accelerated Weathering in a Xenon Arc Weatherometer of Black Exxon Vistalon Ethylene Propylene Diene Methylene Terpolymer. 394

TABLE: 182 Accelerated Outdoor Weathering in Arizona by EMMA and EMMAQUA of Black Exxon Vistalon Ethylene Propylene Diene Methylene Terpolymer. 395

TABLE: 183 Accelerated Weathering in a UVCON and a Xenon Arc Weatherometer of White, Randomly Selected, Strained Ethylene Propylene Diene Methylene Terpolymer. 396

TABLE: 184 Accelerated Weathering in a UVCON and a Xenon Arc Weatherometer of White, Randomly Selected, Unstrained Ethylene Propylene Diene Methylene Terpolymer. 397

TABLE: 185 Accelerated Weathering in a UVCON and a Xenon Arc Weatherometer of Black, Weather Resistant, Strained Ethylene Propylene Diene Methylene Terpolymer. 398

TABLE: 186 Accelerated Weathering in a UVCON and a Xenon Arc Weatherometer of Black, Randomly Selected, Unstrained Ethylene Propylene Diene Methylene Terpolymer. 399

TABLE: 187 Accelerated Weathering in a Xenon Arc Weatherometer of Black Exxon Vistalon Ethylene Propylene Diene Methylene Terpolymer. 400

TABLE: 188 Accelerated Weathering in a UVCON and a Xenon Arc Weatherometer of Black, Weather Resistant, Unstrained Ethylene Propylene Diene Methylene Terpolymer. 401

TABLE: 189 Accelerated Weathering in a UVCON and a Xenon Arc Weatherometer of Black, Randomly Selected, Strained Ethylene Propylene Diene Methylene Terpolymer. 402

TABLE: 190 Ozone Resistance of Exxon Vistalon Ethylene Propylene Diene Methylene Terpolymer. 403

TABLE: 191 Ozone Resistance of Ethylene Propylene Diene Methylene Terpolymer. 404

TABLE: 192 Outdoor Weathering, Accelerated Outdoor Weathering by EMMAQUA and Accelerated Weathering with a Xenon Arc Weatherometer of Black DuPont Neoprene W Neoprene Rubber. 411

TABLE: 193 Outdoor Weathering in Arizona and Florida of Black Neoprene Rubber. 411

TABLE: 194 Outdoor Weathering in Arizona of Black DuPont Neoprene W Neoprene Rubber. 412

TABLE: 195 Outdoor Weathering in Arizona With and Without Water Spray Added of Black DuPont Neoprene W Neoprene Rubber........ **413**

TABLE: 196 Outdoor Weathering in Florida With and Without Water Spray Added of Black DuPont Neoprene W Neoprene Rubber. **414**

TABLE: 197 Accelerated Outdoor Weathering by EMMA and EMMAQUA and Accelerated Weathering in a Xenon Arc Weatherometer of Black DuPont Neoprene W Neoprene Rubber.......... **415**

TABLE: 198 Accelerated Outdoor Weathering in Arizona by EMMA and EMMAQUA of Black DuPont Neoprene W Neoprene Rubber. .. **416**

TABLE: 199 Accelerated Weathering in a Xenon Arc Weatherometer of Black DuPont Neoprene W Neoprene Rubber. **417**

TABLE: 200 Ozone Resistance of DuPont Neoprene Rubber.......... **418**

TABLE: 201 Ozone Resistance of Japan Synthetic Rubber JSR BR Polybutadiene Rubber. **419**

TABLE: 202 Ozone Resistance of Goodyear Natsyn Polyisoprene Rubber. **422**

TABLE: 203 Ozone Resistance of Goodyear Natsyn Polyisoprene Rubber. **423**

TABLE: 204 Ozone Resistance of Goodyear Natsyn Polyisoprene Rubber. **424**

TABLE: 205 Ozone Resistance of Goodyear Natsyn Polyisoprene Rubber. **425**

TABLE: 206 Accelerated Weathering in a Xenon Arc Weatherometer of Polyurethane Rubber. **427**

TABLE: 207 Outdoor Weathering, Accelerated Outdoor Weathering by EMMAQUA, and Accelerated Weathering in a Xenon Arc Weatherometer of White Silicone Rubber. **429**

TABLE: 208 Outdoor Weathering in Arizona and Florida of White Silicone Rubber. **430**

Endnotes to Tables

(a) test method: CIE Lab color scale

(b) test apparatus: Hunter Colorimeter; test note: samples washed with water

(c) test apparatus: Hunter Colorimeter

(d) test method: ASTM D882

(e) test name: static strip test; test results note: Results of the test are reported with two digits separated with a virgule. The number before the virgule indicates the number of quarters of the test strip which showed cracks (total range is 0-4). The number after the virgule indicates the size of the cracks in length perpendicular to the length of the test strip. The range is 0 to 10 using the following scale in millimetres: 1) 0.25-0.5 2) 0.5-0.76 3) 0.76-1.02 4) 1.02-1.52 5) 1.52-2.03 6) 2.03-2.5 7) 2.5-3.8 8) 3.8-5.1 9) 5.1-6.35 10) >6.35.

(f) definition: L= 100 lighter, 0 darker, a= -green, + red, b= -blue, + yellow;
material grade: Novaloy 9000-17

(g) test name: Gardner gloss; test method: ASTM D3134

(h) test method: SAE J545

(i) test name: grey scale; test method: ISO 105A02, DIN 54001

(j) test method: ASTM D1003

(k) specimen type: ASTM D638, type IV; strain rate: 508 mm/min.

(l) specimen type: ASTM D638, type IV; strain rate: 50.8 mm/min.

(m) strain rate: 50.8 mm/min.

(o) test method: ASTM D1003-61 (1988); test apparatus: HunterLab Model D25P-9 colorimeter; test apparatus note: equipped with a pivotable sphere optical unit; test note: coded surface away from source

(p) test method: ASTM E308-66; test apparatus: Beckman 5240 spectrophotometer; test apparatus note: utilizes an integrating sphere; test note: wavelength region measured was from 780 nm to 350 nm, coded surface away from source

(q) name: yellowness index; test method: ASTM D1925-70

(r) name: yellowness index; test method: ASTM E313-73

(s) name: delta E; test description: change in E value comparing the color measurement on the exposed sample versus the original blank sample; test apparatus: spectrophotometer

(t) test apparatus: Glossguard 2-glossmeter

(w) test method: ASTM D882; strain rate: 508 mm/min.; test note: jawspace is 50.8 mm, benchmark is 25.4 mm; test direction: transverse

(x) specimen thickness: 3.2 mm

(y) specimen thickness: 2.03 mm; specimen type: microtensile specimen to ASTM D1708

(z) name: relative tensile modulus; test note: not an ASTM test, strain calculated from grip separation

(ab) test apparatus: FMC-2 Hunterlab; test method: SAE J1545

(ac) test method: ASTM D412

(ae) test method: ASTM D2240

(aj) test note: as viewed through a 5x lens

Reference Index

(2) *Kynar Polyvinylidene Fluoride,* supplier technical report (PL705-Rev4-1-91) - Atochem North America, Inc., 1991.

(16) *The Radiation Response of Udel Polysulfone,* supplier technical report (Number: 101) - Amoco Performance Products, Inc.

(20) *Torlon Engineering Polymers / Design Manual,* supplier design guide (F-49893) - Amoco Performance Products.

(25) *Lupolen, Lucalen Product Line, Properties, Processing,* supplier design guide (B 581 e/(8127) 10.91) - BASF Aktiengesellschaft, 1991.

(26) *Polystyrol Product Line, Properties, Processing,* supplier design guide (B 564 e/2.93) - BASF Aktiengesellschaft, 1993.

(27) *Ultrapek Product Line, Properties, Processing,* supplier design guide (B 607 e/10.92) - BASF Aktiengesellschaft, 1992.

(28) *Ultrason E, Ultrason S Product Line, Properties, Processing,* supplier design guide (B 602 e/10.92) - BASF Aktiengesellschaft, 1992.

(29) *Styrolux Product Line, Properties, Processing,* supplier design guide (B 583 e/(950) 12.91) - BASF Aktiengesellschaft, 1992.

(30) *Luran Product Line, Properties, Processing,* supplier design guide (B 565 e/10.83) - BASF Aktiengesellschaft, 1983.

(51) *Ultem Design Guide,* supplier design guide (ULT-201G (6/90) RTB) - General Electric Company, 1990.

(53) *Noryl Extrusion Resins,* supplier design guide (CDX-265) - General Electric Company.

(62) *Technical Information Evoprene G,* supplier technical report (RDS 028/9240) - Evode Plastics.

(68) *Design Handbook For Du Pont Engineering Plastics - Module II,* supplier design guide (E-42267) - Du Pont Engineering Polymers.

(70) *Vectra Polymer Materials,* supplier design guide (B 121 BR E 9102/014) - Hoechst AG, 1991.

(77) *Victrex PEEK,* supplier design guide (VK2/0586) - ICI Advanced Materials, 1986.

(78) *Calibre Engineering Thermoplastics Basic Design Manual,* supplier design guide (301-1040-1288) - Dow Chemical Company, 1988.

(89) *Foraflon PVDF,* supplier design guide (694.E/07.87/20) - Atochem S. A., 1987.

(93) *Ultramid Nylon Resins Product Line, Properties, Processing,* supplier design guide (B 568/1e/4.91) - BASF Corporation, 1991.

(101) *Engineering Properties Of Marlex Resins,* supplier design guide (TSM-243) - Phillips 66 Company, 1983.

(111) *Hytrel Technical Notes - " Weather Protection of Hytrel with Carbon Black",* supplier technical report (I-48) - DuPont Company.

Reference numbers correspond to our assigned source document number, if you wish additional information, please contact Plastics Design Library.

(112) *Marlex Polyethylene Weatherability,* supplier technical report (TIB3 (78-89 02)) - Phillips 66 Company, 1989.

(113) *Acrylite Plus Acrylic Based Molding and Extrusion Compounds,* supplier marketing literature (1511A-293-5CG) - Cyro Industries, 1993.

(114) *Hyflon ETFE 700/800 Properties and Application Guide,* supplier design guide - Ausimont USA, Inc.

(115) *Kynar Polyvinylidene Fluoride,* supplier design guide (15M-8-88-TR PL705-REV-2) - Penwalt Corporation, 1988.

(116) *Rovel Weatherable High Impact Polymers,* supplier marketing literature (301-622-285) - Dow Chemical Company, 1985.

(117) *Aristech - Unpublished Test Results,* supplier written correspondence - Aristech Chemical Corporation, 1993.

(118) *Cycolac ABS Resin Design Guide,* supplier design guide (CYC-350 (5/90) RTB) - General Electric Plastics, 1990.

(119) *Weathering Data of Santoprene Thermoplastic Rubber (Black Grades) Versus Standard Thermoset Rubbers,* supplier technical report (TCD03787) - Monsanto Company, 1987.

(120) *Weathering Of Santoprene Thermoplastic Rubber Black Ultraviolet Grades,* supplier technical report (TCD00592) - Advanced Elastomer Systems, 1992.

(121) *Parylene Conformal Coatings Specifications and Properties,* supplier technical report - Union Carbide Speciatly Coating Systems, 1992.

(122) *Celanex Thermoplastic Polyester Properties and Proecessing (CX-1A),* supplier design guide (HCER 91-343/10M/692) - Hoechst Celanese Corporation, 1992.

(123) *Upimol Polyimide Shape,* supplier technical report - Ube Industries.

(125) *Solvay Polyvinylidene Fluoride,* supplier design guide (B-1292c-B-2.5-0390) - Solvay, 1992.

(126) supplier written correspondence - Rogers Corporation, 1991.

(127) *Plexiglas Acrylic Sheet General Information,* supplier technical report (PL-1p) - Rohm and Haas Company, 1985.

(128) *Novatec Novaloy 9000,* supplier marketing literature - Novatec Plastics & Chemicals Co.

(129) Gish, Brian D., Jablonowski, Thomas L., *Weathering Tests for EPDM Rubber Sheets for Use in Roofing Applications,* 8th Conference On Roofing Technology, conference proceedings - National Bureau of Standards and NRCA, 1987.

(130) *Elastollan Design And Processing Guide,* supplier design guide - BASF Corporation, 1993.

(131) *Thermal And Other Properties Of Halar Fluoropolymer,* supplier technical report (GHG) - Ausimont.

(132) *Tecnoflon,* supplier marketing literature - Montefluos.

(133) *Weatherability,* supplier marketing literature (E-53525) - Du Pont Company, 1983.

(134) *Resistance to Ultraviolet Irradiation for Surlyn Ionomer Resins,* supplier technical report (E-78693-103520/A) - Du Pont Company, 1986.

(135) *IXEF Reinforced Polyarylamide Based Thermoplastic Compounds Technical Manual,* supplier design guide (Br 1409c-B-2-1190) - Solvay, 1990.

(136) *Bayblend FR Resins For Business Machines And Electronics,* supplier marketing literature (55-D808(5)J 313-10/88) - Mobay Corporation, 1988.

(137) *Terluran Product Line, Properties, Processing,* supplier design guide (B 567e/ (8109) 9.90) - BASF Aktiengesellschaft, 1990.

(138) *Aclar Performance Films,* supplier technical report (SFI-14 Rev. 9-89) - Allied-Signal Enineered Plastics, 1989.

(139) Dupuis, I. C., Cumberland, D. W., *Compounding 'Hypalon' For Weather Resistance,* supplier technical report (E-23070-1 HP-515.1) - Du Pont Company, 1987.

(140) *Ryton Polyphenylene Sulfide Compounds Engineering Properties,* supplier design guide (TSM-266) - Phillips Chemical Company, 1983.

(141) *Physical Properties Kydex 100 Acrylic PVC Alloy Sheet,* supplier technical report (KC-89-03) - Kleerdex Company, 1989.

(142) *Luran S Acrylonitrile Styrene Acrylate Product Line, Properties, Processing,* supplier design guide (B 566 e / 11.90) - BASF Aktiengesellschaft, 1990.

(143) *Luran S Acrylonitrile Styrene Acrylate Product Line, Properties, Processing,* supplier design guide (B 566 e / 10.83) - BASF Aktiengesellschaft, 1983.

(145) *Ardel Polyarylate - The Tough Weatherable Thermoplastic,* supplier marketing literature (F-47141C) - Union Carbide Corporation.

(146) *EVA Greenhouse Film - Pesticide Study,* supplier technical report - Cyanamid Polymer Additive.

(147) *Plexiglas Acrylic Sheet General Information And Physical Properties,* supplier design guide (PLA-22a) - AtoHaas North America Inc., 1992.

(148) *Cyanamid TMXDI (META) Aliphatic Isocyanate,* supplier marketing literature (90-4-849 3K 5/90 - UPT-061) - American Cyanamid Company Urethane Chemi, 1990.

(149) *Capron Nylon Effect Of Exposure To Sunlight,* supplier technical report (842-149) - Allied Chemical, 1976.

(150) *The Enduring Beauty Of Architectural Finishes Based On Kynar 500,* supplier marketing literature (PL500-TR-10M 12-90) - Atochem North America, 1990.

(151) *Physical Properties Acrylite AR Acrylic Sheet And Cyrolon AR Polycarbonate Sheet,* supplier design guide (1632B-0193-10BP) - Cyro Industries, 1993.

(152) *Comparison Of Plastics Used In Glazing, Signs, Skylights And Solar Collector Applications - Technical Bulletin 143,* competitor's technical report (ADARIS 50-1037-01) - Aristech Chemical Corporation, 1989.

(154) Rainhart, L. G., Schimmel, Jr., W. P., *Effect Of Outdoor Aging On Acrylic Sheet,* 1974 International Solar Energy Society, U. S. Section Annual Meeting, conference proceedings (SAND 74-0241) - Sandia Laboratories, 1974.

(155) *Set A New Standard Of Performance For Your Non-Residential Glazing Seals,* supplier marketing literature - Advanced Elastomer Systems, 1993.

(156) *Lower Costs, Increase The Service Life Of Your Expansion Joints And Water Stoppers.,* supplier marketing literature - Advanced Elastomer Systems, 1993.

(157) *Weatherability Of Santoprene Rubber Compared To Other Materials,* supplier technical report (TCD01588) - Advanced Elastomer Systems, 1988.

(158) *Ozone Resistance Of Santoprene Rubber,* supplier technical report (TCD01787) - Advanced Elastomer Systems, 1986.

(159) *Microcal Spa C110S For Polypropylene,* supplier technical report (APP033 PI) - ECC International, 1993.

(160) *Evoprene Super G Thermoplastic Elastomer Compounds,* supplier marketing literature (RDS 050/9240) - Evode Plastics.

(161) *Forprene By S.O.F.TER.,* supplier marketing literature (RDS 049/9240) - Evode Plastics.

(162) Fusco, James V., Hous, Pierre, *Butyl And Halobutyl Rubbers,* reference book - Exxon Chemicals, 1987.

(163) *BROMO XP-50 Optimizing Key Properties,* supplier marketing literature - Exxon Chemicals.

(164) *Bromobutyl Rubber Optimizing Key Properties,* supplier marketing literature - Exxon Chemicals.

(165) *Kodar PETG Copolyester 6763,* supplier technical report (MB-80F/June 1988) - Eastman Plastics, 1988.

(166) *Kodar PCTG Copolyester 5445,* supplier technical report (MB-94/August 1985) - Eastman Plastics, 1988.

(167) *Weathering Of Tenite Butyrate,* supplier technical report (TR-25C) - Eastman Plastics, 1984.

(168) *ECOSTARplus Leads The Way,* supplier written correspondence - Ecostar International L.P., 1993.

(169) *Chapman, G.,* New Technologies And Applications For Starch Containing Degradable Plastics, supplier technical report - Ecostar International L.P.

(170) *Grilamid TR55 Transparent Nylons,* supplier design guide (GR1-104) - EMS-Chemie.

(171) *Tensile Strength And Color Difference After Weathering - Unpublished Test Results,* supplier technical report - EMS-Chemie.

(172) *Outdoor Exposure LPP30/PP Copolymer,* supplier technical report - Ferro Specialty Plastics Group.

(173) *Weatherability Of Noryl Resins,* supplier technical report - General Electric Company, 1992.

(174) *Natsyn Polyisoprene Rubber,* supplier design guide (700-821-980-540) - Goodyear Chemicals, 1988.

(175) *Florida Weathering Of Chemigum TPE,* supplier technical report (TPE 06-0292/498900-2/92) - Goodyear Chemicals, 1992.

(176) *Business Equipment Externals Comparative UV ===dE Data,* supplier technical report (GH-110792/19) - General Electric Company, 1992.

(177) *Weatherability of Cycolac Brand ABS - Technical Publication P-405,* supplier technical report (8203-5M) - General Electric Company, 1982.

(178) *Vestolit BAU For World-Wide Windows,* supplier technical report (1083e/May 1987/bu) - Huls AG, 1987.

(179) *Dutral And The Automotive Industry,* supplier marketing literature - Montedison Specialty Chemicals Ausimont.

(180) *Ultradur Polybutylene Terephthalate (PBT) Product Line, Properties, Processing,* supplier design guide (B 575/1e - (819) 4.91) - BASF Aktiengesellschaft, 1991.

(181) *Ultraform Polyacetal (POM) Product Line, Properties, Processing,* supplier design guide (B 563/1e - (888) 4.91) - BASF Aktiengesellschaft, 1991.

(182) *Topics In Chemistry - BASF Plastics Research And Development,* supplier technical report - BASF Aktiengesellschaft, 1992.

(183) *Ultraform Outddor Exposure - Unpublished Data,* supplier technical report - BASF, 1993.

(184) *Unknown Name,* reference book,.

(185) *Ultramid T Polyamid 6/6T (PA) Product Line, Properties, Processing,* supplier design guide (B 605 e / 3.93) - BASF Aktiengesellschaft, 1993.

(186) *Lupolen Polyethylene And Novolen Polypropylene Product Line, Properties, Processing,* supplier design guide (B 579 e / 4.92) - BASF Aktiengesellschaft, 1992.

(187) *Du Pont Faxed Correspondence,* supplier technical report (P10-2123) - Du Pont Company, 1994.

(188) *UV Stabilization Of Aromatic Pellethane Thermoplastic Polyurethane Elastomers,* supplier technical report (306-00439-1293 SMG) - Dow Chemical Company, 1993.

(189) *Engineering Design Guide To Rigid Geon Custom Injection Molding Vinyl Compounds,* supplier design guide (CIM-020) - BFGoodrich Geon Vinyl Division, 1989.

(190) *BFGoodrich Fiberloc Polymer Composites Engineering Design Data,* supplier design guide (FL-0101) - BFGoodrich Geon Vinyl Division, 1989.

(191) *Geon HTX Polymers Product Data Sheets,* supplier marketing literature - BFGoodrich Geon Vinyl Division, 1991.

(192) *Duracap Vinyl Capstock Compounds,* supplier marketing literature (DC-001) - BFGoodrich Geon Vinyl Division, 1988.

(193) *Geon Flexible Compounds,* trade journal (Bulletin G-36) - BFGoodrich Geon Vinyl Division, 1990.

(194) *Rovel Weatherable Polymers,* supplier technical report (301-621-285) - Dow Chemical Company, 1985.

(195) *Tyril SAN Engineering And Fabrication Guidelines,* supplier design guide (301-665-1085) - Dow Chemical Company, 1985.

(196) *Styron 6000 Ignition Resistant Polystyrene Resins,* supplier marketing literature (301-01673-192R SMG) - Dow Chemical Company, 1992.

(197) *Beetle & Jonylon Engineering Thermoplastics,* supplier marketing literature (BJ2/1291/SP/5) - BIP Chemicals Limited, 1991.

(198) *Estane Thermoplastic Polyurethane Product Data Sheets,* supplier technical report (BFG-15512-J) - The BFGoodrich Company.

(199) *Pulse 1745 Polycarbonate / ABS Resin For Computer And Business Equipment,* supplier marketing literature (301-00425-793 SMG) - Dow Chemical Company, 1993.

(200) *Rynite Design Handbook For Du Pont Engineering Plastics,* supplier design guide (E-62620) - Du Pont Company, 1987.

(201) *Delrin Design Handbook For Du Pont Engineering Plastics,* supplier design guide (E-62619) - Du Pont Company, 1987.

(202) *Alcryn Weather Resistance Guide,* supplier technical report (196240A) - Du Pont Company, 1990.

(203) *Pigmentation And Weathering Protection Of 'Hytrel',* supplier technical report (HYT-303(R1) / E-73191) - Du Pont Company, 1985.

(204) *Engage Polyolefin Elastomers,* supplier marketing literature (305-01995-1293 SMG) - Dow Chemical Company, 1993.

(205) *Tefzel Fluoropolymer Design Handbook,* supplier design guide (E-31301-1) - Du Pont Company, 1973.

(206) *Engineering Thermoplastics For Lighting,* supplier technical report (6m/0387) - General Electric Plastics Europe, 1987.

(207) *Ultranox 626 / 626A Antioxidants,* supplier technical report (CA-243B) - General Electric Specialty Chemicals, 1990.

(209) *Cyasorb UV-531, UV-3346, Combination Concentration Guidelines For PE Greenhouse Film Products,* supplier technical report - American Cyanamid Polymer Additives, 1994.

(210) *Celcon Acetal Copolymer,* supplier design guide (90-350 7.5M/490) - Hoechst Celanese Corporation, 1990.

(211) *Shinko-Lac ASA T Weatherable And Heat Resistant ASA Resin,* supplier design guide - Mitsubishi Rayon Company.

(214) *Bayflex And Baydur RIM Polyurethane Systems For Unparalleled Design Freedom With RIM,* supplier marketing literature (53-D601(10)L) - Mobay Corporation, 1986.

(215) Cloud, Peter, Theberge, John, *Glass-Reinforced Thermoplastics,* Thermal And Environmental Resistance Of Glass Reinforced Thermoplastics, supplier technical report - LNP Corporation, 1982.

(216) *Japan Synthetic Rubber JSR RB,* supplier design guide - Japan Synthetic Rubber Company.

(218) *Optical Property Test Report,* supplier technical report (F188101.020) - Heraeus DSET Laboratories, Inc., 1991.

(219) *Merlon Polycarbonate Design Manual Section VI - Environmental Effects,* supplier design guide - Mobay Corporation, 1986.

(221) *Colo-Fast Spray,* supplier marketing literature - Recticel n.v. - s.a.

(223) Carver, T. Granville, Kubizne, Peter J., Huys, Dirk, *Reaction Injection Molded Modular Window Gaskets Using Light Stable Aliphatic Polyurethane,* Polyurethanes: Exploring New Horizons - Proceedings Of The SPI 30th Annual Technical / Marketing

Conference, conference proceedings - The Society Of the Plastics Industry, Inc.

(225) *"TPX" Polymethylpentene,* supplier design guide (88.06.3000.Cl.) - Mitsui Petrochemical Industries, Ltd., 1986.

(228) *Hostalen GUR - Effects Of Heat And Light Aging,* supplier technical report (HCC Rev 1/90) - Hoechst Celanese Corporation, 1990.

(229) Stein, Harvey L., *Ultrahigh Molecular Weight Polyethylenes (UHMWPE),* Engineered Materials Handbook, Vol. 2, Engineering Plastics, reference book - ASM International, 1988.

(232) *Introducing Superior UV Stability With Good Looks That Last In Business Machine Housings.,* supplier marketing literature (7110) - Monsanto Chemical Company, 1990.

(233) Baseden, G. A., *Compounding Nordel Hydrocarbon Rubber For Good Weathering Resistance,* supplier technical report (E-88779 / 5/87 118 545/A) - Du Pont Company, 1987.

(234) *Hytrel - Resistance To Mildew And Fungus,* supplier technical report (E-84285) - Du Pont Company.

(235) *K-Resin SB Copolymers UV Stabilization - Plastics Technical Center Report #412,* supplier technical report (778-93 K 01) - Phillips Petroleium Company, 1994.

(237) *Engineering Plastics Acetal Copolymer - Iupital,* supplier design guide (M.G.C.91042000P.A.) - Mitsubishi Gas Chemical Company, Inc., 1991.

(238) *Ube Nylon Technical Brochure,* supplier design guide (1989.8.1000) - Ube Industries, Ltd., 1989.

(239) *Mater-Bi - The Latest Plastic Material Introduces The True Value Of Biodegradability. Today, 1991.,* supplier marketing literature - Novamont, 1991.

(240) *Kraton Thermoplastic Rubber,* supplier design guide (SC:198-89) - Shell Chemical Company, 1989.

(241) *We Test Everything Under The Sun,* supplier marketing literature (2.5M114) - Heraeus DSET Laboratories, Inc., 1994.

(242) *Weathering Services and Equipment Catalog,* supplier marketing literature (7M14) - Heraeus DSET Laboratories, Inc., 1994.

(243) Brennan, Patrick J., *UVA -340 Exposure of Automotive Materials: A Status Report,* technical journal (Vol. 12, No.2) - Journal of Vinyl Technology, June 1990.

(244) Monte, Salvatore J., *High Levels of Weatherability and Functionality of Carbon Black, Organic and Inorganic Pigmented Compositions Using Titanate and Zirconate Coupling Agents,* RETEC: Weathering Well with Color and Additives - Color and Appearance Division and Polymer Modifiers & Additives Division, conference proceedings (Oct. 11-13) - The Society of Plastics Engineers, 1993.

(245) Grossman, George W., *Correlation of Laboratory To Natural Weathering,* technical journal (L-824, Vol. 49, Number 633, Pages 45-54) - Journal Of Coatings Technology, 1977.

(246) Brennan, P., Fedor, C., *Sunlight, UV, & Accelerated Weathering,* RETEC: Automotive, conference proceedings (L-822, Nov.) - The Society of Plastics Engineers, 1987.

(247) Ketola, Warren D., Grossman, D.M., *Errors Caused by using Joules to Time Laboratory and Outdoor Exposure Tests,* Accelerated and Outdoor Durability Testing of Organic Materials, conference proceedings (ASTM STP 1202) - American Society for Testing and Materials, 1993.

(248) Fedor, G., Brennan, P., *Correlation Of Accelerated and Natural Weathering of Sealants,* trade journal (May) - Adhesives Age, 1990.

(249) *Hunter Lab,* supplier marketing literature - Hunter Associates Laboratory, Inc.

Trade Name Index

Aclar (Allied Sig.)
CTFE 67

Acrylic (Aristech)
Acrylic 39

Acrylite (Cyro)
Acrylic 39

Acrylite Plus (Cyro)
Acrylic 39

Acrysteel (Aristech)
Acrylic 39

Alcryn (DuPont)
TPO 285

Altair (Aristech)
Acrylic 39

Ardel (Amoco)
Polyarylate 165

Bayblend (Miles)
PC ABS Alloy 263

Baydur (Miles)
RIM PU 277

Bayflex (Miles)
RIM PU 277

Beetle (BIP)
Nylon 6 91
Nylon 66 103

BROMO XP-50 (Exxon)
Butyl Rubber 359

Calibre (Dow)
Polycarbonate 131

Capron (Allied Sig.)
Nylon 6 91

Celanex (Hoechst Cel.)
Polyester - PBT 147

Celcon (Hoechst Cel.)
Acetal Copol 23

Chemigum TPE (Goodyear)
Nitrile TPE 357

Colo-Fast LM (Recticel)
RIM PU 277

Colo-Fast SPR (Recticel)
RIM PU 277

Cycolac (GE)
ABS 1

Cycoloy (GE)
PC ABS Alloy 263

Cyrolon (Cyro)
Polycarbonate 131

Delrin (DuPont)
Acetal 21

Dutral-CO (Ausimont)
EPM 375

Dutral-TER (Ausimont)
EPDM 377

Elastollan (BASF)
TPAU 345
TPEU 345

Engage (Dow)
TPO 285

Estane (BF Good.)
TPU 345

Evoprene G (Evode)
Styrenic TPE 341

Evoprene Super G (Evode)
Styrenic TPE 341

Fiberloc (Geon Co.)
PVC 239

Foraflon (Atochem)
PVDF 71

Forprene (Evode)
TPO 285

Geon (Geon Co.)
PVC 239
PVC Polyol 353

Geon Duracap (Geon Co.)
PVC 239

GPA (Aristech)
Acrylic 39

Grilamid (EMS)
Nylon 12 87

Halar (Ausimont)
ECTFE 61

Hostalen GUR (Hoechst Cel.)
UHMWPE 203

Hyflon (Ausimont)
ETFE 63

Hypalon (DuPont)
CSM 361

Hytrel (DuPont)
Polyester TPE 331

Iupital (Mitsubishi)
Acetal Copol. 23

Ixef (Solvay)
Polyarylamide 125

Jonylon (BIP)
Nylon 6 91
Nylon 66 103

JSR BR (Jap. Synth.)
Polybutadiene 419

K-Resin (Phillips)
Styr. Butad. Copol 235

Kapton (DuPont)
Polyimide 167

Kodar PCTG (Eastman)
Polyester PCTG 159

Kodar PETG (Eastman)
Polyester - PETG 161

Kraton (Shell)
Styrenic TPE 341

Kydex (Kleerdex)
Acrylic PVC Alloy 261

Kynar (Atochem)
PVDF 71

Lexan (GE)
- Polycarbonate 131

LPP30 (Goodyear)
- PP Copol................................... 211

Lucalen (BASF AG)
- PE Acrylic Acid 205
- PE Ionomer................................ 205

Lupolen (BASF AG)
- EVA PE...................................... 201
- PE... 179

Luran (BASF AG)
- SAN 229

Luran S (BASF AG)
- ASA 31

Makrolon (Miles)
- Polycarbonate 131

Marlex (Phillips)
- HDPE.. 185

Mater-Bi (Novamont)
- Starch Synth. Alloy...................... 269

Merlon (Miles)
- Polycarbonate 131

Minlon (DuPont)
- Nylon 66..................................... 103

Natsyn (Goodyear)
- PI 421

Neoprene (DuPont)
- Neoprene 409

Nordel (DuPont)
- EPDM 377

Noryl (GE)
- PPE 81

Novaloy (Novatec)
- ABS PVC Alloy 259

Parylene (Union Carbide)
- Parylene..................................... 129

Pellethane (Dow)
- TPAU 345
- TPEU 345
- TPU 345

Perspex (ICI)
- Acrylic 39

Plexiglas (Rohm & Haas)
- Acrylic 39

Polystyrol (BASF AG)
- GPPS 219
- IPS 223

Pulse (Dow)
- PC ABS Alloy 259

Rovel (Dow)
- OSA 231

Rynite (DuPont)
- Polyester - PET 153

Ryton (Phillips)
- PPS...217

Santoprene (Adv. Elast.)
- TPO 285

Shinko-Lac (Mitsub. Ray.)
- ASA...31

Solef (Solvay)
- PVDF 71

Styrolux (BASF AG)
- Styr. Butad. Block Copol. 237

Styron (Dow)
- IPS 223

Surlyn (DuPont)
- Ionomer.......................................75

Tecnoflon (Montedison)
- Fluoroelastomer 407

Tefzel (DuPont)
- ETFE...63

Telcar (Technor Apex)
- TPO 285

Tenite Butyrate (Eastman)
- CAB 55

Terluran (BASF AG)
- ABS...1

TMXDI (META) (Am. Cyan.)
- PU 277

Torlon (Amoco)
- PAI 171

TPR (BP Chem.)
- TPO 285

TPX (Mitsui)
- PMP 215

Triax (Monsanto)
- ABS PVC Alloy 259

Tyril (Dow)
- SAN 229

Ube (Ube)
- Nylon 6 91
- Nylon 66 103

Udel (Amoco)
- Polysulfone 225

Ultem (GE)
- PEI 173

Ultradur (BASF AG)
- Polyester - PBT 147

Ultraform (BASF AG)
- Acetal Copol. 23

Ultramid (BASF)
- Nylon 6 91

Ultramid (BASF AG)
- Nylon 6/6T 119

Ultramid (BASF)
- Nylon 610 101
- Nylon 66 103

Ultrapek (BASF AG)
- PAEK...177

Ultrason E (BASF AG)
- PES 227

Upimol (Ube)
- Polyimide...................................167

Vectra (Hoechst AG)
- Liquid Crystal Polymer................163

Vestolit BAU (Huls)
- PVC...239

Victrex PEEK (Victrex USA)
- PEEK...175

Vistalon (Exxon)
- EPDM...377

Zytel (DuPont)
- Nylon 66 103

(Exxon)

BIIR 359
Butyl Rubber 359

(DuPont)

CSM 361

(Rogers)

DAP 271

(Exxon)

EPDM 377

(LNP)

GPPS 219

(DuPont)

Neoprene 409

(LNP)

Nylon 6 91
Nylon 610 101
Nylon 66 103
PE 179
Polycarbonate 131
Polyester 273
Polysulfone 225
PP 207
PVC 239